# Changing European farming systems for a better future

### *International Steering Committee*

Jacques Baudry
Jacques Brossier
Benoît Dedieu
Kirsten von der Heiden
Rebecka Milestad
Colin Newsham
Egon Noe
Milan Slavik
Luigi Omodei Zorini

### *Workshop convenors*

Jacques Baudry
Marianne Cerf
Benoît Dedieu
Ken Giller
Kirsten von der Heiden
Andrea Knierim
Cees Leeuwis
Rebecka Milestad
Colin Newsham
Egon Noe
Philippe Perrier-Cornet
Milan Slavik
Sri Sriskandaraja
Chris Stoate
Luigi Omodei Zorini

# Changing European farming systems for a better future

## New visions for rural areas

edited by:

Hans Langeveld

Niels Röling

Wageningen Academic Publishers

ISBN-10: 90-86860-02-8
ISBN-13: 978-90-86860-02-9

First published, 2006

Wageningen Academic Publishers
The Netherlands, 2006

# Preface

We live in an ever changing world, today probably more than ever. Conditions for farmers as well as for consumers and policy makers are evolving continuously. Processes such as trade liberalisation, reducing support for farming and increasingly strict demands on food quality, environmental quality and animal welfare place heavy demands on land use. Land managers need to identify innovative strategies for improved technical and economic performance, as well as for successful communication with non-agricultural stakeholders. Policy makers and scientists, on the other hand, need to find new ways to reconcile often conflicting interests of farmers, consumers and other stakeholders.

It is under these conditions that the contributions of this book have been prepared. Following a call from the organisers of the 7th European symposium of the International Farming Systems Association, researchers, farmers, advisors and policy makers have used their expertise, experience, insights and vision to define what they think successful changes in land management could and should look like.

Wageningen University and Research Centre has played its role in developing innovative land use practices. Our track record builds on decades of experience, both in Europe and elsewhere in the world, representing a huge scientific and policy oriented research experience that has impacted European and development policy and practice.

Over the years, participatory, interactive, systems-oriented research developed into an important element of our research and education activities. Learning from it we have defined the 'Wageningen approach', that is multi-stakeholder, inter-disciplinary, participative and systems-oriented. We now have quite a large number of examples where this approach has worked and has managed to bring people and solutions together.

The Board of Wageningen University and Research Centre finds it extremely important to present this way of thinking to a wide audience. Therefore, we are happy and proud that this IFSA conference takes place here in Wageningen. We feel that it represents an important approach to the development of innovative, sustainable land use. I hope that our Wageningen work in this field makes a useful contribution. I congratulate the organisers with the Conference Proceedings before the conference. It contains over 100 contributions from around the world. I am confident that this book will contribute in the search for new, improved land use practices.

Wageningen, April 2006

*Martin Kropff, Rector Magnificus of Wageningen University*

## Table of contents

# General introduction

*Hans Langeveld[1], Niels Röling[2] and Janice Jiggins[2]*
*[1]Plant Research International, [2]Department of Communication and Innovation Studies, Wageningen University and Research Centre, the Netherlands*

## New arrangements for land use and communication

Rural areas face many challenges. They are posed by processes of trade liberalisation, globalisation, declining commodity prices and reducing support for farming, increasingly strict demands on food quality, environmental quality, and animal welfare and by climatic change. They result in increasing competition for land, space and water and require highly skilled land managers. Clearly these challenges have to be addressed not only by farmers and other land managers but also by consumers, researchers and policy makers.

Farming families, researchers and policy makers, all need to come up with solutions to help maintain and improve rural conditions. How are we to reconcile the different demands on the rural areas? What land use systems are to help realise cleaner, safer and more animal friendly food production, conserve biodiversity, or improve water management and nature development? What arrangements can be made to support decision-making and problem solving in multiple-stakeholder arenas and transform (potential) conflict into concerted action?

To answer these questions considerable efforts are required in research and management. Knowledge of and skills in technical and bio-physical processes, combined with a keen eye for economic performance of the land use systems, are no longer sufficient. In order to be able to be successful, farmers need to identify new, innovative, strategies and communicate them with stakeholders that have may have different priorities. Scientists, advisors and policy makers, on the other hand, will have to learn to collaborate with farmers and consumers, groups that are increasingly educated, informed and organised. Together, all involved need to find ways to interact efficiently in a world where information is abundant but not necessary always correct.

A clear role for research is laid out in the design of new systems and arrangements that are adapted to the changing conditions and create conditions under which stakeholders can learn and implement new ways of cooperation. Different research schools exist, be it with specific scopes and application areas. Despite the many differences, also some similarities can be identified. Various approaches have been developed that combine an integrated view, multi-disciplinarity, and systems oriented focus. In his overview of systems oriented, participative research, Norman (2002) sketches the development of major research orientations from 1960 to present day.

Over the years, participative, systems oriented research, that integrates multi-disciplinary knowledge at different scale levels, has played an increasingly important role in the design and implementations of new, alternative land use systems, as well as in the communication between the actors involved in rural development. Such research can help different stakeholders, including farmers, civilians and policy makers to participate in problem formulation and in identifying effective policies and farming practices, and can assist in the adoption of such practices.

## IFSA

Over a period of 25 years, the International Farming Systems Association (IFSA) has been at the forefront in developing strategies for rural areas in Europe and elsewhere. IFSA has its origins in Farming Systems Research, as it evolved after the Second World War, largely for the purposes of smallholder development in the tropics. Different groups of researchers from many disciplines as well as extension and development project professionals joined forces in a series of annual meetings hosted by Kansas State University in the 1980s. Together, they helped defining a systems researching approach to agricultural development.

In 1989, the Association for Farming Systems Research and Extension (AFSRE) was born, open to agricultural and rural systems researchers and developers of any discipline, hosting members from almost all geographical regions. Regions started organising themselves in the 1990s, hosting the global symposium on a rotating basis. In 1998, the name of the global association was changed to the International Farming Systems Association (IFSA in short). The European group started organising symposia in the 1990s, starting in Edinburgh (1994), followed by Granada (1996), Hohenheim (1998), Volos (2000), Florence (2002), and Vila Real (2004).

Wageningen University and Research Centre has been actively involved throughout the history of IFSA, both at the European and the global level. In 2002 it was decided that Wageningen would be invited to host the 7th European symposium in 2006. The symposium has been organised including five workshops as well as two special sessions. Workshop themes are represented as sections in this book, covering issues of natural resources management and regional development (sections 5 and 2, respectively), perspectives for farmers (section 3), and knowledge systems and learning processes in rural development and land use system innovation processes (sections 1 and 4), while a special session covers the perspectives for systems oriented, participative research approaches (final section of this book).

This book contains a unique collection of papers prepared by researchers, advisors, farmers and policy makers. Over 100 contributions have been selected, originating from over 20 countries located at six continents and ranging from one page (short communication) to a general five pages (full contribution). They refer to insights and experiences from bio-physical as well as social sciences, focus ranging from crop and animal level to landscapes and entire regions. Together, the contributions will be valuable to policy makers, farmers, advisors and all those that are involved – or feel related to – issues of land use in rural areas. Given the broad scope of the papers and diverse background of the authors and contributions, the book can be considered to present a thorough overview of thinking on land use, farming systems and communication between stakeholders involved in rural development.

## References

Norman, D.W., 2002. The Farming Systems approach: A historical Perspective. Presentation at the 17th Symposium of the International Farming Systems Association, Lake Buena Vista, Florida, November 17th-20th, 2002. http://conference.ifas.ufl.edu/ifsa/papers/invite/Norman.doc

## Acknowledgements

Papers in this book have been prepared at the occasion of the $7^{th}$ European IFSA symposium: 'New visions for rural areas. Changing European farming systems for a better future'. The symposium could be organised thanks to considerable efforts by Agrosystems Research of Plant Research International, Communication and Innovation Studies and Animal Sciences Group (all Wageningen University and Research Centre). Additional support has been received by the LEB fonds, programme 'Nieuwe veehouderijsystemen' and Transforum.

Many persons have played a role in the realisation of the symposium. We are extremely grateful to members of the International Steering Committee: Jacques Brossier, Artur Cristóvão, Benoît Dedieu, Kirsten von der Heiden, Loes Heuff, Alex Koutsouris, Rebecka Milestad, Colin Newsham, Egon Noe, Milan Slavik, Luigi Omodei Zorini, who defined the symposium theme and provided huge support to the organisers. Workshops have been elaborated by Jacques Baudry, Jacques Brossier, Marianne Cerf, Benoît Dedieu, Ken Giller, Kirsten von der Heiden, Andrea Knierim, Cees Leeuwis, Rebecka Milestad, Colin Newsham, Egon Noe, Philippe Perrier-Cornet, Milan Slavik, Sri Sriskandaraja, Chris Stoate, and (last but not least) Luigi Omodei Zorini. Together, they have done a wonderful job in reviewing the papers en posters that found their way into this book.

Thanks are further due to members of the Board of Advisors, Johan Bouma, John Dixon, Janice Jiggins, Herman van Keulen and Niels Röling and our collegues at Wageningen University and Research Centre. A special 'thank you' is in place to the members of the Local Organising Committee (Irene Gosselink, Albert Meijering, Conny van der Plas, Ben Rutgers and Anita Wolsing) and many others who have done a great job in organising the symposium (especially Anouk van Gils, Ria van der Hulst, Koos Nijssen, Jet Proost, Jan Verhagen, Willem van Weperen and Han Wiskerke).

# How sustainable can agriculture be in an EU-scenario of free market & social policy?

*P.H. Vereijken[1] and C.M.L. Hermans[2]*
*[1]Plant Research International, PO Box 16, 6700 AA Wageningen, The Netherlands; [2]Alterra, PO Box 47, 6700 AA Wageningen, The Netherlands*

The EU (European Union) has a Common Agricultural Policy (CAP) that is currently in line with a scenario emphasising protected markets and liberal policies. It is argued that during the next twenty years the EU will be driven towards a CAP in line with a scenario of free markets and social policies. Will such a CAP ensure a sustainable agriculture? This question is explored by stating simple indicators for "people, profit and planet" (3Ps), which reflect the opportunities of EU-regions to maintain their agriculture in the price squeeze and technology race caused by the liberalisation of the food market. The results are presented in four maps classifying the more than 160 regions of the current 25 member states from best to worst outcomes. Per region, the forecast outcomes are so variable that an overall map does not make much sense. There is no overall 3Ps answer to the key question. Nevertheless, the four forecast maps reveal enormous differences in opportunities between regions and seem useful for policymakers and entrepreneurs in deciding on long-term policies and sustainable investments in EU-regions.

A major question to the EU-CAP is: "should the EU keep on protecting and subsidising its agriculture, or switch to a more liberalised market and policy?". This paper wants to contribute to the debate from the starting point that policy and market are already being liberalised, so the question is not whether or not to liberalise, but to what extent. Theoretically, the EU could refuse to liberalise, but its partners in the WTO would in their turn refuse to liberalise too, to the detriment of other EU-industries and services, far more important than agriculture in terms of employment and their contribution to GDP. So, complete refusal to liberalise is not in the interest of the democratic majority and will probably be rejected. The recent liberalisation debate interferes with the old sustainability debate. Some argue that liberalisation should be stopped because it renders agriculture less sustainable, notably in developing countries. Others argue that liberalisation can go along with sustainable agriculture, or even that only by further liberalisation an economically and ecologically sustainable agriculture can be achieved.

This paper supposes quite variable effects of liberalisation on sustainability, depending on the agrarian regions, agricultural sectors and aspects of sustainability. Thus it explores the effects of liberalisation per EU-region using overall indicators for the 3 basic aspects of sustainability: "profit", "people" and "planet". As in the rest of the world, the long-term forecasts for market and policy in the EU in general and of agriculture in particular are subject to two sets of antagonistic human drivers:

- nationalism versus cosmopolitanism;
- liberty versus solidarity.

These two sets of drivers create a tension field in which market and policy can interact in four basic scenarios (Figure 1). For each scenario, examples for the world and the EU are added to show that market and policy can indeed vary independently. Thus, the examples show that market liberalisation does not have to imply a shift from a social to a (facilitating) liberal

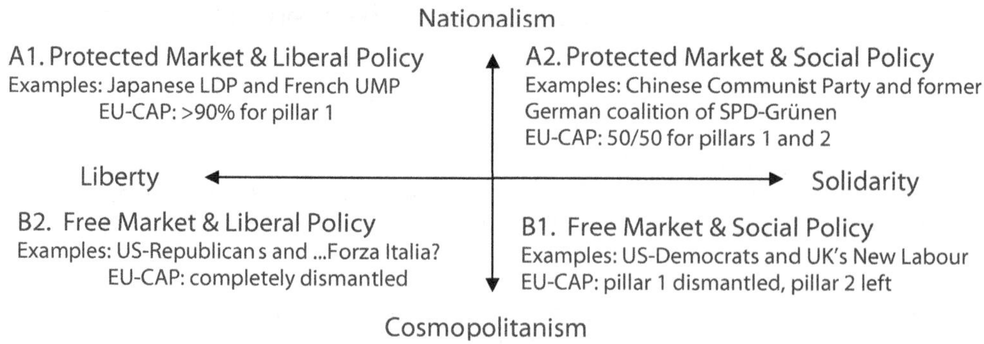


*Figure 1. Future scenarios for Market and Policy of EU in general and the Common Agricultural Policy (CAP) in particular, in the tension field of four basic human drivers.*

policy; it can also combine with a (compensating) social policy. In this scheme, markets are liberalised because cosmopolitanism becomes an ever-stronger driver, to the detriment of nationalism.

Cosmopolitanism seems to become ever more attractive and rewarding than nationalism for especially the younger generation, whether it is driven by individual liberty or international solidarity. It is strongly catalysed by an explosive growth of communication and transport and increasing welfare. As a result, liberalisation of EU-markets and policy is expected to go beyond scenario A2 and reach B1. Therefore, B1 is used as the most plausible scenario for the EU from 2025 on. To explore, in which EU-regions and to what extent agriculture can be sustainable in this free market and social policy scenario; we present forecasts of sustainable agriculture from 2025 onwards at the regional level of the current EU-25. These regional forecasts are based on a limited set of indicators covering sustainability in all its three aspects: economical ("profit"), social ('people") and ecological ("planet").

# Section 1 – Learning as a process

# Introduction to section 1 – Learning as a process: Understanding one's role in the new learning demands of multifunctional land use systems, working with different actors, tools and scales

*Nadarajah Sriskandarajah[1], Marianne Cerf[2] and Egon Noe[3]*
*[1]Unit of Learning, The Royal Veterinary and Agricultural University, Denmark; [2]INRA, UMR SAD-APT, 78850 Thiverval Grignon, France; [3]Danish Institute of Agricultural Science, Department of Agriecology, Blichers Allé, Postbox 50, DK-8830 Tjele, Denmark*

Learning processes associated with research and extension in rural and farming systems have made up a regular and prominent theme in past symposia of the European IFSA network. Theories of learning and their implications for practice, the role of researchers as facilitators of learning in the interplay between farmers and their providers of service, and experiences of using different approaches and tools in such learning-oriented research have featured in workshops under this theme in previous symposia. At the last gathering of the network two years ago in Vila Real, Portugal in 2004, some 33 papers were discussed under the theme of 'Knowing and Learning: labour and skills at stake for a multidimensional agriculture'.

The 2004 workshop on learning identified several challenges facing farming systems practitioners. Some of these relate to their changing roles from experts to participants of new networks, when they have to create the networks and also facilitate meaningful dialogue in them. A new challenge confronting them as part of the new visions for rural areas is that their R&D would now take place in new locations and in new contexts; researchers have to develop new competences; and make learning more explicit for themselves as well as for the multiple stakeholders.

The evolution of farming systems towards multifunctionality ties in needs of sustainability as well as needs of society. The new thrust on rural development demands learning processes in motion, some of which would involve people at the individual farm while others may encompass a whole rural area. Some will involve specific projects; others may take in farmers, citizens and many other local and non-local stakeholders. The future challenge will be about learning processes in open networks and less so in well-defined and often familiar groups. Learning among heterogeneous groups of stakeholders, and among different epistemologies has become one of the most central issues today.

Such a move has to be put in relation with the increasing interest for issues of resources management. If the primary focus in the past was mainly on the building of farmers' capacity to cope with uncertainty and production stakes, it has now moved towards the building of collective action among rural inhabitants who share common resources and they have to agree upon their way to use and keep such resources. Therefore the range of situations in which researchers and developers are involved has expanded, requiring them to appeal to concepts, which would enable them to address problems encountered in such situations.

Irrespective of the situations, the common thread for learning-oriented researchers has been to consider learning as a process, and to agree on the fact that knowing and acting are intertwined. This can also refer to many theories as pointed out in a publication on this theme by the LEARN Group (Cerf *et al.*, 2000).The papers being presented at this workshop have touched on many angles one could take on this learning – knowing – acting theme.

One of these angles is to look at how human interactions can result in or foster learning processes. While understanding how interactions between individuals could support learning at the individual level still remains as a field of interest, communication and dialogical processes have increasingly become meaningful for coping with situations in which people do not necessarily share stakes, epistemologies, practices, or tools (whether symbolic or material artefacts). While some papers leaned on theories of discourse and conversation analysis and elaborate these, others develop some practical tools such as diagramming, use of metaphors or of intermediary objects in order to promote such processes. This is still an open field of questions, and some papers in the current workshop will allow us to discuss those questions further.

A second focus is on identity and capacity building. While understanding practices of reflexivity and the conditions favouring individual stakeholder's practices of reflecting upon their own process of learning and working out how such process can be promoted, increasing interest has developed on how to address collective identity through for instance the development of work on Communities of Practice. This concept was first introduced by Etienne Wenger and then further developed within the theoretical framework of situated learning (see Wenger, 1998).

Finally, a third focus is on the development of collective action and the understanding of public action as a bottom-up process. While aiming at developing approaches that can improve the capacity of stakeholders to be actors in the process of knowing in such situations, it also aims at understanding how knowing, as effective action, operates across scales ranging from the individual to the group, community, organisation and polity. Studies have emerged about social learning, systems social learning, and participatory learning in order to address such questions (Leeuwis and Pyburn, 2002). Social learning refers to (i) convergence of goals, criteria and knowledge leading to agreement on concerted action among interdependent stakeholders where (common pool) resource dilemmas have arisen, (ii) co-creation of knowledge needed to understand issues and practices and (iii) changes in behaviours, norms and procedures arising from mutual understanding of issues. Social learning is thus a feature of doing and knowing within a situation, and an emergent property of the process to transform a situation. Systems social learning insists more on the systemic dimension of the collective action by questioning what lies within a system of interest that enables it to function as a system with the purpose of learning, what lies in its environment that affects or is affected by social learning, what transformation takes place and what the underpinning world views are. Participatory learning places the emphasis on authentic participation by relevant stakeholders at all stages of the exploratory process and their reflective learning on action.

In this section, 18 papers are presented. These take on learning as a process and practice, and its facilitation as a professional role to help meet the new demands. Rural development and the future of rural areas in Europe require formulation of new visions and design of new tools and approaches for enabling learning across the different levels of society. The social, participatory and transformative learning we have been discussing hitherto, when taken in such a collective, societal sense, become 'societal learning', and the new magnified set of epistemic challenges that come with it are, therefore, ours to deal with.

## References

Cerf, M., D. Gibbon, B. Hubert, R. Ison, J. Jiggins, M. Paine, J. Proost and N. Röling (Eds.), 2000. Cow up a tree - Knowing and learning for change in agriculture: Case studies from industrialised countries. Paris, INRA.

Leeuwis, C and R. Pyburn (Eds), 2002. Wheelbarrows full of frogs - social learning in rural resource management, Koninklijke Va Gorcum: The Netherlands

Wenger, E., 1998. Communities of Practice. Learning, meaning and identity, Cambridge University Press, New York

# Where do participatory approaches meet social learning systems in the context of environmental decision making?

*Chris Blackmore*
*Open Systems Research Group, The Open University, United Kingdom*

## Abstract

If popular environmental magazines and journals of 2005 are to be believed there has been an upsurge in participatory decision making backed up by legislation, for instance that arising from the EU's Aarhus Convention. However, there are also many critiques to be found of these participatory approaches and a range of policy makers, researchers and individual authors have suggested that in many situations where there are resource dilemmas, social learning rather than more participation is required. There are many different theories and practices of participation and social learning. How do they differ and how do they coincide? Participatory and social learning approaches are often grouped together but do they always complement each other? What kind of participation is required to bring about social learning as part of environmental decision making? And does it help to think of social learning as systems? This paper will explore some of these questions through use of examples from environmental decision-making situations, mainly from the UK.

**Keywords:** participation, social learning, systems, environmental decision making

## Introduction

The UK ENDS report, included a feature in May 2005 entitled 'A new age of participatory decision making?' that began...

> *"Environmental decision-making is going through a quiet revolution. The public is having more of a say in environmental decisions, and is now consulted on everything from water abstraction to waste collection. This explosion in stakeholder engagement has required both regulators and industry to develop new skills and approaches to decision-making. But what methods are they using to try to involve the public - and how much attention do they pay to the response?"*

Others too have commented on a recent upsurge in public participation but as something making a comeback rather than as something new (Lane, 2005). The ENDS article went on to critique participatory approaches claiming they have many actual and potential benefits but in practice they often also have difficulties, for instance those associated with transfer of power and lack of clarity of purpose, facilitation skills and resources. This kind of critique of participatory approaches is also not particularly new but European Directives such as those arising from the Aarhus Convention on Access to Information, Public Participation in Decision Making and Access to Justice in Environmental Matters and the Strategic Environmental Assessment Directive have recently started to come into effect and have brought an increased focus to issues of participation.

Recognition of the need for learning approaches and social learning as part of environmental decision making has also been gaining momentum. There is perhaps less readily available evidence that social learning, rather than participatory approaches, has taken place though the two are of course related. (I will go on to explore just how, in the final section of this paper.) Many authors and practitioners have framed the development and use of legislation such as environmental impact assessment, and schemes for environmental management systems and corporate social and environmental responsibility as a learning process. (Glasson *et al.*, 2005; McCulloch, 1996; Open University, 1997, 2006). How and where a focus on social learning in the context of environmental decision making has occurred will be discussed later in this paper.

What is meant by participatory approaches varies in terms of who participates in what, with whom, how and for what purpose. But many of these approaches can be linked to the idea of community participation in local governance (Woodhill, 2002; Mayoux, 2003; Gayenta, 2004). Rose (2000) (as cited in Lane, 2005) refers to a host of 'new technologies of governance' ranging through 'governance through communities' and 'Third Way' approaches (Giddens, 1998) to 'public-private partnerships' unified by the need to involve citizens, non-governmental organisations and others in the development and implementation of policy. Social learning also has been considered as a mechanism of governance that may complement others (SLIM, 2004). In order to understand how participatory and social learning approaches differ and coincide in the context of environmental decision making, I will briefly discuss the evolution of each.

## Participatory approaches

Participatory approaches to environmental decision making and analyses of them can be traced back several decades in fields such as planning (Lane, 2005), rural development (Chambers, 1994 ) monitoring and evaluation (Mayoux, 2004; UNDP, 1997) and waste management (Kraft and Kraut, 1985; Kraft and Clary, 1991). Organisations such as FAO, GTZ in Germany, ODA, IIED and IDS in the UK, PRIA in India and CIDA in Canada have long been advocates of participatory approaches, particularly in development contexts (UNDP, 1997). Similar organisational advocates can be found in other parts of the world. Exactly when citizen participation fully emerged, in processes of administration and governance in which environmental considerations were a part, depends on the sector and location under consideration. But some suggest that in western societies it became of concern in public policy making sometime in the late 1960s and '70s (Kraft and Clary, 1991) and that during the 1980s it gained widespread acceptance in effective environment and development work. (Woodhill, 2002). Mayoux (2003) suggests that in the 1990s participatory approaches evolved from a focus at project and programme level to a concern with civil society development and good governance.

This evolution of participatory approaches was evident at a major international conference on community-based sustainable development, hosted by the IIED in 1994 at the University of Sussex in the UK where many examples of long standing and well-established participatory approaches to sustainable development were brought forward as case studies. But there were notably more examples coming from developing countries than from industrialised countries at that time. Tony Gibson and his colleagues' 'Planning for Real' was one of the few examples presented from Europe at that time, though European donor agencies were involved in financing participatory approaches to development in other parts of the world and from the early 1990s participatory approaches were a required component for funding in development programmes.(Mayoux, 2003). Gaventa (2004) suggests that that in the United States and parts of Europe we still have much to learn, regarding participatory approaches, from poorer countries of the world who are in the process of creating new democracies.

## Social learning approaches

Social learning traditions can also be traced back many decades. But what was meant by social learning fifty years ago is only one of its meanings in currency today. Bandura is often credited as establishing the concept of social learning but he studied it in a traditional behavioural manner, which was prevalent in the 1950s and 60s and focused largely on learning through imitation. Danish author Knud Illeris, in his classic text on contemporary learning theory (Illeris, 2002) positioned a wide range of theorists and theories in what he called 'the learning theoretical tension field between the cognitive, the emotional and the social'. Illeris identified many who had brought social and societal factors to the fore of their work and who in so doing had also in his view made contributions to learning theory. They ranged from Marx to Habermas to Wenger to Jarvis, to Giddens, Beck to many others. Examples of social learning theories, developed and used explicitly in the context of environmental decision making and sustainable development, have come from Finger and Verlaan (1995) who developed a conceptual framework for social-environmental learning, Daniels and Walker (1996) who considered collaborative learning and improving public deliberation in ecosystem-based management and Woodhill and Röling (1998) who looked into the human dimension in learning our way to more sustainable futures. This work has since been built on and there are many others who have written about social learning (*e.g.* SLIM, 2004; Leeuwis and Pyburn, 2002; Finger and Asun, 2001; LEARN group, 2000; Wildemeersch *et al.*, 1998).

Hence a very broad range of theories could be considered as theories of social learning. What constitutes a social learning approach depends not just on the questions of who learns, how, with whom and for what purpose (which are similar to those question I raised regarding participatory approaches) but on how social learning is theorised.

In the context of environmental decision making, two different but overlapping social learning approaches have emerged as having a lot of potential for both policy and practice in the context of my research. The first is a process of social learning characterised by (i) convergence of goals criteria and knowledge leading to agreement on concerted action among interdependent stakeholders where (common pool) resource dilemmas have arisen, (ii) co-creation of knowledge needed to understand issues and practices and (iii) changes in behaviours, norms and procedures arising from mutual understanding of issues. This was the social learning approach that emerged from the recent EU project on social learning for the integrated management and sustainable use of water at catchment scale, the SLIM project. (SLIM, 2004). It built on previous work of teams in the Netherlands, France, Italy and UK, including previous work on social learning (*e.g.* Woodhill and Röling, 1998) The second approach I call a social learning systems approach. Systems theories were among those that informed the SLIM project, so these two approaches overlap. What is characteristic of this approach is the explicit use of concepts from various systems traditions such as boundaries, levels, emergence, transformation, resilience and dynamics. The 'systems' dimension of social learning has been commented on by Ison *et al.* (2000) who reviewed theoretical approaches to learning in the context of sustainable agriculture. They noted the frequent occurrence of a cybernetic paradigm in learning process approaches. Work done by people like Lance Gunderson, C.S. Holling and Carl Folke on complex adaptive systems to improve understanding of the dynamics of social-ecological systems could also be thought of as a social learning systems approach. The Open University Systems group (Open University, 1997, 2006) and Etienne Wenger (Wenger, 2000) are among those who have developed and used social learning systems approaches. Keen *et al.* (2005) also include a systemic learning approach as one of nine principles of social learning for environmental management. Some different positions on social learning are taken by these different authors but an advantage in the explicit use of systems concepts is that it helps raise questions of what lies within a system of interest that enables it to function as a system with the purpose of learning, what lies in its environment that affects or is affected by social learning, what transformation takes place and underpinning world views. Systems approaches also offer a range of tools and methodologies to facilitate social learning (Open University, 1997, 2006).

## Comparing and contrasting participatory and social learning systems approaches

Finally, to address the question in the title of this paper: *Where do participatory approaches meet social learning approaches and ideas about social learning systems in the context of environmental decision making?* UK cases of planning, managing water and wastes provide some useful examples to consider. Participatory approaches have certainly been in evidence in these domains, as can be seen from a wide range of initiatives such as workshops, focus groups, consensus conferences, 'planning for real' and citizens' juries (IEMA, 2002). But while in many of these cases there is evidence of collective action, there is less evidence of long term concerted and purposeful action with social learning as an emergent property of a number of inter-related elements. That is not to say that some participatory approaches do not lead to social learning. Some authors include participation as just one facet of social learning alongside others (*e.g.* Keen *et al.*, 2005). Several authors have concluded that localised participatory approaches and relying on interactive emergence are not enough to bring about the kind of changes needed for sustainable development and that this will require social learning, including conducive institutions (Röling, 2002; Woodhill, 2002; SLIM, 2004). Röling (2002) has linked interaction and concerted action with the idea of a sustainable society, which he defines as a society based on agreements to control our own behaviour in the context of 'the eco-challenge'. He argues that "most of all a sustainable society must be capable of concerted action".

So comparing and contrasting participatory and social learning approaches seems to indicate both that there is a rationale for grouping them together and that social learning approaches have the potential to achieve more than participatory approaches alone in terms of environmental decision making in the context of sustainable development (with the caveat that some participatory approaches can lead to social learning if other factors also are present.) But what of participatory approaches and social learning systems? Perhaps the explicit use of systems ideas can help in identifying what participatory approaches are and could be a part of and help to make sense of the interactions, purposes and transformations required in environmental decision making? SLIM (2004), Keen *et al.* (2005), Blackmore (2004, 2005) and others have made a start in this direction and in the process have provided further evidence of the need for social learning systems approaches. Wenger also has focused on the idea of participating in a social learning system noting that the perspective of a social learning system applies to many of our social institutions: our disciplines, our industries, our economic regions, and our organizations with implications at multiple levels (individuals, communities of practice and organisational).

Returning to the suggestion made in the ENDS report quoted at the beginning of this paper, that we may have a new age of participatory decision making, I wonder if and when a need for social learning will become recognised in a similar way? The evolution of social learning approaches seems to me to have had much more of a theoretical focus than participatory approaches, which in my experience have had more of a policy making and practitioner focus. Approaches such as SLIM have started to bring theoretical, policy and practitioner focuses and participatory and social learning approaches together. There appears to be a lot to be gained from continuing in this direction.

## References

Blackmore, C., 2004. From one-off events to learning systems and communities of practice. Pre-proceedings Sixth IFSA European Symposium, April 4-7, Vila Real, Portugal, Universidade de Trás-os-Montes e Alto Douro, pp. 449-458.

Blackmore, C., 2005. Learning to appreciate learning systems for environmental decision making – a 'work-in-progress' perspective. Systems Research and Behavioural Science 22, 329-341

Chambers, R., 1994. The Origins and Practice of Participatory Rural Appraisal. World Development 22, 953-969.

ENDS feature, 2005. A new age of participatory decision making? ENDS report 364.

Daniels, S. and G. Walker, 1996. Collaborative Learning: Improving Public Deliberation In: Ecosystem-Based Management, Environmental Impact Assessment Review, pp. 71-102.

Finger, M. and J.M. Asun, 2001. Adult Education at the Crossroads - learning our way out. Zed Books, London & New York.

Finger, M. and P. Verlaan, 1995. Learning our way out: a conceptual framework for social-environmental learning. World Development 23, 505-513.

Gaventa, J., 2004. Strengthening participatory approaches to local governance: lessons from abroad. National Civic Review, Winter 2004.

Giddens, A., 1998. The third way: the renewal of social democracy, Cambridge, MA: Polity Press.

Glasson, J., R. Therivel and A. Chadwick, 2005. Introduction to Environmental Impact Assessment (natural and built environment series). Taylor and Francis, London

Illeris, K., 2002. The three dimensions of learning. Contemporary learning theory in the tension field between the cognitive, the emotional and the social. Roskilde University Press. (Translated from Danish edition Læering, which appeared in 1999.)

Institute of Environmental Management and Assessment (IEMA), 2002. Perspectives – Guideline on participation in environmental decision making. IEMA:Lincoln, UK.

Ison, R., C. High, C. Blackmore and M. Cerf, 2000. Theoretical Frameworks for Learning-Based Approaches to Change in Industrialised-Country Agricultures In: M. Cerf, D. Gibbon, B. Hubert, R. Ison, J. Jiggins, M. Paine, J. Proost and N. Röling (Eds.), Cow up a tree - Knowing and learning for change in agriculture: Case studies from industrialised countries. Paris, INRA.

Keen, M, V. Brown and R. Dyball, 2005. Social Learning in environmental Management towards a sustainable future. Earthscan, London, Sterling, VA.

Kraft, M.E and B.B. Clary, 1991. Citizen participation and the Nimby Syndrome: public response to radioactive waste disposal The Western Political Quarterly 44, 299 -328.

Kraft, M.E and R. Kraut R, 1985. The impact of citizen participation on hazardous waste policy implementation: The case of Clermont County, Ohio. Policy Studies Journal 14, No 1 September.

Lane, M.B., 2005. Public participation in planning: an intellectual history. Australian Geographer 36, pp 283-299.

LEARN Group (Learning in Agriculture Research Network), 2000. In: M. Cerf, D. Gibbon, B. Hubert, R. Ison, J. Jiggins, M. Paine, J. Proost and N. Röling (Eds.), Cow up a tree - Knowing and learning for change in agriculture: Case studies from industrialised countries. Paris, INRA.

Leeuwis, C and R. Pyburn (Eds.), 2002. Wheelbarrows full of frogs - social learning in rural resource management, Koninklijke Van Gorcum, The Netherlands.

Mayoux, L., 2003. Grassroots action learning: impact assessment for downward accountability and civil society development. (Enterprise Development Impact Assessment Information Service Information Resources www.enterprise-impact.org.uk/pdf/EINJune04.pdf, accessed 31/01/06).

McCulloch, A., 1996. Environmental Assessment, the learning society and the search for sustainable development, Paper for conference on integrating environmental assessment and socio-economic appraisal in the development process, University of Bradford, UK, 24–25 May.

Open University, 1997. T860 Environmental decision making: a systems approach Blocks 2 and 5 Open University: Milton Keynes, UK.

Open University, 2006. T863 Environmental decision making: a systems approach Book 2 Open University, Milton Keynes, UK.

Röling, N., 2002. Beyond the aggregation of individual preferences – moving from multiple to distributed cognition in resource dilemmas. In: C. Leeuwis and R. Pyburn (Eds.), 2002. Wheelbarrows full of frogs - social learning in rural resource management, Koninklijke Va Gorcum, The Netherlands, pp. 25 – 47.

Rose, N., 2000. Community, citizenship and the third way. American Behavioral Scienist 43, 1395-411.

SLIM, 2004. Policy briefings and Framework documents (various) available from Http://slim.open.ac.uk (accessed 31 January 2006).

UNDP, 1997. Who are the question makers? A participatory evaluation handbook. OESP Handbook series, UNDP, New York.

Wenger, E., 2000. Communities of Practice and Social Learning Systems Organization Volume 7(2), SAGE: London, Thousand Oaks, CA and New Delhi, pp. 225–246.

Wildemeersch, D., M. Finger and T. Jansen (Eds.), 1998. Adult Education and Social Responsibility. Peter Lang, Frankfurt.

Woodhill, J. and N. Röling, 1998. The second wing of the eagle: the human dimension in learning our way to more sustainable futures. In: N.G. Röling and M.A.E. Wagemakers (Eds.) Facilitating sustainable agriculture. Participatory learning and adaptive management in times of environmental uncertainty. Cambridge: Cambridge University Press.

Woodhill, J., 2002. Sustainability, social learning and the democratic imperative Lessons from the Australian Landcare movement in Leeuwis, C and Pyburn, R. (Eds) (2002) Wheelbarrows full of frogs - social learning in rural resource management, Koninklijke Va Gorcum, The Netherlands, pp 317–331.

# The action researcher as a reflective partner to a core group

*Dorthe Christensen and Nadarajah Sriskandarajah*
*Section for Learning, the Royal Veterinary and Agricultural University, Denmark*

## Abstract

The EU suggests applying bottom-up, participative learning approaches, such as Action Research, as steering instruments to meet the challenge of multifunctionality and its links with rural development. This paper focuses on the many demanding roles placed on an action researcher when working with rural stakeholders to achieve normatively desirable learning. It is suggested that in order to genuinely qualify the learning process and its outcome for all, the action researcher keeps an adequate balance between being "close to" or "inside" the stakeholder arena and "distanced to" or "outside" this arena. A model for how this balance could be achieved is proposed.

**Keywords:** learning, action research, multifunctionality, rural development

## Introduction

European agriculture is confronted with the challenge of multifunctionality and its links with rural development. The functions currently expected of agriculture are: 1) a competitive production of food and non-food while meeting consumer demands for food safety, quality and variety; 2) stewardship of the countryside regarding pollution and the combination of nature conservation, recreation, cultural heritage and land use; 3) additional services contributing to improving quality of rural life and rural economy (Durand and Van Huylenbroeck, 2003). As a policy response, EU suggests complementing the usual three mechanisms, namely regulatory, market-based and information measures with a fourth governance mechanism. This mechanism, frequently referred to as the "fourth axis" of the rural development programme, cuts across the three principal functions expected of agriculture in relation to rural development. It involves applying a bottom-up, participative hypothesis to achieving sustainable rural development. The LEADER programme introduced and part-sponsored by the EU, is an example of this ethos in operation (Ray, 2000).

Research has in fact demonstrated the advantages of such a fourth governance mechanism, showing how local stakeholder networks can co-create knowledge of local relevance by joint facilitated reflective dialogue and experimentation (*e.g.* Röling and Wagemakers, 1998; Cerf *et al.*, 2000). Yet going beyond the first steps taken by the EU towards complementing top-town expert knowledge with knowledge of local relevance, the idea dominating most of this research is local capacity building as means for sustainable development. This capacity building takes place through the joint learning of rural stakeholder in networks and in a normatively desirable twofold way: 1) learning towards joint action in managing the rural area in a more sustainable fashion and; 2) learning how to learn to meet this end. The notion put forward here is that as rural areas can never be designed to be ultimately sustainable from the "outside", the use of the sustainability concept ought to be a focus of concern about their futures by rural stakeholders themselves. The challenge is to create self-critical systems that are continually learning how to persist, achieve long-term mobilisation and stability through joint learning and learning to learn.

Much of the research referred to above is conducted as Action Research (AR), a participatory research paradigm advocated also in the domains of agricultural development, rural development and environmental management. AR is concerned with improving local situations through the development of relevant knowledge, which is then used to guide informed action. An AR project therefore often proceeds through iterative cycles of planning, acting, and reflecting, each of these being self-critically implemented in an integrated way by participating stakeholders. In addition to giving local stakeholders power to act by generating knowledge of local relevance through a reflective process, AR aims to disseminate the more generic knowledge gained to a wider research community and to people in comparable situations.

## Multiple roles of the action researcher

Participatory processes are far from smooth and the action researcher plays many roles in order to push rural stakeholders towards learning that leads to joint action in managing a rural area in a more sustainable fashion, and towards learning how to learn to meet this end. We identify the following five roles as crucial for this purpose:

1. Providing context-free and contextual knowledge: Implicit in AR is often the assumption that many stakeholder groups feel stuck in particular and/or conflicting views of their situation. In order to open up for alternatives, they are in need of an improved and more widely shared understanding. For this purpose, the role of the action researcher as a provider of context-free knowledge as opposed to the contextual knowledge already existing in the situation is often put forward (*e.g.* Greenwood and Levin, 1998). If the issue at stake is about swine industry for example, it would be desirable if the action researcher knew about the industry, the polluting impacts it may have, alternative views on nature and farming, economic mechanisms *etc.* Unfortunately, many action researchers are unknowing of issues at stake in the situation of concern. By contextualizing the local situation to participants with context-free knowledge, action researchers can assist stakeholders in learning about conflicting viewpoints, the need for change and options for the future. At the same time, the action researcher must seek out the contextual knowledge that participants hold about the local scene and which may also be key resources for change.

2. Facilitating communicative planning and decision-making processes: The action researcher is expected to facilitate the joint communicative learning process, provide the large variety of planning and decision-making tools required and create a cooperative climate. The focus on communication is closely linked to the idea that local stakeholders can reach agreement on a shared definition of both their present and an improved situation as basis for co-ordinating action.

3. Designing for higher levels of learning: In AR joint learning is closely affiliated with improving communication in the way described above. However, this communication should be directed towards the higher "levels" of learning described by a variety of authors (*e.g.* Argyris and Schön, 1996). Single loop learning, which merely improves present routines, is not sufficient if stakeholders shall achieve the transformation of awareness needed to continue as a self-reflective learning system beyond the time the action researcher is participating. Action researchers must therefore design learning processes so that they enable stakeholders to significantly alter and challenge the assumptions underlying their practice (double loop learning) and the way they think about and organize the learning process itself (triple loop learning).

4. Ensuring reflection on boundaries: In systemic AR approaches, it is often the aim that all stakeholders affected by the definition of the improvement participate. The contention is twofold: 1) progress can only be made when all key players reach a shared understanding; 2) wide-spread stakeholder involvement allows the most inclusive, most ethical and most sustainable position on improvement to emerge (Midgley, 2000). Setting boundaries is therefore a central activity in any knowledge generating system that wants to avoid taken-for-granted assumptions flowing into intervention. Boundaries define both the knowledge to be considered significant and the people who generate that knowledge and who represent a variety of different perspectives. Rural stakeholders may be unaware of some sources of knowledge and/or other stakeholders, or the latter may be stigmatised, resulting in their marginalisation. The action researcher should assist stakeholders in appreciating all knowledge sources and/or for example insist on better gender, class or ethnic balance in the group. Since boundary judgements are closely tied to values, the boundary setting activities can also support moral reflection on the purposes of the intervention itself, and thus, higher levels of learning.

5. Facilitating negotiation processes and conflict resolution: Directing communication towards higher levels of learning can overcome conflicts. Yet when rural stakeholders fail to act communicatively, this is not always adequate and participatory development efforts may be best organized as a negotiation process (*e.g.* Leeuwis, 2000). This has implications for the action researcher, who then needs to shift from a fairly neutral role to being someone with a power-base to forge agreements.

A final, but central role of the action researcher is as a scientist, where s/he is interested in generating more generic knowledge beyond local relevance, for a wider research community and to people in comparable situations in society.

## An adequate balance of closeness and distance

In order to genuinely qualify the participatory process with the competences outlined above and in this way qualify the learning outcome for all, we suggest that the action researcher consciously designs the process so that s/he can maintain an adequate balance between being "close to" the stakeholder arena and "distanced to" this arena. This suggestion is in part rooted

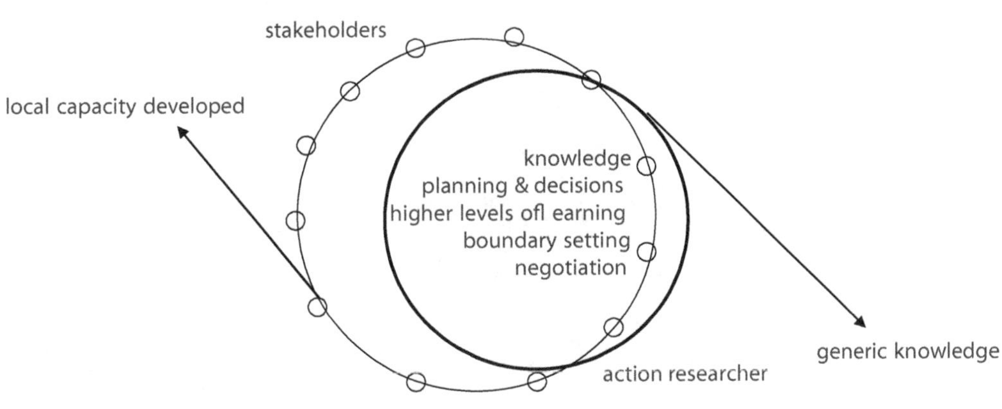


*Figure 1. Multiple roles of the action researcher in relation to stakeholders.*

in queries about the dual synergism between "outsiders" (action researchers) and "insiders" (local stakeholders) – *i.e.* between fundamentally different forms of knowledge input and output, practices and communities. The suggestion also emerges from an enquiry relating to the creation of self-critical systems learning how to persist. Taking the latter first, we worry that the commitment of stakeholders to consciously learn their way into the future, in the way described above, often fades out when the action researcher leaves the situation. We ask if a more balanced action researcher role in relation to "closeness" and "distance" could support continuity towards self-functioning learning systems.

Regarding the dual synergism between outsiders and insiders, we note that Greenwood and Levin (1998) construct an AR model that illustrate how the different learning loops (iterative cycles of planning, acting, and reflecting) in an AR process are carried out partially by the action researcher as an outsider and local insiders together, and to some extent by each part in separation from the other. We ask if the time and space for action researcher activities in separation from insiders, like reflection-in-action and subsequent reflection-on-action and reading up of relevant literature, is adequate to advance own competence development and practice. There is also an ongoing need to make the research explicit and open for the academic community. Through this the action researcher can expose the AR process to critical reflection by fellow researchers. If such "distanced" activities enable the action researcher to better qualify the process with new insights and improved competences ongoing, a more distantly positioned action researcher may be to the advantage of the rural stakeholder situation of concern. Further, we note that it is very demanding on the action researcher to simultaneously manage the classical systematic processes characteristic of scientific inquiry and the dynamic change processes in active cooperation with stakeholders. For example, it is very demanding to facilitate a process and at the same time be an observer, register and collect data. This dual role can also present the action researcher with the dilemma of having to constantly make decisions which may conflict with either the quality of research or the quality of cooperation with stakeholders. We ask if a role where the action researcher is in a more reserved position, however still deeply involved in qualifying the intervention and the process leading to it, may also increase and qualify the more generic knowledge content gained. Figure 2 shows a model for how the action researcher may achieve an adequate balance of closeness and distance.

In the proposed model, the role of the action researcher is as a reflective partner to a core group responsible for coordinating and facilitating the participatory change process. In between the

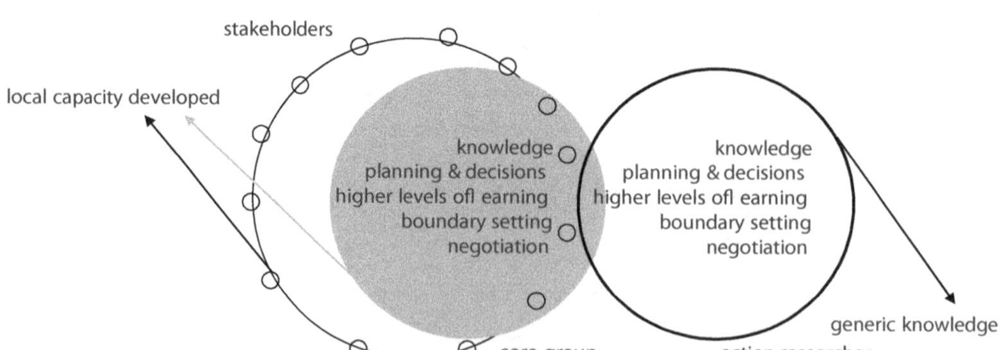


*Figure 2. An adequate balance of closeness and distance.*

different learning loops facilitated, the action researcher has close interaction with this core group. The interaction is aimed at qualifying reflection on the five facilitative roles (now taken by the core group) required to push rural stakeholders in the normatively desired direction. As opposed to the role model shown in Figure 1, the action researcher is not in the same way responsible for the intervention as an active facilitator. This does not mean however that the action researcher becomes an "objective" researcher merely observing the stakeholder situation from the outside. Through the interaction with the core group, and if needed occasionally also with the wider stakeholder group, the action researcher is still deeply engaged, qualifying interventions and the process leading to them with genuine outsider competences. Since the stakeholders from the beginning function as an entity independent from the action researcher, the more distanced position proposed is likely to also support the ongoing self-functioning ability of the system.

The proposed model will be tested as part of a PhD research project conducted in Denmark. The PhD researcher will take the proposed role in relation to 3-4 core groups responsible for coordinating and facilitating in their respective 3-4 existing rural stakeholder networks. In Denmark such core groups typically consist of an extension officer, a local government person, a LEADER coordinator or a local enthusiast - each of them with 1-3 stakeholders. We envision that the PhD research may contribute with valid knowledge on how researchers can actively engage in the "fourth axis" and simultaneously contribute to the generic pool of knowledge on how to meet the challenge of multifunctionality and its links with rural development.

## References

Argyris, C. and D. Schön, 1996. Organisational learning II: Theory, method and practice. Reading, Mass: Addison Wesley, New York. 302 pp.

Cerf, M., D. Gibbon, B. Hubert, R. Ison, J. Jiggins, M. Paine, J. Proost and N. Röling, 2000. Cow up a tree. Knowing and learning for change in agriculture. Case studies from industrialised countries. Institut National de la Recherché Agronomique, Paris. 492 pp.

Durand, G. and G. van Huylenbroeck, 2003. Multifunctionality and rural development: a general framework. In: G. van Huylenbroeck and G. Durand (editors), Multifunctional agriculture. A new paradigm for European agriculture and rural development. Ashgate, Hampshire, pp. 1-16.

Greenwood, D.J. and M. Levin, 1998. Introduction to action research: social research for social change. Sage, Thousand Oaks. 275 pp.

Leeuwis, C., 2000. Reconceptualizing participation for sustainable rural development: towards a negotiation approach. Development and Change 31, 931-959.

Midgley, G., 2000. Systemic intervention. Philosophy, methodology and practice. Kluwer Academic/Plenum Publishers, New York. 447 pp.

Ray, C., 2000. The EU LEADER programme: rural development laboratory. Sociologia Ruralis 40, 163-171.

Röling, N.G. and M.A.E. Wagemakers, 1998. Facilitating sustainable agriculture. Cambridge University Press, Cambridge. 318 pp.

# New futures for traditional regions: A dairy farming systems response to regional development in Australia

*Anne Crawford[1], Mark Paine[1], Howard Smith[2] and Tom Davison[1]*
*[1]University of Melbourne, Victoria 3010, Australia; [2]Dairyfarmers Cooperative, Malanda, Queensland 4885, Australia*

## Abstract

'*GROW*Malanda' is a farmer led regional development program that developed in response to declining returns. In this paper we ask the question 'To what extent can Australian agri-food systems be successfully developed as regional initiatives?' We explore this question using a case study of the *GROW*Malanda project. Within Australia we have focused on the standardisation of our production systems to service commodity markets. The Atherton Tablelands is a unique place to question the relevance of the regional food systems as it is geographically defined, has only one major processor and the farmers are self organised through *GROW*Malanda to develop new ways to farm and assist in the production of valued added products in partnership with their factory. *GROW*Malanda provides an example of innovation across the supply chain, as an integrated research, development and extension project encompassing all aspects from farm to factory.

## Introduction and background

Australian dairy farmers have faced declining returns in recent years as they coped with industry deregulation, higher compliance costs, drought and international market pressures (Dairy Australia, 2004). This was exacerbated in more marginal areas of dairying, where the price adjustment following deregulation was greater, yet there are also higher costs of production due to geographic isolation. These factors combined to reduce regional industry confidence.

In Europe farmers have captured a price benefit by reorganising to produce value-added products or establish 'labels of origin' accreditation schemes such as France's *appellation d'origine côntrolée* (Barham, 2003). Such schemes relate a distinctiveness associated with a specific place or method of production, and have been supportive of rural development, particularly in marginalised areas (Barham, 2003). As a 'new world' country, Australia, at first glance, doesn't have the same extended history of embedded traditions of agriculture and food production, rather is a melting-pot of traditions and methods which have been adapted from a variety of sources.

Dairying has been an important part of the Atherton Tablelands community since last century, and provides a valued 'way of life' (Anderson, 2004) for the 100 farm businesses still operating in the region. These farms are mostly family operated, tropical pasture-based and have an average herd size of 193 cows. However, low industry confidence has seen many farmers exit the industry in the past decade, with a recent survey indicating further exits over the next three years potentially accounting for 27% of current milk production, with 'adequacy of price' as a key factor influencing farmers future plans (Dairy Moving Forward, 2005).

As a region, the Atherton Tablelands has a number of attributes which can be further developed; a unique, highly valued landscape, essentially a single milk processing company (Dairy Farmers Cooperative), a well-recognised geographic location and close proximity to established tourist markets. What were the rural development possibilities to maintain the viability of the dairy industry in this region?

The announcement of a Federal government regional funding initiative catalysed action on behalf of the local milk processing company (Dairy Farmers Cooperative) and the national dairy industry development corporation (Dairy Australia), in partnership with local farmers and the funding organisation (ATSRAC). This process was facilitated by the University of Melbourne.

## Participatory development and social learning

Community participation in development programmes became popular in the mid-1990's (Okali *et al.,* 1994) Conceptual advances in participatory approaches have addressed issues at the level of organisational alignment and the associated function of mediators. At another level participatory approaches have focused on the farm system, and the behaviour of farmers as decision-makers (Paine and Kenny, 2002).

Participatory development and social learning approaches have in large part circumvented many of the adoption problems confronted in traditional approaches to technology transfer by enlisting practitioners in the design of interventions or development of new technologies. The pragmatic orientation of participatory approaches ensures that it is activities (*cf.* concepts and knowledge) that are of central concern to participants. These activities are organised at the level of collective action rather than the individual.

## Developing the proposal

Local farmers had previously mobilised in response to industry deregulation during 2000-2001, undertaking a foresighting task which resulted in the *GROW*Malanda strategy. In October 2003, this was revisited, and further options identified and explored during a two day dairy industry forum. We used an adapted form of the Project Logical Framework (logframe) approach (AusAID, 2000; Farrington *et al.,* 1997) to align the expectations and development criteria of investors with the learning and community development needs of the Malanda region. The logframe approach was used because it provided a process that exposed the similarities and differences in people's preferences and views about the proposal, and orders workshop outputs into the key elements of a business plan that was required by the investors. Our interest in using the logframe was to progress development in the region, not evaluate the methodology for doing this. Our paper is therefore more concerned with the conditions and adaptations that underpinned the use of a logframe approach to a regional development initiative.

The options considered were comprehensive and broad, including niche branding, designer milk and specific breeding programmes, taking into account for each option the market potential, the costs of implementation and change, and its fit with the region. As a first step, farmers worked in five small groups, each with five to seven members, to identify the strengths, weaknesses, opportunities and threats for each option. These options were prioritised and further developed using the logframe approach, capturing local knowledge. All participants had the opportunity to rank each of the four proposed projects, identifying what was attractive about the project, what required further work and providing a rating (low/ medium/ high).

The 'high priority' options were further developed by industry partners into a funding proposal which attracted some $9.82M AUD (€ 5.9M) over two years of federal government and industry investment. The subsequent *GROW*Malanda programme of work is comprehensive and includes farm and factory dimensions, focusing on increased protein production, improved farm profitability and factory value-adding (Table 1). The programme is overseen by a steering committee which includes representation of all the funding organisations and key local farmers.

The purpose of *GROW*Malanda is to help the local dairy community to take charge of its future development by focusing simultaneously on: rebuilding dairy business viability; developing and up-skilling the capabilities of dairy industry businesses to meet existing and future challenges of dairying; and re-establishing regional confidence and capability. The attributes of the options selected were therefore based on these requirements which included:

- reasonably risk adverse - high probability of generating profits in the short term;
- maximised benefits from the unique characteristics of the region; and
- perceived as feasible to local farmers.

The selected development option also needed to avoid introducing more stress and anxiety in terms of additional workload, large learning challenges, high capital outlay etc, on behalf of the farmers. As a consequence, a reasonably conservative option was selected relative to the range of options put forward.

The research, development and extension programme involves the management and implementation of nine project modules operating under two interrelated streams, Farming Systems (Profit through Protein) and Factory (Building Protein Capability) (Table 1). Through

*Table 1. Summary of the Farm and Factory modules.*

| Farm Systems Modules | Objectives |
|---|---|
| FS1 Growing farm businesses | Develop individual farm business plans to improve profitability, based on the findings of the other Farm Systems modules. |
| FS2 Influencing protein fractions | Identify the on-farm factors influencing whey protein levels. |
| FS3 Alternate farming systems | Form farm study group to identify, evaluate and promote alternative farming systems. |
| FS4 Greenfield prospectus | Identify and assess new dairy farm sites. Prepare business prospectus. |
| FS5 Attractive regional employment | Train specialist dairy farm managers and employees. |
| Factory Modules | |
| P1 Protein yield improvement | Increase lactoferrin yield by extraction from cheese milk or salty whey. |
| P2 Testing whey proteins | Develop a test method for measurement of low levels of lactoferrin and other whey proteins in milk, for factory process control and farm milk quality and payment systems. |
| P3 Freezing protein quality | Improve protein bioactivity and solubility through better freeze-drying pre-treatment methods. |
| P4 Proteins for immunity | Build and commission the second stage of the Malanda whey protein plant, production of immunoglobulins. |

this range of integrated capacity building, skills development and technology commercialisation activities, *GROW*Malanda aims to halt the decline of dairying in the region.

## A collaborative learning approach

A focus on farmers accessing higher milk prices by increasing the protein production of their farm business underpins the *GROW*Malanda strategy. As part of this, the project team was interested in achieving a fundamental shift in farmers' thinking from 'our farm's economic performance is due to factors outside my control' to one that says 'it is primarily my responsibility…'. Evaluation was therefore undertaken in real-time to inform the project team members of farmers' attitudes, knowledge and understanding of relevant concepts to allow the programme to be tailored accordingly. A farmer survey conducted just prior to the project commencing indicated a general complacency about managing their farming systems for increased milk protein (QMR, 2005). Our interpretation was that we were starting from a low base in terms of farmers taking responsibility for their performance, as well as providing a focus for the technical team to address methods that enabled the farmer to have more control eg grazing management or feeding strategies. Addressing this requires a greater connection between opportunities to improve protein and impact on milk price, that is, farmers proactively becoming price makers, as well as linking with greater use of business plans to increase confidence as farmers are encouraged and supported to take control of their whole farm performance. Business plans made targets and strategies more explicit for the farmer and advisor to work together and also provide a point of reference for reviewing performance – a reflective learning opportunity.

The survey also revealed to the core team that some farmers required a combination of new skills and knowledge and increased business confidence if they were to increase milk protein production profitably. Issues of farm profitability, farmer confidence, maintaining a critical mass of farmers and production, workload and stress levels on farm had previously been identified (*GROW*Malanda, 2001 unpublished report; Dairy Moving Forward, 2005) and required addressing through an integrated supply chain approach. The project was designed to be comprehensive, drawing on existing and new research and providing a variety of learning experiences; one-to-one support, discussion groups, field days and newsletters. This was undertaken through the established networks of the current dairy advisor.

The project team also recognised that it was important to develop a diversity of future farming system options, as to recommend any single 'solution' would jeopardise a proportion of the milk supplier base. For example, two working groups were established to assist in defining two alternative profitable and sustainable farming systems. These groups included farmers, scientists and dairy advisors, to collect and review information, undertake computer modelling on revised scenarios and present recommendations to all farmers for information and discussion. The two working groups address different questions, namely, an 'easy care' system of milk production for the region, and large milk production farms. The approach taken here was designed to capture the experience and knowledge that all parties contributed to the development of new farming systems.

The project is still underway, but we can say that several lessons have emerged from our experiences to date. These include:

- appreciation of the historical context when entering the field is required to build out of earlier initiatives where possible;
- the importance of co-development with the investors when securing funds;
- the need to include power brokers from both local and federal levels;

- development options require sufficient detail for locals to engage with them;
- the success of an option relies on attracting a critical mass of support;
- a core team was required to follow through on plans and proposals; at the same time there had to be a process to maintain engagement with all stakeholders;
- flexibility and adaptation around the business planning approach is required;
- monitoring and evaluation procedures provide a framework for planning and reviewing progress; and
- the local voice (farmers and advisors) is central to specifying needs and targets.

## Conclusion

We conclude that when resources and conditions support coordination and collective learning it is possible to instigate regional agri-food initiatives. However these initiatives need to work within the historical and socio-economic context of the region. *GROW*Malanda has required a large investment of resources, commitment and activity from farmers, milk processors, regional development organisations and research and extension organisations. To date, all partners would suggest that it has been a rewarding undertaking to be involved in, especially with the concurrent increase in local industry confidence. It provides a valid example of a regional development approach for other dairying regions in Australia, where similar levels of industry motivation exist. Motivation and commitment by industry partners are the essential elements to this initiative.

## References

Anderson, J., 2004. Dairy deregulation in Northern Queensland: the end of traditional farming? Anthropological Forum 14, 269-282.

AusAID, 2000. The Logical Framework Approach. Australian Agency for International Development. The Australian Government, Canberra, Australia.

Barham, E., 2003. Translating terroir: the global challenge of French AOC labelling. Journal of Rural Studies 19, 127-138.

Dairy Australia, 2004. Australian Dairy Industry in Focus 2004.

Dairy Moving Forward, 2005. Dairy 2005: Situation & Outlook. Report to the Australian dairy industry. June 2005.

Farrington, J., C. Thirtle and S. Henderson, 1997. Methodologies for monitoring and evaluating agricultural and natural resources research. Agricultural Systems 55, 273-300.

Okali, C., J. Sumberg and J. Farrington, 1994. Farmer Participatory Research: Rhetoric and reality, Intermediate Technology Publications, London.

Paine, M. and S. Kenny, 2002. Intentional Learning: interplay between farmers and service providers. Fifth IFSA European Symposium, Florence, Italy, 8-11 April.

Quantum Market Research (QMR), 2005. Grow Malanda Baseline Research 2005. Survey Amongst Dairy Farmers in the Atherton Tablelands. Report prepared for *GROW*Malanda.

# Learning from participatory research with organic farmers in the UK and Sweden

*David Gibbon[1], Cheney Longville[1] and Karin Eksvärd[2]*
*[1]Craven Arms, SY7 8DR, Shropshire, United Kingdom; [2]SLU, CUL, Centrum för uthålligt lantbruk, Box 7047, 750 07 Uppsala, Sweden*

## Abstract

This paper is about learning among researchers, advisers and farmers within research projects on organic farming systems in Sweden and the United Kingdom. Both projects involved partnerships between several institutions, natural and social scientists, advisers and many farmers. Scientific data, farmer written case studies, field notes, learning experiences and evaluations have been collected in both countries. The UK research project and the Swedish initiative arose out of a wish to bring organic farmers into a research process which, in the past, had been driven by researcher-defined problem areas and designs. Both developed research in which farmers had an equal role in the definition of the "problem" to be researched and the design, implementation and monitoring of the process over time. Mutual learning among participants has occurred through their engagement with other stakeholders and a willingness to change behaviour within different areas of activity. Participants agreed that the approach can make a very positive contribution to the evolution and credibility of organic farming.

## Introduction

This paper is about two initiatives which are engaged in the development of farmer participatory research within UK and Swedish organic farming systems. The primary focus of the paper is on communication and learning processes among farmers and researchers and in contributing to new ways of conducting research among organic practitioners and researchers. The outputs are discussed in other papers (see Davies *et al.*, 2006; Ögren *et al.*, 2002).

## Background

In both cases, a variety of stakeholders were initially involved, including farmers, advisers and researchers. The projects arose out of a wish to bring organic farmers into a research process which in the past had been driven by researcher-defined problem areas and designs. Both wished to develop a new approach to research with organic farming in which farmers had continuing roles in the definition of the research agenda and in the design, implementation and monitoring of the process over time.

### *The Swedish initiative*

The Swedish research initiative began in 1998 and involved several groups of ecological farmers, who wished to develop a research capability around their common interests. Participatory research groups formed from different motivations and financial support including: - Cereals, Greenhouse Tomatoes, Dairying, Vegetables, Ley seeds, Weeds, Green Manure Systems and Poultry Production. The initiative to start the groups had mainly come from advisors or researchers. Group meetings were usually held on farms. Resources came from educational, research and development grants from the Board of Agriculture, regular research grants, some contributions from CUL, other EU-finance and farmers own time. In all existing groups there was at least one person who had some training in the facilitation of participatory research. During

2003, the Programme for Ecological and Organic agricultural Research, which is the main guide for government and private funding bodies, was rewritten. In this programme, participatory research was not only mentioned but emphasised as one of the major complementary approaches requested for future research.

### *The UK project*

The UK project began in 2002 and was built around a partnership between several research and advisory institutions and farmers. The project was funded by the Department for Environment, Food and Rural Affairs (Defra) who wanted to support the development of farmer-participatory research approaches. The first activity was training in participatory methods for the research team, followed by a large farmer stakeholder meeting during which the actors discussed alternative worldviews, expectations, key problem areas and methods. This workshop has been repeated annually and continues to be a constructive social learning forum. Three focus groups were established on: perennial weeds, farming systems and communication and knowledge. These groups contained farmers and researchers/advisers in equal numbers and they met regularly to develop strategies, plans and action. The other key elements of communication were farm-based field days (30 over 3 years), with mixes of farmers and researchers present, the presentation of papers of work in progress at conferences, the development of a Website which collated all current and past information on weeds and weed management. 60 interviews were conducted with farmers and these can be found on the website (http://www.organicweeds.org.uk/).

## Results

### *Outputs*

In Sweden, the tangible results shared mainly come from the dairy group, the green manure group and the tomato group, but also from other individuals and groups working with participatory research. The dairy group farmers' interests were in finding ways to become self sufficient in fodder and to use only organic fodder. Two dairy farmers have bought small oilseed mills to improve the nutrient content of their fodder. The green manure group worked on learning about qualitative aspects of different green manure systems. The tomato group developed their competence in better nutrient management, reducing labour demands and disease control through a shared experiential learning. New energy initiatives also emerged from this group.

The Swedish groups have published their outputs and learning in magazines and leaflets for farmers, reports published by CUL (Centrum för uthålligt lantbruk) the county administration, posters and presentations at the organic farming conferences in Sweden. Most groups have published some kind of article and three groups have produced larger project reports. There has also been farmer to farmer exchange outside of the groups and advisors applying the results and experiences with other farmers.

In the UK weeds project, there have been many lessons and learning of new and more effective ways of managing weeds within different farming systems (arable, dairy, vegetables and mixed). Much of this has come through farmer-farmer sharing of experiences, but also through the accumulated knowledge from papers and farmer case studies on the website. A great deal of this has been through an appreciation of the need to maintain sound, strategic farm and land planning (rotations), good husbandry, greater flexibility and tolerance to weed presence and an array of control methods which include hand tools and/or the careful and timely management of machinery and grazing animals (wherever possible).

The Website has developed as an accessible reference, a recording system and site for feedback. Also it has become a store of 60 farmer case studies. Journal articles, leaflets, articles in farming papers, conference papers, summary leaflets on key weeds, systems and management strategies have also been produced A low cost book is now planned. The meetings were held as a means of both planning with and updating stakeholders in accumulated knowledge and experience.

### *The learning process*

Within both initiatives, care has been taken to note the processes of learning among participants through conscientious recording of planning processes, meetings, field visits and both formal and informal exchanges. Some of the methods and techniques used in the UK project have been developed during recent European research initiatives on learning (Cerf *et al.*, 2000) and social learning (SLIM 2005). There have also been exercises in monitoring learning experiences among both researchers and farmers. These include: SWOT analyses, short, regular evaluation exercises and the use of several PLA methods such as Venn diagramming, force field analysis and critical reflection on researcher learning (in the UK case).

*Farmers*

Within Swedish groups, new skills have emerged among farmers, notably in facilitation, the recording of meetings and the development of e-mail communication, as they have taken on greater responsibility for planning and driving the process.

In both projects, farmer participants derived the greatest benefits from the communication that emerged from farm-based field visits and from small group meetings in which ideas and experience are shared by those working in similar contexts. Several farmers, in both projects, used words like 'inspirational' when reflecting on the intensive group meeting benefits.

The visibility of results – through documentation, accessible magazine articles and other accessible publications in the Swedish case and through the production of short leaflets and the reference material on the website in the UK case – has been highly valued by farmers.

*Researchers*

In both cases, researchers have had learning experiences through their involvement in participatory work. For many, their normal work environments, include very clear agendas and programmes of work. The need for flexibility in responding to other world views and changing contexts has at times been challenging, but in general, they have recognised that being involved in these projects has been a very positive learning experience for them. Some researchers have expressed how this working mode gives research more relevant meaning and how important it has been to get a broader, systemic picture of how farmers' situations affect their areas of interest.

Additionally, researchers have expressed an appreciation of other researcher skills and knowledge, both in different disciplines and in thinking about more interdisciplinary and systemic ways about farm based problem areas. They also recognized the importance of informal, farm based research which had great relevance when dealing with complex, systemic problems.

*Advisors*

Both advisors and researchers have learned new communication and presentation skills and participatory techniques. In the UK case they have recognized the value of access to large knowledge store and how they will have an important future role in extracting and summarizing information in different ways for a variety of farmers and other clients.

At an Open Space arranged by the Board of Agriculture in Sweden, advisors said that the development of participatory research has provided a positive platform for advisers to explore questions in depth, to adapt their focus to local situations, to be involved in a group which is created to support local decisions and to facilitate iterative action and follow up.

The more open and flexible the project agenda is, the greater the possibility is for all participants to engage. From the work in Sweden there were experiences of how difficult it could be to reach a high degree of genuine participation when the setting of the project had not been participatory in its original format.

## Conclusions

Flexible, responsive planning, careful and accurate recording of process and the provision of platforms for critical reflection by actors, were all considered to be key methodological elements which ensured success in these projects

The multi-stranded approaches that have been adopted here have generated outputs of interest for researchers, advisers and farmers. New scientific findings have been interpreted and communicated, but also everyone involved has engaged in a constructive social learning process.

Both farmers and researchers used almost the same words to express both the importance of this mode of research and the enjoyment derived from engaging with people who are familiar with the subject but who come from different perspectives and knowledge bases.

Most researchers, advisors and farmers who were involved in these research projects recognized that participatory processes are important to include in future organic research and that valuable lessons have been gained from these cases on: - understanding systems of interest, the tools and platforms for stakeholder learning and the role of facilitation for this kind of research to be effective.

## References

Cerf, M., D. Gibbon, B. Hubert, R. Ison, J. Jiggins, M. Paine and N. Röling, 2000. Cow up a Tree: knowing and learning for change in agriculture. Case Studies from Industrialised Countries INRA, Paris. 492 pp.

Davies, G., R.J. Turner and D. Gibbon, 2006. Knowledge development for weed management in organic farm systems in the UK. In: Langeveld, H. and N. Röling (eds.) Changing European farmingsystems for a better future, Wageningen Academic Publishers, Wageningen, The Netherlands, pp. 304-308.

Ögren, E., K. Homman, O.Andersson, E.I. Adhe, K-G. Berglund, K. Eksvärd, U. Engström, A. Mizban, B. Eriksson, L. Gäredal, A-M. Johansson, A.Y. Larsson, L. Bartoft, B-I. Nilsson, D. Johansson, H. Nilsson, J. Björklund, G. Pellas, K.M. Sjöstedt, L.S-E. Wilhelmsson and T. Wålstedt, 2002. Växtnäringsnyttjande i ekologisk tomatodling ett dokumentationsprojekt genomfört under 2002 i Dalarna Gästrikland, Hälsingland, Uppland, Västmanland och Södermanland samt sammanfattning av projektperioden 2000-2002. Länsstyrelsen, Västmanland.

SLIM, 2005. Social Learning for the Integrated Management and Sustainable Use of Water at Catchment Scale. EU Contract No. EVKI-CT-2000-00064 SLIM. Five country project. http://systems.open.ac.uk/page.cfm?pageid=sustdevptS.

# Setting a research agenda to support change in agriculture: Learning processes, stakeholder participation, R&D policy making

*Bernard Hubert*
*INRA, France, on behalf of the LEARNing Group*

## Abstract

The LEARNing Group (Learning in European Agricultural and Rural Networks: Institutions, Networks, Governance) is constituted by researchers from 9 countries (Australia, Belgium, Denmark, France, Greece, Hungary, Poland, The Netherlands & UK), working with the aim of analysing the relationships between technology and society within the agricultural sector. It aims to produce concepts, facilitation processes and tools that support participation mechanisms for collective action at different levels and began sharing experiences in creating spaces for learning, capacity and action building in various agricultural contexts during the 1990s. This paper presents the goals, process and results of an EU funded project carried out in 2003-05.

## Rationale

The relationships among stakeholders in agricultural, food, environmental and rural policy and practice are not optimal in that these relationships are not leading to the desired outcomes such as multi-functional agriculture, safe food, and a vibrant rural economy. New modes of collaborative adaptive management among farmers, scientists, consumers and stakeholders in the rural world need to be developed, as indeed society's organisational capacities to carry out R&D.

The development of the agricultures of industrialised countries has benefited from strong public and private backing to support the continuous productivity gains with a key role for agricultural research in it. But the situation has now changed. No real consensus has been reached on the role of agricultures in European countries, and the funding of public R&D is under question. This creates uncertainty about the models of development appropriate to meet the challenges agriculture has to face. Each country deals with this situation in its own ways according to its agricultural history and model of development.

This situation is challenging R&D professionals and posing new questions to social sciences research. Our project seeks to ground pragmatically the relationships among researchers and practitioners within situations of learning and change. Our main research question was: *How to involve diverse stakeholders in meaningful conversations, so that they engage in new ways of learning to lead to actions and outcomes that move farming and related industries toward a multi-functional agriculture?* With the aim to expand capacity building among stakeholders through their participation in the development of a research agenda for R&D in agricultures of industrialised countries (ICA), a participative bottom-up and comparative procedure was implemented. Each country team undertook a set of small-scale exploratory procedures in order to put heterogeneous actors at work to produce a critical agenda of R&D activities within ICA. A cross-national exchange of practices and agendas became the central feature of a European conference in February 2005 and led to the creation of an Agenda for Learning in ICA.

Our participative bottom-up approach was also seen as a way to promote policy process, enable stakeholders to express the problems they face while implementing R&D, and initiate a dialogue with policy makers. Our comparative approach to local R&D practices enacted at the European conference also allowed us to better understand how to do R&D in a European space by considering the way that R&D practitioners and other stakeholders interpret the heterogeneity of national cultures, institutions and economic objectives towards agriculture.

The process is based on a *"search* and *propose"* procedure using meetings, exchanges of information, round tables and conference. We set up a two loop process: the first loop aimed at involving various stakeholders and us in the design of a new agenda for issues concerning R&D in ICA (WP2, WP4, WP6); and the second aimed at co-ordinating us, as the researchers working on learning in R&D processes in ICA, in order to translate this agenda into a common platform for social science researches on learning processes (WP1, WP3, WP5, and WP7).

## Performing the process

During the first meeting in Paris (WP1 December 2002), LEARNing participants agreed on WP2 tasks in each country and devoted time to developing a common methodology, based on the distinction of what we named (Table 1):

- A "promising configuration", where interested stakeholders are involved and interact to deal with a common issue at stake. It represents a situation in which there is potentially an "arena" to design.
- And an "arena", *i.e.* a closer setting (*"dispositif")* built upon the process identified as a promising configuration to create a situation of conversation between some of the stakeholders and the researchers.

*Table 1. A range of relevant situations.*

| Domain of change | Countries where in it addressed | Main stakes | Partners involved in the arenas |
|---|---|---|---|
| Production systems | Belgium, Denmark, France, Greece | Developing alternative farming models for environmental friendly agriculture | Farmers, rural, environmental and agricultural advisors, consumers |
| Role of agriculture in rural areas | France, Netherlands, UK, Poland | How does agriculture participate in a multifunctional territory? | Local communities, environmental associations, water supply companies, foresters, farmers advisors |
| Agriculture advising | Australia, France, Hungary | Which skills, education, institutions to promote in order to support a higher diversity of farmers and to deal with stakes linked to multifunctionality? | Farmer's representatives advisory agencies public agricultural administration, agricultural education |

## Lessons learned: A renewal of research position in social debates

The Country teams reported a change in the role of research position in the social debates initiated by them. A schematised representation of the relationships between involved stakeholders is shown in Figure 1, with the traditional relationships shown in 1A and the innovative ones, where our research teams were engaged, shown in 1B.

The very classical situation (1A) has been disrupted by recent changes, even crises, that happened in industrialised countries with environmental issues (N pollution, biodiversity erosion, *etc.*), food safety (BSE, dioxin contamination, development of organic farming, *etc.*), changes in urban-rural relationships (housing, employment, landscapes, *etc.*). It needs new forms of research involvement to go with these changes and commit citizens in new forms of action (NGOs with regard to environment or food), emphasize farmers' practices, rush the traditional extension services and push new institutions in charge of new issues (water management, landscape preservation, *etc.*) and shout at a renewal of democracy in public action. Figure 1B represents new ways of practicing research in such complex situations.

***Learning a process to be managed***

We can point out some specific elements related to our research process, based on a strong commitment of partnership with stakeholders, involving them in dialogue on the role of social science in innovation processes and facilitating learning among a range of partners. This put into consideration some crucial elements such as:

*A - How to initiate the conversations and invite participation*

Would stakeholders recognise the rationale and aims of the LEARNing project as legitimate and of compelling interest in terms of their own 'stakes'? According to Country teams, the question had been put explicitly to the stakeholder participants, as a means to legitimate the processes set in train, and in order to refine through discussion the nature of the engagement to which stakeholders were invited.

In each case, different procedures have been followed to introduce the LEARNing project and invite participation. Typically, where new contacts have been made, the protocols have included: construction of systems diagrams, by the research team, of the boundaries of the sub-systems of interest; a more or less formal Stakeholder analysis, of the actors and Intermediary Bodies (IB) in each sub-system, and their existing relationships; preliminary identification by the researchers of the key stakes involved; construction of a preliminary mailing list, crafting and dispatch of a letter of invitation to engage in preliminary conversations; preliminary 'round tables'; feedback reports to round table participants; launching of jointly planned activities.

*B – The representativity of our partners*

According to whom we invite to these exercises, how representative will the emerging R&D agendas be? Or, what interests and stakes will be represented? Different and as yet partial answers can be offered in each case. There has been no attempt to standardise across the nine countries either the actors, or the stakes, in terms of what they might 'represent'. The unity of focus is at a higher level, on the processes by which a wider set of actors can be effectively drawn into R&D agenda setting for the social sciences in agriculture.

*C – From dialogue to inter-action*

The intention is not to assist stakeholders to talk about R&D agendas only, but to elicit agendas through *inter-action*. Scientific institutions, and many scientists, tend to think about 'integrated research' as primarily a communication problem, to be overcome by more explicit and sophisticated use of communication tools and processes. For example, a survey carried out in Australia for leading environmental research funding organisations, elicited six different metaphors to describe how 'integration' could be achieved. In contrast, the research carried out by the LEARNing group over 15 years, has surfaced the central idea that integrated research most effectively occurs when stakeholders collaborate in shared actions around a 'socio-technical mediating object', the material focus of enquiry, the physical embodiment of the issue at stake. But, it is not some essential characteristic of the object that is important but its power to shape the processes of 'stake-holding' (the exercise of one's stake), and the understanding of what is at stake, as actors engage in resolving problems and improving messy situations, through action.

*D – The researcher's role*
Researchers in the LEARNing project are playing (at least) three roles: (i) as researchers, independent of, and external to, the processes observed; (ii) as 'expert resource persons' in a discipline, a theme, or as practitioners of particular research practices; and, (iii) as stakeholders, whose central stake is an interest in researching inter-active learning participants. Some researchers are also playing process management roles in ways appropriate to their own particular suite of activities. For example, in the Polish case, the researchers act as the moderators of the Focus Group Interviews. Each identity confers different strengths and limitations: *e.g.*, the choice of the Dutch team to place itself as a participant in inter-active dialogue, as one stakeholder interest among others, limits their ability to impose formal research instruments on what is monitored. In this case, research techniques will favour those that rely on the formalisation of self-learning and participatory assessments.

*E – The learning process itself*
It could be investigated as: (1) the dimensions of learning: as an object (the content of learning), as an action (the process) and as policy (the agency effect); (2) learning as a process: the role of systems of activities, and more precisely the interdependency of various communities of practice (CoPs), specially when collaborative action for change is the main concern; (3) the place of individual reflections in learning (role of narratives, interplay between individual and social learning processes); (4) the role of intermediary objects and concepts (*e.g.* maps, a herd-book, *etc.*) in the process of knowing: objects and quasi-objects used in translation processes, and consequently transformations of human actors and ability to commit into change.

One important conclusion arose: we noticed how much the issues of collaborative action and learning were interwoven and dependant on situations in which change were happening in ICA. One reason seems to be the need for clarification despite some theoretical difficulties to assess and somehow evaluate learning as an outcome of a process. We concluded that this process was itself not only path-dependant in a course of action but also performed by humans positioned in their own life, and finally affected by pervasive or enhancing structural effects in a given social context. Therefore, we came to raise the question about the causality or at least the co-definition of learning, collaborative action and changes in situation.

## Social Learning: Some questions to follow up

Social Learning is a process that needs time and taking place in concrete situations with regard to who was concerned and willing to interact, what issues were at stake and in which general (political, economic, social) frame. It could be used as a co-ordination mechanism besides market and institutions in situations characterised by uncertainty and complexity of the faced

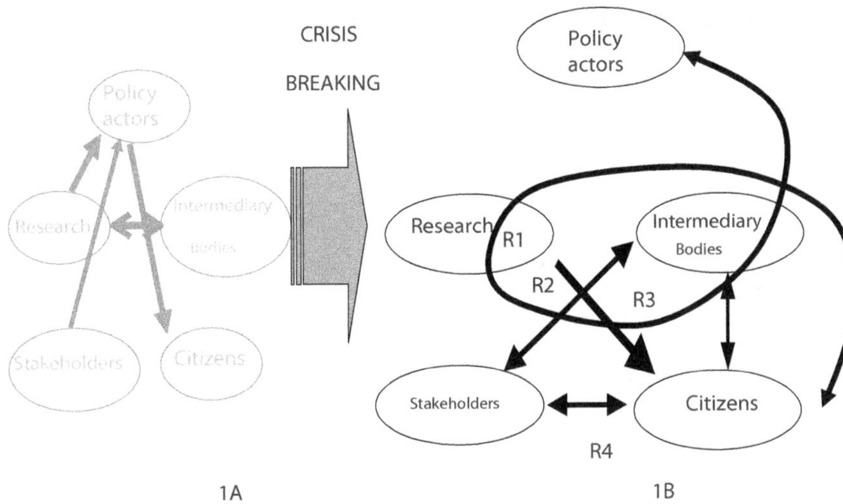


*Figure 1. Traditional relationships (1A) and innovative ones (1B) in which the researchers in LEARNing are engaging.*

issues. It can be a cognitive process, where the researcher's role is to intervene as an agent of change, committed in a collective action which is leading to innovation (technical as well as organisational).

Intermediary Bodies (IB) - institutions that organise action, e. g. advisory bodies, environmental agencies; Stakeholders (SH) - individuals or groups with defined interests and explicit stakes in R&D processes in agriculture; Citizens (C) - members of the public with a general interest in the well-being and sustainability of rural livelihoods, natural resources, food safety and quality; Policy actors (P) with whom IB, SH, and C are invited to engage in the setting of R&D agendas; Researchers (R).

In Figure 1A, researchers are in relation mainly with IB and give advice to policy makers, in a traditional position of "expertise" and a classical 'institutional' way of functioning. They have very few relationships with SH and C, mainly considered as observed objects in their research. Figure 1B represents new ways of practicing research in such complex situations, where researchers are involved in learning processes with a wide range of stakeholders, by creating loops for interactions.

Researchers in this representation are positioned as instrumental to the process and its outcomes, in four specific ways: as researchers external to the process, observing, recording, and analysing methods and processes for the co-creation of inter-active learning (R1), who through the LEARNing project are orchestrating invitations to engage in conversations and actions (R2), and who are looking at the 'objects' around which new actions and relationships among IB, SH, and C, are mobilised (R3), as well as participating in the co-creation of the new actions and relationships (R4).

# Sustainable rural development and communicative learning

*Egon Noe and Anne-Mette S. Langvad*
*Danish Institute of Agricultural Science, Department of Agriecology, Blichers Allé, Postbox 50, DK-8830 Tjele, Denmark*

## Abstract

Functional differentiation within society at large poses a major challenge to practising sustainable rural development. Multiplication of perspectives on sustainability calls for a theoretical position that is based on the integrity of each of the perspectives in play and for an approach that is able to coordinate the various partial perspectives. In this article we present such a theoretical framework for poly-ocular communicative learning.

## Introduction

Agriculture and rural development constitute an entirely new complex of problems acting as a challenge to the theories of learning. Negative rural development is often explained by general societal development traits, such as globalisation, centralisation or heightened demands for efficiency. However, there are other ways in which to approach the rural crisis situation: We hold the view, that due to functional differentiation society has seized being perceivable as local entities. To an increasing extent, local community functions are attached to networks, which have no local rooting and therefore do not contribute to local development with respect to *e.g.* job provision, earnings and social contact.

Non Governmental Organisations representing *e.g.* environmental, nature quality, landscape or out door life perspectives on sustainable rural development have no clear local attachment either. They are *perspectives* organised across geographical and cultural boundaries.

Specialisations into highly differentiated perspectives apply to scientific disciplines and institutions preoccupied with creating sustainable rural development as well. Consequentially, 'sustainability' means something very different to different sciences. Because these multiple understandings cannot simply be added it becomes more and more difficult for the disciplines to look upon rural districts as coherent entities and to take into consideration other perspectives.

Without question, differentiation of perspectives is connected to a general positive societal development in many fields, but as already mentioned it does have its negative consequences, as well. To our view a primary obstacle is the fact that it becomes more and more complicated to integrate across differentiated perspectives and on grounds of rural districts *per se*.

## Proposal of further developments in light of the present challenge

It lies within the concept of 'sustainable development' that it can only take place through a balancing of the various interests, perspectives and considerations involved. From a scientific perspective, rural development can only be made sustainable when research is able to support local stakeholders in balancing their various perspectives and in reaching synergetic and locally bounded solutions. In other words, what constitutes the greatest challenge is to develop a foundation and an approach to support local involvement and coordination of various partial perspectives and conceptualisations in order to generate an integrated rural developmental *practise*.

Within the IFSA network and the LEARN-group in particular, there is a tradition for studying agriculture and sustainability from the perspective of learning (LEARN Group 2000). However, we find that within the tradition there is a demand to make further research in order to be able to cope with rural sustainable development in the future. Our proposition to manage complexity is subsumed in a concept that we call 'poly-ocular communicative learning'. To comprehend what poly-ocular communicative learning means, it is pre-requisite, first to explain what lies in the concept of poly-ocularity and secondly to elaborate upon what is meant by discursive learning.

## Poly-ocularity

If we give up for a moment the classical reductivist perception of knowledge as something that can be added and like bricks or puzzles come to constitute a whole and if, simultaneously, we acknowledge that different disciplinary observations contribute with different understandings to a certain problematic, consequentially we have to find an alternative theoretical foundation to capture multi-perspectivism.

By way of the American anthropologist and philosopher, Gregory Batesons', concept of 'binocularity' or 'perception of depths' we can understand the implication of working with isolated single perspectives as foundations for observation.

By way of two observations slightly displaced, which our eyes represent, we are able to perceive depths, *i.e.* to estimate of relative distance of two objects irrespective of their absolute differences of size. Via our perception of depths, with great accuracy, we are able to pour water into a glass in front of us. The information upon which the estimation of distance is based is neither included within one picture or the other. This we can make sure of by trying to close first one eye and then the other. The third dimension, *i.e.* the distance, is produced within the brain on the basis of another type of information. This other type of information is in fact generated by the little differences among the two pictures, that our eyes produce (Bateson, 1979). Projecting the example onto the differentiated perspectives described in the above section, the single perspective, *i.e.* what one eye observes, becomes more and more nuanced, but at the expense of perception of depth.

Magoroh Maruyama, a Japanese sociologist, has made further developments to Bateson's binocularity. According to Maruyama the difference between bunches of pictures make it possible for us to generate invisible dimensions. Such dimensions do not come into existence simply by adding more perspectives. They are produced cognitively as part of communicative operations other than those that give rise to single perspectives. To this principle Maruyama give the name of poly-ocularity (Maruyama, 1978, 1985, 2004).

According to Maruyama differences between many oculars produce large amounts of so called non-redundant complexity and give comprehensive meaning to an object. The more dimensions that are involved in a single perception by way of sequence in time, the more comprehensive an understanding ensues. But as a negative side effect to this complexity of understanding is the risk of chaos due to information overload.

## Discursive learning

The concept of discursive learning builds on Niklas Luhmann's conceptualisation of discourse. To Luhmann, communication is a self-referential autopoietical process, *i.e.* a discursive process enabling and simultaneously delimiting both the conditions for communication and the way that social relations can accordingly be formed (Luhmann, 1995).

The product of *cognitive* learning can be defined as change that is produced by experience coupled to reflexivity. As a process cognitive learning leads to change or reconstruction in the manner that a person observes his or her surroundings. The changed manner of observation subsequently influences how the person acts.

Comparatively, *discursive* learning can be defined as reflexive communication on the basis of some kind of experience, which leads to a reconstruction of discourse. The way that discourse is reconstructed influences how discursive elements can potentially interrelate, and the way that discursive elements become receptive to the surrounding world of communication.

For instance construction and maintenance of a disciplinary discourse can be seen as discursive learning. Each researcher's contribution to discourse by way of research results can be considered as discursive experience and subsequent scientific debates with respect to results then as reflexive communication. Some times such debate leads to a general theoretical reconstruction of the way in which scientific disciplines observe their objects. When this happens discursive learning has taken place.

Both cognitive learning and discursive learning have practical consequence. A reorganisation of the way that communication proceeds - *i.e.* a change in the rules of what constitutes a better argument - leads to change in the meaning obtained through communication. When this happens the object of communication is practically affected as well.

## Poly-ocular communicative learning

In order to ensure sustainable rural development, it is pre-requisite to simultaneously involve multiple perspectives, and thus to look upon rural districts in a poly-ocular way. This is a central statement of the paper. Because poly-ocularity is something too complex for one person or one perspective to perform we have to manage poly-ocularity by mutual involvement within communication processes. As a learning procedure the aim of poly-ocular communication is to change and to develop our possibilities for enabling sustainable rural development. By communicating poly-ocularly it is possible to both enlarge and understand the possibilities for synergy that can potentially evolve.

Even though poly-ocular learning also proceeds through communication, it is different from discursive learning. Discursive learning is what happens when self-organising disciplines communicate within their own rationality and thereby produce purer perspectives from an internal disciplinary point of view. Characteristic for intra-disciplinary communication is that it settles for the better argument, which by definition has to be communally agreed upon. In contrast poly-ocular communication makes a virtue of having very different points of departure. It seeks the different structures of meaning that different viewpoints produce. Poly-ocular communication takes seriously any viewpoint put forward by the participants within a communication. It readily acknowledges any value of looking upon matters differently. In addition, the principle of poly-ocular communication is the fact, that nothing can ever be satisfyingly described from a single perspective; no single perspective is ever able to capture the matter rightly. Operatively this means that integration has to be met by bringing more perspectives into play within a specific context.

Poly-ocular communicative learning influences individual learning and deliver input to discursive learning as well. Acknowledging that a cause can be looked upon from various angles and to communicatively describe these angles simply make for a change within the participants involved. Through engagement in poly-ocular communication participants come to acknowledge that the object is more complex to participants, than they are themselves able to observe.

## Three levels of communicative poly-ocular learning

Whereas the objective for discursive communication is to generalise and universalise its observations, poly-ocular communication has to be about something concrete, so that various perspectives can simultaneously attach to it. It is imperative to speak of the same object in order to be able to meaningfully communicate and to proceed communicating from different points of view. Still, participants need to be able to observe the object in their own right. There are three essentially different levels of concrete objects for communication. Thus, there are also three levels for poly-ocular communicative learning.

The first level refers to concrete, spatially bounded complexes of problems. The second level refers to objects for communication, which are generalised representations of something objectively real. Finally, at the third level, poly-ocular communication treats of values and rationalities on behalf of which the observers observe and conceptualise groups of problems or generalised objects.

At the first level, poly-ocular communication treats of *questions addressing a concrete complex of problems*. The question of how farms in a farming area can contribute to rural development through interaction with their surroundings could be one example of what is meant by addressing a concrete problematic. The objective of such a communication oriented to application is to develop a spatial understanding of the problematic and thereby to enlarge the number of alternatives to problem solutions or simply the possibilities for action. It is by way of the spatial perspective – *i.e.* the instantaneous understanding of multiple dimensions captured within the problematic - that new visions arise. Such visions cannot be visualised from within a single perspective and they cannot be unfolded by any one of the various organisations that may be enrolled in communication.

Poly-ocular communication can also relate to *concrete generalised representations* of something physical, such as 'soil' for instance. Even if 'soil' refers directly to a physical phenomenon, 'soil' can attain very different meanings depending upon whether it is a biologist, an agronomist, a geologist, a real estate agent or a landscape architect that attaches to the word. Anyway it is possible to communicate about 'soil' as something very concrete. As a consequence of poly-ocular communication it becomes possible to relate various perspectives, and make them relevant to one another. In other words it *e.g.* becomes meaningful to an agronomist to look upon 'soil' as a living entity or as a mineral medium for growth. Multidimensional understandings of a generalised object that are produced by poly-ocular communication can be transferred to any single discipline or interest organisation involved – or it can contribute to the experiential, sense-based repertoire of an individual. During the transfer however, the object becomes less many facetted of character.

The third level for poly-ocular communication is about *values attached to concepts or values attached to real complexes of problems.* Communication within this third level is centred upon aspects such as nature quality or styles of learning. Whereas the second level of poly-ocular communication treats of the object as seen from different points of view, the third poly-ocular level for communication treats of the value-rationality, which is precept when presenting a certain point of view. Different values will be ascribed to nature quality depending upon which view point is in play. To handle different value perspectives is a task of considerable difficulty and abstraction (Noe *et al.*, 2005; Hansen *et al.*, 2006).

Trying to reduce matters to what can be consented is of no point to poly-ocular communication. Due to the differentiation process of disciplines as well as society, single value perspectives do not constitute logical oppositions but are only adding possibilities to an object. During the process of poly-ocular communication many facetted perceptions of multi-functionality arises, which can serve as a firm foundation for any project. Consensus in stead has to be reached by finding ways to include multiple perspectives that may not be perfectly integrated.

## Further on

Agriculture and rural development constitute an entirely new complex of problems acting as a challenge to the theories of learning. Potentially we can draw upon each other's perspectives much more intensely, and by way of intense inter-reliance generate deeper understandings and better descriptions. Not least, by taking into account more dimensions we are able to generate visions of future possibilities that are qualitatively different of character. On the basis of the concept of poly-ocularity this paper has reflected on how to meet future challenges. It has done this in a theoretical and methodical way.

Poly-ocular communicative learning cannot be performed within a single disciplinary discourse. In contrast to disciplinary discourse it does not produce knowledge valid across time and space, but is performed on an ad hoc basis, only when real problems so demand. In this sense poly-ocularity is formal theory of learning that can be made operational only in procedural rules.

A primary challenge for researchers to meet in order to enable sustainable rural development therefore is to frame poly-ocular communicative learning in practise even when research is not directly involved in the communication.

## References


Bateson, G., 1979. Mind and Nature: A Necessary Unity. Bantam Books.

Cerf, M., D. Gibbon, B. Hubert, R. Ison, J. Jiggins, M. Paine, J. Proost and N. Röling (eds.), 2000. Cow up a tree – Knowing and Learning for Change in Agriculture. INRA Paris.

Hansen, L.; E. Noe and K. Højring, 2005. Nature and nature values in organic agriculture. An analysis of contested concepts and values among different actors in organic farming. J. of Agr. and Env. Eth. (In press).

Luhmann, N., 1995. Social systems. Stanford Univ. Press.

Maruyama, M., 1978 Endogenous Research and Polyocular Anthropology. In: Perspectives on Ethnicity. R. Holloman and S. Arutiunov (Eds.), The Hague, Mouton Publishers.

Maruyama, M., 1985. Mindscapes: How to understand specific situations in Multicultural Management. Asia Pac. J. of Man 2, 125-149.

Maruyama, M., 2004. Polyocular vision or subunderstanding? Org. Stud. 25, 467-480.

Noe, E. and H.F. Alrøe, 2005. Combining Luhmann and Actor-Network Theory to see Farm Enterprises as Self-organizing Systems. Cybernetics and Human Knowing (In press).

# Landscapes and sugar boxes for learning in rural areas management

*Richard Raymond, Salma Loudiyi and Sylvie Lardon*
*ENGREF, BP 90054, 63171 Aubière cedex 9, France*

## Abstract

The users and stakeholders of rural areas have often different knowledge and expectations about these spaces. To plan shared goals, it is important to build a common representation of the areas in concern. We observed several meetings between farmers and non-farmers aiming to build a common cognitive framework of collective management of these rural areas. We show that references used are closely related to the social and spatial organisation level. Accordingly, they have different impacts on social learning and shared cognitive framework.

**Keywords:** natural spaces management, speech and references

## Introduction

In our days rural areas are supposed to fulfil multiple functions related to an ensemble of several involved parties with different expectations. In order to achieve the objectives of management of the rural areas' multifunctional role it is necessary to obtain some coordination between these parties to overcome antagonisms of management objectives, of dispersed practices, of inefficiency public measures *etc.* Many authors consider that these forms of organization take the form of networks (Callon *et al.*, 1983; Latour, 2005). Even if we do not always share the same conclusion with these authors who neglect the emergent properties of social systems, we recognize the heuristic efficiency of this analysis model. This is particularly true for innovating processes facing at new stakes (Lorrain, 1998).

Farming system practitioners (FSP) could be involved in networks which bring together stakeholders with different statuses and different expectations. These networks take the form of distributed networks with a weak or without any hierarchical basis or central node (Baran, 1964). But the special role of the FSP in these networks cannot be neglected. Firstly because these stakeholders practise and put in form the major part of rural areas. Secondly, because even if they have to answer to the social expectations they must also preserve their primacy over their production areas in order to control their production tools. So, these stakeholders must participate more actively to their defence and put themselves into the regulation of these networks.

One of the most efficient regulation modes is based on the construction of a common representation of the shared territory and of the problems that have to be solved in a collective way. The acquisition of the world's representation is allied with a learning process. When this representation results from the confrontation and the articulation of knowledge distributed in a social group, it is similar to a social learning: each stakeholder (1) teaches to the others their point of view while at the same time; (2) learns the others point of view; and (3) takes part in a collective that constructs a new common representation of the world. Within this framework, the main goal of FSP is to convince the other stakeholders of the pertinence of their own representation of the rural world. By doing this, they reinforce the legitimacy of their position in the common shared territory.

In this paper, we propose a mode of apprehension of speech's heuristic dimension. After having presented the framework of our analysis, we approach three types of speech developed in relation with the place of agriculture in French suburban areas. The heuristic dimension of each type of speech is revealed and afterwards we draw the conclusions from the deployment of those types of speech and their juxtaposition.

## How to apprehend the heuristic dimension of a speech?

In our point of view a line of arguments has more chances to be credible when the used arguments are considered as comprehensible and objective. While human minds are not simple computers and a speech cannot be just a simple series of symbols. For being the support of a learning process speech must be associated with an intentionality (Searle, 1980). Hilary Putnam deals with this very question of the intentionality that is related to words in his work about references (Putnam, 1981). In the present article the same approach as Putnam's of "making a reference" has been adopted. That indicates a univocal or of a little ambiguity relation between a word and something that really exists. We have to keep in mind that making references allows putting a speech under the test of observed or observable facts. The objectivity of these facts supports the objectivity of the associated speech and reinforces the credibility of the references. We allocate the status of objectivity to a fact following John Searle's theory: either this fact belongs to an external reality, or to the social reality that underlies an ensemble of stabilized conventions (Searle, 1995). On the other side, Donald Davidson analyzes the processes of mutual comprehension and highlights the principle of charity. According to this author during a conversation we allocate to our interlocutor the same beliefs, the same world as to ourselves (Davidson, 1984). If this hypothesis is true, then a new representation of the world must be in adequacy with this ensemble of already instituted beliefs for being integrated by a social agent. So it seems that a speech is more credible, and therefore a possible vector of learning, when it is based on a group of values or beliefs already shared out by the receivers or when it refers to an ensemble of facts that can be tried and tested.

Our analysis relates to a group of speeches related to agriculture in suburban areas of two French agglomerations. The first that is located in the North-West of the Ile-de-France area, is the Vexin Français. The other is the suburban area of the agglomeration of Clermont-Ferrand in the Massif Central that is a zone of medium-size mountains in the centre of France. Within this intention, we observed several formal or informal scenes of construction of these cognitive frameworks of rural territories' collective management. These observations attempted to seize the means by which the various developed speeches gained the support of their listeners. We were particularly interested in the references used during this conviction process.

Our analyses are based on two types of data. The first one results from the observation of various scenes of discussions or meetings between involved stakeholders to the management of the shared territory. The second type of data is based on the analysis of the documents that formalize or present the public policies or the collective management of these areas. We noted down the references used by the speakers and we focused on their allocated properties and on the logical consequences that could be deduced from these pairs {references, properties} for the management of rural areas and the orientation of agricultural practices.

The analysis of these references reveals several levels of organization. Three types of speech are distinguished: the first corresponds to a very broad level of organization; these speeches are diffused all around the French society. The second one concerns the speeches addressed specially to the stakeholders involved to the specific territories. Finally, the third type of speech refers to a more precise level of organization. This one brings together the farmers on their activity spaces: their exploitations and their plots of land. Three heuristic aspects correspond to those three levels of organization.

## What can we learn from a speech without univocal references?

The first type of speech, concerning the management of rural areas and the place of agriculture in these areas, is part of a very general level. In both field cases, the used references belong to very general categories whose limits and properties remain ambiguous. These speeches mention the importance of rural and suburban areas for the life quality of their inhabitants and underlie their natural character. These rural areas are described only by polysemous words and ambiguities such as rural, natural, inheritance, green, agglomeration's lungs... without detailed precisions of the indicated facts or objects. Only, a model of ring organization of the space, limited to the regional administrative (case of the Ile-de-France) or institutional limits (case of the agglomeration Community of Clermont-Ferrand) indicates that these rural areas are the peripheral spaces, coloured in green on the maps.

These ambiguous references encourage the participation of social partners to convincing discussions. In fact, everyone gets something out of it. The exact meaning of these words remains uncertain but the social agents interpret them in a way that every reference remains comprehensible and in line with their system of values. On this purpose they follow the principle of charity developed by Donald Davidson. The main lines of management of these rural areas are fixed by tacit agreement between all the social agents and the decision makers: the rural areas must be protected and the agriculture which maintains them must be protected. No definition is been given in connection with these rural areas or with the type of agriculture, agricultural practices or modes of production are used in them, so every reference could be transposed without great modifications to all the French suburban areas.

Thus, these main lines remain very far away from concrete action. The used references do not allow giving any direction to special agricultural practices but they remain more in the sphere of ideas and values. So, the learning that results from these speeches is connected more with a consciousness raising. It concerns less ways of action or future practices' orientation than great and general principles which would or could have be translated on other levels of organization.

## Which are the consequences of a repeated speech on the landscapes?

Another type of speech concerning these rural areas and the agricultural activities in these areas can be identified. These speeches are more definite because in this case the used references by the social stakeholders are referred to a specific perceived or perceptible area. Even if these references are numerous there is a level of space organization and a mode of perception of these references that dominate: it concerns landscapes and elements, forms or associations related to them. This role given to landscapes is consolidated by an ensemble of largely diffused images. In the same way, the application of many public regulations is accompanied by reports emphasize on the attention that has to be given to the rural and agricultural areas and to landscapes. The epistemic objectivity of these references becomes stable.

These landscapes are above all picturesque images of the concerned areas. In this way, they fulfil the principal aesthetic representation of the countryside common in the French society (Luginbühl, 2001). However, this model of landscape is adapted to local situations. In the suburban area of Clermont-Ferrand, the volcanoes and the meadows which cover them are pointed out. In the Vexin Français a special emphasise is given to the curves of the plateau. Nevertheless, many elements refer to the picturesque model of the XVIII$^{th}$ century. So the references that can be found on these speeches are more precise and less ambiguous than previously. The allocated properties belong to two types: either they remain primarily aesthetic or they are based on a confusion between the appearance of a thing and the thing itself.

Thus, the discovered area embodies the common representations of what should be a beautiful rural area. However, glancing at these spaces means also glancing at cultivated spaces. In this case the agriculture is presented like the guarantor of these landscapes. An alliance is then proposed between FSP and the local populations concerned about the safeguard of their environment. It is both a question of maintaining these landscapes and shielding them from the urban threat.

The social learning carried by these references joins together the local society and the farmers. They well-sustain arguments of an aesthetic conception of these areas and of the agricultural activity that gives them shape. The rural areas are likened to a scenery. That tends to divert the attention from productive grounds which are perceived on a scale that does not permit to focus on technical management even if these grounds remain elements of these landscapes.

## How do the Vexin Français FSP teach what their practices are?

Finally, the third type of speech refers to areas of agricultural techniques. It brings together, around everyday or more technical objects, the FSP and the rest of the local society. We present an example developed during a biannual meeting between the FSP of the Vexin Français and the rest of the residents of this area (Raymond, 2003). Because they were infuriated by the suspicion that goes often with the large-scale farming concerning its impacts on environment, some FSP took the initiative in presenting their job to "teach" to the local society what the agriculture is really. We recount the principal elements of the developed argumentation and of the used references in order to convince of its truthfulness.

At the beginning, some notice-boards present various techniques of soil analyses and of crops needs... They show the rigour and the objectivity of the know-how of the agricultural practices. Presentations illustrate the processes of products transformations coming out from large-scale farming, from grains of wheat to bread, from beet to sugar; from barley to beer... forge the bonds between large-scale farming crops and everyday consumption products. So the main objective of the large-scale farming crop is established in an undeniable way: all of us need everyday the food provided by it. Other references are also used. The farmers chose to convince about the truthfulness of their environmental concerns by using objects which visitors can check out in a very simple way. So, several trial plots of land are presented. On every plot a small notice-board indicates what has been made or rather what has not been made: here they have not used any nitrogen fertilizer, further they have not used any herbicide or any fungicide... The visitors can observe the consequences. In front of each plot, a farmer gives explanations. On a table, several elements show out what the farmer explains. A wooden frame, one-meter-square big, represents an easy to conceive surface unit. Next to it, there are some Petri dishes containing granules. They represent the amount of input usually used for this culture for this specific surface. A bag of grains or a box of sugar represents what is produced using this amount of granules...

These objects are the mouthpiece of the relations between farmers and their cultivated lands. All of them express the same thought: that the farmer controls his crops and pays a lot of attention to the protection of the common environment. These references form a clear speech on which is based the lesson given by the farmers as well as the information given to the local society about the agriculture. This transmission of points of views renews farmer's legitimacy in order to be presented as an expert (and sometimes as the only expert) of the management of their production area. These demonstrations had a great success. They contributed to reinforce farmers' position in the local networks as well as their importance in the definition of management orientation of this rural area (Raymond, 2004).

## Conclusions

The analysis of the references of each argumentation developed in relation with the management of rural areas and with the role of the agricultural activity in these territories permits to distinguish the different impacts of these speeches. Even if these references seem to refer to the same categories, they have different implications as it concerns the division of cognitive frameworks. We show that when the level of a certain organization is close to the action, the used references for argue and reinforce the credibility of the speech must be univocal. Nevertheless, it seems that the effectiveness of these references is reinforced when the stakeholders (to whom these references are being aimed at) can check out these references in a simple way. In this way they can also test easily the truthfulness of speeches. On the other hand, the diffusion of action's purposes and the constitution of a community of common interests can be achieved thanks to a conscious raising based on ambiguous references.

The necessary but non sufficient conditions for convincing and increasing the participation of the involved partners in the management of rural areas and the choice of the used references depend on the objectives to be reached. After all, the three identified organization levels and the references that correspond to each of them are undoubtedly complementary. The collective learning, which is the base of the collective management of these areas, supports probably each of these organization levels. Broadest levels will permit to establish the backgrounds in which references of more precise speeches will be interpreted.

## References

Baran, P., 1964. On distributed communications: introduction to distributed communications networks. United States Air Force Project Rand. 37 pp.

Callon, M., J.P. Courtial, W. Turner and S. Bauin, 1983. From Translation to Problematic Networks: an Introduction to Co-word Analysis", Social Science Information 22, 191-235.

Davidson, D., 1984. Inquiries into Truth and Interpretation. Oxford University Press (2001, 2nd edition). 320 pp.

Latour, B., 2005. Reassembling the Social: An Introduction to Actor-Network-Theory. Oxford. 316 pp.

Lorrain, D., 1998. Administrer, gouverner, réguler. Gouvernances. Les Annales de la recherche urbaine 80-81, 85-92.

Luginbühl, Y., 2001. La demande sociale de paysage. Conseil National du Paysage, 21 pp. www.environnement.gouv.fr/telch/2001-t3/CNP_luginbuhl.pdf.

Putnam, H., 1981. Reason, Truth and History. Cambridge University Press. 224 pp.

Raymond R., 2003. La concertation sur l'espace cultivé et la nature dans le Vexin Français. Economie Rurale 273-274, 169-183.

Raymond, R., 2004. La nature à la campagne: identification sociale et argument pour la gestion d'un territoire rural partagé. Le cas du Vexin Français, PhD thesis, Departement of Geography, Université Panthéon Sorbonne, Paris. 593 pp.

Searle, J.R., 1980. Minds, Brains and Programs. The Behavioral and Brain Sciences 3, 417-457.

Searle, J.R., 1995. The Construction of Social Reality. New York, The Free Press. 241 pp.

# Farmer stable schools: A fruitful way for Danish organic dairy farmers with the goal to phase out antibiotics from their herds

*Mette Vaarst*
*Danmarks Jordbrugs Forskning, Denmark*

## Abstract

The Farmer Stable School concept developed when a large group of Danish organic dairy farmers faced a situation of having a common goal to phase out antibiotics from their herds. This was a complex goal which could be reached in several ways, but at the same time, there was no previous experience in Denmark of explicitly having this goal in a group of farmers. In order to establish a good common learning environment the concept of Farmer Field Schools (FFS) known from developing projects in African and Asian countries was taken in. To fit a Danish organic farming context, what adaptations are needed? In this project, we build on ideas from experiential learning and action research (med mindre action research er formuleret som grundlag for FFS i den anvendte FFS artikel). The farmers formed small learning groups and used an organic dairy husbandry consultant (one of the authors) as their process facilitator. The groups took their starting points in their own farms, each farmer learning by taking action on his/her farm and the farmers learning together by exchanging experiences and giving mutual advice while visiting each others' farms. The results from the herds indicate that crucial changes took place during the project period. We suggest that the farmers' ownership over the common goal in combination with advice from the group based on the articulated goals of each farm both are crucial factors in this process. We identify the farmers' change process towards the common goal as an equal common learning process. Therefore we conclude that Farmer Stable Schools as a modified version of FFS are very useful in the process of learning how to phase out antibiotics in Danish organic dairy herds. In the light of the educational background and position of Danish organic dairy farmers (a North Western European context), the concept of empowerment may however be understood differently as compared to been used in the context of developing countries.

**Keywords:** experiential learning, context analysis, empowerment, advisory service, Danish organic dairy farming

## Introduction

Farmer Field Schools (FFS) is a concept for farmers' learning and empowerment through knowledge and experience exchange, which has been developed and used in Indonesia and after this, in a number of developing countries, based on an innovative, participatory and interactive learning approach as a sustainable way of learning and developing farming for small-scale rice farmers (Gallagher, 1999). The FFS approach has been widely used in Africa in various forms in order to empower farmers to take responsibility and control over their own lives, be active participants in their local communities and to improve their knowledge and skills in relation to their farming businesses. In Denmark, the so-called 'Farmer Experience Exchange Groups' have been used through decades. They are often groups of 10-15 farmers from quite similar farms (*e.g.* dairy farms with a certain housing system and/or breed), which meet on regular basis shifting between their private farms. The whole group is normally run by an agricultural advisor, who is a kind of coordinator and professional expert in the field. Often, an external expert within the field (*e.g.* in farm economy, buildings, feeding, *etc.*) will be invited and give a lesson on a certain topic.

In this project, a group of organic dairy farmers had a common goal of phasing out the use of antibiotics from their herd through improved herd management, health promotion and disease preventing actions. As a group, the farmers had on their own initiative defined this rather complex goal and therefore had a strong sense of ownership and accordingly a strong motivation to do an effort to reach it. In this presentation, we focus on the process of creating and working in farmer groups following a modified approach to the FFS concept as a contrast to the model of Farmer Experience Exchange Groups in the learning approach. The concept of Danish Farmer Stable Schools is described as a case of practical learning experience and discussed in relation to aspects of common learning, improved farming practices and empowerment.

## Identifying the method of phasing out antibiotics as a common goal

In Denmark, a group of 23 organic dairy farmers initiated a project in March 2004 with the aim of phasing out the use of antimicrobial drugs from their herds, in collaboration with the Danish Organisation of Organic Farmers and Danish Institute of Agricultural Sciences. Inspired from personal experiences with an FFS Approach modified to Ugandan context gained by one of the project researchers, four so-called 'Stable Schools' were established among the project participants. This approach was taken for the following reasons:

1. It was a goal set commonly by a group of farmers. It seemed logical that the farmers should reach the goal together as a group.
2. Not much existing knowledge could support the process, but a lot of various more fractionated pieces of knowledge. Knowledge therefore had to be built up and made explicit along the process.
3. It was a complex goal, which could be reached in several different ways. The most appropriate approach would therefore be to take the starting point in each farm and its individual goals and practices.
4. Using this method, farmers could communicate about the initiatives taken at each farm in order to reach this goal, and to learn from each other.

## The Danish context

In Denmark, all farmers are well educated. Organic farming has been developing in at least two major movements: the biodynamic movement from mid-1950s and the so-called modern organic movement basically starting in 1970. These two movements are now operating under the common term and legislation of 'organic farming'. Organic dairy farming has followed the general development of farms in Denmark, where the size of farms and herds has increased dramatically over the past decade, with the consequence that the number of farms has decreased as dramatically and many farmers have felt that they had the choice between increasing the size of their herd or closing their farm. From the mid-1990s, a majority of farmers, who decided to convert their farms to organic farming, were mainly driven by economical incitements in terms of subsidies, premium prices and expectations that converting the farm to organic farming and at the same time increasing the herd size and improving the farm structures and practices accordingly could be a way of ensuring the future of ones farm. Most of the participating farmers in this project typically went through a 'mental conversion process' after the so-called 'technical conversion process', whereas the first groups in the 1950s and 1970s onwards built their technical conversion on a long process of mental conversion where they gradually gained insight and consciousness about environmental issues and sustainability of farming. Few of the participating farmers in this project had been through this process. Furthermore, organic farmers throughout Europe have very little support in terms of advisory assistance and veterinarians, who are specifically skilled and knowledgeable about organic farming. This serves as an argument for claiming that organic farmers compared to other groups of farmers have gone through several learning processes, where not only their technical skills but also their awareness towards a huge number of issues like environmental care and animal welfare has been challenged, and where they also often have had to be innovative (because of lack of *e.g.* veterinary advice suited for their particular situation) and to defend themselves and their organic approaches to other parts of the farming environment. In this way, organic farmers have a special position of being supported by groups of consumers but not by others, sometimes being isolated in relation to farmer colleagues who feel that their organic approach is threatening the non-organic farming methods, lacking support from the conventional advisory system, and being empowered by the processes of conversion, innovativeness and gradually gaining consciousness and insight to professional issues of farming as well as societal and environmental issues. These processes are valuable elements of experiential learning processes (Kolb, 1984).

## The concept of 'Stable Schools' developed

Built on the awareness of the above mentioned factors, it became clear that many Danish organic farmers are generally well-educated and very reflective persons, conscious about their own situation, and have taken many active choices on which they have based their farming practices. The step to set the goal of phasing out antibiotic treatments from their herds furthermore added to the perception of ambitious and generally self-confident farmers. We therefore concluded that the approach of these Stable Schools should be adopted as equal common learning, each farmer actively contributing by reflection as well as by action, thus being more participative than in the traditional FFS setup. The exchange of experience and knowledge in these groups should be based on the participants' own herds and farms. Since these farmers emphasised that they were busy in their daily lives, monthly meetings were found more appropriate than *e.g.* weekly meetings. This partly pointed towards groups of 5-6 farmers, so that two meetings could take place at each farm over one year. The grouping of farmers was based on geographical conditions. Consequently, the herds in each group were mixed with regard to size, breed, housing system, milking system, *etc.*

## Conduct of meetings and the role of the facilitator

The facilitator sent out an agenda 1-2 weeks before the meeting in the group based on a consultation of the farmer, on whose farm the next Stable School meeting should be held. One success-story and two problem areas as perceived by the farmer were taken up at the meetings. Key data from the herd (from the Danish central cattle data base) were included in the letter in order to let the whole group gain insight to the herd in focus, and to prepare them for the discussion. The group meeting was built of a ½-1 hours walk with free discussions in the buildings and outdoor when relevant, and a session of 1½-2 hours around a table indoors. Crucial to the process was that the facilitator kept out of giving advice. Apart from facilitating the discussion and the process in the group, the facilitator also recorded brief minutes, basically giving the mutual advices and the main conclusions of the meetings.

## Results of the process in terms of development in the participating herds

The results reflecting the development within the participating herds in this project indicate that during the project period, the number of antimicrobial treatments was reduced by 50% in total, and selected animal health parameters in the herds remained more or less unchanged or improved. The results point to the fact that actual improvements of the herd health situation–in some herds quite dramatic ones –took place as a concrete and measurable consequence of the work and learning process of the farmers. The farmers stated in the final focus group interviews as well as in the individual semi-structured qualitative interviews that their consciousness and awareness of animal health and welfare promotion had been improved, and that the Stable School groups had been crucial in this development.

## Overcoming barriers of differences by working on a common goal

The type of the farm improvements varied widely in each group. Some farmers built a new housing system at their farm, and others improved the hygiene when milk feeding calves, improved the feeding or changed milking routines. Despite the differences between farms in general and between focus areas for the meetings, all farmers seemed to gain much from the meetings. At all farms, the individual goals of the farmer were articulated and included in the group discussions of the Stable School, which made all discussions relevant for the host farmer. The fact that the farmers had the same overall goal, namely phasing out antibiotics from the herd, made all discussions relevant to all participants. The combination of a common goal and herd individual goals vividly illustrated the fact that the common goal could be reached in a number of different ways, which again stimulated the innovative thinking in the group. All farmers came from their own complex herd situation, which enabled them to take many different factors into account also when discussing solutions to a specific and apparently limited animal health related problem. In conclusion, the Stable School approach seemed a valuable way to promote animal and herd health in cases, where complex farming situations are in focus. Problems can be identified and changes or solutions proposed based on a context analysis oriented approach in a group with an identified common goal.

## A common goal as crucial basis for common learning

We suggest that the many changes which took place in the herds during the project, could happen because of the consistent, continuous and common learning processes and exchange of experience and knowledge between farmers based on the identification and ownership of a common goal in combination with individual farm based analyses and goals. This process demands more than listening and thinking, and the aim to reach the common goal stimulates to changed practices. This is the main difference between the Farmer Stable Schools and the well-established concept of Danish Farmer Experience Exchange Groups. At all meetings, a case of success at the host farm was presented to the group. This always gave a good, positive and encouraging perspective at the meetings. The focus on both success cases and problem areas encouraged all farmers and gave other farmers innovative ideas to adjust to their own farm.

## From being taught to being empowered through experiential learning

A fundamental part of the FFS concept is common learning based on practical examples and 'hands-on' experiences. This was also the case in projects in Uganda where a modified model of FFSs were implemented in groups of smallholder dairy farmers. In the classical FFS approach, the learning often takes place in few 'learning farms' or plots of land, where experiments are performed and all farmers can see and learn from this (Gallagher, 1999). This did not seem relevant in a Danish context. In FFSs in African and Asian countries, where the group members are the rural poor, 'empowerment' is a crucial goal of the common experiential learning process. Danish farmers face a rapidly changing farming environment, many explicit and implicit expectations from the surrounding society, including consumers and other stakeholders, and many of them express doubt whether they can continue as dairy producers, and if so, then under which circumstances. In this sense, the group of organic farmers definitely benefited from common learning and action as well as the strong support of each other, on the personal level and on group level, as well as in relation to other groups in society, and through the whole development of this concept and action research project (Checkland, 1991). In this sense, 'empowerment' also takes place in groups of farmers, who in many ways are well educated and do not seem 'powerless'. A relevant question could be whether the farmers, who joined the Stable School groups, did this because they were capable, had ambitions and articulated goals and expectations, and were empowered enough to dare jump into a process, which would challenge their daily practices - a process that was chosen due to interest and possible benefit rather than driven by poverty or lack of life income. In relation to the commonly used 'expert advisory system' in Denmark, the processes in these group could be characterised as that of an alternative advisory service, where farmers built up and accumulated common knowledge in an exchange process of individual knowledge contributed to by each farmer in a professional dialogue. The characteristics of a process of empowerment of farmers in a North Western European context may differ substantially from the empowerment of rural poor in an African or Asian context.

## References

Checkland, P., 1991. From framework through experience to learning: the essential nature of action research. In: H.E. Nissen, H.K. Klein R. Hirschheim (Eds.). Information systems research: contemporary approaches and emergent traditions. Elsevier, Amsterdam.

Gallagher, K.D., 1999. Farmers Field Schools (FFS): A Group Extension Process Based on Adult Non-Formal Education Methods. Http://www.farmerfieldschool.net/document_en/FFS_GUIDe.doc.

Kolb, D.A., 1984. Experiential learning. Prentice Hall PTR, Englewood Cliffs, New Jersey.

# The changing perception of expert advice in nature conservation: A case of the Drentsche Aa

*Severine van Bommel*
*Wageningen Universiteit, Leerstoelgroep Bos en Natuurbeleid en Communicatie en Innovatie Studies, Wageningen, The Netherlands*

## Abstract

The paper shows that during the last decade, governance has created the setting for a new role for nature conservation experts in the policy process in the Drentsche Aa area. For decades before, expert knowledge was used in nature conservation policy as experts were asked by policy makers to give advice on matters on which they were considered to be knowledgeable. During the last decade, the role of experts has changed. Experts have become just another stakeholder among the number of private and public stakeholders that influence the outcome of the policy process and the value of 'scientific' knowledge is put into perspective, because other types of knowledge, such as local knowledge, are increasingly valued. This means that the experts-of-the-past and their claimed expertise are loosing credibility. As a result, experts-of-the-past are confronted with competing claims on expertise by 'new' stakeholders. Boundaries and identities with regard to expertise have to be renegotiated and redefined.

**Keywords:** experts, expertise, governance, Drentsche Aa, nature conservation

## Introduction

Societal processes such as globalisation and decentralisation have caused a shift in policy coordination mechanisms from traditional modes of 'government' to new modes of 'governance'. Governance has increasingly received attention as a research object from many different disciplines. The main emphasis in research on governance has been on identifying and describing the shift in policy coordination and the mix of co-ordination types. So far little attention has been paid to the changing role of experts.

In the Netherlands, governance has created the setting for a new perception for nature conservation experts and expertise in the nature policy process during the last decade. In the traditional modes of 'government' expert knowledge in nature conservation policy was used as experts were asked by policy makers to give scientific and technical advice on matters on which they were considered to be knowledgeable. *Vice versa* nature conservation expertise was also political as it was shaped by political preferences and considerations to protect certain species and certain types of nature (Turnhout, 2003).

Nowadays, the perception of nature conservation experts seems to be changing. There is now increasing citizen awareness and involvement in most Dutch political issues. Policy makers can no longer ignore the opinions of the people interested in and affected by the implementation of nature policy (Van de Poll and Glasmeier, 1997; Van der Windt *et al.*, 1997). 'Governance' is thought to give legitimisation to the policy process by providing a broad social basis of public support (Aarts, 1998). The perception of experts can be expected to have changed as nature conservation experts have become just another actor among the number of private and public actors who influence the outcome of the policy process (Hisschemöller *et al.*, 2001; Turnhout, 2003).

Several authors (Hisschemöller *et al*, 2001; Turnhout, 2003) show that the perception of experts is contingent on the context. This means that the perception of expertise can be expected to differ in different governance arrangements. It can be hypothesised that a shift from 'government' to 'governance' may have implications for the perception of experts and expertise. This paper sets out to explore this changing perception of expert advice in nature conservation.

## Concepts

Science studies problematises the boundary between science and non science (including policy) and argues that these boundaries are not based on any essential criteria but are better understood as the outcomes of strategic negotiations in certain (policy) contexts (Gieryn, 1995).

Hoppe (2002) states that traditionally science and non-science are conceptualised as different domains or cultures in which actors draw boundaries between themselves and others. These boundaries demarcate science. According to Gieryn (1995) there are always people who contend for, legitimate, or challenge the cognitive authority of science – and the credibility, prestige, power, and material resources that attend this position. This is what he calls 'boundary work'. As the essential features of science are then provisional and contextual results of successful boundary-work, what is considered science and what is not then becomes the outcome of a negotiation.

This paper will use the notion of boundary work to study the boundary between experts and non-experts in different modes of governance. It will analyse the claims that actors make about what should and should not be considered expertise. It will also study how actors carve out a domain of cognitive authority of their expertise.

## Methodology

The paper will follow a social constructivist perspective. This has two important implications: the actor oriented perspective of expertise, and the contingency of practices.

As Turnhout (2003) has argued, the relation between science and policy is complex and perception of expertise in nature conservation policy is contingent on context. The complexity of the relation between science and policy is the result of the ambiguous nature of expertise. Although theoretically appealing, it is empirically problematic to establish the difference between science and policy. Therefore, instead of using a predefined notion of expert and expertise, an actor oriented perspective of who is considered to be an expert and what is seen as relevant expertise (scientific as well as lay) in which circumstances and by whom were used to understand the perception of experts and expertise.

The context dependency of the perception of experts in nature conservation requires an in-depth case study of the perceptions of experts and expertise. Therefore, the changing perception of expert advise in nature conservation was studied by means of a case study of the National Landscape the "Drentsche Aa" in the Netherlands where 75 semi-structured interviews (1.5 hrs) were carried out.

## Context

The Drentsche Aa is the name of a system of small streams that originate on the plateau's of glacial and eolian sands in the Province of Drenthe, in the North of the Netherlands. Together they constitute one of the last relatively unspoilt river systems on the North German Plain. The area has been farmed for some 4000 years. The DRA is, therefore, not a pristine nature area, but an antique farmed landscape. The unique characteristics of the landscape are the result of generations of farmers cleverly trying to eke out an existence on very poor sandy soils. From a point of view of landscape, natural beauty, and bio-diversity, the area is considered unique. Brook meadows border the small streams. The meadows have always been very wet and were mainly used by farmers for hay making during summer. As a result of the high water table and the continuous off-take of organic material over the centuries, a very rich and diverse herbal flora has developed.

## Results

### *The role of experts in a government mode of policy coordination*

In the sixties, the landscapes and the biodiversity that the old type of farming had generated were threatened by agricultural modernisation. At the time, the late Harry de Vroome and Edgar Stapelveld, who were both regional managers of the State Forest Service, took the initiative to fight this destruction in the DRA area. Their good relationships with the then 'Commissioner of the Queen' in the Province of Drenthe allowed them to effectively promote a conservation plan with the Ministry of Culture and Spatial Planning (Mr. Modderkolk, one of the authors of the conservation plan, Min. of CRM, pers. comm.). At the time, purchase of land from farmers was the only option for conservation. The State Forest Service, which already owned some state forests, got the Ministry on their side and was charged with the management of these lands. Farmers and other stakeholders were not involved in the decision making about the creation of the reserve areas at all. An upheaval followed during which the farmers protested against the hierarchical and top-down formulation of the plan but to no avail. The State started buying meadows from 1965 onwards and soon State Forest Service became the largest landowner in the area. This was a rather 'government' approach to coordinating policy.

During interviews, when asked who was considered to be 'the expert' in the past, the majority of my respondents replied that they considered the people of the State Forest Service to be the experts. This can be illustrated by the response of a farmer:

> *"I think experts are, what the nature conservationists used to be in the past."*
> *(farmer)*

This means that the people of the State Forest Service had effectively managed to carve out a domain of cognitive authority of their expertise and convincingly defend its boundary.

### *The role of experts in a governance mode of policy coordination*

In the nineties, a situation occurred which is similar to the situation of the sixties. This time, however, a 'governance' approach to policy coordination was chosen. In the beginning of the nineties, the Ministry of Agriculture, Nature Conservation and Fisheries (Min. of LNV) was charged with the creation of (additional) National Parks in the Netherlands. As the Ministry considered the Drentsche Aa being a potential National Park, they asked the Temporal Commission for National Parks (VCNP) to study the possibilities for assigning this status to the area. The Temporal Commission for National Parks encountered strong resistance by local farmers and their Union, the NLTO. Due to this resistance, they advised against the creation of a National Park. Subsequently, in 1998, the Secretary of State for Agriculture, instituted a Regional Advice Commission in which all the major stakeholders with claims on the area were represented. This Regional Advice Commission was charged with studying the way forward. They finally agreed on the creation of a 'National Landscape'. In a National Park, all land use is dedicated to 'the function Nature', while, in a National Landscape, multifunctional land use is accepted to a point. The National Landscape formula was the outcome of a compromise. Its report, the BIO Plan, *i.e.*, the plan for Design, Management and Organisation of the National Landscape was accepted and on December 4, 2002, the Minister for Agriculture, Nature Conservation and Fisheries officially declared the area a National Landscape ("National Brook and Village Landscape, the Drentsche Aa"). The Multi Stakeholder Platform that prepared the plans was virtually the same as the one that now guides the implementation of the National Landscape.

During interviews, when asking who is considered to be 'the expert' now, my respondents refer to a number of other stakeholders, among whom the State Forest Service is one. This means that the claim on cognitive authority by State Forest Service is loosing credibility. They are currently having difficulty convincingly defending their boundary against other claims. This can be illustrated by the response of a villager:

> *"The State Forest Service used to be the expert, but they are not the experts anymore. I sometimes think that they are still stuck in the past. If you are looking for expertise, talk to the older farmers in the area. They know beautiful stories about the Drentsche Aa area." (villager)*

We can also observe that other stakeholders are challenging these claims and putting their own counter claims. This can be illustrated by the response of a farmer:

> *"The farmers here say:' all the money goes to the State Forest Service. We can also do this nature conservation stuff. The State Forest Service gets paid for every hectare, why can't I do that being a farmer? If I carry out the management, I will get that money.' Of course that will have consequences for the type of nature that you will finally get." (farmer)*

This means that the boundaries that demarcate expertise and the boundaries that demarcate the identities of experts have to be renegotiated by the State Forest Service and the other stakeholders.

## Conclusion

This paper set out to explore the changing perception of expert advice in nature conservation in different modes of governance. Its point of departure was that the perception of experts is contingent on the context and therefore the perception of expertise can be expected to differ in different governance arrangements.

When analysing the 'government' way in which the conservation plan (Gedachtenplan) was formulated and implemented in the sixties, we can conclude that the nature conservationists of the State Forest Service were perceived to be the experts. We can conclude that a 'clique' of nature conservationists and policy makers, consisting of the regional managers of the State Forest Service, the Commissioner of the Queen and the Ministry were involved in the policy process. In this clique, the role of nature conservationists consisted of advising the policy makers at the Ministry. Their expert advice legitimised the Ministry's nature conservation policy.

When analysing the 'governance' way in which the BIO-plan was formulated and implemented in the nineties, we can observe that it is no longer the expert advice that legitimises nature conservation policy but it is the involvement of all major stakeholders in the policy process. This has consequences for the perception of experts. Their claims become contested as they have now become 'just' a stakeholder among stakeholders, instead of being *the* expert. With this, the value of their scientific and technical expertise is put into a different perspective as other types of expertise (*e.g.* local expertise) enter the political process. As a consequence, nature conservationists have become 'boundary workers' who are confronted with competing claims on expertise by 'new' stakeholders. Boundaries and identities with regard to expertise have to be renegotiated and redefined.

## References


Aarts, N., 1998. Een kwestie van natuur. Een studie naar de aard en het verloop van communicatie over natuur en natuurbeleid. Wageningen, Wageningen University.

Gieryn, T.F., 1995. Boundaries of science. In: S. Jasanoff, G.E. Markle, J.C. Petersen and T. Pinch (editors), Handbook of science and technology studies. London and New Delhi, Sage publications.

Hisschemöller, M., R. Hoppe, P. Groenewegen and C. Midden, 2001. Knowledge use and political choice in Dutch environmental policy: a problem structuring perspective on real life experiments in extended peer review. In: M. Hisschemöller, R. Hoppe, W.N. Dunn and J.R. Ravetz (editors), Knowledge, power and participation in environmental policy analysis and risk assessment. New Brunswick, Transaction Publishers.

Hoppe, R., 2002. Rethinking the puzzles of the science-policy nexus: boundary traffic, boundary work and the mutual transgression between STS and Policy Studies. Paper presented at the EASST 2002 conference: Responsibility under uncertainty, York.

Turnhout, E., 2003. Ecological indicators in Dutch nature conservation. Science and policy intertwined in the classification and evaluation of nature. Vrije Universiteit, Amsterdam.

Van de Poll, N.E. and A. Glasmeier, 1997. Natuurontwikkeling: waarom en hoe? Platform wetenschap en ethiek. Verslag van een debat. Den Haag, Rathenau Instituut.

Van der Windt, H.J., J.A.A. Swart and R. Rabbinge, 1997. De wortels en dilemma's van natuurontwikkeling. In: N.E. van de Poll and A. Glasmeier (editors), Natuurontwikkeling: waarom en hoe? Platform wetenschap en ethiek. Verslag van een debat. Den Haag, Rathenau Instituut.

# Lessons learned about instigating change towards an AKIS/RD in Brandenburg, Germany

*Kirsten von der Heiden*
*AfoReg, Germany*

## Abstract

Manifold radical structural changes have formed the present Agricultural Knowledge System - AKS in the post-socialist German Federal State of Brandenburg. Organisational and institutional structures have been in transition, compelled to search for new functions and roles in the free market system. Parallel to this re-structuring process, EU CAP-reform set a readjusted pattern for farming in Europe. The second pillar of agricultural policy funding structures, as well as the instruments of environmental policy on landscape development and natural resource management, offer new challenges in the transformation towards future Agricultural Knowledge and Information Systems for Rural Development - AKIS/RDs.

Experiences in applied research activities on integrating environmental aspects into agricultural extension, from the author's current doctoral thesis, are blended with theoretical considerations on instigating change in post-socialist states. With regard to the Brandenburg privatised extension system and aiming at developing a participatory AKIS/RD model for Brandenburg, important lessons learned about how to get key-stakeholders to alternate and negotiate between top-down and bottom-up approaches are pinpointed.

**Keywords:** post-socialist conditions, privatised extension, tracks for change, interpretive scheme

## Transformation processes in Brandenburg

The adaptation processes in the agricultural sector at the time of German reunification (1989) differ from organisational transfer of other sectors: in the academic research or health care sector, *e.g.*, eastern German collective actors lacked strategy-capacity to the advantage of western German players with high institutional consensus (power asymmetry). In the agricultural sector, on the other hand, power asymmetry has been of little importance in negotiating new sectoral objectives and structures. Following Lehmbruch *et al.* in Czada *et al.* (1998), obstructive goals set by western German associations and administrations for the future of agricultural structures in a united Germany (*e.g.* the favoured family farm model) have even been counteracted by the eastern German players, who were capable of acting and finding strategic solutions. Their skill and experience in playing political games enabled them to lobby continually and effectively. Although those conditions are tending towards market mechanisms, the endogenous adaptation processes under changed institutional framework conditions, *e.g.* new legal forms, can be characterised by farmers' adherence to and lobbying for collective farming with centralised decision making and large firm structure.

The decision for a privatised agricultural extension system followed, again driven by farmers' demands in a mixed decision making panel, made up of farmers, officials and founders of advisory services. The extension system of the German Federal State of Brandenburg was completely privatised in 2002, in the sense of Rivera's classification (2001), which means that all farmer

- advisor contacts are determined by direct contracts (Von der Heiden, 2006). This is combined with a lack of knowledge exchange based on public goods issues (Von der Heiden, 2004).

Radical political and structural changes have been made in post-socialist agriculture. There was no clear competence for supplementary structural implementation of nature protection and landscape care in the former GDR and during the first years after reunification (Hübler *et al.*, 1993; Auster *et al.*, 1998). A new system of landscape planning and rural development instruments has been built up, driven by individual academics committed to nature protection, namely Müller-Helmbrecht, Freude, Jeschke, Knapp, Meßner and Succow (vice-Minister for Environment in 1990), volunteers in associations and administrative bodies. Instruments of *e.g.* agricultural planning (*i.e.*, Agrarstrukturelle Vorplanung - AVP) and landscape planning (*i.e.* Landschaftsrahmenplanung - Lrp) still overlap. There was much conflict between agricultural and resource management representatives. The fusion of the Brandenburg agricultural and environmental ministries was a positive step.

In view of such fundamentally changing environments of framework conditions, constellations and linkages between AKS meta-system and the three major groups of actors - the sub-systems extension/ education, research, farmer - the survival of an organisation has been dependent on its own voluntary institutional transformation. The need for change has often conflicted with resistance to it. Each organisation which is still in existence today has coped with transition successfully - so far.

## Lack of integration of public goods issues into the Brandenburg AKS

During this period of re-balancing in a changed political system, and competition for survival, the decision was reached on AKS meta-system level to improve farming and rural development interactions. Sustainable rural development (the preservation of natural resources and a diversified farm structure) was established as the explicit aim of national German and of EU agricultural policy. The future paths of integration, most recently with the CAP-reform and the Brandenburg specific "Integrated rural development - ILE", present new challenges to AKS sub-systems searching for balance. Those top-down directives to integration theoretically involve new role definitions and divisions at Brandenburg federal and local level, *e.g.* of former (GDR) integrated local farm development functions (municipal infrastructure and social tasks), which must be taken over by potential organisations. The assumption that rural actors would continue to take charge of this rural development process and also environmental responsibilities, complemented by agricultural support programmes, was criticised by Nagel *et al.* (2003) as poor leadership in times of financial weakness. With full privatisation and an almost total disruption of the government - advisor linkage, the state government has relinquished an important direct way to drive a hard bargain with the farmers and advisors in view of its rural policy. Under present conditions, the only conceivable way for Brandenburg to achieve an Agricultural Knowledge and Information Systems for Rural Development - AKIS/RD (specified in Von der Heiden, 2006) is in mostly new socialised and largely informal linkages, dependent on goodwill and high personal commitment of actors. This is potentially unstable, particularly with regard to activities of societal interest.

Regional feedback via interviews, feedback workshops, plenary discussions, personal reference over the period of the GRANO applied research on transferring agro-environmental issues into training programmes and farmer-extension-researcher events have shown some possible solutions in practice (BMBF funded research programme 1998 to 2003, specified in Nagel *et al.*, 2003; Von der Heiden, 2004). The trials of the GRANO sub-group "Agri-environmental Extension" aiming at improving the knowledge exchange on agro-environmental issues in the AKS comprise:

- arranging practical field days in co-operation with scientific experts;
- organising seminars for farmers and advisors in co-operation with local training organisations;
- elaborating demand driven extension material on agro-environmental topics, addressing farmers.

I realised that most of the blockages and restraints from AKS actors the GRANO sub-group had to face were from the independent advisor stakeholder group and the Farmers´ Unions. The latter kept an eye on the scientists, to ensure that ecological ideas did not gain too much importance for research goals and policy advice. Throughout the GRANO project, we had difficulty particularly integrating private advisory firms into the running trials (except the qualification seminar trial) and implementation process. A consideration of the first survey based on interviews (personal consultations of the author during the project run) with advisors soon revealed hints concerning the reluctance of most advisors to introduce innovative material to farmers and on dialogue and networking activities of the project. Advisors were worried about losing the farmer's trust and acceptance, particularly by integrating extension topics, which they had not specifically requested. Most advisors felt that the scientists owed them on-the-spot support, to avoid wasting time during (workshop) dialogues. Some absent themselves from research and development activities because they are overwhelmed with unspecific research inquiries or in the case of GRANO trials are not ready to risk introducing innovative knowledge.

A tentative comparison with Czech experiences of agro-environmental topics for extension and education, in the study of Rolls *et al.* (2003), shows the same as the few existing studies in Brandenburg, Germany in one important point. Agro-environmental topics are not seen as most important information by the respective interviewees, in particular farmers and advisors; it is mostly even one of the lowest ranked information, compared with *e.g.* economics, production technology and national policy obligations. Interviewees in Rolls *et al.* (2003) ranked the use of such "external effects" information as useful. Also in Brandenburg, farmers and extension workers evaluate those topics as useful as preparation for the future (see CAP-reform, national policy demands, federal regulations). That means that the top-down demand, change as a matter of survival, is comparable to the one directed towards Brandenburg's organisations of the AKS. As we learnt from the Brandenburg example, that does not necessarily express a "need" which will be followed by action.

## Theoretical consideration, reflection and lessons learned

Taking Norman Long (1989) into consideration, communication of innovations is a process of negotiation between social actors. If relations among stakeholders have been disrupted, as described for the Brandenburg transformation situation in general, much care has to be taken to weave new networks. To be successful, the importance of regional diversities must be adequately considered, closely linked to the subject of identity - need identification combined with a survey of organisational culture(s) and identities of participating stakeholders is consequently an essential starting point for initiating change towards sustainable organisational transformation. Thus changes, *e.g.* towards new settings in AKS, such as the AKIS/RD approach, need to be facilitated by a non-partisan, disinterested or more precisely, by an open minded intermediary who represents all sides and all interests. The GRANO sub-group took over an intermediary role during the action research process and compiled processing results. The sustainable implementation of such stimulus remains to be investigated, in particular if one of the addressing key actor groups - the advisors - has not been convinced to take over an intermediary role on the subject in focus.

Provided that the integration of public goods issues in AKS is jointly desired in consensus, we may ask how to inspire, to attract decision makers and how to empower rural communities in Brandenburg to follow bottom-up approaches going hand in hand with top-down demands. Going one step further, we may ask how to provide the impulse for organisational changes towards a sustainable AKIS/RD. Nagel *et al.* (2003) found that intensive and sustained communication in multi-stakeholder groups can bring positive results in informal co-operative solutions in implementing important linkages towards an Agricultural Knowledge and Information System for Rural Development - AKIS/RD, posing the problem of new role settings. The applied research project GRANO showed that these intensive communication opportunities are a "luxury which public extension can afford but for which private extension would have to charge dearly" (Nagel *et al.*, 2003).

I hypothesise that the key actor group of independent advisory services in the GRANO trials, was not "tracked" (in the sense of Laughlin 1991, organisation researcher) by us on a level which would enable sustainable changes without the continuation of such stimulation. Because of the non-participation during GRANO of this key actor group, which habitually acts as intermediary in AKSs, we as applied researchers took on a double role - the intermediary role and the researcher role, reflecting processing options towards AKIS/RD transition. No intermediary organisation was found to implement AKIS/RD processing, from participatory theory development to reality. Including the theoretical thoughts and excellent overview Clark *et al.* (1995) gave on organisational transformation, the GRANO project group accomplished lots of impulses towards the most visible and tangible aspect of participants´ organisations (their so-called sub-systems, not to be equated with the term in AKS theory). Following Laughlin (1991), the sub-systems are always expressions of deeper-lying characteristics. "The generic concept of an organisational form is constituted by three interrelated components: the sub-system (tangible organisational elements), its structure (organisational architecture) and culture (interpretive schemes)" (Laughlin, 1991: 211). "Organisational culture emerges and consolidates historically over time and provides a sense of purpose and meaning for all organisational activity and behaviour." (Clark *et al.*, 1995).

In Brandenburg, the need for change towards improved agro-environmental behaviour, as mentioned above, often coincides with resistance to change. There is an internal conflict of social actors on "cultural unacceptability" and "an inbuilt systemic bias towards evolution within the prevailing organisational logic or paradigm" (Clark *et al.*, 1995), particularly applied to our Brandenburg case, shortly after the process of transformation of the political system.

"The assumption is", following Laughlin (1991: 209-210), "that organisations, for possible psycho-social reasons are naturally change-resistant, with a strong tendency to inertia and will only change when forced or kicked, or disturbed into doing something". Stating that the organisations that make up a promising AKIS/RD need to rethink their functions and roles, we have to consider suitable 'tracks' (Greenwood *et al.*, 1988 in Laughlin, 1991: 213) or 'kicks' (Morgan, 1986 in Laughlin, 1991: 213) environmental (in the sense of surrounding) disturbances may take. Therefore we have to take into account that "organisations contain certain tangible elements about which inter-subjective agreement is possible (...) and two less tangible dimensions which give direction, meaning, significance, nature and interconnection to these more tangible elements and about which inter-subjective agreement is very difficult" (Laughlin 1991: 211).

The so-called "evolution" pathway (or paradigm) is a morphogenetic change, thus involving major shifts in the interpretive schemes. Most important for the intermediaries facilitating such processes is to assume that the decision has been discussed and freely accepted by all the organisational participants. We might summarise that within a privatised extension setting, especially in former socialist states, effective and durable co-operation requires networking based on a deeper understanding of the highly intangible "organisational architecture" and "interpretive scheme" of the participating organisations, which have emerged and consolidated historically. The assembling of informal linkages thus has to be developed over a long period of time, aiming at instigating change towards a new "dynamic balance". Culturally unacceptable changes will be rejected by AKS organisations, even if entrepreneurial survival strategies are promised as sort of stimulation within environmental policies.

## References

Auster, R. and H. Behrens, 1998. Naturschutz in den neuen Bundesländern - ein Rückblick. In: IUGR (editors): Umweltgeschichte und Umweltzukunft VI, Forum Wissenschaft, Studie Bd 45, Hb I + II. BdWi-Verlag, Marburg.

Clark, E. and A. Soulsby, 1995. Transforming Former State Enterprises in the Czech Republic. In: J. Child (editor): Organization Studies 1995, 16/2, Walther de Gruyter & Co. Berlin, New York, pp. 215 -242,

Hübler, K.-H. and H.-J. Cassens (editors), 1993. Naturschutz in den neuen Bundesländern: Chancen für die Landesentwicklung, Bewertung der Naturschutzpolitik, Stand der Gesetzgebung in den neuen Bundesländern und in Berlin. Blottner Verlag, Taunusstein.

Lehmbruch, G. and J. Meyer, 1998. Kollektivwirtschaften im Anpassungsprozeß: Der Agrarsektor. In: R. Czada and G. Lehmbruch (editors), 1998. Transformationspfade in Ostdeutschland - Beiträge zur sektoralen Vereinigungspolitik. Schriften des Max-Planck- Instituts für Gesellschaftsforschung, Köln Bd.32, Campus Verlag Frankfurt/M.

Laughlin, R.C., 1991. Environmental disturbances and organizational transitions and transformations: some alternative models. Organisation Studies 12/2, Walther de Gruyter & Co. Berlin, New York, pp. 209-232.

Long, N. (editor), 1989. Encounters at the interface: a perspective on social discontinuities in rural development. Wageningen, Agricultural University.

Nagel, U.J. and K. v.d. Heiden, 2003. Privatising Extension in Post-Socialist Agriculture - The Case of Brandenburg, Germany. In: G. Alex and W.M. Rivera (editors), Vol. 2, World Bank 2004. Extension Reform for Rural Development - Case Studies of International Initiatives. http://lnweb18.worldbank.org

Rivera, W.M., 2001. The invisible frontier: The current limits of decentralization and privatization in developing countries. In: Brewer (editor), Agricultural Extension: An International Perspective, 2001, Chapter 17, Erudition Press.

Rolls, M.J. and M. Slavík 2003. Change in Information Systems in Czech Agriculture, CUA, Department of Education, Ministry of Agriculture CZ and British Council.

Von der Heiden, K., 2004. Missing links towards bi-directional agri-environmental knowledge transfer - the knowledge systems approach of Brandenburg, Germany. In: Proc. on the Intern. Symp. ´Education – way for the prosperity of future generations´, Sept. 2004, Department of Education, CUA, Prague, pp. 63-70.

Von der Heiden, K., 2006. Chances of success in investing all AKIS/RD's efforts in informal linkages under privatised extension conditions. In: Langeveld, H. and N. Röling (eds.) Changing European farmingsystems for a better future, Wageningen Academic Publishers, Wageningen, The Netherlands, pp. 370-375.

# Building the capacity of the Australian dairy RD&E continuum: How do we create a learning platform?

*Jane M. Weatherley[1] and Mark S. Paine[2]*
*[1]School of Agricultural Science, University of Tasmania, GPO Box 252-54 Hobart, Australia; [2]Faculty of Land and Food Resources, University of Melbourne, Royal Parade, Melbourne, Victoria, Australia*

## Abstract

This paper analyses how the Australian dairy industry research, development and extension (RD&E) continuum including funding organisations, scientists, extension practitioners, interact and learn around farming systems projects (farmlets). We argue that there is an imperative for those in the RD&E continuum to be active in their own learning. A case study is used to explore the construct of learning platforms. The case provides an explanation of how a learning platform was created through the interaction of various project teams to develop and implement a model for dairy farming systems projects. The paper then concludes with a summary of how a learning platform emerged from the interactions between three key stakeholder groups.

## Introduction

Studies that have focussed on learning processes in agriculture have come to the fore over the last 10 years. This is, as stated by Hubert *et al.*, (2000), because *"learning is central to improvements in agricultural systems and farmers' ability to cope with change"*. As for farmers and their practice, researchers' tools and the extension practitioners' approaches are adequate for certain technical eras, however with the presentation of new challenges, the tools and approaches become insufficient in their capacity to deliver the required interventions. Adaptation to changing circumstances is thus central to learning across the RD&E continuum. This paper aims to provide a perspective on learning and bring into debate some understanding of how practitioners in an RD&E continuum work and learn together to improve their practices and increase their capacity to deliver new management approaches.

The context for this paper is the Australian dairy industry. Through the industry funding organisation Dairy Australia, significant levels of investment have been allocated to coordinating and encouraging learning across the RD&E continuum that operates at the level of regional farming systems programs. Workshops held in 1999 and 2000 resulted in farming systems project stakeholders recognising the need for national coordination and learning across projects. To do this, the National Dairy Farming Systems (NDFS) project was initiated in 2001, with a coordinator being appointed to network and facilitate learning across Australian dairy farming systems projects.

Along with the NDFS project emerged the concept of learning platforms. A PhD project was initiated to explore the concept of learning platforms. By studying how practitioners in the RD&E continuum interacted and learnt together it emerged that a learning platform arises when a group of people agree on an objective and share physical and intellectual resources to manage adaptation.

Platforms for learning have previously been described by Woodhill and Röling (1998) as part of a social ecological system. Specifically, they refer to platforms for social learning as an element of a holistic approach to environmental management. Röling and Jiggins (1998) used a construct of platforms for resource use negotiations, similar to that of Woodhill and Röling's (1998) 'platforms for decision making'. Platforms were referred to here as a collection of people taking action to manage resources. Most recently, Boxelaar (2004) developed the construct of platforms for change. An alternative to the previous platforms concepts, Boxelaar explored issues surrounding the convergence of various stakeholders in land management under different context or sites of integration. These sites of integration, including a community, a problem or issue, a practice, a material or substrate, or landscape, are where the construct of a platform for change can be created.

Platform constructs then are providing focus on how groups of people come together and work for a specific purpose, rather than focussing on achieving objectives around a technical issue. Learning platforms aim to build on the existing platform constructs, taking the focus and understanding on how groups of stakeholders share physical and intellectual resources for the purpose of learning.

## Studying learning platforms in the Australian dairy industry

A case study approach, using DairyMod, provided the empirical basis to this paper. DairyMod is a biophysical model that simulates Australian dairy systems. The model was originally developed for exploring principles of grazing management as applied to dairy pastures. The model developers' aim was to develop a tool for research that could look at the effect of manipulating grazing management inputs for example stocking rate, and the interactions between various climatic conditions and pasture species. Data for the case study was gathered using semi-structured interviews with key participants and participant observation at workshops. Data was analysed using thematic coding with Nvivo qualitative research software. Analysis focused on events such as:

- How stakeholders came together.
- What was the objective/purpose of coming together - how was objective constructed.
- What events transpired around trying to achieve the objective.
- Interactions that occurred during stakeholder interactions and in isolation from the group.
- How physical and intellectual resources were used to foster adaptation.
- Emergent learning outcomes including factors encouraging learning and those inhibiting it.

The development of DairyMod coincided with a proposal from the NDFS project that modelling play a more significant role in dairy farming systems RD&E. It was proposed that modelling would assistance with the design, implementation and analysis of farming systems projects, improving the cost-effectiveness of systems experiments and enhancing the predictive capacity and utility of research results. The first stage of DairyMod development had been completed before the case study commenced. This paper analyses the period from model refinement through to legitimisation for use in Australian farming systems research projects. It includes the work of the NDFS extension leader and the DairyMod team to initiate joint activities.

## Discussion: emergence of a learning platform

Key events that provided the environment for a learning platform to emerge are presented in Table 1. Joint activities commenced with the DairyMod and NDFS teams meeting to negotiate

*Table 1. Timeline of sharing intellectual and physical resources around DairyMod.*

| Activity/time frame | DairyMod team | NDFS team | Farmlet teams |
|---|---|---|---|
| April, 1999 - DairyMod development begins. A multidisciplinary team of dairy farming systems specialists and a professional modeller come together to create a model that could be used to explore the principles of grazing management as applied to dairy pastures. | Learning process begins for DairyMod team in terms of model building requirements. | No exposure to DairyMod at this stage as the NDFS project is in development phase. | No exposure to DairyMod. Modelling tools are being used in some farmlet projects. |
| April, 2001- NDFS team and DairyMod team come together to explore the opportunities for the two projects to work together. Negotiation process for potential model integration begins. Negotiation begins by both teams introducing their projects objectives and opportunities for project integration with the farmlets identified. The model is presented at this stage as a research tool. | Building almost complete. Team looking for utility of the model and keen to explore dairy farmlet projects. | NDFS project seeking opportunities for learning across projects DairyMod explored. | No exposure to DairyMod. Modelling being used in some farmlets. |
| June, 2001 - First DairyMod project completed – model up and running. The model is at a stage where it needs to be implemented and farmlet projects are an option. At the first NDFS workshop DairyMod is presented as a possible tool for farmlet teams. All stakeholders are now in the negotiation process and are all learning about DairyMod and its capabilities. | Model utility and refinement stage. Basic model capabilities presented to the teams. | Workshop used as a platform for launching the opportunity of DairyMod. | Farmlets teams first introduction to DairyMod. Agreement to explore model. |
| October, 2001- First workshop with all farmlet stakeholders to use the model and establish the role it might play for their projects. Farmlets have the opportunity to use the model and begin working with farmlet data. DairyMod team provides support and obtains feedback on further requirements of the model. | Learning for DairyMod team remains largely in the area for refining the model. | Learning for NDFS team: how models can be used to link farm systems teams. | Learning begins on model structure, utility and where it fits in practice. |
| October, 2002 - Second DairyMod workshop with farmlet stakeholders, specialist groups for animals and pastures developed. Farmlet teams that are interested in using the model participate in the workshop and can run simulations. Farmlet teams request one to one sessions with the model developer to learn more about how to use the model in their region. | Model refinement process continues. Support to teams if requested. | Linkage opportunities being sought with individual farmlet teams. | Limited use of DairyMod though interest regenerated at the workshop. |
| December, 2002 - Meeting with the Mutdapilly Research farm team as a case study to assist with implementing DairyMod into their project. Other regional teams use of the model remains limited. | Specific integration requirements. | An example for other projects. | Full utility of DairyMod explored. |

how to bring the farmlet teams together and to learn about the model and consider its potential applications in their projects.

***Stakeholders coming together***
The leadership and initiative of the NDFS project team to fund and coordinate stakeholders provided the foundation for a learning platform. Individuals in regional farming systems teams did not have the resources, mandate nor inclination to bring stakeholders together. However the NDFS initiative to build a partnership between the DairyMod team and the farmlets created possibilities for mutually beneficial learning. Not all farmlet projects participated, with teams that didn't work with models in their everyday practice or that had a negative perception of models electing to not be involved.

***Setting objectives***
Setting the objectives for team interaction is a key aspect of stakeholder engagement. In the first instance it was the objectives of the DairyMod and NDFS teams that were largely being addressed through the collaboration. Farmlets were not consulted and engaged until the objectives of the other two teams were aligned because if the NDFS investment. This meant some of the farmlets participated because they felt they had to, rather than seeing their participation as an opportunity. Many of the farmlets had not incorporated modelling as part of their project and as such it was an additional activity on top of already busy work schedules. Farmlet teams interest in the model increased with workshops and one to one interactions. For these teams there was a steady progression in learning how to use it and establishing the model utility in their practice. Their commitment came with time, in association with refinements to the model that made it more relevant to their needs.

***Sharing physical and intellectual resources***
DairyMod provided an interface for NDFS, DairMod and farmlet teams to work together. To achieve the goals of each team, there was an interdependency created, requiring each team to share their physical and intellectual resources. The learning platform emerged through this sharing. Effective sharing of these resources was achieved through different mechanisms of contact and communication. For example the workshops provided a group environment where all the teams were together and were able to establish how the other operated and how they might assist their learning. One to one contact between farmlet teams and the DairyMod team enabled indepth understanding of the model to emerge for both teams.

The farmlet teams providing feedback on the model to which the DairyMod team responded in real time. This on-going checking and re-checking of the utility of the model between the project team and farmlet teams kept the learning processes going, making the model more relevant and likely to become a part of the practice of the farmlet teams. This adaptive process contributed significantly to the learning of all teams. Conversely, the model was developed to a beta test level by the DairyMod developers in isolation of the end users. The beta test model was complex, having detailed parameters which only the model developers had a clear handle on. The significant time spent by farmlet teams to get up to speed on the model was a learning process that could have been fast tracked if the intellectual resources of the farmlet teams were shared earlier with the DairyMod team.

## Conclusion: creation of the learning platform

While the model acted as an interface that brought each of the stakeholders together, it was the farmlet teams sharing their knowledge around the model that allowed the learning platform to emerge. The emergence of the learning platform was dependent on how the resources (*i.e.* the model and farmlet knowledge) were shared, and the utility of these resources to the practices of individual stakeholders. The sharing approach catalysed the learning process, with workshops and one on one methods providing different avenues by which resources were exchanged. The timing of exchanges and maintaining the continuity in communication between events were additional elements that affected the emergence of a learning platform.

## Acknowledgements

The authors would like to thank Dairy Australia for funding this project, along with Anne Crawford the NDFS extension leader for her contribution to the research.

## References

Boxelaar, L., 2005. Diversity and convergence in platforms for change: Building social capability for land management. PhD thesis. Institute of Land and Food Resources. The University of Melbourne.

Hubert, B., R.L. Ison and N. Röling, 2000.The problematique with respect to industrialised country agricultures. In: Cow up a tree: knowing and learning for change in agriculture. Case studies form industrialised countries. Institut National De La Recherche Agronomique, Paris France, pp. 13-31.

Röling, N. and J. Jiggins, 1998. The ecological knowledge system. In. Facilitating sustainable agriculture. Participatory learning and adaptive management in times on environmental uncertainty. Cambridge University Press, pp. 283-307.

Woodhill, J. and N. Röling, 1998. The second wing of the eagle: the human dimension in learning our way to more sustainable futures. In: Facilitating sustainable agriculture. Participatory learning and adaptive management in times on environmental uncertainty. Cambridge University Press, pp. 46-69.

# The learning potential of farmers´ market

Lotten Westberg[1], Johanna Björklund[1], Ulrika Geber[2] and Rebecka Milestad[1]
[1]Department of Urban and Rural Development, SLU, Sweden; [2]Centre for Sustainable Agriculture, SLU, Sweden

## Abstract

This paper is a preliminary exploration into learning in Farmers´ Market (FM) in order to discuss its contribution to a more sustainable food production. The focus is consumers´ opportunities to learn due to the interaction that is taking place between consumers and producers of FM. The learning possibilities are explored by viewing FM as a community of practice (CoP), that is, a community that serves as an important locus for learning through the very practice that is developed by its members. According to our results so far, the CoP of FM includes farmers only. Consumers cannot be regarded as members since they are not involved in the practice giving coherence to the community. FM is arranged for a certain kind of standardised conversation between consumers and farmers that does not seem to support learning of the kind that we are interested in. Consumers and producers have different perspectives on the products for sale, but these do not need to meet in order for the participants to fulfil the goals of the trading activities. On the other hand, the practice of FM creates places where recurrent consumers and farmers can develop good relations. These relations lead to activities and conversations that do not follow the standard procedures and seem to be built on a feeling of trust. The consumers are less interested in learning about the conditions of farming. Instead they seem to be satisfied with having found farmers to whom they may delegate this judgement with trust. The good personal relations that are developed may be of importance for a more sustainable food production in the long run, though Wenger´s theory could not help us to describe the supposed value of them.

## Introduction

Local food systems are regarded by many as important for the development of more sustainable food production. The arguments are ecological (Swift *et al.*, 2004), social (Gale, 1997) and economical (Van der Ploeg and Renting, 2000). Based on the understanding that sustainable food production should be viewed as an ongoing learning process, rather than as ready-made criteria to impose (see *e.g.* Pretty, 1995), there is one additional argument: Local food systems, with their diverse opportunities of face-to-face interaction between consumers and producers, may be of importance for such learning (Feenstra, 2002). From a general point of view, the learning process need to include farmers, consumers and other stakeholders of the food system and support learning in the instrumental as well as normative dimension (Mezirow, 1991). The stakeholders may learn about *e.g.* biophysical conditions that affect/restrict food production as well as about norms and values held by those involved in the system. This may lead to a deeper understanding of the complexity of the food system as such, to more well-founded judgements about what aspects of sustainability to define and strive for, (Proost and Röling, 2000) and how to act in order to support production and consumption patterns congruent with these judgements.

Our hypothesis is that local food systems enable the groups of actors that are involved to learn from experiences and perspectives of each other. By communicating, consumers may *e.g.* get direct insights in the conditions of farming and farmers may get insights in the preferences of consumers. Besides, the communication patterns of local food systems enable flexibility and tight feedback loops (Sundkvist *et al.*, 2005), implying that those involved may act according to their growing understanding and values as well.

This paper is a preliminary exploration into learning in Farmers´ Market (FM) by tentatively viewing it as a community of practice (Wenger, 2003). It is based on an ongoing research project aimed at exploring the learning potential among actors within local food systems in Sweden with regard to a more sustainable food production. The project has been carried out for slightly more than one year and will be completed in another two years. It contains three cases of "local" food systems: Farmers´ Market; Community Supported Agriculture; and a food store (with large supply of locally produced food). Since there are hardly any institutional avenues for consumers to take part in the policy development of food production and consumers generally are alienated from the production, our focus in the paper is on consumers. The question we ask is: Does the interaction that is taking place between consumers and producers of FM enhance consumers understanding of the conditions of farming and help them to reflect upon, and make judgements about, food production and how to act in order to support a more sustainable production?

## Method

Two Farmers´ Market organisations have been selected for the study. One is located in the capital of Sweden and the other in a middle-sized town. The data for this paper is based on interviews and observations made so far:

- Interviews with four farmers, two from each Farmers´ Market organisation. Two of the farmers are producing/selling vegetables, the third cereals/bread and the fourth farmer is raising/selling sheep/meat. The interviews were semi structured (Kvale, 1996) taped, and took around 2 hrs each.
- Four observations of market occasions, two in each town. Each observation lasted for about 3 hrs during which we worked two by two and took notes on the trading-conversations that our farmers were involved in.
- Phone interviews with 12 consumers that we randomly selected during the market observations. These interviews (structured) took around half an hour and were taped.

All the data has been transcribed and briefly analysed. The interviews and observations will be repeated and the analyses made more profound over the next two years.

## Communities of practice – a locus for learning

Wenger (2003) places learning in the context of our lived experiences of participating in society, in the communities and organisations to which we belong. Learning is not viewed as a separate activity, but as an integral part of our lives. Learning starts from the ability of humans to experience the world and their engagement in it as meaningful. Significant learning: *"... is what changes our ability to engage in a practice, to change the understanding of why we engage in it, and the recourses we have to our disposal to do so."* (Wenger, 1998: 95-96). Learning is gaining habits, skills and memories and the formation of our identity. Recognising the importance of participation, Wenger has identified certain social constellations as crucial locus for learning. These are called communities of practice (CoP).

CoPs are everywhere and we all belong to several, at work, with a group of tight friends, within a political movement *etc.* By associating practice with community, it is possible to get a more distinct picture of what kind of communities the theory is focusing on. The practice of a CoP is giving coherence to a community in three dimensions:

- *Mutual engagement* – the members of a CoP maintain dense relations of mutual engagement which are organised around what they are there to do.
- *Joint enterprise* – working with others who share the same conditions is a central factor in defining the enterprise that the members engage in. This gives rise to relations of mutual accountability and shared responsibility among those involved.
- *Shared repertoire* – by their mutual engagement and joint enterprise the members over time create a shared repertoire including routines, words, styles etc, which have become part of their practice and which express their identity as members.

## Farmers´ Market as a community of practice

The idea of FM, with its special regulations (*e.g.* a maximum farm distance of 250 km from the market place; farmers allowed only to sell products from their own farms and own processing), was tested for the first time in Sweden about five years ago. Since the start the practice of FM has been continuously developed by the farmers involved. This practice shapes what the farmers do as well as their meaning (interpretation) of what they do. Through the interviews and observations, we have identified characteristics of *mutual engagement* as well as *joint enterprise* among the farmers that have supported our view of FM as a CoP. (However, we have not caught any evident signs of a *shared repertoire* yet.) When the members interact, do things together, negotiate meanings with each other, the possibilities to share some significant learning arise. Participating in FM shapes a practice and an identity different from practices and identities that the farmers have as members of other CoPs though there are important overlaps.

The idea of viewing FM as CoP in order to explore the learning potential, developed gradually from our observations. At this time we spontaneously included the consumers in the supposed CoP of FM and associated the connections between farmers and consumers to learning opportunities. This seemed evident since the whole idea of FM is to invite consumers and by visiting the market, they get access to the practice of the farmers. Not until we explored the activities going on at the markets and the role of the consumers more thoroughly, we realised that they cannot be viewed as true members of the CoP of FM. They are visitors of the market and not involved in the practice that is developing mutual engagement or joint enterprise among the farmers. The interaction between consumers and producers that are created through the practice of FM, seem standardised. The conversation that we have registered is mostly restricted to phrases like: "Can I have one of those please?", "How much?" *etc.* So the question is, does the role given to/taken by the consumers enhance learning of the kind that we are interested in?

## Markets and products as boundary objects

Wenger defines the process through which we give and take meaning as *negotiating of meaning*. Understanding what other members of a CoP mean, with for instance a process or an expression, and widening/deepening/changing our own meaning as a result of this, is embedded in the practice of a CoP and thereby important for learning. At FM, farmers and consumers can have different perspectives on (meanings of) the products for sale, but these do not need to meet and bridge in order for the participants to fulfil their goals with their trading activities. The products that are for sale on the market (as well as the market place itself) can be viewed as *boundary objects* (Wenger, 2003). These are forms of reifications around which members of a CoP can organise their interconnections with the rest of the world. The objects may serve lots of perspectives at the same time, but without them needing to meet. In other words, the products for sale serve as objects that can be reified quite differently by consumers and farmers, without disturbing the trading. The consumers do not have to know anything about how the vegetables that they buy were produced in order to be able to prepare and eat them. The markets are places where consumers and farmers can meet and talk without actually learning from (negotiate their meanings of the products with) each other. The whole practice of FM is organised in order to sustain this, as it is founded on the idea that the official regulations of FM are sufficient information for consumers to decide whether they want the products for sale or not.

According to our preliminary findings, FM does not generally develop the consumers' understanding of, or support their supposed interest in, the conditions of farming. In other words FM does not seem to support their ability to make reflections and judgements related to more sustainable food production. In the sense that this kind of learning is important for the development of a more sustainable food production, FM does not seem to make a contribution.

## "Learning" through good relations?

The consumers in our study seem satisfied with the market situation described above. Though they show no or little interest in the way the products are produced, meeting farmers face-to-face is of great importance to them. The farmers serve as guarantors (reifications) for good quality and freshness of the products. Besides, knowing that the products are locally produced is of importance. This may contribute to the consumers' identity of themselves as supporting the local farmers as well as organic farming (there are no requirements of organic farming in the regulations of FM though some of the consumers seem to take for granted that everything on the market is organic). In cases where consumers ask farmers questions, it is to learn about unfamiliar products and how to prepare them. This kind of conversations seems to be part of the FM practice (thought they are rare). The farmers are willingly giving the consumers ideas of new dishes, they use tips that they get from consumers and deliver them to others. Sometimes consumers are coming back to tell they have tried a new recipe and how it tasted. In the interviews the farmers tell us stories about how they have learned to recognise some of the recurrent consumers and their specific preferences. They use to put aside the coveted products for them in case they are to be late and the supply is running short. One of the farmers told us that he, several years ago, observed that he was the only one that had a certain kind of pumpkins for sale and that a group of Japanese women were the only ones buying them: *"So now I know who is going to have them already when I grow them."* This kind of activities does not follow the standard procedures and phrases of the market and seem to develop good relations built on a feeling of trust. The consumers are not interested in learning more about the conditions of farming in order to be able to judge whether the products that they buy are produced in a more or less sustainable way. Instead they seem to be satisfied with having found farmers to whom they trustfully may delegate this judgement. In case they regard a more sustainable food production as important, the farmers whom they have chosen to trust are reified as guarantors, not only for good quality, but for a sustainable food production as well. According to our understanding of Wenger´s theory, these relations do not seem to give rise to any obvious opportunities of significant learning.

## Discussion

The question to be discussed in this paper is if the interaction between consumers and producers of FM enhance consumers understanding of the conditions of farming and help them to reflect upon and make judgements about sustainable food production? By using the social theory of Wenger (2003) to explore the learning opportunities, we have not found that the communication taking place at FM is supporting consumers to reflect upon food production related to aspects of sustainability. On the other hand the concept of FM as such seems enough "sustainable" to support consumers' idea of sustainable food production. These consumers prefer locally, often organically produced food and due to FM they have an opportunity to take responsibility for their standpoints and express their appreciations to the farmers themselves. The good personal relations that are developed between farmers and consumers may be of importance for a more sustainable food production in the long run, though Wenger´s theory could not help us to describe the supposed value of them.

We recognise that our results are highly tentative. During the coming years, when we go deeper into the case of FM and other cases of the research project we will explore the possibilities for consumers and other stakeholders to learn due to their participation in local food systems. By this we hope to be able to discuss what kind of practice, what kind of relations and learning qualities would be important in order to support a more sustainable food system, viewed as a learning process.

## References

Feenstra, G., 2002. Creating space for sustainable food systems: lessons from the field. Agriculture and Human Values 19, 99-106.

Gale, F., 1997. Direct Farm Marketing as a Rural Development Tool. Rural Development Perspective 12, 19-25.

Kvale, S., 1996. InterViews. An introduction to qualitative research interviewing, Sage Publication, London. UK.

Mezirow, J., 1991. Transformative Dimensions of Adult learning. Jossey-Bass, San Francisco. California.

Pretty, J., 1995. Participatory Learning For Sustainable Agriculture. World Development 23: 1247-1263.

Proost, J. and N. Röling, 2000. Desperate to Get of the Treadmill. Renegotiating Arable Farming in a Densely Populated Industrialised Country. In: J. Proost and N. Röling (eds.). Cow up a Tree. Knowing and Learning for Change in Agriculture. INRA. Paris.

Sundkvist, Å., R. Milestad and A. Jansson, 2005. On the importance of tightening feedback loops for sustainable development of food systems. Food Policy 30, 224-239.

Swift, M.J., A.-M.N. Izac and M. van Noordwijk, 2004. Biodiversity and ecosystem services in agricultural landscape - are we asking the right questions? Agriculture, ecosystems and environment 104, 113-134.

Van der Ploeg, J.D. and H. Renting, 2000. Impact and potential: A comparative review of European rural development practices. Sociologia Ruralis 40, 529-542.

Wenger, E., 2003. Communities of practice. Learning. Meaning and Identity (second edition). Cambridge University press, Cambridge.

# Project setup and learning processes in participative systems oriented research initiatives

*J.W.A. Langeveld[1], A. Crawford[2], M. Paine[2], S. Pinheiro[3], W. de Boef[4], I.S. Kristensen[5], J. Hermansen[5], B. Dedieu[6], P. Hildebrand[7], V. Cabrera[8], D. Jansen[1] and J. Dixon[9]*
*[1]Agrosystems Research Unit, Plant Research International, Wageningen University and Research Centre, Wageningen, The Netherlands; [2]University of Melbourne, Victoria 3010, Australia; [3]Empresa de Pesquisa Agropecuária e Extensão Rural de Santa Catarina S.A (EPAGRI), Florianópolis, Santa Catarina, Brazil; [4]Department of Crop Science, Centre for Agricultural Sciences (CCA), Federal University of Santa Catarina (UFSC), Florianópolis, Santa Catarina, Brazil; [5]Dept. of Agroecology, Danish Institute of Agricultural Sciences, Research Centre Foulum, PO Box 50, DK-8830 Tjele, Denmark; [6]INRA TSE, UMR Metafort, 63122 St Genès Champanelle, France; [7]Food and Resource Economics, University of Florida, McCarty Hall, Gainesville, USA; [8]Rosenstiel School of Marine and Atmospheric Science, University of Miami, 256 Rogers Hall, Gainesville, FL 32611-0570, USA; [9]CIMMYT, Apdo. Postal 6-641, 06600 Mexico D.F., Mexico*

## Introduction

Conditions for farming are changing rapidly. Development of new, responsive farming systems requires effective interaction between farmers, researchers and other stakeholders. Participative, systems oriented approaches may help to design, test and disseminate appropriate systems. Farming Systems Research (FSR) involves (groups of) actors and stakeholders using formal and non-formal knowledge. Learning in FSR results from interaction between actors with different roles and back-grounds using different types of knowledge. We analyse the design of a variety of participative research projects including dairy farming (Australia), watershed development (Brazil), systems research (Denmark), regional development (France), sustainable production (Netherlands), dairy modelling (USA) and coffee production (Vietnam) and its effect on the learning process. Details on the projects are summarised in Table 1.

*Table 1. Characteristics of the research projects.*

| | Major Objective[1] | Actors Involved[2] | Information sources[3] | Activities[4] |
|---|---|---|---|---|
| Australia | En El | F S B A I | Iv Mt Md Er Of | Ar De Gd |
| Brazil | Gr En Mk R | F B A | Mt Er | Ar De Gd FF |
| Denmark | El L | F S B A | Ds Mt Md Er Of | De Gd |
| France | En El L R | F S B A L | Iv Mt Md Er | De Gd |
| Netherlands | En El | F B A | Ds Mt Md Er Of | De Gd |
| United States | En El | F B A P | Iv Mt Md Er | De Gd |
| Vietnam | Gr En El Mk | F B M | Iv Mt Er | De Gd FF |

[1] Development of economic (En), ecological (El), or marketing (Mk) performance, group (Gr), landscape (L) or rural (R) development
[2] Farmers (F), social (S) or bio-physical (B) scientists, advisors (A), marketing (M), local (L), or policy oriented (P) stakeholders, investors (I)
[3] (Ds) desk study, (Iv) interviews, (Mt) monitoring, (Md) modelling, (Er) experiments, (Of) on-farm research
[4] (Ar) action research, (De) data exchange, (Gd) group discussions, (FF) farmer field schools

## Project setup and learning processes

The projects aim at farm economics, ecological or supply chain performance, sometimes referring to farmer group or farmer-marketing supply chain interactions. Large numbers of actors and information sources are involved in a process with strong emphasis on (internal, external) communication and cooperation. Most projects show similarities to classical FSR. Generally, conditions are favourable for effective learning as there is attention to adequate information exchange, effective interaction and building mutual trust. Differences in project design will favour different types of learning. Number and types of actors involved (types of information), use of information sources (information availability and exchange) and types of activities (information exchange, interaction, building of mutual trust) will determine the character of the learning process. Relevant issues for discussion are character of information exchange and the extent of mutual trust in the exchange process.

# Constructing new appreciative frameworks: Self inspiring from multi-agent systems

*H. Morales[1], J.F. Tourrand[2] and P. Bommel[2]*
*[1]Instituto Plan Agropecuario, Uruguay; [2]CIRAD TERA, France*

Livestock production is an important activity in Uruguay. Cattle were introduced in the territory by the Europeans before the arrival of the 1$^{st}$ European settlers. By the end of the nineteenth century a well established meat industry to supply foreign markets was already established. In the second half of the twentieth century a national program with the aim of increasing the productivity was launched. The main objective was to make pastures more simple and productive but until now cattle production depends on rangelands. The climatic, commercial and financial instability make difficult to livestock farmers to adapt and survive as farmers. At the same time they have to be aware of the requirements of other stakeholders, locally and internationally.
The extension system usually supported operational decisions, but this is not enough when thinking about farms sustainability, so strategic decisions should also be explored. Nowadays, facing stakes such as being - as an industry and not only as a farm - environmentally friendly and safe for people, the extension system should contribute to promote collective actions. This means that in order to attain social acceptability and subsequently, financial viability, many points of view must be taken into account. Continuous learning is necessary at different scales, farms, regional councils, farmers' unions and institutions.
We have explored multi-agent systems (Morales *et al.*, 2005) as a tool to enrich our understanding of the fundamental process ongoing. We have modelled livestock farmers differing in their strategic decisions and simulated it using the Cormas platform. The model is rather simple and transparent for different stakeholders because we wanted to use the model and its results as a tool for discussion about the farms' trajectories and for this it is important to be confident that we can understand – and that we are able to explain – everything that goes into the model (Axelrod, 1997).
We presented it to different stakeholders, mainly livestock farmers and their leaders. Nobody was able to advance the result of the experiment although they faced a very familiar situation. But when they know it, they said that it represents what has happened in reality. So, how can we explain that they – neither we - were not able to advance the results, but after seeing them accept them as valid? The model and its functioning was capable to put in order many known facts and to propose new statements about how things happen. New information brought about a restructuring of mental models, and as a consequence a different kind of interaction with farmers - using UML activity models- was proposed and is being used in a project called "Integrating Knowledge". From this point of view the model has been a good tool in order to learn.

## References

Axelrod, R., 1997. The complexity of cooperation. Agent Based Models of Competition and Collaboration. Princeton University Press.

Morales H., P. Bommel, J.F. Tourrand, 2005. Modelling Livestock Farmers' Strategies in the Uruguayan Pampa. In: A. Zerger and R.M. Argent (eds.) MODSIM 2005 International Congress on Modelling and Simulation. Modelling and Simulation Society of Australia and New Zealand, December 2005, pp. 2340-2345. http://www.mssanz.org.au/modsim05/papers/morales_grosskopf.pdf.

# Labor market competencies and agricultural extension and education undergraduates

*Reza Movahedi and Uwe Jens Nagel*
*Institute of Extension and Communication, Humboldt University of Berlin, Germany*

## Theoretical framework

One of the most important problems faced by the higher agricultural education system in Iran is that subjects studied in the agricultural extension education departments are academic and usually have little relevance to the student's future workplace as well as labor market needs. There is a realization at all levels that students need to be taught practically to meet the requirements of the labor market. One of the most important basic initiatives to address this is to shift from knowledge-oriented to competency-based education, as observed elsewhere. The core research problems to be addressed by this study are lack of relevant competencies of agricultural extension education graduates and students for the labor market and the difficulties of university to conduct relevant curriculum development basis on competency based education model for students.

## Purpose and goals

Improving competencies of agricultural extension education undergraduate students for employment. Improving linkage between education and labor market. Improving curriculum development for student based on CBC approach.

## Outputs and results

Identifying career and vocational competencies for undergraduate students of higher agricultural extension education needed by labor market. Determining level of vocational competencies needed by labor market. Addressing the quality of competencies and skills acquired by agricultural extension education graduates. Recognizing factors involved in strengthening linkages between education and work.

## Methodology

This study utilized both quantitative and qualitative research methods. The quantitative research method used to collect data was descriptive, survey research. The qualitative research method used was interview, discussion group, workshop and focus group. The target population includes employers, undergraduate students, graduates and curriculum developer. Employers include supervisors and heads of extension education working both in ministry of agriculture and private centers (firms) from 32 provinces in all over Iran. Students include some degree level students majoring in agricultural extension education in public universities of Iran. Graduates include some alumni from the aforesaid departments. Curriculum developers include all faculty members, department's heads and educational planners for agricultural extension and education at university level in Iran. As the structure and nature of curricula which are taught in these departments are the same and due to limitations in time and resource three departments have been selected for in-depth studies.

# Building effective learning networks among researchers and stakeholders in the Danish context

*Nadarajah Sriskandarajah[1], Dorthe Christensen[1], Sofie Kobayashi[1] and Henrik Sønksen[2]*
*[1]Unit of Learning, The Royal Veterinary and Agricultural University, Denmark; [2]Viventes Consult, Denmark*

The current trend within EU policy sees agriculture as a provider of public goods in its environmental and rural functions and in the development of rural areas. Researchers and their extension counterparts in traditional agricultural institutions are facing new challenges in fulfilling their functions effectively under this new trend. LEARNing (Learning in European Agricultural and Rural Networks: Institutions, Networks, Governance) was a nine country project funded by EU. Its premise was that in the light of the current trend in policy, existing relationships between researchers and practitioners in the agri-food system are not optimal. The way researchers, extensionists, farmers, government, food-industry, NGOs and consumers engage with each other (if at all) was not leading to the desired outcomes of multifunctional agriculture and a vibrant rural economy. We explored ways of involving stakeholders in meaningful conversations about different aspects of the agri-food system with the focus on creating new learning networks as means of improving the existing system and of understanding the barriers to their effectiveness. This poster outlines the context of this project in Denmark, processes used and some questions that emerged from stakeholder interactions.

Two research groups, one in plant protection and the other in farming systems, met with other stakeholder counterparts (farmers, advisors, policy makers, agencies, citizen groups) to discuss the above premise. In a final Round Table event, a cross section of people representing the two interests met around the theme 'agriculture in the context of rural development in the Danish landscape'. The multiple aims of the event included: continuing to build a network of meaningful conversations, placing R&D in agriculture in the new and broader context of rural development, formulating a research agenda and, developing relevant tools and facilitating learning of new skills among participants.

Participants at the round table acknowledged a need to move beyond researchers and farmers and be inclusive of diverse stakeholders from rural areas. Other emergent themes were: that farmers were moving into a new narrative, including environmental protection, while still living in the old narrative of production; that farmers could no longer survive on the approach of producing food but instead think in terms of making a living, producing food and maintaining a cultural landscape also as purposes of agriculture; that of the many perceptions of landscape stakeholders in society hold, whose would be important in formulating the dominant narrative about landscapes; that scientists needed to find a common language to communicate with other stakeholders; and that it was not so much a question of whether research should be conducted with the stakeholders, but how. Defining the values held by rural people and their urban counterparts about the Danish countryside and ways of facilitating communication between them were emphasised.

# Use of the nutrient cycle tool in developing sustainable dairy production systems

*Willem van Weperen*
*Adviesgroep Nederland, ETC, Leusden, The Netherlands*

Over 100 dairy farmers in Drenthe Province tested new management practises using the nutrient cycle tool in a study group setting. The tool creates awareness among dairy farmers and results in improved N efficiency in farms. Despite the clear advantages of the tool, its use is not yet widespread.

## Background and objectives

During the last 40 years the Dutch dairy sector changed into a high input, high output and intensive production system. The level of concentrate feeding at one time surpassed 2500 kg/cow/year and recommended fertiliser applications on pasture reached 400 kg nitrogen (N) /ha (Anonymus, 1993). In this high input-high output dairy system, the efficiency level went down dramatically. This inefficient use of inputs leads to N leaching and high levels of ammonia emission into the air. The objective of the Bedreven Bedrijven Drenthe project is to show that introducing PMOV management guidelines on dairy farms, increase N efficiency at farm level and improve environmental conditions.

## Materials and methods

A group of dairy farmers and researchers organised themselves in platform PMOV to promote 'eco-technological' farming; amongst others low crude protein and high crude fibre content feeding, feeding less concentrates and limiting the fertiliser applications on pastureland to 150 kg N/ha. These management guidelines were tested in a study group setting in Drenthe Province and their effects monitored using the nutrient cycle learning tool. Nitrate levels are monitored by Drenthe Province. Ammonia emissions are calculated by the NMI, based on the rations fed to the dairy animals.

## Results and conclusions

The use of the tool, increases awareness and understanding of the interrelationships between the various management measures on dairy farms. It also enhances group learning by, providing feedback to the study group members on alternative management measures. This, together with the financial benefits, encouraged farmers to continue the practises. N efficiency at the project farms over time increased from 29% to 32%. Nitrate and ammonia reduction was 20-25%, while milk production and costs were maintained at the same level. Dutch dairy farmers show increasing interest in the PMOV guidelines. The N efficiency tool is however still not used on a wider scale, because it lacks support from policy makers.

## References

Anonymus, 1993. Handboek Rundveehouderij. Informatie en Kenniscentrum Veehouderij, Publication no. 35, Lelystad.

# Section 2 – Agriculture, land use and sustainable development

# Introduction to section 2 – Agriculture, land use and sustainable development

*Artur Cristóvão[1], Ken Giller[2], Luigi Omodei Zorini[3] and Philippe Perrier-Cornet[4]*
*[1]Universidade de Trás-os-Montes e Alto Douro, Portugal; [2]Plant Production Systems, Wageningen University, The Netherlands; [3]Department of Agricultural and Land Economics, University of Florence, p. le delle Cascine 18, 50144 Florence, Italy; [4]INRA-ENSAM - UMR MOISA, 2 Place Viala, 34060 Montpellier cedex 1, France*

This section focuses on the keywords agriculture, land use and sustainable development.

## Agriculture

Agriculture should be considered according to its widest meaning, as a complex productive activity with a multi-functional purpose. The representation of the countryside only as the site of a productive sector appears limited; nowadays the countryside is assuming, or re-assuming, an increasingly multi-functional role, especially in the more developed economies. This development reconfirms the countryside's crucial role in the general socio-economic system.

New functions have emerged in addition to the traditional one of a direct producer of foodstuff. These are: a) a supplier of many raw materials for an increasingly complex agro-food industry, b) an important market outlet for products of the manufacturers of some means of production, c) an occupational buffer to the employment-unemployment cycle of industrial origin (*e.g.*, multi-activity families), d) an autonomous, diffuse, directly productive activity aimed at subsistence, e) an essential contribution to the protection and conservation of the existing land use, f) an opportunity for old people to continue to feel active and useful for the family, g) a therapeutic function for social categories undergoing urban *stress* (*e.g.*, urban population), or finding themselves vulnerable or at risk (disabled, drug-addicted).

In addition, people in many Europeans regions consider agriculture to be a fundamental element of the territorial specificity and identity, particularly in the regions that have a high environmental, cultural (*e.g.*, historical remains and literary associations), and landscape (*e.g.*, location, climate, *etc.*) value and that are suitable for high quality niche products. In these areas, the consumer is offered a full basket of agricultural products, local crafts, sightseeing and cultural services. These situations are deeply rooted in the public's imagination and lead to the identification with the products not only within their natural , but also within their social and cultural contexts. Their development is linked to the capacity of *creating a system* among agricultural and non-agricultural enterprises, institutions and the local population, united by a specific local or *contextual knowledge*, which is not only technical-sectorial, but also cultural and related to territorial resources. The demand for these products comes from outside (mainly from city dwellers) and often takes the form of agri-tourism.

In this kind of *"rural local system"*, the social and economic environment is characterised by a set of productive activities, strongly linked to the valorisation, conservation and reproduction of renewable natural and historical resources, which offer a basket of goods and services able to satisfy a very wide cluster of needs. In this case, the demand by external consumers results mainly in a complex competition among various territories each one of which offers a peculiar basket of goods and services. Every rural local system shows an association between agricultural producers and the whole community, which allows for sharing the costs and benefits of production. In this case we can no longer speak of agriculture only, but must include environment and rural society, whose production of typical food products, specific to that land use, occurs through synergy among craft, agri-tourism, nature tourism, catering, traditional and cultural events, and land use management, both in the open and in inhabited centres.

The first group of papers deals with the integration between agricultural and non-farming activities at the territorial (Theodoropoulou and Kyriakopoulou) and farm level (Contini *et al.*) as well as the conservation of traditional practices and culture (Dneboská *et al.*). The multifunctionality of agriculture is also examined. Methodology and case studies are presented.

## Land use

The demand for these "*complex goods*" creates a network of relationships between urban and rural areas. In fact, it is networking with urban people, who had become alienated from the countryside that represents the second of the novelties of our times.

Until a few years ago, the disadvantages and hardships of living in the country were such that the basic needs of the rural population could not be satisfied. With the improvement of the means of communication, both within the rural areas and between these and the nearby urban areas, with the spread of manufacturing activities in the countryside or in small communities, and with the consequent increase in the average income of rural families, the hardships of country life gradually decreased. Furthermore, if we consider the growing burden, not only in strictly economic terms, of the life in the city (lack of green space, traffic congestion, air pollution, *etc.*), we can better understand the change of attitude towards living in the countryside. This change is highlighted by the well-known and wide-spread phenomenon of urban families who move to the countryside.

The growth of a new rural pride, developed among old and new inhabitants of the countryside, is not unrelated to this phenomenon and radically alters the earlier sense of shame for rural origins. Some factors contributing to marginalisation yesterday paradoxically have become competitive advantages today.

Living in the countryside, particularly in small centres, whether permanently, periodically or occasionally, corresponds, in fact, to a fuller satisfaction of a cluster of needs linked to the search for greater welfare. This "contamination" between urban and rural people , that once represented a lever for migration to the town, has become a complementary relationship, in which urban dwellers rediscover their subdued or corrupted, but never lost cultural roots, and rural dwellers break their isolation and valorise their style of living. The tourist, or at least the person who does not work in the fields, is all the same physically and psychologically immersed in the local production atmosphere, somehow sharing, even if only observing, the "cultural climate". Most agricultural practices, in fact, take place in the open air in everybody's full view (and smell), for better or worse, so the perception of the agricultural typical product by the consumer or tourist includes, at least in part, that specific "cultural climate" that determines the properties of the product.

On the other hand, the lifestyle of the rural population is not immune from contamination by the urban one: it is sufficient to think of the intense social and institutional relations that even rural dwellers entertain with the near urban centres, or to think at the recent development of telejobs that can be carried out at home, creating a dense network among different places and social groups, beyond physical and cultural boundaries.

In conclusion, the relationship between the city and the countryside is no longer conceived as a relationship marked by mutual estrangement, cultural contraposition, or worse, as a relationship marked by economic and cultural dominance of the former upon the latter, but as a dialectic relationship in which both environments interact.

An analysis of the relationship between rural and urban environment (Stobbelaar *et al.*) and the role of the institution (Röling and Jiggins) and "social capital" (Cristóvão, Baptista and Miranda; Schermer) in the development of the territory are presented in the second group of papers. The promotion and enhancement of typical, quality food products and their supply chain in the form of complex goods which are able to satisfy a cluster of needs is also identified as a way to preserve small-scale farms and their livelihood within the territory (Cristóvão *et al.*).

## Sustainable development

The process of sustainable development can ensure our and future generations continuity of the use of resources to satisfy their complex needs. It is centred on:

- an economic aspect in terms of income for a rural population;
- a social aspect dealing with the style of living and welfare;
- an ecological aspect concerning the preservation of the dynamics of natural resources and ecological services.

The fact that a local system can survive and develop in a capitalistic world only if it appropriately preserves its place in the general cultural and economic panorama means that no place can be considered separately from what happens to the rest of the world.

For their survival and development, local systems need to keep the pace of the incessant evolution of needs and technological innovations that occur in the global market, at the same time safeguarding their natural and cultural identity. The future of local systems depends to a large extent on their capacity to use their productive know-how to respond to the challenges of global cultural, economic and ecologic dynamics.

What is to be preserved is not only a cluster of goods, but a cluster of goods based in a whole life style, i.e. an interrelated complex of resources, goods, activities, values and institutions. This is strongly related to an overall, say national, culture, and normally, but not necessarily, is dependent on distance and communication facilities. The general significance of the proliferation of this kind of local systems is an intensification of the network of relationships between cities and countryside, in other words, a restoration, through new forms, of an intimate relationship between city and country, destroyed by the earlier capitalistic invasion of the land use.

Several papers deal with these themes and focus especially on water management (Buijze and Van Lamoen), biodiversity (Wiersema and Van Alebeek), non-timber forest certification (Wilsey), environmental benefits as silvoarable agroforestry (Palma *et al.*), sustainable milk producing techniques (Chataway *et al.*), biomass production, transformation and allocation (Rapidel *et al.*) and availability of manure, food security and poverty in Mali (Keita *et al.*).

Whether and how different farming systems can mobilise capacity to adapt to this situation and become 'local rural systems' in the sense discussed above, and whether and how they can succeed in combining the three key words in the title of this workshop, is investigated and discussed in the papers of this section.

# Towards sustainable farming: Regional pilots on changing land use under the European Water Framework Directive

*Sarie Buijze and Frank van Lamoen*
*Region of North-Brabant, Ecology Division, PO Box 90151, 5200 MC 's-Hertogenbosch, The Netherlands*

## Abstract

The European Water Framework Directive is a driving force behind the transition of agricultural land use in the region of North-Brabant (the Netherlands). The regional government, water boards, farming community and the drinking water company cooperate in regional pilot projects. The aim of these projects is to develop instruments to ensure both good water quality and good perspectives for sustainable farming.

**Keywords:** Water Framework Directive, regional pilots, instruments

## Introduction

The European Water Framework Directive (WFD) sets a challenge for the improvement of water quality. Remediation programmes will include physical conversion of waterways into a more natural state, sanitation of pollution sinks and diminishing pollution sources. The authorities responsible for implementing the ambitious WFD-goals in the region are the regional governments (Provinces), water boards and municipalities. And - important factor - the result is what counts for 'Brussels' in the year 2015.

Within the Netherlands, the region of North-Brabant has major problems with water quality in rural areas. This is due to a combination of intensive agricultural land use and a vulnerable soil and water system, resulting in high levels of nitrate, phosphate and pesticides. To tackle these problems, regional authorities search for new instruments. This will affect agricultural land use on both an individual and a regional scale. Inevitably farmers will have to change the way they manage their land. At the same time, these changes are part of a transition in which the rural landscape as a production space makes place for a space for consumption and recreation (Horlings, 2004). Part of the transition is autonomous. But concerning issues of public interest – such as rural economy, landscape protection and social structure - this transition needs guidance. Closely connected with water issues, this makes transition of rural land use a major topic in regional governance.

In the case of reducing agricultural emissions, regional authorities are experienced in putting forward either stimulation or spatial instruments (Figure 1). The WFD-challenge is high. In Brabant, regional measures are necessary above the national minerals and pesticides legislation. The currently used regional instruments - to periodically stimulate optimisation of the present agricultural practices - may not be enough. Yet, a more structural approach by dividing functions in spatial planning may be too costly and not attractive for any of the stakeholders in the region. We may need new instruments, for the grey section in-between. Although stretching the limits of 'stimulation' is possible – as currently undertaken in regional pilots - this may imply a more systemic change or transition process.

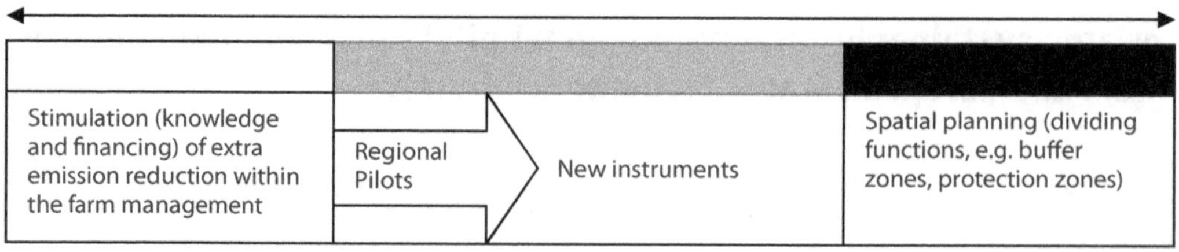


*Figure 1. Possible regional policy instruments.*

## Transition process

For a successful transition process it is necessary that all parties involved share a sense of urgency and a will to cooperate. One of the main prerequisites to start a transition process is to understand each others interests and a will to explore each others motives. The regional Government should recognise that farmers are, above all, entrepreneurs whose income is under stress. The sector deals with market liberalisation and income reductions. Any new threat to their economical validity will – understandably – not be met with great enthusiasm. On the other hand farming is still an important bearer of landscape and nature values. Farmers are well aware of this fact – and proud of it – but do not welcome new costly claims on behalf of the water quality.

In this delicate balance, both the regional Government and the farming community share a problem and are in need of realistic solutions. Yet, individual farmers are not fully aware of the challenge and may not be ready to cooperate. So it is up to the regional Governments to 'seduce' farmers into making joint plans for integrated water management in order to secure both good water quality and good perspectives for sustainable farming.

We describe two cases of regional pilot projects on this matter. Both cases show a process, based on mutual profit and direct influence.

## Case 1. Regional pilots on water quality improvement

The Regional Government of North-Brabant and the Water Boards, in cooperation with the farmers' organisation (ZLTO), carry out five regional pilot projects to improve water quality (mainly concerning nitrate, phosphate and pesticides and sometimes heavy metals) in rural areas. The five areas are more or less hydraulically isolated and represent different types of land use. The farming communities involved are not considered particularly innovating but reflect the 'average farmer'.

Aim of the pilots is to generate 'hands on' experience with cost-effective measures and to create a 'community of practice' for sharing and propagating experiences. The community of practise involved in the process is broader than the pilot areas and includes the total of Brabant (Figure 2).

Central questions are:

- What measures can be taken?
- How effective are these measures for improving water quality?

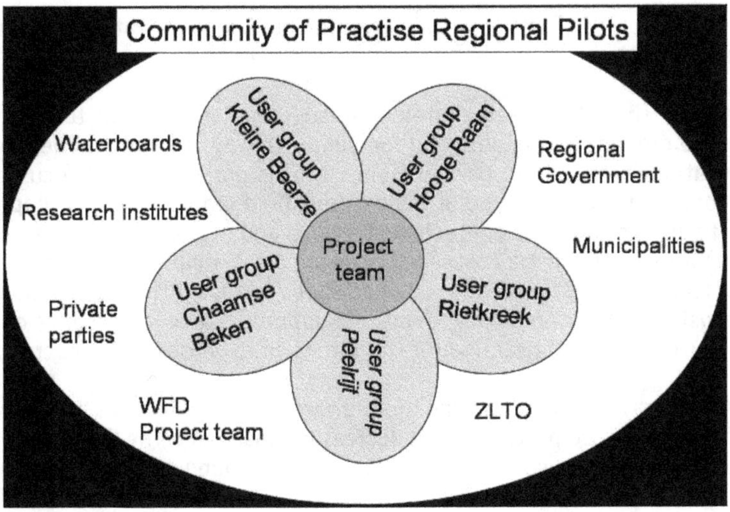


*Figure 2. Parties involved in the community of practise.*

- What are the costs – and who is going to pay?
- Can measures be applied outside the pilot areas?

The focus of the pilots is on actual improvement of water quality. Participation of individual farmers in the pilot projects is voluntary, but needs commitment. Plans are drawn up in close, active cooperation with local farmers groups.

In four pilot areas, planning is done and implementation has started. The different areas show different approaches, ranging from tailor-made measures for individuals, to standardised measures for all farmers in the region. This generates a broad spectrum of experiences to be shared in the community of practise. So far the selected sets of measures seem beneficial for all parties involved, but ex ante evaluation indicates they may not in all cases meet WFD-goals. For this a second set of measures is needed, based on (technical) innovations or even systemic transitions.

## Case 2. The 'Clean Water' approach reduces pesticides leaching

Drinking water in the region of North-Brabant is produced out of groundwater in 40 areas. The regional Government aims for multifunctional land use: drinking water production and agricultural production (the main form of land use) going hand-in-hand. In 8 of these areas (9,000 hectares), drinking water is produced from freatic groundwater. These are indicated as 'Extra vulnerable groundwater protection zones' and of special interest under the WFD (Min. V&W, 2005). Leaching of chemical pesticides is (next to nitrate) a problem here. The regional Government, responsible for the groundwater policy, considered to issue a prohibition in these protection zones on all the problematic pesticides. A quick investigation however learned that that would mean a drastic restriction on the number of crops possible (Boland *et al.*, 2001). This is a critical point for the farmers' perspectives, so their protest was load and clear. On the other hand, the Government is not eager to pay for the consequences. This triggered a joint 'Clean Water' project of the regional Government, Drinking Water Company and farmers (individuals and Union).

Central strategy in the 'Clean Water' project is to stimulate farmers to innovate the best practices for good groundwater quality and for their own way of farming. The project fits under the left (stimulation instruments) part in Figure 1. This project did not focus on 'innovators' or 'early adaptors', but on every type of grower (out field crops), who just happened to be in these zones. The project offers them technical advice and a small stimulance-fee (max. €325 per participant) for eventual extra costs. Practices are often simple (choose pesticides with low leaching risks, reduced doses, mechanic weed control), but need time to be adopted. Technical innovations boost the effects (släpduk, decision support systems) (Bosker *et al.*, 2005). Many practices are neutral or even attractive for the farmer in expenses or in practice.

The effects on groundwater quality in the groundwater protection zones are shown in Figure 3. The use of most problematic pesticides has decreased drastically. The Clean Water practices are workable as cost-effective measures to implement the groundwater quality standards in the WFD and the EU groundwater guideline.

For a few substances we still search for innovative and more cost-effective solutions. When they cannot be found by stimulation, we may need other instruments, in the middle (grey) or right (spatial instruments, such as installing protection zones) part of Figure 1.

## Discussion

The pilots described offer farmers the opportunity to prepare their business to a new set of demands with the Water Framework Directive as a driving force. In the end these demands can either be met by voluntary changes in farming systems or by external legislation and spatial restrictions. First results show that when offered this choice, farmers are willing to explore new ways of farm management. This process can be guided by the regional government by taking up

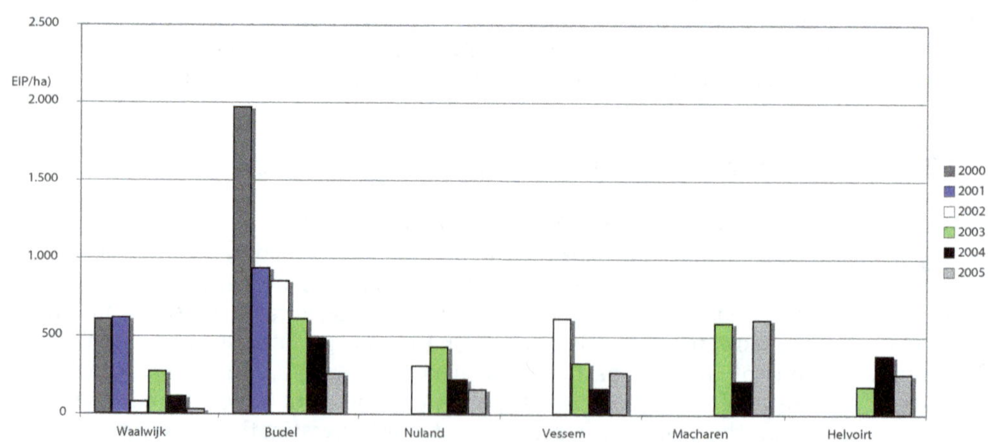


*Figure 3. Effect of the Clean Water project (Leendertse* et al.*, 2006), indicating the risk of leaching, in terms of 'environmental impact points' (CLM, 2005) per area per year. The groundwater quality standard is 500 eip. Number of participants was 36 in 2000 and 250 in 2005.*

a stimulating role. Stimulation may come in the form of knowledge or – temporary – financial compensation. A key role in this process is played by installing a community of practise in which information and experiences are exchanged between user groups in the pilot areas and shared with stakeholders in the total of Brabant.

Expectations are that not only farming practise will change but also the farmers' attitude towards external demands. Actively meeting these demands by taking measures themselves reduces the need for repressive measures by the government. This makes farmers better entrepreneurs, generating long term perspectives for their business by taking up a more sustainable way of farming.

But what if there is need for a more radical systemic change and new instruments. Is this going to undermine the will to cooperate? The outcome of the WFD-implementation at this moment is still uncertain, but parties involved for certain will insist on having influence. In years to come we may have to explore the edges of cooperation. Time will tell whether the instrument will stretch or snap.

## References

Boland, D. and P. Leendertse, 2001. CLM 2000. Bestrijdingsmiddelen in grondwater in Noord-Brabant: risicostoffen en knelpunten. CLM, Culemborg (NL).

Bosker, T., E. van der Wal, M. Boer, P.C. Leendertse and B. Aasman, 2005. Schoon Water rapport 2003-2004. http://schoon-water.nl.

CLM, 2005. Environmental indicator for pesticides. www.milieumeetlat.nl.

Horlings, I., 2004. Inspelen op veranderingen; de bijdrage van de Agenda Vitaal Platteland aan de transitie van landbouw en platteland. Sociaal-wetenschappelijk Tijdschrift voor de Landbouw 19-2.

Leendertse, P.C. and E. van der Wal, 2006. Schoon Water rapport 2005. CLM, Culemborg (NL) (in press).

Ministry of Transport, Public Works and Water management, 2005. Karakterisering Nederlands Maasstroomgebied. www.kaderrichtlijnwater.nl.

# Sustainable dairy farming systems for a deregulated market in northern Australia

*R.G. Chataway, R.G. Walker, M.N. Callow and H.L. Todd*
*Department of Primary Industries and Fisheries, M/S 825 Peak Crossing, Qld 4306, Australia*

## Abstract

The northern Australian milk market was fully deregulated in July 2000 resulting in a decline in milk prices received by farmers of *c.* 20%. It was anticipated that median farm production would need to at least double over the next 5 years for sub-tropical dairying enterprises to remain internationally competitive. At a farm level these changes were expected to have significant socio-economic and environmental impacts. A farming systems research project was established to address the impact of market deregulation on three main parameters; farm economics, the social well being of farming families and the farm's natural resource base. Five production systems were identified, through desktop modelling and iterative review, as having the economic capacity to provide a way forward for faming families in the new deregulated environment. To further assess these systems, field investigations were conducted over a four year period using physical models (farmlets) linked with commercial farms.

Adverse climatic conditions were experienced during the period of field investigations, compounding the impact of reduced milk price on farm profitability and farmer confidence. Three of the five farming systems demonstrated the potential to achieve satisfactory economic goals under these environmental conditions. However all three required significant investment in dairy specific plant and equipment to achieve goals in a manner that was socially and environmentally acceptable. Sociological studies showed an unwillingness of farming families to make these investments during this period post deregulation due to uncertainty they would continue in the industry, and, that the value of these dairy specific investments would not be able to be realised if they withdrew. In conclusion, in intensifying farming operations, incremental changes are relatively easy to manage but large rapid changes, still necessary at particular junctures in the expansion process, carry inherent risk. Contributing significantly to this risk is the historically variable rainfall regime of northern Australia.

**Keywords**: dairy farming, subtropics, internationally competitive, lifestyle

## Introduction

The northern Australian dairy industry is located in a subtropical environment between the latitudes of 17 and 32°S. It produces about 10% of the nation's milk and comprises mainly family farms with a typical milking herd size of 150 to 200 cows. Cattle are typically grazed year round on improved pastures or sown forage crops and supplemented with grain, industry by-products and conserved forage. Rainfall is summer dominant, often intense and of moderate to high variation (Webb *et al.*, 1997). In contrast to southern Australia, production in the north has traditionally focused on the domestic, fresh milk market which has demanded low seasonality of production. Until the year 2000, regulation of the fresh milk market by the State was significant with interventions that included price control at all levels of the marketing chain and statutory production controls (quotas) at the farm level. Farmers received a large price differential between quota (fresh milk) and above quota milk (manufacture milk). This price differential encouraged farmers to produce a level of milk not greatly over quota albeit at a higher cost of production. In contrast, dairy farmers in southern Australia and New Zealand have traditionally produced milk on a seasonal basis primarily for the export market with minimal government regulation of milk marketing and lower cost of production (Parker *et al.*, 2000).

With complete deregulation of the milk marketing chain in the year 2000, farmers in northern Australia experienced the introduction of a single, and lower overall milk price from *c.* A$0.40/L to $0.32/L (Bake *et al.*, 2002) but more flexibility to vary production throughout the year. The challenge for dairy farmers was how to respond to this new lower price coupled with greater production flexibility. Analyses of present production systems showed only modest potential to reduce costs (Hoekema *et al.*, 2000), however, past analyses of farms has shown a generally high potential for increased milk output (Kerr *et al.*, 2000) and cows and infrastructure are often used well below their potential. An initial assessment by Hoekema *et al.* (2000) was that farm production would need to increase by *c.* 100% to manage the change in milk price. This paper describes a series of activities, and the interim findings from these activities, to address the issue of rapid decline in terms of trade that resulted from deregulation of the milk market.

## Analysis of current farming systems

Building on previous reviews conducted of northern Australian farming systems (Ashwood *et al.*, 1993) a project team comprising experienced research, extension and farm management practitioners, reviewed current farming systems. The team identified five production systems, each applicable to a significant number of producers in northern Australia, that would enable substantial increases in milk output without having to expand the farm land resource base. However, to achieve these production increases required substantial increases in both milking cow numbers and in the level of feeding of purchased supplementary feed (Callow *et al.*, 2005) (Table 1).

*Table 1. Characteristics of proposed intensive sub-tropical dairy farming systems.*

| | Farming System | | | | |
|---|---|---|---|---|---|
| Variable | Tropical pasture dominant 1 | Tropical pasture dominant 2 | Forage crops 3 | Temperate pasture dominant 4 | Feedlot 5 |
| Rainfall (mm/yr) | >800 | >800 | 600-800 | 800-1000 | <600 |
| Irrigation | Nil | Limited | Limited | Extensive | Extensive |
| Purchased feed | | | | | |
| - hay (t/cow) | 1 | 1 | Nil | Nil | Nil |
| - grain (t/cow) | 3 | 3 | 3 | 3 | 3 |
| Cows milked/yr | 280 | 360 | 360 | 360 | 900 |
| Stocking rate (head/ha) | 1.9 | 2.8 | 1.4 | 3 | 3.9 |
| Milk (L/cow/yr) | 7 000 | 6 500 | 7 300 | 7 100 | 9 650 |
| Milk Total (ML/yr) | 1.8 | 2.2 | 2.3 | 2.4 | 7.5 |
| Labour units | 3 | 3.5 | 4 | 4 | 9.5 |

Such intensification of the farming system has many implications, particularly for dairy farm families inexperienced in rapid farm development. Callow *et al.* (2005) outline some of the issues involved, including business risks such as debt levels and cash flow, challenges to family well-being, degradation of the natural resource base as a result of the higher number of animals and increased use of fertilizer and supplementary feed on the farm, and greater demands on the management skills of the owner/operator. There were no successful commercial models of such rapid development from which to learn, and consequently there was a need to assess the proposed intensive farming systems in practice. Such an assessment needed to include production and profitability goals of each system, and also environmental and sociological robustness. This was done through the development of physical models of the proposed farming systems and the use of commercial companion farms.

## Physical models and companion farms

On a research station in south east Queensland (27°46′ south, 152°40′ east; altitude 40m; mean annual rainfall 800 mm) a physical model of each farming system developed in the desk top study was established. Managing each farmlet was a team of agency staff and farmers with a particular interest in that system, while overseeing the whole project was a common steering committee. Paired with each farmlet was at least one commercial farm using a similar farming system. These commercial farms, known as 'companion farms' enabled some of the whole farm system implications of enterprise intensification to be more fully investigated. For example, social and labour management implications of farm intensification cannot be adequately studied through farmlets. The commercial farms also enabled systems to be evaluated across a broader geographical base in terms of resource endowment (rainfall, temperature, soils, topography). Selection of companion farms was based on farmers interest in the goals of the project, being a suitable match for one of the five farming systems, and being a participant in a regional dairy farm economic benchmarking scheme (Bake *et al.*, 2002). For these farmers, involvement in the project provided an opportunity to engage with other farmers and agency staff on relevant livelihood issues. Project staff envisaged that both the farmlets and companion farms would provide meeting places for interactive engagement between scientists, extension agents and farmers.

To determine the sustainability of these proposed farming systems, investigations took place under five separate themes; production, water use efficiency, environmental sustainability, business management and sociological factors (Table 2).

The investigative teams formed to address each theme had dual roles. The first was to contribute their expertise to the overall management of farmlets to assist in the implementation of best management practices; and secondly, to monitor the five systems for differences between, and within systems, over time. Leaders of the investigative teams met weekly to share findings and make recommendations on management practices to the project management team. Field studies continued for four years.

## Key findings

The modelled farming systems generally achieved milk production targets. However, higher than predicted costs for purchased feed, due to drought conditions across Australia in 2002-2003, and lower forage production due to reduced rainfall over the whole study period (670 versus 800 mm/annum) resulted in those systems heavily reliant on rain grown forage and purchased

*Table 2. Themes of investigation and their focus.*

| Theme | Investigative Focus |
|---|---|
| Production | Forage production, ration formulation, milk volume and milk components, cow condition, reproductive performance. |
| Water Use Efficiency | Species selection and management to maximise water use efficiency in terms of forage production and milk yields. |
| Environmental Sustainability | Fertiliser, effluent, irrigation and tillage. management and monitoring of deep drainage, nitrogen leaching, whole farm nutrient balance, soil cover, soil chemistry changes. |
| Farm business management | Evaluation of each system for profitability (operating profit, return on assets) and cash flow. |
| Social | The family-business interaction and their dependency on each other. Reviewing issues considered by families before making farm adjustments and the implications arising following changes. |

supplements (Systems 1 and 2) (Table 1) achieving unsatisfactory economic results; assessed in terms of return on assets, operating profit and cash flow.

It was found that rapid intensification of farming systems imposed a great deal of strain in terms of cash flow and level of indebtedness on farming businesses. Rapid expansion is sensitive to the cost of inputs, supply management, herd factors and acquisition of new management skills. These problems were exacerbated with intensification coinciding with a period of prolonged low rainfall as occurred during this study. Predictions that climate change will increase interannual rainfall variability, reduce effective rainfall through higher temperature regimes and place more pressure on water resources (Pittock, 2003) has particular implications for the northern dairy industry.

Sociological investigations found that farmers considered the new milk price received to be low and unpredictable. This engendered a general lack of confidence in the industry, and while farmers were willing to invest in assets that were not dairy specific, such as land, or were easily realisable, cattle, there was a resistance to invest in dairy-specific infrastructure such as milking parlours, feed pads, forage conservation and effluent management systems. Without these dairy-specific resources, farm intensification increased the demands on family labour and made environmental management harder due to the impact of increased cow numbers. Labour management was also an issue for farmers inexperienced in employing non-farm labour. These farmers continued to avoid employing staff, even during or after a period of farm expansion. Farming families considered the view of the farm as a business that is highly focused on profit maximisation (Hoekema, 2003) as being too narrow. Issues such as labour management, the intentions of the next generation, time demands, living with debt and the desirability of the farm and its environs as a place to live were all strong influences on the reaction of farm families to the process of rapid business growth.

With respect to environmental management, intensifying the farming system required greater investment in specialist plant and equipment to reduce the risk of point source pollution. Whole farm nutrient balances showed that with higher levels of supplementary feed inputs, fertiliser recommendations need to be reviewed and often reduced. On farms where cropping is routinely practiced, a benefit of intensification is an increased cropping frequency which has positive outcomes in maintaining higher levels of soil cover, maintaining or increasing levels of soil organic matter, and reducing the risk of nutrient and water loss below the root zone.

Supplementary irrigation had a very positive impact on forage productivity. The benefits measured were probably amplified by the rainfall shortfall during the investigative period. While tropical grasses were found to have a high rate of water use efficiency (kg DM/L water) compared to temperate species, this efficiency was offset by their lower nutritive value. This needs to be considered in forage selection.

## Conclusions

Intensifying current farming systems through increased stocking rate, greater forage production and utilisation and increased use of purchased feeds does achieve higher milk output and economies of scale. However appropriate phasing and sizing of the development process is necessary to match labour and budgetary constraints. Unpredictable climatic conditions increase the risk of business failure during periods of rapid intensification. A reduction in both effective rainfall and its' reliability, due to climate change, poses a major threat to dairying systems not well buffered against low rainfall. Uncertainty of the business future due to a more open and competitive milk pricing environment and/or continued poor seasonal conditions leads to an unwillingness of farmers to invest in dairy-specific plant and equipment which is required if intensification is going to be managed successfully from a lifestyle and environmental perspective.

## References

Ashwood, A., D. Kerr, R.G. Chataway and T.M. Cowan, 1993. Northern Dairy Feedbase 2001. 5. Integrated dairy farming systems for northern Australia. Tropical Grasslands 27, 212-228.

Bake, G., G. Busby, G. Hetherington, R. Itzstein and L. Thorburn, 2002. Queensland Financial and Production Trends 2001. Department of Primary Industries, Queensland, Brisbane.

Callow, M.N., N. Gobius, and G. Hetherington, 2005. Development of profitable milk production systems for northern Australia: an analysis of intensification of current systems. Farm Business and Farming Systems Management 2, 25-39.

Hoekema, M., 2003. Milk Business - a process of analysing your business, planning changes and monitoring progress, Training Program, Tropical Dairy Research Centre, University of Queensland.

Hoekema, M., M. Jelbart, D. Clark and A. Worsley, 2000. 'Future Farming Systems - options for a deregulated market.' Final report to the Dairy Research and Development Corporation:August 2000.

Kerr, D.V., P.M. Pepper and R.T. Cowan, 2000. Estimates of achievable milk production on subtropical dairy farms in Queensland. Australian Journal of Experimental Agriculture 40, 805-811.

Parker, A., K. Woodford and E. Woods, 2000. Deregulation and restructuring in the Queensland dairy industry. University of Queensland, School of Natural and Rural Systems Management, Occasional paper 7, 1-28.

Pittock, B., 2003. Climate change - An Australain Guide to the Science and Potential Impacts Australian Greenhouse Office, Australian Government, Canberra.

Webb, A.A., M.J. Grundy, B. Powell and M. Littleboy, 1997. The Australian sub-tropical cereal belt: soils, climate and agriculture. In: A.L. Clarke and P.B. Wylie, eds. Sustainable crop production in the sub-tropics: an Australian perspective. Department of Primary Industries, Queensland, Brisbane, pp. 8-23.

# Network analysis of farming and non-farming activities for sustainable development

*Caterina Contini, Luigi Omodei Zorini and Roberto Polidori*
*Department of Agricultural and Land Economics, University of Florence, p. le delle Cascine 18, 50144 Florence, Italy*

## Abstract

The aim of our work was to establish a methodology to analyse the relationships between the different economic activities in the territories where there are strong ties between agriculture, tourism, the landscape and traditions as well as to analyse the impact that rural development policies may have on the local economic system. The study has shown how the ability of a territory to trigger mechanisms of development by enhancing the typical features derives from the ability to create a system on the part of all the economic and social-institutional players. Regarding development policies, these considerations underscore the need to overcome a limited sectoral vision and to envision a wide-reaching system to be achieved by implementing territorial measures aimed at reinforcing local synergies.

**Keywords:** rural tourism, sustainable development, input output analysis

## Objectives

Integrated territorial policies have attributed growing importance to local plans for development. Consequently, it is very important to ascertain the possible impact of policies being implemented to enhance the value of local specifities on the entire economic system of the territories involved. The aim of our work was to establish a methodology to analyse the relationships between the different economic activities in the territories where there are strong ties between agriculture, tourism, the landscape and traditions as well as to analyse the impact that rural development policies may have on the local economic system.

## Areas studied

The study was conducted in three Italian areas: the Etna Park in Sicily, the Mountain Community of Val Pellice in the northern Cozie Alps, neighbouring with France, and the area of Bassa Valdelsa in Tuscany. These three zones differ geographically, social-institutionally and economically. However, they are all characterized by the fact that they are agricultural regions where agriculture is both a productive activity as well as the foundation of the social texture. The areas in question also share a marked potential for tourism.

## Methods

An analysis of the official statistics made it possible to define the areas studied according to the degree of development and the economic structure. Interviews conducted with key informants made it possible to identify the weak points of the systems which formed the object of the study as well as the sectors which drive development. Based on the indications obtained, various measures were considered to enhance local resources. The application of an input output model allowed assessment of the effects that these measures could have on the local economic system. For this purpose, we calculated the production multiplier type I $(I\text{-}A)^{-1}$

and type II $(I-A-cv)^{-1}$, where I is the identity matrix, A is the matrix of technical coefficients, c is the vector of coefficients of the propensity to consume and v is the vector of the added value coefficients. The type I multiplier assesses the direct and indirect impact of demand on the local economic system, while the type II multiplier makes it possible to identify the effect induced by introducing demand in the model; this is a function of the income of those who participate in the productive process. For the three areas, calculations began with the input output tables which were produced by the Regional Institute for Economic Planning in Tuscany (IRPET); they refer to the year 2001 and are made up of 30 sectors. This calculation required the solution of two types of problems. Firstly, the lack of local tables and secondly, the lack of sufficiently detailed tables. In the case of the Etna Park and Val Pellice, it was a matter of solving the first type of problem since the tables available were the regional ones (Sicily and Piedmont). Consequently, it was necessary to assess the matrix of technical coefficients relating to the two areas being studied. For this purpose, the FLQ formula, put forth by Flegg (Flegg *et al.*, 2000) was applied. This formula was chosen because recent studies have shown that it is capable of providing a satisfactory assessment of regional multipliers (Bonfiglio *et al.*, 2004; Thomo, 2004; Bonfiglio, 2005). The formula estimates the location coefficient as a function of three variables: (i) the relative size of the sector which produces the final resources j; (ii) the relative size of the sector supplying inputs i; (iii) the relative size of the region.

$$FLQ_{i,j} = \frac{RE_i / NE_i}{RE_j / NE_j} \times \lambda^* \qquad (1)$$

where: $RE_i$ and $RE_j$ respectively indicate the workers employed in the local firms in the area studied based on the sector supplying the input and on the sector which produces the final resources. $NE_i$ and $NE_j$ respectively indicate the workers employed in the local firms of the region based on the sector supplying the input and on the sector which produces the final resources. $\lambda^*$ is derived from the following formula:

$$\lambda^* = \left[\log_2\left(1 + TRE / TNE\right)\right]^{\delta} \qquad (2)$$

where: TRE and TNE are respectively the workers of all the local firms in the area studied and in the region.

$\delta$ is a coefficient greater or equal to 0 and inferior to 1. In our case, we assumed that $\delta$ is equal to 0.0878, based on the results of Thomo (2004) and Bonfiglio (2005). These results demonstrate that with this value the multipliers obtained are very close to those which would have been obtained if the input output tables had been constructed by means of direct investigation. When the location coefficients ($FLQ_{ij}$) are inferior to 1, the technical coefficients for the area being studied are reduced, compared to the regional coefficients, by multiplying the latter by the relative location coefficients. In case the location coefficients are greater or equal to 1, the technical coefficients for the area studied are equal to the regional ones. In the case of Tuscany, it was not necessary to estimate the local technical coefficients since the table available was the table pertaining to the local system which formed the object of the study. In this case, our calculations involved introducing the agro-tourist sector in the matrix (sector which drives the local system). The cost structure and the organization of work in this sector differs from the "restaurant and hotel" sectors within which it was incorporated. For this purpose, a direct investigation was conducted with a representative sample of agro-tourist firms in the area in order to: (i) describe the costs incurred by the agricultural entrepreneurs for the purchase of intermediate goods; (ii) define the final value of production; (iii) verify the number of workers employed in the agro-tourist sector. Lastly, a direct investigation was conducted to assess the average tourist spending. This was based on presences in the three areas under study and on the distribution into the sectors of the input output matrixes which were used. For the Etna Park and Val Pellice, the investigation involved all the tourists while in the case of Bassa Valdelsa, the interviews were limited only to the guests of the agro-tourist facilities.

## Results

The Etna Park is a marginal area. It is characterized by low wages, low levels of education and by a high rate of unemployment. Agriculture plays an essential role from the economic, social and landscape point of view. Typical productions of value are found in the area; these are agriculture and local crafts. Combined with the natural resources, they are the strong points of the local system. Development of this system cannot be based on a single process but rather on an agglomeration of processes and on environmental sustainability. Local products must be linked to the natural, social and cultural context within which they are produced (Becattini *et al.*, 2003). Therefore, local development is dependent upon the local territory since consumer appreciation of the product is linked directly to the experience of a stay in the area. For this reason, tourism seems to represent a privileged vector which makes possible to promote and attach value to the local supply, thus having a direct bearing on development of the system. Tourist arrivals in the year 2003 were 84,130 while the presences were 252,492. This is mainly summer tourism which peaks during the month of August (Statistics Office of the Tourist Board in Catania). Average tourist spending per presence is equal to 74.17 € of which, 78% (57.89 €) was spent within the park. Tourist spending within the park in the year 2003 produced an overall income of 18,508,334.66 €, thus activating 347 workers.

As to the distribution of work, approximately $^1/_3$ of the workers are employed in the hotel and restaurant sector, ¼ in the agricultural sector; $^1/_6$ in the commercial sector and 5% in the agro-industrial sector. Especially noteworthy is the relationship between tourism and agriculture. This is confirmed by the high number of workers employed in the agricultural sector as a result of tourist demand. This ratio certainly is a distinctive feature of local tourism which underscores the importance of tourism in developing agriculture within the park's territory. Synergies between tourism and the territory, as identified in the survey, could be strengthened by promoting a type of tourism which is closely linked to the natural environment, the culture and the typical local products. This can be achieved by reinforcing both the association between tourism and local resources and by increasing tourist flow. In order to assess the impact that a policy targeted at promoting tourism tied to endogenous resources might have on the local economic system, we assumed an increase in presences during the low season months which then reach an average level during the month of May. As to the spending vector, and taking into account considerations which emerged during meetings with local experts and representatives of the local institutions, we assumed a 10% increase in the "agricultural" and "agro-industrial" sectors. The increase in the per capita spending would be 0.50 € per day; that is to say, 1.50 € for the entire duration of one's stay in the area. The results of the simulation conducted with the input output models show an increase in income and employment equal to 21%. The sectors most involved are agriculture and agro-industry. In the agricultural sector, income went from 897,076.67 € to 1,100,065.19 € while employed workers rose from 87 to 107. In the food industry, income rose from 1,319,702.82 € to 1,602,917.33 €; employed workers increased from 16 to 20.

In Val Pellice the main problem has to do with depopulation related to the industrial crisis. On the other hand, the territory's strong features are to be identified in the natural resources and in the historical cultural heritage; more specifically, the "valdese" culture which is a strong tourist attraction. The main weakness of the local tourist system has to do with a lack of coordination as well as the lack of an adequate organization of the territory's different tourist products. Official arrivals in the year 2003 were 20,504 for a total of 83,982 presences which were concentrated in the summer period (data obtained from the Province of Turin). Average spending per presence is 53.08 €, 80% of which (42.43 €) occurred within the valley. Overall tourist spending in Val Pellice in the year 2003 was estimated to be equal to 2,807,197.13 €. This produced an income of 2,822,656.33 € and employed 54 workers. As to the distribution of employment which was created through tourist spending, more than half of the work force is employed in hotels and restaurants (31 workers), 12% in the commercial sector and another 12% in the agro-industrial sector. If one compares the average spending in the valley for each presence to spending estimated by the Federazione Italiana Pubblici Esercizi, in 2002 on a national level (equal to 63 € for the Italian tourists and 83 € for foreign tourists), one can conclude that spending is relatively low in Val Pellice. This data would seem to confirm the main limit of the tourist system which emerged during interviews with sectoral experts. It is the result of a lack of coordination and of an adequate organization of the territory's products. This limits the visibility of the offer for tourists and thus reduces the level of spending. Consequently, the need arises to make it easier to enjoy the goods and services available in the territory. Information must be gathered and the offer organized so as to provide the tourist with a range of options. This would create new spending opportunities for the tourist which in turn would increase the impact of tourism on the valley's economy while not modifying the tourist flow. The input output model was applied to estimate the impact on the local system of measures aimed at improving the organization of the tourist offer. Based on the results which emerged during research and on the remarks made by the local operators, daily per capita spending within the valley was increased up to 50 €. Sectors involved in an increased demand were "agriculture", "food industry", "wood industry and wood products" and "transportation, storage and telecommunications" which include tourist assistance activities such as guides and escorts. Spending in these sectors was increased by 63% compared to spending as found by the direct survey. The impact on employment within the valley was estimated as equal to 63 workers which corresponds to an increase of 16%, compared to the current situation; the impact on income was estimated to be 3,310,184.96 € with an increase of 18%.

The Bassa Valdelsa is characterized by its tourist vocation. The data was obtained from the Province of Florence and pertains to the year 2002. They indicate the number of tourist presences as equal to 411,333. The ratio between presences and the resident population is among the highest in the province. The tourists who, in the year 2002, stayed in agro-tourist facilities represent 15% of the total. The data provided by the Official Registry on Agro-tourism in Tuscany reveal a continuous increase in the number of firms in the area. This increase underscores the growing interest in agro-tourism both on the part of tourists and local operators. The results of the input output multiplier indicate that for agro-tourism the income and employment multiplier is greater than the multiplier for the "hotel and restaurant" sectors. An important feature is also the strong multiplying component of the agro-tourist sector which remains within the local economic system. This is proof of the strong relationships between this sector and local economy. As to tourist spending, it was estimated to be equal to 61.43 € for the presences; 79% of this (46.73 €) was spent within the area. The overall impact of spending on local economy, on the part of tourists lodged in the agro-tourist facilities in 2002, in terms of income was 4,313,382.93 € with 101 workers employed. Besides agro-tourism, the sectors most involved were commerce and restaurants where the tourist directly expresses his or her requests, as well as agriculture. In this case, the effect on the added value and on employment partly depends upon spending made in the sector and partly on spending in agro-tourist enterprises which provide refreshment facilities and which are defined by strong ties to agriculture. A comparison was made of the average daily spending per capita to the estimates of the Tourist Study Centre which conducted a survey of the entire province of Florence during 2001. It was found that spending in our study area was markedly lower than spending on the part of tourists who vacationed in other rural areas of the Province. In particular, in the Chianti region, a neighbouring area with characteristics similar to the area we studied, the average spending per presence was calculated to be 98.29 €. According to these data it is reasonable to speculate that in Bassa Valdelsa an increase in the impact of tourism on local economy might be derived from an increase in average tourist spending per presence. This could be achieved by expanding the territory's range of goods and services in the sectors of food and wine, local crafts and local initiatives. The results of the simulation conducted on the input output model for the reference area, based on tourist daily per capita spending in the Chianti area show an impact on the economic system which is more than double the current one; this is both in terms of added value (891,9331.37 €) and in terms of employment (213 workers).

## Conclusions

The study has shown how the ability of a local system to trigger mechanisms of development by enhancing the territory's typical features does not depend on agriculture alone. It derives from the ability to create a system on the part of all the economic and social-institutional players. If synergies are created between institutions, farmers, artisans, restaurateurs, businesses and organizers of traditional and cultural events, the intrinsic value of the production in rural areas might be able to emerge even if its quality is not yet sufficiently appreciated by the consumers. In this kind of context, the costs and benefits of production are shared by agricultural producers and the rest of the community. Therefore, we can no longer speak only of agriculture; rather we must speak of a rural territory. Regarding development policies, these considerations underscore the need to overcome a limited sectoral vision and to envision a wide-reaching system to be achieved by implementing territorial measures aimed at reinforcing local synergies. In light of this, it would seem opportune that the political and financial commitments on the part of the European institutions be used for the territories and for the promotion of rural development on the whole rather than for sectoral policies aimed only at sustaining agriculture.

## References

Becattini, G. and L. Omodei Zorini, 2003. Local identities and globalisation. In: L. Omodei Zorini and A. Cristovao (editors). Proceedings of the Fifth IFSA European Symposium, 8-11 april 2002, Firenze, ARSIA, Firenze.

Bonfiglio, A., 2005. Can non-survey methods substitute for survey-base models? A performance analysis of indirect techniques of estimating I-O coefficients and multipliers, Quaderno di ricerca n. 230, Dipartimento di Economia, Università Politecnica delle Marche, Ancona.

Bonfiglio, A. and F. Chelli, 2004. An Impact Analysis of SAPARD in Rural Areas by Alternative Methods of Regionalization, Quaderno di ricerca n. 218, Dipartimento di Economia, Università Politecnica delle Marche, Ancona.

Flegg, A.T. and C.D. Webber, 2000. Regional size, regional specialization and the FLQ formula. Reg Stud. 34, 563-569.

Thomo, T., 2004. New developments in the use of location quotients to estimate regional input output coefficients and multipliers. Reg Stud. 38, 43-54.

# Local organizations in the Douro-Duero region of Portugal and Spain: Towards a pluralistic local animation system

*Artur Cristóvão, Alberto Baptista and Raquel Miranda*
*Universidade de Trás-os-Montes e Alto Douro, Portugal*

## Abstract

Many organizations work today in rural areas attempting to implement strategies able to promote the local economy and sustainable livelihoods. These organizations – public, private or third sector related – work in diverse fields, such as education and training, social and economic development, environmental protection or cultural promotion. The paper presents a critical and comparative analysis of organizational dynamics in the Douro areas of Portugal and Spain, identifying problems, issues, initiatives and challenges to their sustainability. This research was inspired by the recent theoretical reflections on local development, the importance of social capital and territorial competitiveness, the State-civil society relationships, the new models of participation and participatory democracy, and the creation of partnerships and networks as an asset to development work and animation processes.

## Introduction and theoretical orientation

Today, the border territories of Portugal and Spain are, more and more, spaces of cooperation and joint construction of develpment paths (Rodríguez-Spiteri, 2001). The EU INTERREG[1] Initiative has contributed to this objective, and an example is the Project under which this research was conducted, that united the Universities of Trás-os-Montes and Alto Douro (Portugal) and Salamanca (Spain). This Project, entitled "Douro-Duero Século XXI", aimed at the promotion of rural sustainable development, by defining strategies and actions based on local resources and inter-institutional partnerships. In its frame, the organizational issues deserved close attention, given its crucial importance in a time citizens are confronted with multiple challenges, particularly in peripheral rural areas.

This research was inspired by recent theoretical reflections on globalization and local development, the importance of social capital and territorial competitiveness, the State-civil society relationships, the new models of participation and participatory democracy, and the creation of partnerships and networks as an asset to a continued development work and animation processes.

In many rural territories, the globalization process directed by financial and business interests led to marginalization and population exodus, as a result of the concentration of attention and resources in areas considered to be more productive and profitable. On the other side, pressure upon the State has reinforced this trend, leading to policies that favour the zones and activities more capable to reproduce capital or those that have more population pressure (Moreira, 2001). These "global social mutations", as stressed by Bryden (1998), also provide new perspectives to rural areas. This author mentioned the growing importance of the notions of territory and local identity, and identified a set of alternative opportunities, such as the access to new markets,

[1] INTERREG is a EU Initiative aiming at the development of social and economic cohesion, and the promotion of border, trans-national and inter-regional cooperation.

the creation of new activities in the service and environment sectors, the attraction of new residents, and the development of cultural, leisure and tourism activities in rural areas.

For those working at the field level, in fragile areas like the Portuguese-Spanish border of the Douro valley, corresponding to the regions of Trás-os-Montes & Alto Douro and Castilla & León, the challenge is to explore the margins to manoeuvre, in order to fight the observed marginalization trends and to promote people and community-centred development projects that value endogenous resources and consider the competitive advantages of the territory. In such a process, the notions of people-centred development (Korten, 1990), empowerment (Pretty *et al.*, 1995; Shor, 1992), and social capital (Putnam, 1996) are critical. Local organizations become important elements, as part of the social capital, as instruments for power (re)distribution in society, and as appropriate spaces to educate for democracy and active citizenship. Besides, in present times, they tend to fill the intervention gaps of both the public and the private sectors (Clark, 1995) and contribute to the construction of a pluralistic local animation system.

## Objectives and project construction

Many organizations operate today in the rural areas, attempting to implement strategies able to promote the local economy and sustainable livelihoods. These organizations – public, private or third sector related – work in diverse fields, such as education and training, social and economic development, environmental protection or cultural promotion. In the frame of this project, the research looked at the organizations concerned with local development, with the aim of comparing dynamics across borders and stimulationg more cooperation. In more specific terms, the objectives were: (1) to analyse and compare organizational processes and dynamics related to rural-local development; (2) to identify obstacles to a better performance of these organizations and challanges to its sustainability; and (3) to strengthen partnerships and networks related to rural-local development.

The work was initiated in October 2003, with exploratory field contacts, in order to discuss the project aim and objectives, and to design the local institutional map. At the end of this phase a Seminar was held in the region, to involve a variety of local actors, by presenting the project, and talking about the organizational issues and methodologies to further study them. These activities were followed by document analysis and two moments of data collection: (1) a mail survey questionnaire sent to 115 organizations (90 in Portugal and 25 in Spain), of which 42 were returned (29 from Portugal and 13 from Spain, representing a response rate of 37%). This questionnaire included a set of questions on: organization identification and history; number and type of members; area of intervention; objectives and fields of activity; services provided; sources of funding and involvement in EU or national programs and projects; human resources; and cooperation in the border space; (2) a series of case studies of different types of organizations activelly involved in local development animation work, prepared through field visits and open-ended interviews with directors and technical staff. These interviews looked at the following issues: history of the organization and major motivations to create it; objectives, approaches and activities; structure, leadership and resources; relationships with other organizations and involvement in parnerships and networks; changes or impacts produced; and organizational sustainability. A total of 16 cases were prepared, 9 in Portugal and 7 in Spain. At the end of the Project, in June 2005, a second Seminar was organized, to share the results and debate the major conclusions, issues and recommendations.

## Organizational dynamics in the Douro-Duero

The Portuguese-Spanish border of the Douro-Duero, characterized by its rurality, relative isolation, and low population density, is a living space in organizational terms. This work allowed us to understand that local development organizations are numerous, have very different histories, some being relatively old (particularly those related to farming and agriculture development) and others novel, have interventions in diverse domains and in different territorial scales, from the parish or municipality to broader areas. All organizations share a common preoccupation: to defend the territory, fighting against problems and negative trends, and attempting to value resources, develop the local potential, and take advantage of opportunities to conquer competitiveness. Summarizing, they act to reverse the socio-economic decline and the tendencies shown by most studies and prognostics. If it is clear that many of these organizations look at implementing alternative development paths in their territories, it is questionalble if they work under a "territory-project" logic (Melo, 2003), that is, with a strategic view and in articulation with other actors. In fact, many operate in a more or less isolated fashion, implementing activities that are not part of a global plan.

Talking about programs, most organizations, in their plans and daily practices, show strong relationships to the EU, national, regional or local policies. The relationships with the local and regional adiministrations are evident, as are the importance of the EU Initiatives, LEADER and INTERREG in particular. In the Portuguese case, it is quite common the involvement in programs managed by the public bodies related to Training & Employment, Agriculture & Rural Development, and Environment & Nature Conservation. These programs facilitate progress, but also generate problems, for instance in terms of coping with bureaucracy, endebtment (due to late reimbursements) and dependency (on policies and public funding).

The obstacles to the work of local organizations are many and diverse. Besides the difficult work context, marked by negative socioeconomic trends, organizations have to face an administation often not sensitive to problems and not prepared to act in a concerted way and with a territorial perspective. The population is ageing, young people run away from agriculture, entrepreneurs are scarce, parnerships are difficult to maintain, the State is distant and slow, funds arrive late, and the organizations manage their activities in cycles of progression and retraction, facing the lack of financial resources and the instability of the technical teams. The conclusions of Moreno (2003), in a broader study of local development organizations and initiatives in Portugal, were confirmed: the major difficulties have to do with a strong financial dependency from the State, that generates a context of uncertainty and jeopardizes the possibility of acting regularly and sustainably with a strategic sense and according to a plan.

Citizen and community participation seem to be difficult to implement. At the organization level, it is not easy to mobilize people to assume leadership positions, and the same persons tend to be in power for long periods. The presention of alternative lists in elections is quite rare. This problem, understandable in a context of population scarcity, can also derive from the fact that many organizations depend on a nucleous of elements with more power and willingness to participate in community matters. In this situation, the challenge is to assume a pro-active position of power redistribution, aiming at the involvement of the hard-to-reach and difficult to involve. If this possibility is not considered, it is likely that more power concentration and exclusion of the less participative will happen. Regarding the participation of members, usually very reduced, the challenge is to break the common logic of accomodation and passivity, assuming that mobilization and animation activities are needed. These activities are, in a way, the possible path to create spaces of encounter and communication, promoting the exchange of ideas and collective learning. Mobilization and animation are, as well, a form of creating the neccessary spirit of identification and sense of belonging to the organization, that seem to be a problem in many cases.

In places with low population density and development problems, and in situations where many local organizations exist, interinstitutional cooperation is an irrefusable option. Agreements and contracts, networks, partnerships and other mechanisms to cooperate, formally or informally, strengten the density of the institutional tissue and permit it to gain power and to reinforce the intervention, accomplishing the perspective of "territory-project". In the studied area there is a long way to go to build the network, gain relational density and acquire negotiation power. When talking about partnerships and networks, it is essential to mention the cooperation across borders. The EU Initiative INTERREG has spread out cooperation seeds, and many organizations are involved in projects, in different fields related to local and regional development. The same could be said about the LEADER program. In general, only the stronger organizations, better equiped in human resources and with better technical capacity, are engaged. The smaller and sector-oriented ones, tend to remain outside such cooperation initiatives. In matters of border cooperation, the idea emerged that it is still very incipient and based on projects that lack continuity. If some organizations show some maturity in this respect, most others have no experience at all. However, it is crucial that the institutional tissues of Douro-Duero get crossed and gain density, as development issues are the same or similar, and the paths to explore are common.

The results of development interventions are disseminated in the territory and related to a broad range of aspects. We could see advances in the territorialization process, gains of competence and self-esteem, growth in power and capacity to attract resources, increased consciousness about the value of local resources and the possibility to use them sustainably. The interviewed actors are aware of these changes, but question the possibility to sustain them, given the socio-economic context and their limitations in terms of financial and technical autonomy. The sustainability of local development organizations is an issue. In fact, most of them rely on public funds to work and survive. Although providing services of public interest, clearly needed by citizens and communities, they don't see this effort dully recognized by the State. Such situation leads to the issue of State-civil society relationships, and to the construction of instruments capable of facilitating symetric partnerships and assuring less bureaucracy, favourable fiscal systems and adequate financial support (ANIMAR, 2003; Cristóvão *et al.*, 2004).

## Three major ideas for a sustainable future

As stressed before, the challenges to local development work and a better performance of the studied organizations are multiple. The following three (complementary) ways should be considered (Amalric, 1999; Cristóvão, 1999; Stahl, 1999): (1) to defend the sustainability of local development organizations, that requires: improved leadership and technical capacities; promotion of a culture of democracy and participation; better visibility of the work and accomplishments, at the local, regional and national levels; continuous construction of networks and partnerships, at variuos spacial levels; (2) to promote a new type or style of intervention, both in the State institutions (from the national to the local level) and local development organizations, actor-oriented, interdisciplinary, and considering the social energy of the territory, in contrast with the normal intervention style, top-down, essentially sectorial, unidisciplinary, and technicist; and (3) to build intelligent territories, with the capability to learn by generating dynamic processes of social or collective learning, namelly through: active exchanges of information; continuous human resource training and development; open exchange of experiences; permanent dialogue with other territories; creation of interfaces between actors (networks, parnerships, linkage committees, *etc.*), active involvement of research and other knowledge related centres; stimulus to innovation and creativity; in a word, open mobilization of the territorial social capital. These measures will contribute to strengten the already existing framework of a pluralistic local animation system.

## References

Amalric, F., 1999. Natural Resources, Governance and Social Justice. Development 42, N° 2, 5-12.

ANIMAR, 2003. Teses da V Assembleia do Desenvolvimento Local, Manifesta 2003. Lisboa: ANIMAR.

Bryden, J., 1998. Novas Perspectivas para a Europa Rural: Tendências Globais e Respostas Locais. LEADER Magazine 18, 4-12.

Clark, J., 1995. The State, Popular Participation and the Voluntary Sector. World Development 23, N° 4, 593-601.

Cristóvão, A., 1999. Para a Valorização dos Recursos Naturais do Vale do Douro. Douro – Estudos e Documentos, Vol. IV (8), pp. 19-31.

Cristóvão, A., Melo, A., Moreno, L., Monteiro and A.E.J. Brás, 2004. Relacionamento Estado-OCS/OIDL, Bases para o Debate sobre uma Carta de Relacionamento. Documento de Trabalho apresentado no Encontro Nacional de ODL – Os Novos Desafios do Desenvolvimento Local. Santárém: ANIMAR.

Korten, D., 1990. Getting to the 21st Century: Voluntary Action and the Global Agenda. West Hartford: Kumarian Press.

Melo, A., 2003. As Organizações da Economia Social e Solidária na Construção de uma Democracia Participativa e de um Desenvolvimento Sustentável. Texto de Conferência da ULHT. Lisboa: ULHT.

Moreira, M.B., 2001. Globalização e Agricultura. Zonas Rurais Desfavorecidas. Oeiras: Celta Editora.

Moreno, L., 2003. Guia das Organizações e Iniciativas de Desenvolvimento Local. Lisboa: ANIMAR.

Pretty, J, I. Guijt, I. Scoones and J. Thomson, 1995. Participatoy Learning and Action: A Trainer's Guide. London: IIED.

Putnam, R., 1996. Comunidade e Democracia – A Experiência da Itália Moderna. Rio de Janeiro: Editora da Fundação Getúlio Vargas.

Rodríguez-Spiteri, J., 2001. Espanha e Portugal: Balance de una Amistad. In Fundação Rei Afonso Henriques, Espanha-Portugal Horizonte 2010, pp. 33-27. Zamora: FRA.

Shor, I., 1992. Empowering Education: Critical Teaching for Social Change. Chicago: The University of Chicago Press.

Stahl, T., 1999. Stimulating Regional Innovativeness: The Learning Region. In CEDEFOP, Towards the Learning Region: Education and Regional Innovation in the EU and USA. Thessaloniki: CEDEFOP, pp. 51-60.

# Is there a place for small-scale production and local food systems?

*Artur Cristóvão, Manuel Luís Tibério, Sónia Abreu and Alberto Baptista*
*Universidade de Trás-os-Montes e Alto Douro, Portugal*

## Abstract

The paper presents an overview of the study conducted in a border area of Portugal and Spain, in the Douro-Duero valley. This is a peripheric rural territory, characterized by deep depopulation and low population density. Farming has been declining, but there are several small-scale productions with recognized quality and potential to contribute to greater economic diversification and revitalization of the local economy. The objectives of the study were: (1) to analyze the "filiéres" of a set of small-scale productions; (2) to assess the contribution of local distribution and restaurants to support the promotion of local products; (3) and to assess the tourists and visitors' interest and demand concerning these products. In general terms, the study looked at a major question: is there a place for small-scale production and local food systems?

## Introduction

Local agri-food products, in the past neglected by researchers, extension agents and policy makers, are today looked as an essential instrument in the strategy to develop depressed or marginalized rural territories. Those responsible for rural development consider that local or traditional agri-food products are important to fight depopulation, and those in charge of tourism development consider them as instrumental to structure and preserve the landscape (Bérard and Marchenay, 1996). This phenomenon has been described as a "re-discovery of small-scale traditional agri-food products" (Bérard and Marchenay, 1996; Bernat, 1996; Carvalho, 1994; Ribeiro and Martins, 1996; Tibério and Cristóvão, 1998; Tibério, 2004) and emerged from a set of circumstances (Tibério and Cristóvão, 1998; Tibério, 2004): (1) the crisis of the productivist agricultural development model; (2) the difficulty of agriculture in some marginal areas to compete with other types of farming; (3) the integration of tradition agri-food products in the EU rural development and food quality policies; and (4) the growing consumers' mistrust regarding mass production, particularly in terms of quality and safety.

With these ideas in mind, the study presented in the paper was developed in a border area of Spain and Portugal, the Douro-Duero valley, including two Natural Parks: "Douro Internacional" (PNDI) and "Arribes del Duero" (PNAD). This is a peripheric rural territory, characterized by deep depopulation and low population density (see Box 1). Farming has been declining, but there are several small-scale productions with recognized quality and potential to contribute to grater economic diversification and revitalization of the local economy. In general, access of these productions to the market is very incipient, but the growing number of tourists and visitors opens up perspectives for future development. The objectives of the study were: (1) to analyze the "filiéres" of a set of small-scale productions; (2) to assess the contribution of local distribution and restaurants to support the promotion of local products; (3) and to assess the tourists and visitors' interest and demand concerning these products.

***Box 1. The Douro-Duero Region of Portugal and Spain***

The studied region corresponds to a border area along the Douro river in the Portuguese province of "Trás-os-Montes e Alto Douro", and the Spanish region of "Castillla y León". It is characterized by a deep population regression, ageing, and low population density (total population of about 60.000, 15 inhabitants/km$^2$ in the Portuguese side and between 8 and 15 habitants/km$^2$ on the Spanish one). Most villages are inhabited by elderly people. Farming is the major economic activity, but the number of farm holdings has been decreasing and the pattern of agriculture activities changing, with less annual crops and more permanent crops (olive and almond orchards, vineyards). Beef cattle, sheep and pigs are important animal productions. Industrial activities are scarce and services are concentrated in the small towns. In the area there are two twin Natural Parks, one in Spain and one in Portugal. Tourism activities, especially rural and eco-tourism, have great potential and have been growing in the past decade.

## Research methodology

The study was initiated in October 2003, with exploratory contacts with the Directors of the two Natural Parks, agriculture and rural development agents, and local leaders. An interview guide was prepared, including the following aspects: local agricultural potential; reasons for agricultural decline; strategic agricultural production considering the new touristic demand; and measures and actions needed to add value to small-scale production. The information collected, complemented with document analysis, allowed us to have a good grasp of the situation. A Seminar was held in the early stages of research, to present and discuss the preliminary results, and to gather ideas on the methodological procedures to be implemented in the next research phases.

In the following phase, the research team prepared a list of local food distributors, restaurants, markets, local fairs, specialized food shops, and places commonly visited by tourists and visitors. Given the study objectives, three interview/survey guides were prepared: (1) restaurants (type of restaurant and major services, sources of food products, major dishes served, motivations to use (or not to use) local products, types of clients); (2) distributors of agricultural products (type of distributor, sources of products commercialized, motivations to sell (or not to sell) local products, importance of local products in the total volume of transactions, types of clients and demands); and (3) tourists and visitors (profile, motivations to visit the area, knowledge and consumption of local products). Additionally, a guide to observe local markets and fairs was also elaborated.

In the case of tourists and visitors, the survey guides were distributed in tourism information places, hotels, rural tourism houses, restaurants and local food shops: A total of 123 responses were gathered, 56 from Spanish visitors and 67 from Portuguese. In the case of restaurants and distribution places, 45 interviews were done, 28 in Portugal and 17 in Spain.

In parallel, six traditional small-scale products were studied: almonds; olives; chestnuts; figs; oranges; and vegetables (horticulture). The wild mushroom case was also analysed, as it constitutes a recent development with growing economic importance and some specific challenges. In each case the following aspects were studied: socio-economic interest; production system, production and productivity; "filiére" organization; and value added to each production and local retention of benefits. This information was obtained through interviews with key informants, from statistics and other documentation. A Seminar was held at the end of the project, with about 40 participants from Portugal and Spain, to share and debate the major results.

## Major conclusions

In spite of the agricultural decline, a considerable diversity of crops and animal products still exists in the region. However, traditional crops like cereals or potatoes have been replaced by permanent or temporary pastures, forest or shrub areas, as well as by non-cropped parcels. Some local productions, like almonds, table olives, figs, oranges, honey, sheep and goat cheese, and meat products from sheep, goat and beef are protected under the PDO or PGI label, according to the specific EU Regulations.

For each selected small-scale production a SWOT type analysis was done (Freire, 2001), as exemplified by the case horticulture in Table 1. As a general summary, the following aspects seem to be crucial: (1) these products are declining but are still important to the local economy; (2) production systems tend to be traditional; (3) some products benefit from the PDO or PGI collective label, but often are not present in the market; (4) other products, like vegetables or wild mushrooms, have no institutional support or particular recognition from public services; (5) the processes leading to add value are very incipient, and the productions are mostly sold at the farm to local intermediaries; (6) some productions are simply not perceived as market oriented and, as such, considered not strategic by producers and development agents; (7) some are demanded by tourists and visitors; and (8) some are gaining recognition and sold at local fairs, specialized food shops and tourism events.

The promotion and sustainability of a local food system depends upon the food distribution sector and local restaurants. In this regard, the following conclusions are worth noting: (1) the supply is not organized to answer for the commercial requirements of distribution, especially in terms of quantity, image, regularity of supply and invoicing practices, (2) small distributors (grocery stores, small supermarkets) show little interest for these products; (3) some grocery stores sell their own farm productions; (4) the questions of invoice issuing and control are an important obstacle to the acquisition of local food products by distributors and restaurant owners; (5) traditional markets no longer serve as structures supporting the commercialization of local food products; (6) the new fairs and specialized shops are providing an impulse to small-scale quality productions; (7) gastronomy is seen as an excellent way to add value to local

*Table 1. Horticulture in the Douro-PNDI: Strengths and opportunities, weaknesses and threats.*

| Strengths and opportunities | Weaknesses and threats |
|---|---|
| Great diversity of products | Most production is for self consumption |
| Special taste, due to local varieties and production practices | Producers don't regards horticulture as an income making activity |
| Production system follows traditional practices, such as the use of animal manure | Distribution circuits are not adequate |
| Production places ("hortas") are close to the villages | Technicians don't regard horticulture as strategic to local development |
| Some products are used to prepare traditional gastronomy dishes | Local varieties are close to being extinct |
| Products are increasingly present in local fairs and events | |

products, and some restaurants are using them to prepare typical dishes, desserts and preserves; (8) tourists are looking more and more for local food products, particularly wine, bread, olives and olive oil, almonds, dry figs, smoked sausage, cheese, beef and honey, but the supply is not prepared to respond positively to such new demand.

The tourists and visitors tend to be highly educated and come usually in groups, with family and friends, motivated by the region's landscape, gastronomy and historic heritage. The large majority buy and consume local products, as previously indicated. Restaurants are seen as key players in the promotion of local small-scale products. However, the Portuguese like to buy directly from farmers and the Spanish prefer restaurants, grocery stores and small shops.

## Strategic orientations

Even in relatively peripheric areas, like the Portuguese-Spanish border close to the Douro-Duero Valley, agriculture and the food system were deeply transformed in the last decades. The traditional agricultural systems declined and some of its products are no longer valued or tend to be marginal in quantitative terms. Producers are ageing and their knowledge and know-how is at risk to be lost, along with the local genetic heritage and biodiversity associated to farming. Local products and markets were progressively integrated into the larger framework of the global food market. Distributors, restaurant owners and consumers in general rely more and more on external food products. However, the quality of existing products is recognized and a new demand is emerging and gaining scale. A place still exists for small-scale production and local food systems, but a considerable effort needs to be done in order re-establish them.

The work developed and understanding of the situation led to the development of a set of strategic orientations related to four major aspects:

1. The institutional scene – local development institutions should integrate farming into their action strategies and work needs to be done across borders, perhaps through a "permanent platform" (integrating the two Natural Parks and other public and private institutions) and "product committees" to stimulate synergies between producers, distributors and other "filére" agents.

2. The attitudes of producers and other agents – a new consciousness needs to be developed concerning the value and importance of small-scale farming, as a way to promote local agriculture, preserve biodiversity and regional landscape, and respond positively to the new demands associated to local products, gastronomy, nature and rural tourism.

3. Commercialization and marketing – product processing and commercialization need to be structured and attention must be given to safety, quality improvements, direct sales, better use of the collective labels PDO and PGI, introduction of organic production, and involvement of young farmers, local farmers' organizations, distributors, restaurant owners and population in general. 4. Research and professional training – a comprehensive strategy requires closer attention to innovation, particularly in the fields of processing, marketing and commercialization, and valuable progresses can only be done trough continued research and professional training efforts. Local farmer organizations, together with the higher education and research institutions located in the region, have a particular responsibility in these respects.

## References

Bérard, L. and M. Philippe, 1996. La construcción social de los productos de la tierra. Agricultura y Sociedad nº 80-81, 31-56.

Bernat, E., 1996. Los 'nuevos consumidores' o las nuevas relaciones entre campo y ciudad a través de los productos de la tierra. Agricultura y Sociedad nº 80-81, 83-116.

Carvalho, B., 1994. Produtos Agro-alimentares de Qualidade e de Tecnologia Caseira. Utopia? Semente n.º 8/9, 20-22.

Cristóvão, A., M. Tibério and M. Teixeira, 2003. Propostas de Estratégias de Valorização. "Terras de Bouro: Estudo e caracterização de produtos locais". Câmara Municipal de Terras de Bouro.

Freire, A., 2001. Estratégia: Sucesso em Portugal. Editorial Verbo. Lisboa/S.Paulo.

INE, 2000. Recenseamentos Gerais da Agricultura. Dados comparativos 1989-1999 (Versão 1.01). Lisboa: Instituto Nacional de Estatística.

LEADER, 2000. Comercializar os produtos locais: circuitos curtos e circuitos longos. www.rural-europe.aeidl.be/rural.pt.

Ribeiro, M. and M. Conceição, 1996. La certificación como estrategia de valorización de productos agroalimentarios tradicionales: la alheira, um embutido tradicional de Trás-os-Montes. Agricultura y Sociedad nº 80-81, 313-334.

Tibério, L. and A. Cristóvão, 1998. Produtos tradicionais: importância sócio-económica na defesa do mundo rural, 1ªs Jornadas de Queijos e Enchidos - Produtos Tradicionais, IAAS, EXPONOR, Abril. 12 pp.

Tibério, M.L., 2004. Construção da Qualidade e Valorização dos Produtos Agro-Alimentares Tradicionais - Estudo da Região de Trás-os-Montes. Tese de Doutoramento, UTAD, Vila Real. 542 pp.

# Marvão polyculture mosaic's fate: Searching the way for a rural area in transition

*Milena Andrade Dneboská, Teresa Pinto-Correia and Patricia Pedro-Rêgo*
*University of Évora, Colégio Luís António Verney, 7000-67 Évora, Portugal*

## Abstract

All throughout the Mediterranean, European agriculture undergoes deep structural transformations, related both with new production conditions, but also with new demands and requirements for the rural space and the functions that agriculture displays in it. Remote rural areas are normally more reluctant towards these new demands, especially when these concern other issues than production, but nevertheless the attitude toward the landscape understanding and its values has been shifting also here.

At the foot of Serra de São Mamede uplands, the confrontation of the new agriculture and society demands with the small-scale farming is observed. As the most conflicting influences on landscape seem to be:

- The complex modification of rural society and its lifestyle in last three generations.
- The possibility of Marvão being accepted as world heritage by UNESCO.
- Intellectual and economic impact on area form tourist and new-settlers.

Intensification and rationalisation of agriculture are weak. While the oldest generations are still mainly responsible for the practical field-work, maintaining the traditional polycultures; the younger ones are leaving in search for better future or work in other professions. On the other side of the border the situation is different, the intensification and specialisation of agriculture has already taken place here. Through other political supports, this area with similar biophysical characteristics has developed in a different way.

The aim of this paper is to analyse the probable future of these traditional farming systems considering the area capacity for trans-border interaction, which may contribute to their renovation.

**Keywords:** rural landscape, trans-frontier relations, polyculture farming

## Introduction

While the most valued cultural landscape in rural Europe are decaying, their appreciation and interest in them increases (Pinto-Correia & Vos 2004). But this interest comes mainly from other parts of society than the one that preserves these values through specific land use systems. In general terms, more detailed mosaics are disappearing while more specialised land-uses are developed and less productive areas are left fallow, both processes leading to landscape simplification. Together with this simplification, its accessibility and whole complex of its meanings is decreasing (Antrop 2005).

In the specific case of Marvão, this mosaic is created by an unique complex of intensive polyculture in fertile valleys and forest of *Quercus suber* and *Q. pyrenaica* with chestnut and olive orchards on poor hillsides. The peaceful hilly landscape with detailed polyculture mosaic is already highly valued by its visitors, but its non-productive functions are only slowly gaining importance among the locals.

Together with the transformation of values among the rural society, landscape appearance, functions and its structure are being transformed. However, policies and related instruments are still much involved in the maintenance of traditional farming systems, without focusing clearly enough in these new functions and the new needs of landscape management. This study aims at pointing out interesting possibilities of good practices and management, which might be the strong points for the future.

## Case study and its trans-frontier context

The Portuguese municipality of Marvão (Figure 1) has strong connections with its Spanish neighbour, Valencia de Alcántara, in geomorphologic and natural terms as well as similarities in historical, social and economic development. However the reactions of each rural area on specific historic and political impulses led towards particular characteristics of social, physical and political context in each case. We focus on each area development trajectory, its particularities and viability.

Both areas have been occupied by humans since prehistoric times and are rich in related monuments, especially megalithic burial places "antas". The climate is dry Mediterranean, with dry long summers and mild winters. The poor soils of mainly granite origin are covered by a combination of open pastures, mixed agro-silvo pastoral system and by more closed pine forests. This territory is rich in birdlife, especially of birds of prey.

In Marvão, the elderly population with poor education is dominant. A rather small variety of professions distributes the population among services (63%), industry (27%) and agriculture (10%). In Spain the percentages are similar with less people employed in the secondary sector (23%) and more in the primary sector (14%). Also here the increase of elderly population is significant.

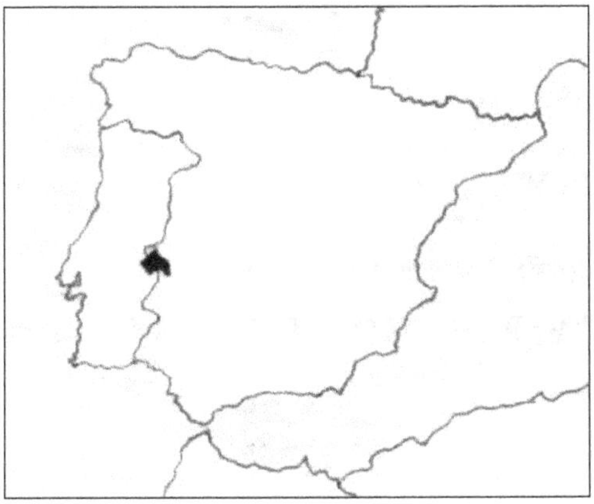

*Figure 1. Location of the study area.*

There is a high number of very small farms on both sides of the border. Meanwhile in Spain they are followed by the large and very large farms, in Portugal the second most represented group are small properties (Figure 2). Most of their owners work only their own land and are mainly retired.

Their production is especially for self consumption (they rarely sell more than chestnuts production once a year) as a system of subsistence and of complement to the low pensions. The humble life conditions of most of these elderly farmers reflect hard work and deep relation to the land, but also the lack of other opportunities.

Four main farm types of farms can be identified in Marvão: two types of small size property with more detailed mosaic, on the South, and two types of larger estates, less diversified in terms of land use, in the North:

- Type 1, South: Very small farms, of less than 5 ha, generally divided in various scattered parcels. The typical polyculture farm unit of the valley includes irrigated potato plantations under the olive trees' crowns and mixed vegetable plots with some solitary chestnut trees.
- Type 2, South: Small to medium holdings from 5 to 20 hectares. Together with the above mentioned uses these would work also several scattered parcels of olive and chestnut orchards on the slopes and forested areas of pine or cork oak on the hilltops and rocky slopes.
- Type 3, North: Farms of medium size from 5 to 20 or 50 hectares. Combination between horticulture on small fields bordered by stone walls and montado with cork production on rocky slopes mixed with pastures mainly for the sheep.
- Type 4, North: Large properties, over 50 or even several hundreds of hectares, with extensive agriculture on rather poor and shallow skeleton soils. Their primary use is cattle breeding and secondary cork from montado over the granite slopes.

In Spain, we find a significant percentage of farm units without agricultural land, mainly intensive chicken and pig breeding. As a second kind of farms, small size properties, up to 5 ha, close to the Portuguese border can be identified. Their domain is polyculture with important component of pasture areas. Their number and productivity is fast decreasing as the age of farmers increase. Third and the biggest type of farm units are the extensive cattle breeding

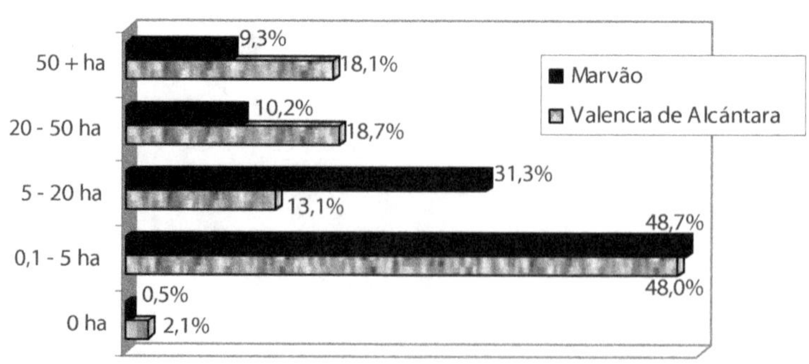


*Figure 2. Distribution of farm-units in relation to their Utilised Agriculture Land = UAA.*

farms with size between twenty and several hundreds of hectares. Their owners are normally younger farmers with specific agricultural education.

## Methodology

The initial statistic data was obtained from agricultural census from 1979, 1989 and 1999; and population census from 1991 and 2001. This data was completed by the interviews to farmers and technical staff corresponding to decision making agents – in the Portuguese and in the Spanish study area.

The aims of the interviews were: 1) to characterise the farm units and their main products, main problems and potentialities; 2) to understand the relation of Portuguese farmers with their Spanish neighbours and the possible influence of the border proximity on the agriculture; 3) to identify the non-productive uses of the landscape, the most preferred or valued functions outside agriculture; 4) to evaluate the importance of these functions within the local landscape and own property, as well as the personal attitude and value assigned to each one of them; particular importance was given to the tourism, hunting and local patrimony; 5) finally, to assess the vision for future development of the landscape and particularly of agriculture, in the area.

Interviews were made to 30 farmers and nine staff member of technical organizations in Marvão, between June and October 2005. The Spanish technical staff members and representatives of different kinds of farm units were interviewed in December 2005 and January 2006. Together with comparison of Portuguese results, the characterization of the situation in Valencia de Alcántara was searched.

## Farm units and their owners

Marvão's agriculture keeps many of its traditional land uses, however new tendencies and attitudes towards the agricultural profession can be detected. The farm units between 20 and 50 ha are the most diversified, having many separate parcels with different and mixed cultures, what is generally considered as a problem. 30% of interviewed farmers have three different crops, 20% have four, and 20% explore two different crops. 10% have five crops and another 10% have six.

The traditional land use systems still influence a lot management decisions for the different parcels among elderly farmers with small farm units. These small size properties mean very small credit and economical capacity of re-investment, which do not allow important longer-term innovation plans. However an important part of the income of many of these small farmers depends on subsidies, the amount of support often does not justify the bureaucratic effort required.

The younger generations of farmers (mid forties till mid sixties) are more influenced by the actual economic context – production supported by subsidies – without many demands for changes in the farming style. This group is rather sensitive to the concrete and practical subsidies and measures. The youngest farmers (till mid forties) are in general open to technical innovation. Especially the part-time farmers among them might, with some support, have an active role as innovators and promoters of newly valued products and services.

## Relation between both sides of the border

According to the interviews, Portuguese farmers are seen as less collaborative and less open to the dialogue about the management practices. The strong personality involvement reveals essential in activities such as cooperatives or exchange of information.

The Spanish side presents larger capacity of cooperation among farmers – also supported by several state or regional measures (economical facilities for shared machinery acquisition, *etc.*). The Spanish farmers are in general seen as more progressive in terms of innovation techniques and stronger in competition.

Another difference is the ownership of agricultural land and persistence of small properties. The Spanish side presents high rates of rented land. Especially the beef producing farm units can afford, due to high subventions for the extensive cattle production, to work mainly over the rented land. Marvão's farmers mainly inherited and own most of the land of their farm units.

Most of interviewed on both side of the border (100% of Spanish and about 80% of Portuguese farmers and 90% of the technical staff) declared a much faster disappearance of very small farms in Valencia de Alcántara. Small farm units in Marvão are also diminishing but their number decreases slowly. One of the reasons might be the creation of a special Extremadura Law N° 19/1995 for the modernisation of farm units. This law considers as "priority farms" those, which can employ at least one and as maximum four men full time, being the concept based on official calculations of demands of working hours per year for each kind of land-use. Due to special support given to these priority farms, many farm-units have been changing their productive orientation or even size, aiming at fulfilling the size criteria.

## Other uses of territory

The importance of non productive functions of landscape is increasing. 60% of the farmers consider landscape as the most valuable feature in the area and 90% of them see in it as an important resource for the tourism. The architectonical and archaeological heritage was considered less important; however 86% of farmers and 100% of technical clerks consider it as an integral part of landscape. The tourist impact on the area is being understood in a rather ambiguous atmosphere. 23% of respondents consider it as an important economical income for the local community, while more than 50% of them express the lack of return from this income to the landscape and the farming systems. Another worry is connected with the rising of land prices, due to demand from foreigners or Lisbon dwellers for the retirement or secondary residence, making more difficult establishment of local young population.

## Potentials for the multifunctionality

The tourist and new-settlers flow connected with Marvão's historical patrimony as well as with the natural and cultural richness of its landscape, influences the locals' understanding of their own environment and in some cases also their action. Nevertheless, the increase of alternative uses within the farm units does not always result positively on the landscape mosaics' diversity. In most cases of rural tourism, the agricultural production of farm unit diminishes or even disappears. The area closest to the residence house is kept in an almost garden-like treatment, while the more distant parcels are rented or forested. As exceptions can be seen very large farms in Spanish side, where both of activities can be maintained in parallel and estates with parallel dry-fruits production in Portugal. But still, this modality, with more careful regulations, might be one of the promoters of multiple landscape values.

The increasing economic exploitation of hunting is seen as another alternative use of the territory on both sides of the border, however developed more in Spain. The appreciation of denser native vegetation by hunters might preserve parts of the local ecosystems and connected activities might promote other values of the area such as traditional products. Nevertheless, due to restricted access to the hunting lodges (only available for paying guests), these are not much appreciated by the local population and in social terms bring neither many working opportunities nor life to the area.

Although the traditional products are already highly valued and their certification has been in Portugal strongly implanted, their larger commercialisation seems rather difficult. The chestnut is the most specific product of the Marvão area, which may be seen as a possible bridge towards a more complex landscape valuation, care and understanding. Recently the irregularity of production and lack of farmers' collaboration has been making larger marketing rather difficult. But the collaboration with Spanish cooperatives, mainly for the improvement of management techniques and commercialisation seems to be promising.

## Conclusion

Agriculture in Marvão seems to face strong limitations, also when compared with the more modernized farm units on the very similar and adjacent Spanish region. But more detailed examination shows that there is potential for a multifunctional use of the region, which has been preserved due to the resistance of traditional farming. The actual task is to wake up carefully these latent resources and develop them with well formulated management plans towards the future combining production with appreciation of the rural services and environmental quality.

## References


Antrop, M., 2005. Why landscapes of the past are important for the future? In: Landscape and Urban Planning 70, 21-34.

Blas Morato, R., 2004. Valencia de Alcántara: Estudio de Planeamiento Estratégico; final degree thesis - University of Extremadura.

Brandt J. and H. Vejre, 2004. Multifunctional Landscapes – motives, concepts and perceptions. In: J. Brandt, H. Vejre (Eds.). Multifunctional Landscapes. Vol. 1: Theory, Values and History. WIT Press, Boston.

INE (Instituto Nacional de Estatística de Portugal). Recenseamento Geral da Agricultura, Dados comparativos 1989 e 1999,

INE (Instituto Nacional de Estatística de Portugal). Censos Populacionais 1991 e 2001.

Pinto-Correia, T. and R. Mata Olmo, 2005. Desafios para a Multifuncionalidade da Paisagem. In: Proceedings from Colóquio Ibérico de Geografia, Évora IX, 2005.

Pinto-Correia, T. and W. Vos, 2004. Multifunctionality in Mediterranean landscapes – past and future. In: Proceedings from conference: From knowledge of landscape to landscaping action, Bordeaux.

# Organic manure availability, food security and poverty relationships at farm level in the old cotton basin of Mali

*M.S. Keita[1], J.F. Bélières[2], J. Coulibaly[3], B. Rapidel[4] and R. Poccard-Chapuis[5]*
*[1]Programme Economie des filières, IER, BP 1813, Bamako, Mali; [2]CIRAD/IER, Bamako, Mali; [3]IPR-IFRA, Katibougou, Mali; [4]CIRAD, Montpellier, France; [5]CIRAD/IER, Sikasso, Mali*

## Abstract

In the cotton growing areas of Southern Mali, further food security improvement and populations income rise will depend on productivity increase through biomass recycling and/or diversification of activities. The objective of our study is to identify and analyze the relationships existing between practices of biomass management and the performance of farmers in terms of food security and agricultural income. In the cotton growing area, achieving security in cereal production remains the main purpose of production for farmers. The levels of crop yields and production means (equipment, labor, cultivated area, herd size) determine land rotation and therefore the share of cotton area in medium to large farms (10-40 ha). This is no longer true in the case of smallholders (less than 7 ha) which in some cases produce cotton solely to gain access to input supply. The use of mineral fertilizers, although not fully homogenous, does not vary much, unlike the quantity of organic manure available in the farm, which affects the crop yields. These manure quantities are linked to the herd size and to the practices of manure production. Food security is generally obtained in normal years except for the 2004/2005 growing season, a particularly bad one.

Agricultural income varies quite a lot among households. These disparities are often linked to the practice of biomass management and to the means of production's availability. However, these incomes are globally low and below the poverty line in Mali. These overall poor economic performances, even for well-equipped households, raise questions on the socioeconomic sustainability of the system.

## Introduction

The increase of agricultural production in the cotton growing region of Southern Mali has occurred over the past 15 years through the increase of acreage (Club du Sahel/OCDE, 2005). However, in the north of this region, all agricultural lands are now under cultivation: fallows are no longer observed while demographic pressure remains high (Breman and Sissoko, 1998). Further improvement in food security and income thus passes by the increase of agricultural productivity and/or diversification of productive activities (Deveze *et al.*, 2005). The produced biomass is often quoted as a factor limiting the farming systems intensification and sustainability (De Ridder *et al.*, 2004).

Therefore, socio-economic sustainability is linked to the improvement of productivity and to the evolution of biomass production and management.

Socio economic sustainability of agricultural households in the area can be evaluated in terms of (i) food security regarding the cereal crops; (ii) and the level of monetary income to satisfy the consumption needs of the household members. Households failing to afford the minimum in terms of food and money have no other alternative than rural exodus of some of their members.

The main objective of this paper is to elucidate the relationships between the level of poverty and food security in agricultural households of the saturated cotton growing areas of the southern Mali and the practices of organic manure production. These relationships, associated with the flow of biomass within the same villages (Poccard-Chapuis *et al.*, 2006), will be used to calibrate a simulation model (Rapidel *et al.*, 2006).

## Materials and methods

The study was conducted in the Koutiala « Circle » in the old cotton basin of Mali. The farming systems of the region depend mainly on cereal and cotton crops as well as livestock (Sissoko, 1998). During the past two decades, steep increases were observed in the following variables: population density, livestock, cotton acreage, oxen driven equipment of the farms. Consequently, the pressure on lands and natural pastures increased. The region is typified as poor (in 2001, 82% of the people lived below the poverty line). Nevertheless, when the access to social services is taken into account, the regional indices almost reach the national average.

The scale used as basis is the village, but lower scales (plot, farm) should also be considered.

We worked in two highly constrained villages with contiguous territories, Dentiola and Nankorola. Agriculture there suffers from frequent delays in the onset of the rains with important adverse effects on cotton cultivation. The population is dense, and land reserves are few, thus limiting the availability of land per active.

A survey was conducted from May to July 2005 on 87 farms, randomly chosen among the total 305 farms, based on then results of a pre-survey conducted earlier. The practices of manure and residue management were studied, along with the incomes and the level of food security. It addressed a whole year (from June 2004 to May 2005). Organic manure production was assessed by inventory of the different techniques, and by asking the number of manure charts produced and applied to the fields. The average weight of a manure chart was set to 250kg, as proposed by Kanté (2001).

The results are presented with respect to the typology currently used in this region (Kébé *et al.*, 1999), adapted in our case by regrouping the two classes of the less equipped farms; (i) type A farms: big size farms with two pairs of plowing oxen, one full equipment and a total of at least 10 oxen (plowing oxen included); (ii) Type B farms: medium size farms with at least a complete harness; (iii) Type C and D farms: small, poorly equipped farms.

## Results and discussions

### *Farming systems and agricultural production in 2004*

The overall characteristics of the farms in the two villages are approximately at the same level as the regional means (CMDT, 2005), although the proportion of big farms seems to be lower in the two villages. The cultivated areas per worker vary between 0.97 and 1.29 depending on the farm type, bigger farms showing the lower areas. Nevertheless, bigger farms also have higher productivity per area and per worker, and this over compensates the lower area cultivated. The mineral fertilizers are the same between types, so the differences can be allocated either to the organic manure used, or to the better initial fertility of the land (data not shown).

The 2004-05 climatic year was dry (total 2004 604 mm *vs.* 1985-2004 mean 846 mm), particularly at the end of the wet season. The yields of the crops were low compared to the inter-annual mean, particularly for maize (-1000 kg/ha), but also for other cereals (pearl millet and sorghum, -500 kg/ha) and, to a lesser extent for cotton. Consequently, the smaller farm types produced hardly enough grain to cover their theoretical needs (225 kg/pers/year). Among the type, the variability is high, and numerous actual farms went through food shortage before the harvest of the following year.

### *Farm incomes*

The farm incomes were low in 2004 (Table 1 and annex): 66,600 FCFA per person per year, and 94,000 FCFA per worker per year. During this year, 97% of the population had an income that located them below the poverty line.

Part of revenues coming from cropping activities is important but fluctuates between 44% and 63%. The cotton share is particularly high, reaching 32% of the total in the bigger farms type, but much less in the smaller farms type. The cotton place is linked to the cereal's low yields. The livestock share is low. Other activities are important for small and big farms.

The sustainability of the system is at stake from three points of view: the high share of agriculture in the global income, the lack of diversification of sources, and the overall low income. This supports the findings of Djouara *et al.* (2006), stating that farm income are very low in this region and depend strongly on climatic conditions.

### *Organic manure production in 2004*

Usually, the farmers apply manure preferentially to the cotton fields, or, to a lesser extent to the maize fields. The mean production of organic manure per farm is 30 tons. It varies greatly, from 0 to 190 tons (coefficient of variation 90%). It corresponds to 2.7 tons per ha of cultivated area (coefficient of variation 55%), although the actual doses in the fields are much higher on small surfaces. Few farms (5%) apply less than 1 ton per cultivated ha. At the end of the 90s, the doses were lower with a mean value of 1.5 ton/ha (Kanté, 2001, Sissoko, 1998).

The organic manure production is linked to the farm type, though the herd size, workforce and equipment (Table 2).

*Table 1. Income composition for the farm types, year 2004. The income was calculated by using the mean actual price at which the farms sold their products. It does not include provision for depreciation.*

| | Big farms (A) | Medium farms (B) | Small farms (C&D) | All |
|---|---|---|---|---|
| Total income (FCFA) | 2,089,185 | 909,696 | 297,490 | 1,062,766 |
| Crops (excl cotton) | 26% | 35% | 31% | 31% |
| Cotton | 32% | 28% | 13% | 29% |
| Livestock | 15% | 15% | 12% | 15% |
| Services sale | 4% | 1% | 2% | 2% |
| Gathering | 2% | 4% | 4% | 3% |
| Work sale | 3% | 4% | 10% | 4% |
| Others activities | 19% | 14% | 27% | 16% |

*Table 2. Correlations between organic manure (OM) production and structural characteristics of the farms.*

| | Workers | Charts | Bovines | Total OM | OM / ha cultivated | Cultivated area | Total income |
|---|---|---|---|---|---|---|---|
| Charts | 0.56** | 1 | | | | | |
| Bovines | 0.65** | 0.59** | 1 | | | | |
| Total OM | 0.70** | 0.58** | 0.58** | 1 | | | |
| OM/ha cultivated | 0.04 | 0.22* | 0.13 | 0.62** | 1 | | |
| Cultivated area | 0.85** | 0.69** | 0.72** | 0.74** | 0.08 | 1 | |
| Total 2004 income | 0.728** | 0.72* | 0.81** | 0.58** | 0.08 | 0.83** | 1.00 |
| Income 2004/person | -0.1357 | 0.26* | 0.19 | 0.05 | 0.09 | 0.09 | 0.40** |

* and **: Respectively, significant correlation at the 0.05 and 0.01 levels (2-tailed).

The produced organic manure depends on the farm size, the equipment and the herd size. The mean dose of organic manure applied per ha is correlated to the number of charts only, which appears as a determinant factor of intensification.

The methods used to produce manure have been described elsewhere and are the same in this case.. All farms have a heap for domestic wastes, and only 2% do not apply the compost produced to the fields. No farm has improved manure pit, although this technique has been thoroughly recommended during years. Most farms (83%) have a cattle ground and a ground for minor animals (55%). 88% of the cattle grounds are improved, that is, residues are applied for the animals to speed their decomposition.

The organic manure comes from domestic waste composting (43%), ground residues and faeces (44%) and only 14% from compost.

***Relationships between organic manure production and income***

The total production of organic manure produced by a farm is logically linked to its size, equipment and herd (Table 2). On the other hand, the mean organic manure dose applied per cultivated area is not linked to farm size, nor to farm income. The incomes (total or from agriculture) and gross margins of cropping activities are correlated to the total organic manure applied, but we could not find any direct relationship with the mean organic manure dose per ha. By contrast, there is a clear relation between cotton and maize yields and the total organic manure produced.

## Conclusions

The organic manure production does partly explain the crop yields. These quantities are linked to the farm size, the herd size and the equipment.

The livestock contribution to the income is low in this case, far behind agricultural crops, and particularly cotton. The food sufficiency, which relies mainly on cereal crops, was hardly reached during the bad 2004 climatic year.

High income differences exist between farms. These disparities are related to the biomass management practices, and to the available production means. However, the incomes found in this study are low, and inferior to the poverty line defined for Mali. These poor economic performances put at stake the socioeconomic sustainability of the farm, including the richer ones.

## References

Breman, H. and K. Sissoko (Ed.), 1998. L'intensification au Sahel: vouloir c'est pouvoir. Harmattan Paris

Club du Sahel/OCDE, 2005. La concertation sur la crise du coton en Afrique de l'Ouest et du Centre: chronologie des événements clés 2001 - 2005. Paris Septembre 2005. 39 pp.

CMDT, 2005. Annuaire statistique 04/05. Bamako 2005. 86 pp.

De Ridder, N., H. Breman, H. van Keulen and T.J. Stomph, 2004. Revisiting a 'cure against land hunger': soil fertility management and farming systems dynamics in the West African Sahel. Agricultural Systems 80: 109-131.

Deveze, J.C. and D. Halley Des Fontaines, 2005. Le devenir des agricultures familiales des zones cotonnières africaines. AFD. Paris Octobre 2005. 85 pp.

Djouara, H., J.F. Bélières and D. Kébé, 2006. Les exploitations agricoles familiales de la zone cotonnière du Mali face à la baisse des prix du coton graine. Cahiers Agricultures 15, 64 -71.

DNSI, 2004. Enquête malienne sur l'évaluation de la pauvreté (EMEP), 2001. Principaux résultats. Ministère du Plan et de l'aménagement du territoire / Banque mondiale. Bamako juin 2004. 46 pp.

Kanté, S., 2001. Gestion de la fertilité des sols par classe d'exploitation au Mali-Sud. Wageningen, Pays Bas, Université de Wageningen, Documents sur la gestion des ressources tropicales n° 38. 237 pp.

Kébé, D., B. Fomba, M.C. Sidibé and H. Djouara, 1999. Le conseil de gestion aux exploitations agricoles: un outil de vulgarisation. IER/ESPGRN. Bamako Mali Mars 1999. 30 pp.

Poccard-Chapuis, R., M. Thales, A. Venturieri, M.-G. Piketty, B. Mertens, J. Bastos da Veiga and J.-F. Tourrand, 2006. The livestock commodity chain: A lever for monitoring pioneer dynamics in the Brazilian Amazon Region? In: Langeveld, H. and N. Röling (eds.) Changing European farmingsystems for a better future, Wageningen Academic Publishers, Wageningen, The Netherlands, pp. 430-434.

Rapidel, B., R. Poccard-Chapuis, J.-F. Bélières, B.S. Traoré, J. Coulibaly, D. Diallo and M. Lessnoff, 2006. Biomass production, transformation and allocation at the village scale in Southern Mali: Development of a conceptual model. In: Langeveld, H. and N. Röling (eds.) Changing European farmingsystems for a better future, Wageningen Academic Publishers, Wageningen, The Netherlands, pp. 144-148.

Sissoko, K., 1998. Et demain l'agriculture? Options techniques et mesures politiques pour un développement agricole durable en Afrique Sub Saharienne. Cas du Cercle de Koutiala en zone sud du Mali. Wageningen, Pays Bas, Université de Wageningen, Documents sur la gestion des ressources tropicales n° 23. 179 pp + annexes.

## Annex

### *Main characteristics of the farms in the survey*

| | Big farms (A) N=17 (19.5%) | | Medium farms (B) N=59 (68%) | | Small farms (C&D) N=11 (12.5%) | | All farms N=87 (100%) | |
|---|---|---|---|---|---|---|---|---|
| | Mean | Std Dev. | Mean | Std Dev. | Mean | Std Dev. | Mean | Std Dev. |
| Persons | 30.53 | 20.36 | 13.44 | 7.92 | 6.91 | 2.07 | 15.95 | 13.29 |
| Workers | 18.94 | 13.86 | 8.19 | 4.74 | 4.09 | 1.14 | 9.77 | 8.57 |
| Total cultivated area (ha) | 17.82 | 7.42 | 9.78 | 5.32 | 5.07 | 1.78 | 10.76 | 6.66 |
| % cotton area | 32 | 9 | 27 | 9 | 25 | 6 | 28 | 9 |
| Cultivated area/worker (ha)** | 0.97 | 0.36 | 1.23 | 0.47 | 1.29 | 0.37 | 1.13 | 0.44 |
| Draft oxen | 6.47 | 2.70 | 3.63 | 2.35 | 0.91 | 0.94 | 3.84 | 2.77 |
| Others bovines | 22.29 | 24.43 | 2.93 | 5.76 | 0.27 | 0.65 | 6.38 | 14.02 |
| Ovines & caprines | 26.00 | 16.44 | 10.81 | 12.52 | 1.91 | 2.02 | 12.66 | 14.45 |
| Fertilizer on cotton (kg/ha) | 191 | 22 | 192 | 44 | 187 | 78 | 191 | 46 |
| Fertilizer on maize (kg/ha) | 209 | 82 | 185 | 89 | 189 | 160 | 190 | 98 |
| Cotton yield (kg/ha) | 1,104 | 344 | 853 | 276 | 468 | 223 | 853 | 332 |
| Maize yield (kg/ha) | 1,495 | 750 | 1,281 | 703 | 974 | 994 | 1,284 | 758 |
| Pearl millet yield (kg/ha) | 585 | 241 | 612 | 402 | 298 | 252 | 567 | 372 |
| Sorghum yield (kg/ha) | 521 | 389 | 344 | 368 | 147 | 219 | 354 | 369 |
| Total organic matter (kg) | 55,090 | 40,733 | 26,210 | 17,194 | 9,623 | 7,703 | 29,756 | 26,501 |
| Organic matter/area cultivated 2004*** (kg/ha) | 3,091 | 1,486 | 2,679 | 1,384 | 1,898 | 1,090 | 2,766 | 1,431 |
| Total income 2004 (FCFA) | 2,089,185 | 1,097,490 | 909,696 | 577,887 | 297,490 | 164,384 | 1,062,766 | 867,004 |
| Income/person (Fcfa)* | 68,432 | 36,060 | 67,682 | 33,158 | 43,058 | 22,668 | 66,614 | 34,257 |
| Income/worker (FCFA)** | 92,650 | 56,828 | 97,743 | 54,710 | 67,993 | 42,272 | 94,239 | 55,294 |

* weighted by persons; ** weighted by workers; *** weighted par area cultivated.

# Modelling environmental benefits of silvoarable agroforestry in Europe

*J.H.N. Palma[1], A.R. Graves[2], R.G.H. Bunce[3], P.J. Burgess[2], R. de Filippi[1], K.J. Keesman[4], H. van Keulen[5], F. Liagre[6], M. Mayus[7], G. Moreno[8], Y. Reisner[9] and F. Herzog[1]*
*[1]Ecological Controlling Research Group, Agroscope FAL Reckenholz, Reckenholzstrasse 191, 8046 Zurich, Switzerland; [2]Institute of Water and Environment, Cranfield University, Silsoe MK45 4DT, United Kingdom; [3]ALTERRA – Alterra Green World Research, PO Box 47, 6700 AA Wageningen, The Netherlands; [4]Systems & Control Group, Wageningen UR, PO Box 17, 6700 AA Wageningen, The Netherlands; [5]Plant Production Systems Group, Wageningen UR, POBox 430, 6700 AK Wageningen, The Netherlands; [6]Assemblé Permanent des Chambres d'Agriculture, 9 AvenueGeorge V, 75008, Paris, France; [7]INRA, Centre de Recherche de Montellier, 2 Place Pierre Viala, 34060 Montpellier Cedex 1, France; [8]Universidad de Extremadura, Centro Universitario de Plasencia, 10600 Plasencia, Spain; [9]Swiss College of Agriculture, Laenggasse 85, 3052 Zollikofen, Switzerland*

## Abstract

Increased adoption of silvoarable agroforestry (SAF) systems in Europe, by integrating trees and arable crops on the same land, could offer environmental benefits compared with conventional agricultural systems. Soil erosion, nitrogen leaching, carbon sequestration and landscape biodiversity were assessed for a stratified random sample of 19 landscape test sites in Mediterranean and Atlantic regions of Europe using computer models developed in the Silvoarable Agroforestry for Europe (SAFE) project. At each site, the effect of introducing agroforestry was examined by simulating the growth of one of five tree species (hybrid walnut *Juglans* spp., wild cherry *Prunus avium* L., poplar *Populus* spp., holm oak *Quercus ilex* L. subsp. *ilex* and stone pine *Pinus pinea* L.) at two tree densities (50 and 113 trees ha-1) in combination with up to five crops (wheat, sunflower, oilseed, grain maize and silage maize). Across the 19 landscape test sites, SAF had a positive impact on the four indicators with the strongest effects when introduced on the best quality land. The computer simulations showed that SAF could significantly reduce erosion by up to 65% when combined with contouring practices. Nitrogen leaching could be reduced by 21%. With agroforestry, predicted mean carbon sequestration, over a 60-year period, ranged from 5 to 179 t C $ha^{-1}$ depending on tree species and location. Landscape biodiversity was increased by introducing SAF by an average factor of 2.

**Keywords:** alley cropping, erosion, nitrogen leaching, carbon sequestration, landscape diversity, agri-environmental policy

## Introduction

Silvoarable agroforestry (SAF) is a form of multi-cropping which involves combining at least one woody-perennial species with a crop which results in ecological and economic interactions. Such systems are typically associated with a variety of environmental benefits and although agroforestry systems were common in Europe (Eichhorn *et al.*, 2005) they have been greatly reduced because of agricultural intensification (Herzog, 1998).

In Europe, environmental benefits are expected from new land-use systems (Baldock *et al.*, 1993). However, investigating the environmental performance of SAF through field experiments is expensive and time-consuming because trees take decades to mature. Computer models provide one method for overcoming these problems. They can yield results more rapidly and they can be used to extrapolate research results to new combinations of biophysical and management conditions that are too complex to be studied in field experiments (Mobbs *et al.*, 2001). However, the majority of research on temperate agroforestry systems has been undertaken to evaluate their productivity, whereas information on their environmental performance remains scarce.

Palma *et al.* (submitted to Ecological Engineering) therefore developed a modelling approach to assess the environmental performance of SAF systems. This included examining the impact of SAF on soil erosion by water (hereafter called erosion), nitrogen leaching, carbon sequestration and landscape biodiversity, and uses tree and crop yields derived from a biophysical model called YieldSAFE (Van der Werf *et al.*, submitted to Ecological Engineering).

This paper summarizes the results obtained from modelling the potential environmental performance of SAF in representative climatic conditions of southern Europe (Mediterranean Spain), western Europe (France), and northern Europe (the Netherlands) (Palma *et al.*, submitted to Ecological Engineering).

## Material and methods

Based on an environmental classification of Europe (Metzger *et al.*, 2005), randomly landscape test sites (LTS) were selected in Spain, France and the Netherlands (Figure 1). In each site tree and crop yields were modelled on hypothetical farms for SAF at two densities (50 and 113 trees ha$^{-1}$) on 10 and 50% of the total agricultural area, starting with either the best and worst quality land. Current agricultural land use was also modelled to provide a comparison with the status quo.

YieldSAFE, which dynamically predicts site-specific long-term tree and crop yields under competitive conditions on the basis of solar radiation, temperature, precipitation and relevant soil physical characteristics (Van der Werf *et al.*, submitted to Ecological Engineering), was used to generate crop yields for typical crop rotations at each LTS over a 60-year time horizon. The

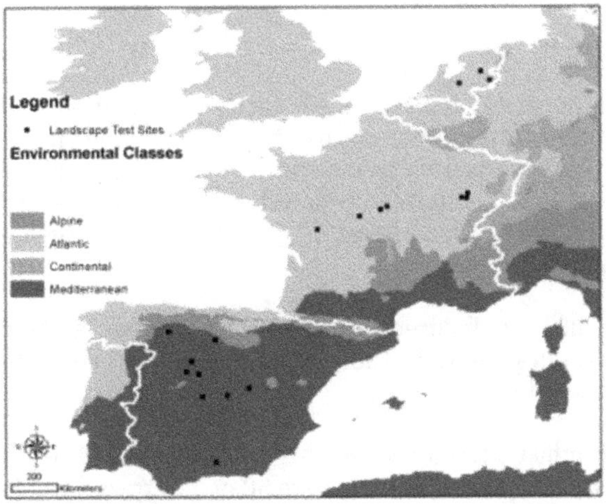


*Figure 1. Landscape test sites based on the European environmental classification by (Metzger* et al.*, 2005).*

same crop rotations were then used in SAF systems that included holm oak (*Quercus ilex* subsp. *ilex* L.) and stone pine (*Pinus pinea* L.) in Spain, and hybrid walnut (*Juglans* sp), wild cherry (*Prunus avium* L.) and poplar (*Populus* spp) in France and the Netherlands.

Geographical and statistical data was gathered during field surveys and source acquisition and processed for further analysis. Landscape test sites (LTS) were subdivided into up to four distinct land units (LU) using a cluster analysis. The LU were considered as homogenous with respect to soil properties and climatic conditions. The environmental assessment was performed in each LU assuming a 60-year rotation for the agroforestry system (three 20 year rotations for poplar).

Because crop yields within an agroforestry system decline over time as the trees increase in size and compete with the crops, it was assumed that farmers would stop arable cropping when it became unprofitable. The cut-off point was estimated from a five-year moving average of profitability as described by Graves *et al.* (submitted to Ecological Engineering).

Erosion was modelled with the revised universal soil loss equation, where SAF was considered to mimic strip cropping which could be implemented with or without contour farming. Nitrogen leaching was assessed based on an annual water exchange factor in the soil and the excess nitrogen potentially available for leaching.

Annual excess nitrogen was estimated from tree and crop productivity, assuming optimized nitrogen fertilization. Carbon sequestration estimations were based on the Intergovernmental Panel on Climate Change (IPCC, 1996). Landscape biodiversity was estimated using a habitat index, which assesses the share of habitats available to wildlife in an agricultural landscape, using broad distinctions between habitat and non-habitat.

Each environmental assessment for each LU in each LTS was then used to calculate a weighted mean at the farm scale, based on the proportion of land occupied by each LU within each hypothetical farm. These were then aggregated to provide an overall assessment of environmental effect for each scenario at each LTS.

## Results and discussion

The results of the environmental assessment for each scenario at each LTS is explored in detail in Palma *et al.* (submitted to Agriculture Ecosystems & Environment). Table 1 summarizes an average calculated for all LTS enabling an overview on the benefits of SAF trough Europe. Mean annual values are indicated to facilitate interpretation, although the environmental parameters evolve over the 60 years time horizon because of variation in weather, crop rotations, and the growth of trees in the SAF systems.

In interpreting the results, we focused on the relative differences between scenarios, rather than on the absolute values. However, absolute values have been tabulated to indicate the magnitude of the computed values.

Soil erosion was reduced by up to 65% when SAF was combined with contouring on 50% of the farm, on the best quality land, whether there were 50 or 113 trees ha$^{-1}$ (Table 1). Similar relative reductions in erosion rates have been found following introduction of hedgerow intercropping, where soil erosion was reduced by up to 90% on gentle slopes in Nigeria, and by 45-65% on steep slopes in maize systems in Colombia (Young, 1989). However, RUSLE does not account

*Table 1. Results of environmental effects of different densities of trees in Silvoarable Agroforestry (SAF) systems implemented on W: worst quality land (W) as compared to Best quality land (B). For detailed results, see Palma* et al. *(submitted to Agriculture Ecosystems & Environment).*

| | | SCENARIOS | | | | | | | | | | | | | | | | |
|---|---|---|---|---|---|---|---|---|---|---|---|---|---|---|---|---|---|---|
| | | | non contouring | | | | | | | | | contouring | | | | | | |
| | | A** | SAF50 | | | | SAF113 | | | | A | SAF50 | | | | SAF113 | | |
| | | | 10%SAF | | 50%SAF | | 10%SAF | | 50%SAF | | | 10%SAF | | 50%SAF | | 10%SAF | | 50%SAF | |
| | unit | | W | B | W | B | W | B | W | B | | W | B | W | B | W | B | W | B |
| Soil Erosion | ton $ha^{-1}$ $a^{-1}$ | 2.5 | 2.5 | 2.4 | 2.4 | 2.0 | 2.5 | 2.4 | 2.2 | 1.9 | 1.6 | 1.5 | 1.4 | 1.0 | 1.0 | 1.4 | 1.5 | 1.0 | 1.0 |
| Nitrogen Leaching | kg $ha^{-1}$ $a^{-1}$ | 96 | 93 | 93 | 88 | 88 | 90 | 89 | 77 | 76 | na | na | na | na | na | na | na | na | na |
| C Seq (after 60 years) | ton $ha^{-1}$ | - | 8 | 11 | 28 | 43 | 11 | 16 | 40 | 60 | na | na | na | na | na | na | na | na | na |
| Landscape diversity | $I_{hab}$* | 25 | 32 | 32 | 65 | 65 | 32 | 32 | 65 | 65 | na | na | na | na | na | na | na | na | na |

* Habitat index; ** Status quo.

for gully erosion. In fact, implementation of SAF without contouring practices could increase the probability of gully erosion along the tree strips, due to the greater erosivity of water drops under the tree canopy (Young, 1989). The orientation of the tree lines along the contour lines would avoid this negative impact.

The nitrate leaching scenario comparisons were restricted to the test sites where nitrogen leaching exceeded 10 kg $ha^{-1}$ $a^{-1}$ (France, The Netherlands). The overall mean shows a maximum decrease of about 21% when SAF is implemented with 113 trees $ha^{-1}$ on 50% of the best land (Table 1). The predicted reduction in nitrogen leaching under SAF is significantly lower than values of 40 and 75% reported by (Nair & Graetz, 2004) for temperate agroforestry systems. However, our modelling approach does not account for the potential of tree roots to recover nitrogen from below the crop root zone nor for the possibility of reducing fertilization due to increase of organic matter in the soil as consequence of tree leaf fall.

Sequestration varied in dependence of the tree species selected for each LTS. Although the modelled situations ranged from 5 to 179 t C $ha^{-1}$depending on tree growth rates,total sequestration was generally between 8-60 t C $ha^{-1}$ (Table 1) which matches other sequestration studies (Dixon *et al.*, 1994, Van Kooten *et al.*, 2002). Lower tree densities result in higher biomass per tree, which explains the small difference found between tree densities. Better results were consistently found in the best land.

The habitat index ($I_{hab}$) expresses landscape biodiversity by relating the share of natural and semi-natural habitats to the total area of a given landscape. The introduction of rows of trees in homogeneous arable areas increases the structural diversity of the landscape, which potentially increases its species diversity (Smart *et al.*, 2002). The introduction of SAF increased $I_{hab}$ for each LTS, with the largest increase in areas where existing natural or semi-natural habitat was low. The current $I_{hab}$ of the LTS had a mean of 32% and increased in general by a factor of 2 (Table 1), even reaching a factor of 4 in some cases.

## Conclusions

The results for 19 randomly selected LTS in Spain, France, and the Netherlands indicate that adoption of silvoarable agroforestry systems can potentially lead to reduced soil erosion and nitrogen leaching and increased carbon sequestration and landscape biodiversity. The extent of the modifications depends on the characteristics of each site and the management of the SAF system selected for each location. Predicted environmental benefits were highest when SAF was implemented on large areas (*i.e.* 50% of the farm) on high quality land, where current agricultural practices are most intensive and thus associated with higher levels of soil erosion and nitrogen leaching.

Agroforestry systems are highly diverse. We only examined five tree species in combination with five crops but, of course, many more tree species and crop types can and should be considered. All those options can only be fully explored with modelling approaches. Such models, however, must be validated on the basis of experimental data from these systems and such data are scarce.

In Europe, new land-use systems should (also) yield environmental benefits. The results presented here increase our understanding of the environmental benefits that can be expected from modern agroforestry systems and complement the economic analysis by Graves *et al.*

(submitted to Ecological Engineering) for the same LTSs. Next we will concentrate on an integrated economic and environmental analysis of the benefits provided by SAF systems.

## References

Baldock, D., G. Beaufoy, G. Bennet and J. Clark, 1993. Nature conservation and new directions in the EC common agricultural policy. HPC, Arnhem.

Dixon, R.K., J.K. Winjum, K.J. Andrasko, J.J. Lee and P.E. Schroeder, 1994. Integrated Land-Use Systems - Assessment Of Promising Agroforest And Alternative Land-Use Practices To Enhance Carbon Conservation And Sequestration. Climatic Change 27, 71-92.

Eichhorn, M., P. Paris, F. Herzog, L. Incoll, F. Liagre, K. Mantzanas, M. Mayus, G. Moreno Marcos and D. Pilbeam, 2005. Silvoarable agriculture in Europe - past, present and future. Agroforestry Systems (in press).

Herzog, F., 1998. Streuobst: a traditional agroforestry system as a model for agroforestry development in temperate Europe. Agroforestry Systems 42, 61-80.

IPCC, 1996. Revised 1996 IPCC Guidelines for National Greenhouse Gas Inventories: Reference Manual. Available at: http://www.ipcc-nggip.iges.or.jp/public/gl/guidelin/ch5ref3.pdf.

Metzger, M., R. Bunce, R. Jongman, S. Mücher and J.W. Watkins, 2005. A climatic stratification of the environment of Europe. Global Ecology and Biogeography 14, 549-563.

Mobbs, D., G. Lawson and T. Brown, 2001. HyPARv4.1 Model for Agroforestry Systems User Guide. Centre for Ecology and Hydrology, Edinburgh. Available at: www.nbu.ac.uk/hypar/.

Nair, V.D. and D.A. Graetz, 2004. Agroforestry as an approach to minimizing nutrient loss from heavily fertilized soils: The Florida experience. Agroforestry Systems 61, 269-279.

Smart, S.M., R.C.H. Bunce, L.G. Firbank and P. Coward, 2002. Do field boundaries act as refugia for grassland plant species diversity in intensively managed agricultural landscapes in Britain? Agriculture, Ecosystems & Environment 91, 73-87.

Van Kooten, G., S. Shaikh and P. Suchánek, 2002. Mitigating Climate Change by planting trees: The transaction costs trap. Land Economics 78, 559-572.

Young, A., 1989. Agroforestry for Soil Conservation. CAB International, Wallingford, UK.

# Biomass production, transformation and allocation at the village scale in Southern Mali: Development of a conceptual model

*Bruno Rapidel[1], René Poccard-Chapuis[1], Jean-François Bélières[1], Bouba S. Traoré[2], Jacob Coulibaly[2], Drissa Diallo[2] and Matthieu Lessnoff[3]*
*[1]CIRAD/IER, Mali; [2]IPR, Mali; [3]CIRAD/ILRI, Kenya*

## Abstract

The cotton zone of West Africa undergoes a progressive saturation of available agricultural space and a difficult economic situation. The produced biomass is often cited as a factor limiting the farming systems intensification and sustainability. Thus, to analyze the ecosystem's sustainability, it is necessary to understand the production, transformation and allocation of biomass. The biomass management decisions are taken at the farm level, but depend on processes at the village scale, which must be considered.

Scientists of various disciplines developed a conceptual model to represent the biomass flows in two typical villages of the cotton producing area in Southern Mali. The objective was to represent the current situation, and to simulate possible future evolutions.

This model consists of four interlinked modules: The first module considers the biomass production at the field scale and the use of transformed biomass in the fields. The second module describes the biomass uses and transformations of the farm types. The third module represents the operation and biomass production of the common assets, including areas not individually owned and the tree production of the farms' fields, in open access. The fourth module describes the economic operation of the farms -stocks, products valuation, margin calculation- and the workforce balance.

The model will be formalized in a simple numerical model. Surveys are underway, to estimate its many parameters and compartments sizes, in such a way that it will be able to reproduce the current state. We will then assess its aptitude to simulate the farm functioning and the biomass balance at village scale in the medium term. We want to use the model to consider the introduction of innovations that interfere with biomass management in a village territory.

**Keywords:** West Africa, sustainability, biomass, modelling

## Introduction

The cotton zone of West Africa undergoes a progressive saturation of available agricultural space and a difficult economic situation that does not favor the use of imported fertilizers (Devèze and Halley des Fontaines, 2005; Djouara *et al.*, 2005). An efficient use of natural resources will be a key factor to maintain food sufficiency in the region. Since the introduction of the cotton crop after the Second World War, the farms have been able to accumulate production means, mainly through a better association between agriculture and livestock production (Pelissier, 1977). But this association seems to be reaching its limits in some regions, and the forage to maintain the herd near the villages becomes scarce. Some herds are sent southward during the dry season, other are reduced in number, thus reducing the farms capabilities to produce manure and to maintain the soil fertility.

The produced biomass is often quoted as a factor limiting the farming systems intensification and sustainability (De Ridder *et al.*, 2004). To analyze the ecosystem's sustainability, it is necessary to understand the production, transformation and allocation of biomass. The biomass management decisions are taken at the farm level, but depend on processes at the village scale, which must be considered (Landais *et al.*, 1991).

Progress on integration between agriculture and livestock is necessary, through the introduction of innovations dealing with biomass production and management at various levels: cover crops, forage crops, green manure, fences, *etc.* The management of space and livestock is a key issue in the villages in Western Africa, rules have been established long ago and are difficult to change. Useful innovations have been poorly adopted in the past (Bationo *et al.*, 1998; Enyong *et al.*, 1999).

The objective of this work was to build a conceptual model of the biomass production, transformations and uses in a typical village of Southern Mali that helps in the following: assessing the actual biomass transfers and their sustainability, visualizing the possible effects of the adoption of various innovations, and focusing the research efforts on biomass management between disciplines and scales.

We built the conceptual model based on the knowledge of the biomass uses in two connected villages on the northern edge of the ancient cotton basin in Southern Mali. Details on the biomass uses and economic issues in theses villages are presented elsewhere in this volume (Bélières *et al.*, 2006; Poccard-Chapuis *et al.*, 2006).

## Modeling principles

We tried to apply four basic principles when conceiving the conceptual model: (1) the precision used in the description of elemental processes was kept as low as possible to maintain the whole model simple, while being sensitive to the management decisions and innovations; (2) the precision was kept equivalent throughout the four different modules; (3) the users may run and test each module independently; (4) the originality of the model lays in the scale considered, not in the treatment of each elemental process. We will use already developed modules to formalize the numerical model, when available.

## Description of the conceptual model

The conceptual model consists in four interconnected modules. The first module (Figure 1) determines the biomass production in the farm's cultivated fields. Biomass production depends on: rainfall -two typical years will be considered, and the production response will vary with the soil type; the cropping system –intensification level and artificial fertilizer quantities; the species; and the soil mineral availability. A specific soil module calculates soil nutrient availability. It considers the initial soil status, the nature and quantity of the organic manure added and the doses of mineral fertilizers. Each crop is simulated in each soil unit. Nevertheless, the organic matter evolution is considered as an average over the whole area cultivated by the farm type in this soil unit (regardless of the crop that was grown this particular year).

The second module (Figure 2) sums up the quantities of biomass recollected by the farmers, and establishes their transformations. Biomasses come from either the farm fields or the commons. Two units transform the biomasses, the family and the livestock. Biomass can also be added directly to some residue compartments. Each residue compartment produces organic manure

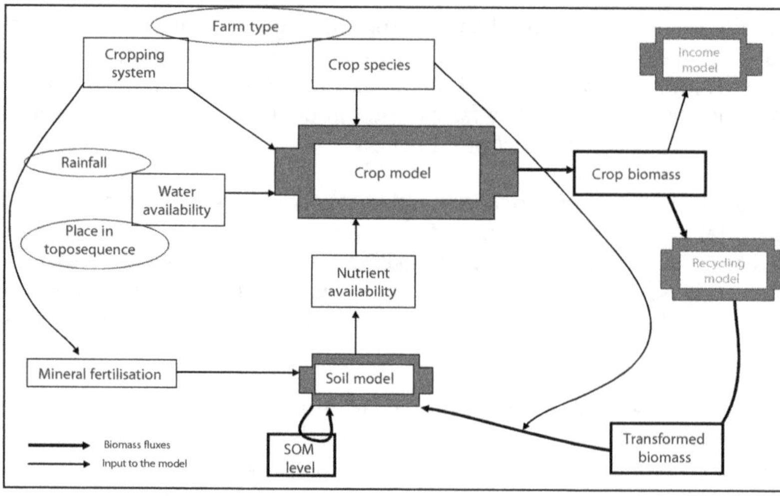


*Figure 1. Module 1, the crop model.*

of a known quality. The livestock does not change in number, but, depending on the forage sufficiency; it is considered to stay the whole year on the farm, or to be sent outward for some months. In this case, the module reduces the biomass transforming capacity of the farm.

The third module (Figure 3) calculates the wood and forage production of the commons, *i.e.* the areas managed collectively. These areas are the collective assets and the trees of the individually owned fields. Wood production depends on the existing biomass. Its uses depend on

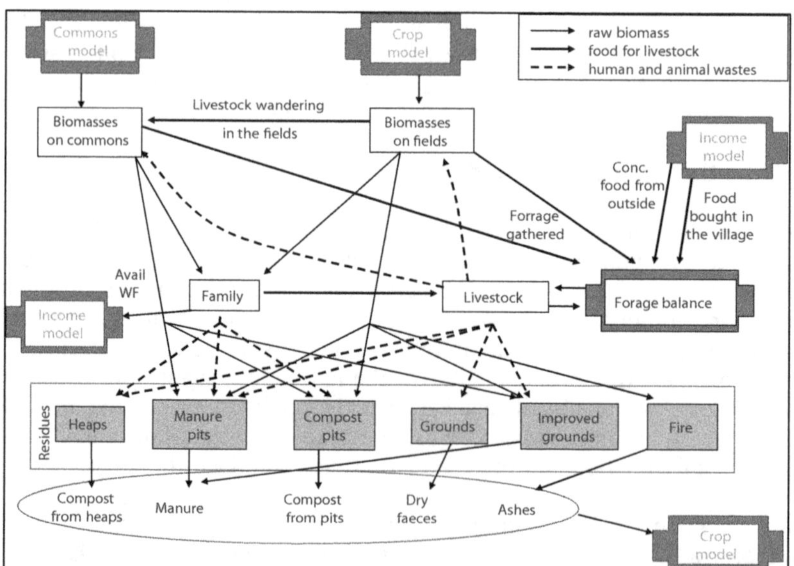


*Figure 2. Module 2, the recycling model.*

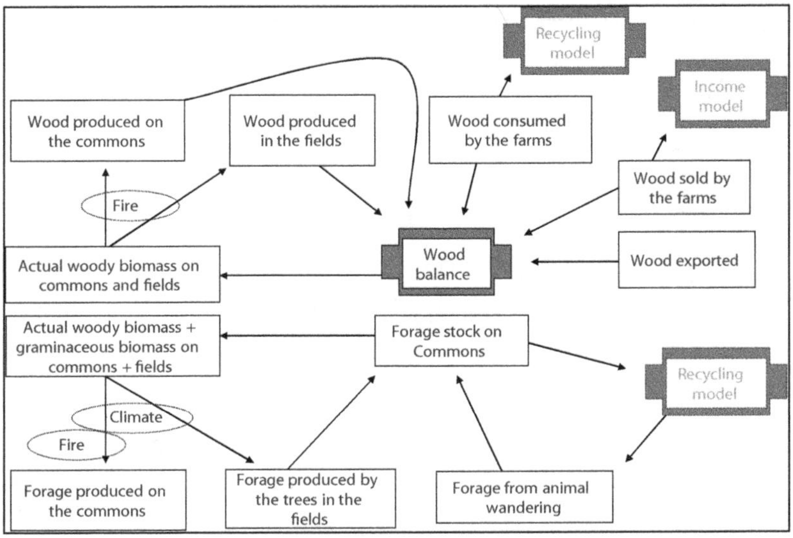


*Figure 3. Module 3, the commons model.*

the population of the village and on possible trade. Consumption is compared to the production to determine the remaining standing biomass, and thus, the production for the following year. The forage production depends on the standing woody biomass; the grazing area; the climate, and the management of this grazing area by fire.

The fourth module (Figure 4) calculates the incomes of each farm type. It considers some social issues (gifts) and incomes from financial transfers from abroad, based on data collected during the 2004/2005 survey. It also calculates the workforce balance.

Food sufficiency assessment relays on two indicators: the comparison between grain production and family needs, and the stock level. The workforce balance is also an indicator, apart from its contribution to global income. We consider three economic indicators: global income, income from the plant production and income from livestock management. For future simulations, we will assess the farm sustainability by two main indicators, global income and percentage of the family needs of grain covered by production. If these indicators are below a threshold, people from the family will have to emigrate.

We built a typology to segregate the farms on the basis of their 2004-2005 practices for biomass management. Modules 1, 2 and 4 are applied to farm types, to sum up the whole production of the villages. The allocation of soil units among farm types is extracted from a geographical information system of the villages, but spatial information is not further considered in our model.

## Perspectives

The development of this conceptual model was an opportunity to compare the perception of scientist from diverse disciplines on a key issue of sustainability. This exercise will continue with the filling of each compartment when integrating the data from the surveys. Then, the model will be formalized into a numerical model, using already developed modules as much

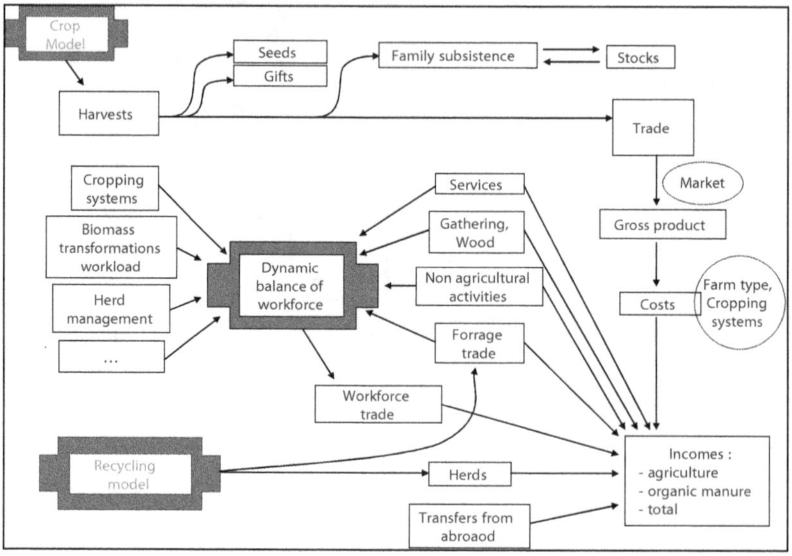


*Figure 4. Module 4, the income model.*

as possible. Modification of the practices by innovation adoption will be introduced in future simulations to test their effect on the biomass fluxes at village scale.

## References

Bationo, A., F. Lompo and S. Koala, 1998. Research on nutrient flows and balances in west Africa: state-of-the-art. Agriculture Ecosystems & Environment 71, 19-35.

Bélières, J.-F. *et al.*, 2006. Manure availability, food security and poverty in cotton farms of the old cotton production basin in Mali, 7th European International Farming System Association, Wageningen, The Netherlands.

De Ridder, N., H. Breman, H. van Keulen and T.J. Stomph, 2004. Revisiting a 'cure against land hunger': soil fertility management and farming systems dynamics in the West African Sahel. Agricultural Systems 80, 109-131.

Devèze, J.C. and D. Halley des Fontaines, 2005. Le devenir des agricultures cotonnières; Cas du Mali, AFD, Paris.

Djouara, H., J.-F. Bélières and D. Kébé, 2005. Les exploitations agricoles familiales de la zone cotonnière du Mali face à la baisse des prix du coton graine. Cahiers Agricultures (in press).

Enyong, L.A., S.K. Debrah and A. Bationo, 1999. Farmers' perceptions and attitudes towards introduced soil-fertility enhancing technologies in western Africa. Nutrient Cycling in Agroecosystems 53, 177-187.

Landais, E., P. Lhoste and H. Guérin, 1991. Systèmes d'élevage et transferts de fertilité. In: C. Piéri (Editor), Savanes d'Afriques, terres fertiles? CIRAD et Ministère de la Coopération et du Développement, Montpellier, pp. 219-270.

Pelissier, P., 1977. Competition and the integration of agriculture and cattle raising in Sahelian and Soudano-Sahelian Africa, Proceedings of interantional symposium on rainfed agriculture in semi-arid regions (17-22 april 1977). University of California, Pasadena, CA.

Poccard-Chapuis, R. *et al.*, 2006. Biomass management in livestock and cropping systems in Southern Mali: how to improve cotton farms sustainability? 7th European International Farming System Association, Wageningen, The Netherlands.

# Institutions for anthropogenic nature: Case study from Brabant

*Niels Röling and Janice Jiggins*
*Communication & Innovation Studies, Wageningen University and Research Centre, Wageningen, The Netherlands*

## Abstract

The paper analyses the institutional dynamics required to maintain nature in spite of competing claims on the resources and ecological services involved. The paper draws on a case study of water management in a densely populated area in which 'wet nature pearls' are maintained amidst intensive industrial farming. Network coordination practices that emerged over time as a result of stakeholders learning to 'give and take' are challenged by pragmatic adaptations in institutional behaviour required by the ecological imperatives embodied in 'green' European legislation.

## Introduction

Institutional innovation in water management is of particular concern in the deltaic conditions of the Netherlands, in which the management of the level and quality of surface, ground, and soil water is determinant of the options for drinking, recreation, farming, and nature conservation. Water management typically operates in conditions of a *resource dilemma* (SLIM, 2004):

- Water is a *common pool resource* where use is conditioned by (1) *sub-tractability* -the use by one subtracts benefits from others; (2) *high transaction costs* incurred by excluding individuals or groups from using the resource; and (3) *high risk* of degrading hydrological systems.
- *Multiple stakeholders* make different claims on the resource, from recreational fishing, to abstraction for industrial purposes, from use as drinking water to a medium for carrying away farm effluent.
- There is *interdependence*: stakeholders can realise their own objectives only through others.
- There is *controversy*: stakeholders hold strong but divergent values and perceptions about what is at stake.
- There is *complexity*: scientific data cannot resolve the issues because they arise from multiple causes and have multiple effects, with different expression in space and time, and the irreducible value dimensions of 'the problem' cannot easily be measured or modelled.
- There is *surprise* that arises irreducibly from complexity.

Hydrological systems in the Netherlands are multiple-scale and multiple-use common pool resources that span national and other borders. Their ecological status is the outcome of the collective impact of the activities of millions of stakeholders acting locally in diverse contexts. Their management demands coordination mechanisms that go beyond the typical 'command and control' hierarchical measures, backed by regulation and legislation, and beyond the operation of the market, because the former rapidly run into problems of non-compliance (with all that entails in terms of surveillance and enforcement, legal costs, social injustice, and eventually political embarrassment), and the latter because it is incapable of dealing with the social and environmental harm and costs that are 'externalised' in competitive market conditions. Table 1 shows that a third coordination mechanism can usefully be recognised (Powell, 1994). Resource

*Table 1. Coordination mechanisms.*

| | Hierarchy | Market | Network |
|---|---|---|---|
| Operational mechanism | law, regulation | invisible hand | inter-dependence<br>negotiated agreement<br>social learning |
| Intervention | policy implementation | liberating market forces | process facilitation |
| Dynamics | cause-and-effect<br><br>social engineering<br>technical measures<br>using force | rational choice driven by individual preferences | co-construction of knowledge<br>reciprocity<br>cooperation<br>negotiation |
| Outcome | directed action | aggregation of individual choices | concerted action |
| Resource | power | financial capital | social capital |

dilemmas are usually considered to require network coordination because hierarchy and market fail to solve them (Röling, 2002).

The European Habitat Directive introduced a framework for the protection of specific species in designated areas that are to be managed as habitats for those species. The Water Framework Directive (WFD), introduced in 2000 aims to provide a common, legally enforceable framework for the governance of Europe's surface and groundwater under River Basin Management Plans. The WFD's way of thinking is radical in that it reifies water as a living entity that has the force to affect the *human* habitat. As a 'life milieu' it is recognised as essential for life. Furthermore, water as an 'ecological agent' is positioned as acting on terrestrial ecosystems, and thereby as capable of shaping and constraining human opportunity within water catchments. The WFD's normative dimension requires that river systems should achieve water of 'good ecological status' by 2015. The procedural dimension of the WFD demands adoption of indicators, standards, and monitoring capacity that can demonstrate improvement, and that are scientifically based, unambiguous, and quantitative. The WFD also specifies explicit sanctions for failing to achieve the ecological quality standards formulated in the river basin management plans, while Article 14 calls for citizen involvement in implementation. By specifying the regulatory framework, demanding 'right prices', and stipulating participatory approaches, the WFD represents a mix of all three coordination mechanisms.

## Case study: *De Heide,* nature as an artefact

A heath is an area of open, uncultivated land, especially on acid, sandy or peaty soil, covered by heather (*Calluna* and *Erica* species) and related plants. Heaths are the remnants of a 5000-year old form of agriculture endemic to Western Europe's mild climate. The former vast areas of heath served as pasture for sheep, which at night were kept in folds where they produced the manure that was applied to small arable fields. Forty hectares of heath were required to maintain one arable hectare. This age-old system of agriculture collapsed with the advent of chemical fertilisers. In the Netherlands, most of the heath subsequently was converted into productive farmland. Maintaining the remnants as nature reserves requires maintaining the old agricultural system; the heath is always trying to resume its identity as forest - if you leave it, it will be covered with trees in 30 years time.

The case is about *De Heide*, a remnant in the south of the Netherlands that has been designated as a nature reserve, and is mainly maintained through a flock of sheep of a variety that had nearly become extinct but was rescued by volunteers. The shepherd is a salaried man. The heath is further maintained through mechanical mowing, sod harvesting, burning, shifting cultivation, and tree removal by volunteers. A major threat to its survival has been the atmospheric deposition of sulphuric acid and ammonia from intensive pig industries in the area. These additives threatened to turn the heath into grass until the deposition was cut by half (ammonia) or more (sulphuric acid), as a result of various 'manure laws'.

*De Heide* harbours a fen that now is the largest oligotrophic fresh water lake in Europe, with a *Littoralion* plant community that is uniquely adapted to the slightly acid, nutrient-poor environment. The lake was restored at a cost of 1.7 million Euro in the early 1980s. However, the re-emergence of the Habitat Directive species around the lake could be achieved only by diverting the *Peelrijt*, a ditch dug by farmers in the early 1800s to drain the heath upstream from *De Heide*. Over the years, this *Peelrijt* had become an open sewer. The heavily fertilised sandy fields leaking large amounts of potash and nitrates were a major contributor to the pollution of the fen. Diversion of the *Peelrijt* meant disconnecting it from one river catchment, managed by one water board, and connecting it to another catchment, managed by another board. The original agreement to divert the *Peelrijt* was signed by both boards in 1987; the arrangement was simplified by the fact that at the time the same man served as director of both.

However, as soon as the WFD came into force, the mixing of the waters of the two catchments became a huge technical and institutional problem. Would one water board have to cede jurisdiction of the entire catchment area that is drained by the *Peelrijt* to the other? Would the EU Commission accept an arrangement by which water is diverted across a natural watershed? Was it fair that the water board that now received the sewer water from the *Peelrijt* became at a stroke responsible for cleaning it up? Faced with this uncertainty, the two water boards agreed that the *Peelrijt* had to be returned to its original course and in 2004, the water boards and the nature conservation agencies responsible for *De Heide* signed a covenant agreeing to 'system restoration'. However, this led to other dilemmas.

In 2005, one of the nature conservation bodies discovered a Habitat species, Floating Water Plantain, *Luronium natans*, in the swampy heath downstream from the fen. Reconnecting the *Peelrijt* would certainly cause its demise. The water from the *Peelrijt*, even if restored to the MTR quality - as the water boards had promised under the 2004 covenant - would still have a phosphate content of 0.36 P mg/litre, while the *Litteralion* and *Luronium natans* can survive only in conditions of 0.02 P mg/litre. The nature conservation agency decided to withdraw from the covenant.

Indignation erupted all over the institutional map. The nature conservation agency in question was called 'unreliable': 'This is not the way we deal with each other'. Here we see a clash between the absolute values emanating from ecological imperatives, and the values of negotiated agreement that have emerged in this very densely populated area of the country, among parties that had become adept at handling the competing claims on natural resources in an interactive (network) manner. Here also is evident the latent clash between the Habitat Directive and the Water Framework Directive. What to do?

The conflict could not be handled by the multi-stakeholder platform, made up of the two water boards, provincial authorities, three nature conservation agencies, farming interests (5 representatives!), and the drinking water companies, that usually handles water problems to do with *De Heide*. It was handed over to a specially convened provincial board, made up of the elected provincial delegate responsible for the portfolio of agriculture and nature, the directors of the two water boards, and the regional director of the nature conservation agency involved. That is, it was made up of senior administrators who were expected to ensure that business was done in the public interest. At a meeting in January 2006, the representative of the nature conservation agency agreed once more to abide by the 2004 covenant. The WFD seemed to have superseded the Habitat Directive. But, at the same time, the water boards promised to ensure that the quality of *Peelrijt* water will be improved, far beyond MTR quality promised in the 2004 covenant, in order to meet the requirements of the Habitat Directive. They also promised to collaborate in a joint search for ways of infiltrating the *Peelrijt* into the sub-soil so that only relatively little of it ends up as surface water flowing into the fen and *De Heide*.

## Institutional arrangements (rules of the game) and ecological imperatives

There is a very clear distinction between the outcomes of negotiated agreement and ecological imperatives. For example, the population of Atlantic Cod does not heed the negotiated agreements reached by fisheries ministers, however fair these are judged to be. It can crash, whatever the negotiated outcome, if only the needs of people and not also those of the fish are taken into account. The negotiations about Cod quotas tend to neglect the latter. Obviously the interests of Cod represented by fish ecologists loses out against the interests of fishermen and their power to make the lives of ministers miserable.

The Habitat and Water Framework Directives are interesting ways by which ecological imperatives enter into multi-stakeholder processes. Nature conservationists and water managers can bolster their claims by referring to these European legislations, in effect relying on hierarchy for coordination of multi-stakeholder behaviour. The fact that these European Directives became laws is itself interesting. The WFD, for example, was strongly supported by the 'green' hydrologists of the Dutch Ministry of Agriculture who were involved in its formulation. It is only when it was too late that the agricultural lobby woke up, especially after a research report that showed that two-thirds of Dutch farms would have to stop operations if the WFD were to be enacted. Since then, the Ministry of Agriculture has sought a '*status aparte*' for the Netherlands on the grounds that it is an artificial country to begin with (according to the WFD, artificial waters can get away with lower quality criteria). In a way, the European legislative capacity took people by surprise and provided a 'by-pass' around national legislative processes that would never have allowed such strong ecological imperatives to be enacted into law.

The difficulty for those trying to bring the intentions and provisions of the European directives into effect rests on the fact that the designation and boundaries of river systems, catchments, and habitats, and of their desirable states, are purely human choices. The definitions of ecological standards and habitats cannot be derived purely from science. The frameworks thus do not so much resolve resource dilemmas, *let alone* remove them, as cast them in another form, as a challenge to existing institutional arrangements, and they introduce a new rationale for the management of the uncertainties that arise.

The case shows the extent to which rules for negotiated agreement have become accepted for resolving resource dilemmas. But the word 'negotiation' with its suggestion of arriving at a low common denominator does not do justice to what has been achieved. In fact, the multi-stakeholder platform responsible for *De Heide* has gone out of its way to divert a flow of maximally 1.3 $m^3$/sec to an entirely different catchment. Similarly, and also at great cost, farms in the catchment of the *Peelrijt* have been connected to the sewer. These measures were taken with ecological imperatives, especially the *Litteralion* plant community, in mind. They represent more than mere negotiation (as in the case of Cod fisheries).

What is interesting is that the introduction of the WFD forced the water boards to come back on their original 1986 agreement to divert the *Peelrijt*. They somehow got the nature conservation bodies to agree to the 2004 covenant. When the nature conservation body, in turn, 'pulled the plug' out of that covenant, the partners were indignant and the conservationists had to eat their words. Strong pressure was applied by appealing to the norms and values of voluntary agreement. Yet, notwithstanding the conservationists' 'loss' (from a strict negotiation point of view), the partners still agreed to an improvement of water quality beyond the MTR. Thus ecological imperatives prevailed. One could argue that this was entirely based on hierarchy: the WFD will allow the conservations to sue water boards if they do not adhere to the quality stipulated in water management plans. On the other hand, given the artificial nature of Dutch water, the chance that management plans would stipulate 0.02 P mg/litre is very small. We conclude that over time, the multi-stakeholder platform responsible for the management of *De Heide* has come to accept that the ecological imperatives should determine the outcome of its interaction, beyond what mere negotiation based on power would have required.

## Conclusion

Western society is dominated by coordination mechanisms based on hierarchy or rational choice. We are less familiar with the arrangements for network coordination. That type of coordination seems more appropriate to resource dilemmas. However, especially under the impact of European directives, stakeholders engaged in multi-stakeholder processes move beyond the rules of negotiation to respond to ecological imperatives.

## References

Powell, W., 1994. Neither Market nor Hierarchy: Networks Forms of Organisation. In: G. Thompson, J. Frances, R. Levavic and J. Mitchell (editors). Markets and Hierarchies and Networks: The Coordination of Social Life. London: Sage, pp. 256-277.

SLIM, 2004. Social Learning for the Integrated Management and Sustainable Use of Water at Catchment Scale. European research in support of WFD implementation. Policy Briefings. http://slim.open.ac.uk.

# The construction of a coherent vision for territorial development

*Markus Schermer*
*Institute of Sociology, University of Innsbruck, Austria*

## Abstract

The new GAP views rural spaces increasingly from a territorial view as opposed to the traditional sectoral view of rural development. It thus allocates partly funds under the measurements of the "second pillar" activities for integrated rural development, not only connected to agriculture. This requires a new positioning of agricultural actors.

For territorial development to be effective, collective action and a shared vision for the common future of a given region is needed. The paper looks into the processes of how a coherent vision for local and regional development is formed. In doing so it combines the theoretical approaches of the theory of diffusion of innovation and social capital theory in analysing an empirical case study from Austria.

The stages of diffusion of innovations (Rogers, 1983) are combined with various forms of social capital needed for successful development (Woolcok, 1998). The resulting framework allows monitoring how a territorial view on rural development is constructed and can become a commonly shared vision.

After developing the theoretical argument, it is used to analyse the example of the eco-region ("Bioregion") Murau in Austria. Eco-regions can be described as attempts to transfer the value system of organic agriculture across economic sectors on a territorial scale. Different kinds of social capital (bonding, bridging and linking) are necessary in the various stages of the diffusion process to create a shared vision.

**Keywords:** territorial development, social capital, diffusion of innovation

## Introduction

Territorial development is a new approach to rural development which focuses on the development of a territory rather than of an economic sector. The shift towards territorial development can be noted in numerous reports of many OECD countries during the 1980s and 1990s (Pezzini, 2000).

The first rationale for this territorial approach results out of the fact that rural is not agricultural any more (Van der Ploeg *et al.*, 2000). This requires also a new positioning of farmers within the rural sphere. They have to look for new allies to safeguard their interests and to design new strategies to influence the further development of their region.

The traditional territorial policies were concerned with the equitable geographical distribution of resources which is not an appropriate answer to new conditions engendered in globalisation processes (Pezzini, 2000). Rural areas are not homogenous with uniform problems and opportunities opposed to urban areas. Often high numbers of commuters blur a distinct urban rural divide. But even if we stick to rural areas with low population density and significant agricultural land use, the development patterns, based on geographical location, proximity to markets, climate, history and endowment of human, social, cultural and physical capital, are different. Therefore the new focus has to be on mobilising local resources to find tailor made solutions for the variety of regional problems. This implies also a shift from supporting economic sectors to a place based, territorial view.

Another important fact is the change in the political power balance between regional, national and supernational levels. It appears that within the EU the national level has lost power, not only to the apparently growing forces on supernational level (from WTO to EU), but also to a regional level. This becomes visible in diverse developments like the creation of a Committee of Regions on EU-level, the devolution of government in countries like the UK, the creation of an "European Region Tyrol" and the existence of new institutional structures of local development like Regional Nature Parks in France, LEADER groups of Local Agenda 21 activities (Pezzini, 2000; Ray, 2001).

For territorial development to be effective, collective action and a shared vision for the common future of a given region is needed. While there is some literature on the effects of synergies through coherence on a regional scale (see for example Brunori and Rossi, 2000), little is known about the processes leading to coherence in development approaches. The paper provides some preliminary findings how a shared vision for local and regional development is formed. It combines the theoretical approaches of the theory of diffusion of innovation and social capital theory for the analysis of an empirical case study on one Eco-Region approach in Austria.

The notion of eco-regions ("Bioregionen" in German) forms an example for such territorial development endeavours. The idea of eco-regions emerged in recent years in the Austrian context of regional rural development as well as of organic farming. It was perceived as an important chance for organic farming, but also for the regional development especially of mountain areas. In the long run eco-regions aim to extend the philosophy, inherent to the organic farming *value* system, of sustainability, to other actors and sectors in a region and apply them territorially. (For a more detailed description of the approach see Schermer, 2006).

## Material and methods

The analysis of the case study uses a combination of the theory of diffusion with social capital theory.

While the seminal work of Everett Rogers (1983) on the diffusion of innovations has been influential for the analysis of technical innovation processes, it has been largely neglected in the context of regional development. However the construction of territorial coherence can be understood as a social innovation, following largely similar patterns to technical innovations. Rogers account on how an innovation is passed on from those who invented it to wider user groups until it becomes a general practice can thus be applied on the construction of a common vision for the future development of a territory.

In his account Rogers lists five categories of adopters with the following characteristics:

- *Innovators* are eager to try out new ideas; they are connected more with likeminded innovators in other regions than with members of their own social system. Still they play an important role in the diffusion process by launching a new idea, importing it from outside of the systems boarders. Thus they have a gatekeeper function in the flow of new ideas into a social system.
- *Early adopters* are more integrated into their social system than innovators. They are more "localites" while the innovators are more "cosmopolites". Their crucial contribution to

diffusion is opinion leadership. They act as "brokers" conveying a subjective evaluation for others to adopt the innovation. In the terminology of Rogers they are more "homophilious" (meaning similar in their attitudes) to the majority than innovators.
- The *early majority* believes into the evaluation of the peers because they have frequent contact with them, but do not hold leadership functions themselves. They provide interconnectedness in the systems network.
- Persons belonging to the *late majority* are sceptical and adopt out of economic or social pressure the idea. They act only if the systems norms are in favour of the innovation.
- *Laggards* are most localite, near isolates. Their point of reference is the past.

Being a social innovation the construction of a commonly shared territorial vision is connected to the existence of social capital. The notion of social capital has gained importance in recent years. The academic discourses on social capital started form such diverse sources like Bourdieu, Loury, Coleman and Jacobs (Haugg, 1997) and have influenced different scientific disciplines like sociology, economy and political science (Panter, 2002). On a more applied scale especially the World Bank has promoted social capital as a concept for rural development. One of the most influential contributions in this respect comes from Michael Woolcock (1998). He tries to combine the different strands of social capital theory into a coherent approach and distinguishes between three types of social capital: bonding, bridging and linking. While bonding social capital refers to a shared value system within a group, bridging social capital is necessary to exchange ideas between different groups. Linking social capital finally refers to the relations between social groups and the institutional sector as It is necessary to develop systems norms in favour of the innovation.

For an innovation to push through it needs to have a mix of relations within the system, which provide the social backing to take up new ideas. This backing is connected to the social embeddedness (Granovetter, 1985), on one hand as well to the existence of "weak ties" (Granovetter, 1972) to extend innovations to other networks. Certain persons have to act as "brokers" between rather closed networks in order to extend the diffusion of new ideas. In essence, within each network the stages of the diffusion of innovations are recurrent: An innovator, who is more oriented towards social groups outside the local network, brings in the idea; opinion leaders take up the idea and act as testimonies for others to follow until there is a social pressure for the whole network to take over the innovation. At the end only some rather isolated laggards remain sceptical.

The empirical data used to apply the theoretical findings of the theory of diffusion and social capital are derived form a case study attached to two research projects: a national project on "Eco-regions as model for sustainable development" and the EU- funded project "Implementing Sustainable Agriculture and Rural Development in Alpine Mountains, IMALP". In the course of these projects workshops, interviews and the analysis of minutes of internal meetings have been used to picture the dynamics in the construction of a shared vision for territorial development.

## The case study

### *The area*

The district of Murau is located in the centre of Austria in the province of Styria. It shares borders with the province of Salzburg in the west and Carynthia in the south. The district encompasses 35 communities with about 31.100 inhabitants on an area of 1.384 km². The population is slightly decreasing over the last decades due to the lack of adequate employment facilities. About 11% of the total labour force is still engaged in farming activities on 1600 farms. About two third of them are run by part time farmers. About 25% of the farms are managed organically and most of the others participate in the Austrian environmental program ÖPUL. The regional dairy runs organic and conventional production lines and supplies one of the major Austrian discounters with dairy products. Besides grassland based dairy farming forestry contributes to a high degree to farm income, especially by providing a source of renewable energy. While the agricultural sector as well as the industrial sector has decreased in importance, in recent decades the tertiary sector with tourism increased. Especially the Natural Park Grebenzen offers possibilities to develop forms of ecotourism in conjunction with farming.

The strategic direction of the regional development is discussed in the "ARGE Bioregion". The core team of this consortium ("ARGE") consists out of the leaders of different sectoral working groups which deal with agriculture, economy, ecology and renewable energy. The development goals are based on the creation of trans-sectoral partnerships to increase the regional value added chains.

***Description of the process***

The first ideas to promote actively territorial development emerged when two business persons, a baker and a plumber took part in an Agenda 21 process, specially designed for business people, in 2000. The baker, who had already been an activist against the use of GMO in agriculture before, introduced the idea of an organic food cluster and the notion of Eco-Region (Bioregion) The plumber supported the idea of focussing on renewable biomass heating systems among colleagues and challenged the farmers to organise the necessary supply logistics. Both teamed up with a change agent employed by the rural development agency who was also a farmer to promote the idea. In June 2001 the three proponents launched a workshop discussing the possibility of establishing a so called "Bioregion" ("Eco-Region") in Murau as a tool for sustainable rural development. A consortium of proponents from different economic sectors was installed to implement the concept. Especially the cooperation between a farmers's cooperative providing wood chips for heating and a newly formed consortium of the major plumbers of the region became influential. It took further the topic of renewable energy which resulted in the "Energy Vision 2015" proposing autonomy with primary energy by 2015. The plumber companies of the region agreed to favour biomass heating systems and to stop installing oil based heating systems after 2005. The initiative was awarded with the "energy globe" an international prize for sustainability. Various other co-operations (between farmers and gastronomy, tourism, Natural Park, leader projects *etc.*) emerged. Finally 2005 two broadcastings on the Eco-Region Murau on national TV increased the recognition of the initiative inside and outside of the region.

***Analysis of the process***

The first proponents were not part of the farmers' social network; they were business people (baker, plumber). Their basic ideas came form an Agenda 21 process. Thus the innovators acted rather cosmopolite, searching for connections from outside. They teamed up with the regional rural development agent who is also a peer in the local farmers' network. Being a farmer himself he is more homophilous to the target group and thus more "localite". On the other hand as a change agent he had many links outside of the farmers' network. Thus he could make use of his bridging social capital to introduce the new ideas to the rather closed network of farmers. First he introduced the idea to other functionaries in the farming sector (conventional and organic) in the region. At the same time the plumber started to organise her professional network and involved her colleagues into a group which called themselves the "eco-plumbers". The common name and goals led to the formation of bonding social capital which is unusual among competing business people. By joining together farmers and business people the idea of closing regional product and energy cycles could grasp. From there the idea of an "eco-region" developed and started to embrace actors of other sectors (like gastronomy, tourism, natural park *etc.*) which had rather separate closed networks before. Finally the proponents (also the baker but especially the change agent) are well integrated into the regional institutional network; they are familiar with regional politicians and also national politicians coming from the region. This allowed mobilising support from the institutional level in form of funds for projects and recognition on official occasions. Thus the diffusion of the idea is horizontal between networks as well as vertical within each network.

While the innovators used their bridging social capital to introduce new ideas into their own professional groups, they would not have succeeded to circulate the innovation further without the change agent who is firmly rooted in the rather closed farmers' network able to build mainly on bonding social capital. Finally the links to institutional representatives and regional politicians could make the integration from an "avant-garde idea" to the "mainstream" possible. This official recognitions is however crucial to convince a broad regional majority.

## Conclusions

The combination of Rogers's theory of diffusion of innovations combined with social capital theory provides a grid for analysing the complex processes involved in the construction and extension of coherent visions for territorial development.

One of the crucial findings is that diffusion of a social innovation on territorial level needs persons which deploy all three forms of social capital. They can be incorporated in one person or shared between an initiating group. Furthermore the innovation is propagated by actors on a middle level who can also link between grassroots and bureaucracy. The challenge for the multiplicators is to work downwards to the actors on the ground (farmers, tourist enterprises, environmental activists *etc.*) as well as upwards towards the institutional levels (regional chambers of agriculture, regional chambers of commerce, regional administration, *etc.*).

## References


Brunori G. and A. Rossi, 2000. Synergy and coherence through collective action: some insights from the wine routes in Tuscany. Sociologia Ruralis 40, 409-423.

Granovetter, M., 1972. The strength of weak ties. AJS. 78, 1360–1380.

Granovetter, M., 1985. Economic action and social structure: the problem of embeddedness AJS 91, 481-510.

Haug S., 1997. Soziales Kapital. Ein kritischer Überblick über den aktuellen Forschungsstand. Arbeitspapiere Arbeitsbereich II/15 MZES Mannheim.

Panter S., 2002. Sozialkapital. Entwicklung und Anatomie eines interdisziplinären Konzepts. In: W. Ötsch and S. Panther (editors) Ökonomik und Sozialwissenschaft. Metropolis, Marburg, pp.155-176.

Pezzini, M., 2000. Rural policy lessons from OECD Countries, Econ. Rev. Federal Reserve Bank of Kansas City, fall issue, pp. 47-57.

Ray, C., 2001. Culture economies, a perspective on local rural development in europe. CRE press, Newcastle. 151 pp.

Rogers, E., 1983. The Diffusion of innovations. Third Edition, The Free Press, N.Y., 453 pp.

Schermer, M., 2006. Regional development: through organic territory: ecoregions in Austria. In: G. Holt and M. Reed (editors). Sociological Perspective of Organic Agriculture: From Pioneer to Policy CABI (in press).

Van der Ploeg, J.D., H. Renting, G. Brunori, K. Knickel, J. Mannion, T. Marsden, K. de Roest, E. Sevilla-Guzman and F. Ventura, 2000. Rural development: from practices and policies towards theory. Sociologia Ruralis 40, 391-409.

Woolcock, M., 1998. Social capital and economic development: towards a theoretical synthesis and policy. Theory and Society 27, 151-249.

# Urban oriented agriculture: The case of Almere (The Netherlands)

*Derk Jan Stobbelaar[1,2,4], Marleen Warnaar[1,4], Jan Eelco Jansma[3] and Walter A.H. Rossing[1]*
*[1]Wageningen University and Research Centre, Biological Farming Systems Group, Marijkeweg 22, 6709 PG Wageningen, The Netherlands; [2]Wageningen University and Research Centre, Plant Production Systems Group, PO Box 430, 6700 AK Wageningen, The Netherlands; [3]Wageningen University and Research Centre, Applied Plant Research, Lelystad, The Netherlands; [4]Wageningen University and Research Centre, Rural Sociology Group, PO Box 8130, 6700 EW Wageningen, The Netherlands*

## Abstract

The discussion on new functions for the rural area, urge for a definition of the role that agriculture can play in this transition. A challenging new form in this respect is urban oriented agriculture; agriculture that is aiming at strengthening the ties with citizens. Research on the conditions for the success of this type of agriculture is urgently needed because of its profound benefits for both farming and society as a whole. Therefore our central research question is: what are the demands of citizens towards urban oriented agriculture? We investigated this question in a survey in the outer suburbs of Almere, a mid size town in the centre of The Netherlands. We used the touristic experience typology to analyse and generalise the data of the 342 respondents that replied on the survey. This typology runs from superficial amusement to existential connection.

The results of a survey show that there is indeed a citizen's willingness to join several farm activities. The most preferred functions are walking trails, cycling tracks, farm shop and educational activities. These functions are connected with rather low intense modes of experience, like amusement and change. Physical features (landscape, assortment, distance, *etc.*) of the farm are more important than social features (contact with farmer). Less preferred but still wanted by a large group, are the medium intense types of experience like interest. This can be found in farm functions like guided tours, information and education. Here contact with the farmer is equally important with the physical conditions of the farm. The least preferred functions are self picking and co-operation on the farm. These functions are connected with very intense types of experience like rapture and dedication. Social contact with farmer and other consumers is very important, next to the physical features.

**Keywords:** multifunctional agriculture, consumer demand, survey, The Netherlands

## Introduction: New functions of the rural area

There is much debate about the future of the rural area in Europe. One of the elements in the discussion is that the rural area should fulfil more and more diverse functions and provide meanings for rural and urban inhabitants (Raad voor het Landelijk Gebied, 2000; Kinsella *et al.*, 2000; Banks and Marsden, 2000). The rural area thus emphasizes its role as the contramal of the cities, where people find rest, space, recreational activities and an attractive landscape. In order to maintain public support, agriculture should have more attention for these new demands (Raad voor het Landelijk Gebied, 2002; Marsden, 1999). Delivering on public expectations will be challenging, not the least because many of these demands are contradictory. Hellerstein *et al.* (2002) note the absence of information on what the public really wants.

Agricultural entrepreneurs have reacted on these new demands from society by starting a process of rural development. Within this at least Europe-wide paradigm shift (Van der Ploeg *et al.*, 2002) – away from the modernisation paradigm with its ever increasing costs and stagnating gross value of production – we can find farmers that are relatively closely situated to cities. Some of these farmers choose to see this position as an advantage, rather than a disadvantage as it is seen in the modernisation paradigm, due to all kinds of (legal) restrictions. This Urban Oriented Agriculture, as we call it, can be seen as an example of how to match the demand and supply of functions of the rural area. In Urban Oriented Agriculture farmers are in close contact with consumers. It offers the opportunity to citizens to be more directly involved in food production and use and management of the rural area. On the other hand, it gives farmers the opportunity to learn about the wishes of its customers, which will possibly result in additional income for farmers.

Research about the conditions for the success of this type of agriculture is urgently needed because of its profound benefits for both farming and society as a whole. Therefore our central research question is: *what are the demands of citizens towards urban oriented agriculture?*

## Methodology

For a consumer the countryside can have a variety of meanings. The touristic experience typology of Cohen (1979) and Lengkeek *et al.* (1997) can be used to express the relevance of the countryside for the consumers. This typology distinguishes 5 types of experience ranging from superficial amusement to existential connection. These intensities of experience can be associated with some specific farm activities that fulfil the wishes of consumers:

- Amusement: a pleasant trip to the country side, with enjoyable distraction and busyness. Suitable functions for the rural area / farms are: restaurant, tea salon, farm shop, farm market.
- Change: to flea from the busy city. Rest and relaxation. Suitable functions: quiet and beautiful surrounding, walking and cycling tracks.
- Interest: Actively getting to know the unknown farming practises. Suitable functions: guided tours through interesting farm land (mixed farming), education, information (meeting the farmer, farm news letters, and info panels), and adoption of farms (in the Netherlands adoption of apple trees, hens or cows).
- Rapture: Adventure en challenge on farm, self discovery. Suitable functions: self picking, free entrance and transparent farm management.
- Dedication: farm life and the rural area are becoming a part of your own life. Suitable functions: co-owner, club activities, participation and contribution to farm work.

We conducted a survey in order to get a representative picture of the wishes of the inhabitants of the fringe of Almere and combined the results with the above touristic experience typology. Almere is a medium-size city in the very Dutch open polder landscape of the province of Flevoland, which was reclaimed from the IJssel Lake in the 60-ies and 70-ies. Almere has been designated to grow from 175,000 inhabitants in 2005 to 400,000 in 2050. Agriculture has to find its place in the shifting fringe of Almere.

The respondents were sent a letter to announce the telephonically taken questionnaire with closed questions. The core questions were about the wanted on-farm activities and features of urban oriented farms. We asked the respondents to estimate the chance that they would make use of a certain activity. Next to this there were questions about the background features of the respondents that enabled us to link these features with their wishes and to distinguish consumer groups.

The survey was selectively aiming at the residential areas in the fringe of Almere. The idea was that these inhabitants are the first group to contact when starting with urban oriented agriculture. In total 342 respondents replied on the survey. The data were processed and analysed statistically using the Mann-Whitney test and the Kruskal-Wallis test in SPSS to evaluate correlations between respondent characteristics and their demand for UOA functions.

## Results: Citizens demand of urban oriented agriculture

The first part of the questionnaire was dedicated to the desired farm functions (see Table 1). 68 percent of the respondents stated to like or to like very much visiting farms that offer several products and services. Walking trails and cycling tracks are the most wanted function, followed by visiting a farm shop. The largest interest was thus for the less intense types of experience 'amusement' and 'change', with little commitment with the farms.

The second most wanted types of functions – preferred by almost two third of the respondents - were guided tours and educational activities for children. These functions are closely related to the 'interest' type of experience. A city farm could deliver to these needs, especially if it is a varied mixed farming system. Activities from the 'rapture' and 'dedication' types that ask for more commitment, like self picking and doing farm activities (consumer co-operation) were less popular, but still had a potential clientele.

The second part of the questionnaire was dedicated to the desired features of urban oriented agriculture (see Table 2). Although all features that were mentioned in the survey are important large parts of the respondents, animal welfare and a beautiful farm surrounding were the most wanted. Contact with the farmer (social distance) is the least important feature. This means that physical features of agriculture are more important than social features.

Based on that information we divided the respondents according their gender, age, education, and household, to see whether these groups would give different answers. No differences in preferences were found between the groups except for two significant effects. Compared to regular consumers organic consumers were significantly more willing to do self picking

*Table 1. Opinion of respondents concerning farm activities.*

| Required farm activities | Will probably to almost definitely make use of that function (type of experience) | |
|---|---|---|
| Walking trails and cycling tracks through the rural area | 90 % | (Change) |
| Visiting a on farm shop | 86 % | (Change/Interest) |
| Educational farm activities for children | 66 % | (Interest) |
| Visiting a Pub and/or a restaurant at the farm | 65 % | (Amusement) |
| Guided tours at the farm | 59 % | (Interest) |
| Pick your own fruit and vegetables | 32 % | (Rapture) |
| Stay overnight at the farm | 15 % | (Change) |
| Voluntarily co-operating at the farm | 13 % | (Rapture/Dedication) |

*Table 2. Percentage of respondents that gives weight to certain farm features.*

| Required features of urban oriented farming | Finds feature important to very important for farming |
|---|---|
| Attention to animal welfare | 89 % |
| Farm in a beautiful surrounding | 70 % |
| Farm on cycle distance | 51 % |
| Small scale farm | 47 % |
| Learning new things | 46 % |
| Organic farming | 36 % |
| Contact with the farmer | 33 % |

(p= 0.002) and to visit farm shops (p= 0.0001). Organic consumers valued all mentioned features in Table 2 higher than regular consumers. Secondly, compared to men, women found it significantly more important that farms were small scale (p= 0.009), organic (p= 0.009), and cared about animal welfare (p= 0.0001).

## Conclusion and discussion

There is indeed a citizen's willingness to join several farm activities. The most preferred functions are walking trails, cycling tracks, farm shop and educational activities. These functions are connected with the low intensity types of experience 'amusement' and 'change'. Physical features (landscape, product assortment, distance *etc.*) of the farm are more important than social features (contact with farmer). Promising activities would be combinations of walking / cycling and visiting a farm shop or a farm restaurant. Less preferred but still mentioned by a large group, were the medium intensity types of experience such as 'interest'. This can be found in farm functions like guided tours, information and education. Here contact with the farmer is equally important as the physical appearance of the farm. The least preferred functions are self picking and co-operation on the farm. These functions are connected with very intense types of experience like 'rapture' and 'dedication'. Social contact with farmer and other consumers is very important, as well as the physical features.

We may see urban oriented agriculture as a concept for studying the potentials of rural development as a whole. The interaction between farmer and consumer is tight, which makes it possible to study preferred connections rather easily. Literature shows an increasing interest in the production – consumption linkages (for example Fine, 2004; Goodman and DuPuis, 2002). Research of Warnaar *et al.* (in preparation) shows that even within the field of urban oriented farming a wide range of strategies exist. In studying and understanding new forms of linking production and consumption, this diversity should have a central role, because our research shows it determines the degree of success in matching supply and demand. Studying and discussing this linkage could also have wider implications for rural development as a whole, because it sheds light on the consumer demand and the way agriculture could respond on this.

An important next question for research is whether and how activities related with urban oriented farming fit in daily urban life, for example which existing activities can be replaced by visiting a farm (visiting shops or nature/leisure areas) and the place the regular farm visits can have in daily life (after school or in the weekend).

## References

Banks, J. and T. Marsden, 2000. Integrating agri-environmental policy, farming systems and rural development. Tir Cymen in Wales. Soc Rur 40, 466-480.

Cohen, E., 1979. A phenomenology of tourist experiences. The journal of the British Soc Ass 13, 179-201.

Fine, B., 2004. Debating Production - Consumption Linkages in Food Studies. Soc Rur 44, 332-342.

Goodman, D. and E.M. DuPuis, 2002. Knowing food and growing food: beyond the production - consumption debate in the sociology of agriculture. Soc Rur 42, 5-22.

Hellerstein, D., C. Nickerson, J. Cooper, P. Feather, D. Gadsby, D. Mullarkey, A. Tegene and C. Barnard, 2002. Farmland protection and the role of public preferences for rural amenities. Agricultural Economic Report. USDA Economic Research Service, Washington DC. Agricultural Economic Report. 815, pp. 66.

Kinsella, J., S. Wilson, F. d. Jong and H. Renting, 2000. Pluriactivity as a livelihood strategy in Irish farm households and its role in rural development. Soc Rur 40, 481-498.

Lengkeek, J., J.W. Te Kloeze and R. Brouwer, 1997. The multiple realities of the rural environment: the signification of tourist images for the countryside. In: H. Haan and N. Long (editors). Images and realiteis of rural life: Wageningen perspectives on rural transformation. Van Gorcum, Assen, The Netherlands, pp. 324-343.

Marsden, T., 1999. Rural futures: the consumption countryside and its regulation. Soc Rur 39, 501-520.

Raad voor het Landelijk Gebied, 2000. For farmers, townspeople and countryfolk. Advice on the importance of socio-cultural developments for the rural area. (In Dutch) The Council for the Rural Area, Amersfoort, The Netherlands. 132 pp.

Raad voor het Landelijk Gebied, 2002. European integration and regional diversity. A challange for the Dutch ministry of agriculture. The Council for the Rural Area, Amersfoort, The Netherlands. 112 pp.

Van der Ploeg, J.D., A. Long and J. Banks, 2002. Living countrysides. Rural development processes in Europe: The state of the art. Elsevier Bedrijfsinformatie, Doetinchem, The Netherlands. 231 pp.

# The contribution of non-farm small-medium enterprises to rural development

*H. Theodoropoulou and S. Kyriakopoulou*
*Harokopio University of Athens, Greece*

## Abstract

Rural areas are affected significantly by social and economic changes. Specifically, the economy of rural areas is becoming more diversified as the service sector grows significantly at the expense of agriculture. The dynamics of rural changes observed are the subject of many studies for policy formulation and program implementation. Devising a rural development policy involves making political choices to ease community adjustments to structural economic changes. The development of these policies requires information on rural trends on economic and societal conditions. The objective of the present study was to examine the contribution of the non-farm small-medium enterprises in social and economic changes in the rural area of Paiania in Greece. The results of the present study indicate that the development of non-farm small-medium enterprises contributes to the local development of the rural area of Paiania in Greece.

**Keywords:** small-medium enterprises, rural areas, local development

## Introduction

At the beginning of the 20$^{th}$ century it was believed that small-medium enterprises contributed to the increase of employment and income. Schumpeter in 1911 said that small-medium enterprises introduce in the market new products and new methods of production as well as they are able to locate new sources in order to obtain primary supplies. In the mid of the 20$^{th}$ century in the developed countries it was believed that small-medium enterprises were connected with low economic development because it was observed that they had incompetent management, inefficient organisation and they used old technology. Therefore, at that time small-medium enterprises were not included in the governmental policy for industrial development (Mason and Harrison, 1990; Stanworth and Gray, 1991). According to Storey (1991) that period was characterised by big industry development located in specific places all together. After WWII intellectuals at that time had persuaded politicians that the future economic development relied on the establishment of big enterprises (Thurik and Wennekers, 2001). It was not until late 1970's, that economic analysts and policy makers started to recognise the importance of small-medium enterprises in the economic development (Moha and Beal, 2000). One important reason for that change was the publication of the Bolton Report in 1971, which provided evidence for the contribution of the small-medium enterprises in economic development (Curran, 1999).

At the beginning of eighties there was a big change in policy formulation for small-medium enterprises, such as less strict rules and regulations and supportive policy for founding small-medium enterprises. Industries prefer to be divided in to different branches and each branch to be located in different areas. Nowadays, small-medium enterprises belong to the economic core of a country (Stanworth and Gray, 1991). Resent studies show that small-medium enterprises are more flexible to adapt to the new markets demands in comparison with big enterprises, they employ more employees per unit of capital, they contribute to a more fair distribution of income and they affect in positive way to the local development (Moha, 1999; Audretsch and Thurik, 2000).

In European Union (E.U.) every year more than 1,000,000 new small-medium enterprises are founded. In 1996 there were 19 million small-medium enterprises in E.U. with more than 110 millions employees. Ninety three percent of these enterprises employed less than 10 employees while only 1% employed more than 50 employees (Storey, 1994, ENSR, 1997). It is considered that small-medium enterprises contribute greatly to the European economic development since they supply 60% to 80% of the new positions annually (Department of Trade and Industry, 1998).

In Greece, most of the small-medium enterprises employ less than 10 people and they are mostly family business. Furthermore, it is more affordable for most of these small-medium enterprises to be launched in rural areas, and there is a tendency for people to abandon farming which is a difficult occupation and to search for employment opportunities in non-farm small-medium enterprises. Therefore, rural areas are affected significantly by those social and economic changes and the economy of rural areas is becoming more diversified as non-farm small-medium enterprises grow significantly at the expense of agriculture. The objective of the present study was to examine the contribution of the non-farm small-medium enterprises in social and economic changes in the rural area of Paiania in Greece.

## Materials and methods

The statistical frame of the survey was based on the census of 392 small-medium enterprises registered with the Municipality of the target area of Paiania. A stratified random sampling of small-medium enterprises by size was used. Specifically, the total population of 392 registered small-medium enterprises were classified into three groups based on the size of the number of employees. The first group comprised of 313 small enterprises with less than 10 employees, the second group comprised of 57 medium enterprises with 10 to 49 employees, and the third group comprised of 22 enterprises with more than 50 up to 249 employees. The size of the sample used in the study was 235 small-medium enterprises, which covered 60% of the total population of registered small-medium enterprises in the target area. The sample included 50% (156) of the enterprises in the first group and all the enterprises in the second and third group (79).

Data on the characteristics of the employees and the owners of the enterprises along with their business management practices were collected through a questionnaire survey during the summer of 2004. Investigators on location completed the questionnaires by visiting the registered small-medium enterprises and directly interviewing the owners in order to avoid misunderstanding in the completion of the questionnaires.

The data collected were analysed by using descriptive statistics for calculating the means and standard deviations of continuous variables and the frequencies and percentages of discrete variables.

Measurements of the small-medium enterprises annual gross income were analysed against three groups of continuous variables. The first group consisted of variables indicating the demographic characteristics (Age of owners, Number of employees). The second group consisted of variables indicating the economic characteristics (Quantity of exported goods). The third group consisted of variables indicating the characteristics of the enterprises (Use of ISO, use of materials produced locally, legal entity, promotions, bonus and prim).

The measurements on the annual gross income were distributed closely to normal probability law and therefore were analysed by logistic regression model having annual gross income as the depended variable and all others as independent. The final model formed by backwards selection was examined for its capacity to explain the variations of the depended variable $Y_i$ (Neter, 1980; Agresti *et al.*, 1979) as follows:

$$\text{Int}Y_i = \beta_0 + \beta_1 X_{1i} + \beta_2 X_{2i} + \beta_3 X_{3i} + \beta_4 X_{4i} + \beta_5 X_{5i} + \beta_6 X_{6i} + \varepsilon_i$$

$Y_i$: Depended variable = Annual gross income
$\beta_0, \beta_1, \beta_2, \beta_3, \beta_4, \beta_5, \beta_6$: Regression model parameters
$X_{1i}$: number of employees
$X_{2i}$: exports
$X_{3i}$: promotions
$X_{4i}$: use of ISO
$X_{5i}$: use of materials produced locally
$X_{6i}$: legal entity
$\varepsilon_i$: Error term

The model parameters were estimated as $\beta_0$=11.32, $\beta_1$=0.01, $\beta_2$=0.88, $\beta_3$=1.71, $\beta_4$=0.93, $\beta_5$=0.009, $\beta_6$=1.72, where the overall capacity of the model to explain variations in annual gross income was measured to be 60.7% ($R^2$=0.607). Furthermore, the model was tested against the basic assumptions. The $\varepsilon_i$ was found to be identically and independently distributed with N $(0, \sigma^2)$, while no signs of heteroscedasticity and multicollinearity where detected.

## Results

The results from the descriptive statistics showed that non farm small-medium enterprises in the area of Paiania are mainly family operated (65%) and owned by locals (51%), so the income generated by these enterprises remains and it is consumed in the community (56%). Also, 33% of these enterprises had employed 1 up to 10 employees. Furthermore, 50% of these enterprises had hired local employees and so they contribute to the decrease of the local unemployment. Finally, some of these enterprises have export activities (18%), which help to increase local production.

The profile of the owners of the enterprises that emerged from this study was that of a man (70%) aged 30-44 years old (50%). The educational level of the men owners was mostly high school (26%) and university (22%) while for women owners was mainly high school (29%). The annual gross income for 43.6% of the enterprises was up to € 100,000, for 28.6% of the enterprises was more than € 1,000,000, and for the rest was between € 100,000 and € 1,000,000. Most of the owners answered that their clients are locals (53%). Also, most of the owners (53%) and the employees (69%) said that they desired the development of a better infrastructure in the area, that will ensure easier transportation of people and goods, which will contribute to the growth of the enterprises.

Also, the analysis of various generalised linear models showed that the significant factors explaining the variation of local development measured as the log average annual gross income of the non-farm small-medium enterprises in the area were the following: age of owners, number of employees, exports, promotions, use of ISO, use of inputs from local sources and the legal entity. These variables were able to explain 60.7% of the variation of local development. In addition, it was found that enterprises, which receive E.U. funding, and their owners are of a young age are more capable to export goods. Finally, enterprises whose owners were knowledgeable on new technologies had higher annual gross income and contributed more to local development.

## Conclusion

Rural areas are continuously in search of new opportunities for economic development. This study examines the role of non-farm small enterprises as a part of livelihood rural system, which revitalizes rural areas and prevents their desertion. The results of the present study indicate that the development of non-farm small-medium enterprises contributes to the local development of the rural area of Paiania in Greece. In addition, the study outlines the involvement of the local population in enterprises activities in order to become better financially and improve their well being. The existence of non-farm income sources and assets is of a great importance for a successful local development.

Non-farm small enterprises, as a part of livelihood system within the rural sector, can be an appropriate 'tool' for integrated local development in socioeconomic rural areas. While rural areas are being diversified, there must be an attempt to avoid the autonomous development of non-farm small enterprises and to combine non-farm enterprises with the primary sector, in order to give the locals the ability to become financially stronger and to become socially reinstated, without the risk of degenerating the rural environment. As far as the relationship between sustainable livelihoods of rural areas and non-farm small enterprises is concerned, non-farm small enterprises are a desirable policy objective, which can be one of the profitable alternatives to improve livelihood security and to raise living standards.

As far as the role of non-farm small enterprises in the local development this study showed that small enterprises contribute to the local development of the area of Paiania, because these enterprises are mainly family operated and owned by locals, so the income generated by these enterprises remains and it is consumed in the community. Also, some of these enterprises have export activities, which help to increase local production and furthermore, since they hire local employees, contribute to the decrease of unemployment. Furthermore, small enterprises and their local owners are more willing to have local clients and there is less risk for small enterprises to be relocated in comparison with large enterprises that employ many employees. Also, the owners and the employees in these enterprises wanted the development of a better infrastructure and the use of new technology in the area that will ensure easier transportation of people and goods, which will contribute to local development.

However, in order to ease community adjustments to structural economic changes, it is of a great importance to devise a rural development policy that involves making political choices. The development of these policies requires information on rural trends on economic and societal conditions in order to choose the right tax policy for small-medium enterprises, as well as to easy the administrative hassles for the establishment of new small-medium enterprises. Finally, information about new technology and possibilities for the use of national and European funds should be easily accessible and provided by the local authorities. Consequently, small-medium enterprises with the appropriate governmental policy can contribute to the local development.

## References

Agresti, A. and B.F. Agresti, 1979. Statistical Methods for the Social Sciences. Dellen Publishing Company, San Francisco, USA.

Audretsch, D. and A.R. Thurik, 2000. Capitalism and democracy in the 21st century: from the managed to the entrepreneurial economy. J. Evolutionary Econ. 10, 17-34.

Curran, J., 1999. What is small business policy in the UK for? Evaluation and assessing small business policy. Int. Small Bus. J. 18, 36-50.

Department of Trade and Industry, 1998. Our competitive future: Building the knowledge driven economy. Stationery Office, Cm 4176, London.

ENSR, 1997. The European Observatory for SMEs, EIM. Zoetermeer, Netherlands.

Mason, C. and R. Harrison, 1990. Small firms: phoenix from the ashes. Western Europe: Challenge and Change. Belhaven Press, London.

Moha, A., 1999. The accessibility of the government sponsored-support programmes for smalland medium-sized enterprises in Penang. Int. J. Urban Regional 16: 83.

Moha, A. and T. Beal, 2000. The strategic contribution of small and medium enterprises to the economies of Japan and Malaysia: Some comparative lessons for Malaysian SMEs. University Sains Malaysia, Victoria Universiy of Wellington, New Zealand.

Neter, J., W. Wasserman, N.H. Kutner, 1980. Applied Linear Statistical Models, 3rd ed., Irwin Press.

Schumpeter, J., 1911. The theory of economic development. An inquiry into profits, capital, credit, interest, and the business cycle. Translated by R. Opie, Oxford: Oxford University Press (1963 (1934)).

Stanworth, J., and C. Gray, 1991. Bolton 20 years on: The small firm in the 1990s. Paul Chapman Publishing, London.

Storey, D., 1991. The birth of new firms-Does unemployment matter? A review of the evidence. Small Bus. Econ. 3, 167-178.

Storey, D., 1994. Understanding small firms. Routledge, London.

Thurik, R. and S. Wennekers, 2001. A note on entrepreneurship, small business and economic growth. Erim Report Series Research in Mangement and Center for Advanced Small Business Economics, Faculty of Economics, Erasmus University, Rotterdam.

# Agrobiodiversity as driving force of sustainable rural development

*Menko Wiersema[1] and Frans van Alebeek[2]*
*[1]Province of South-Holland, PO Box 90602, 2509 LP Den Haag, The Nethelands; [2]Wageningen University and Research Centre, Applied Plant Research, Wageningen, The Netherlands*

## Abstract

The initiative "Biodiversity in the Hoeksche Waard: for and by citizens" is a success. The stakeholders introduced a natural pest suppression method based on biodiversity. Through the initiation of field margins, 'beetle banks' and walking paths, the Hoeksche Waard changed into a biodiverse, environmentally sound and attractive area. How has it been realised? Can we export it to other regions?

**Keywords:** functional biodiversity, stakeholders, natural pest suppression, Hoeksche Waard

## Introduction

The Hoeksche Waard (population 83.000) is an island south of Rotterdam in the Province of South Holland in the Netherlands (Zuid-Holland, 2006). The Hoeksche Waard consists of polders that have been reclaimed gradually after the Saint Elisabeth flood of 1421. As a result, dikes dominate the landscape. On these dikes, farms were built, with trees in between. The dikes lined with houses and trees are an important feature of the landscape. The open landscape in between dikes and the relative silence contrast strongly with the densely populated cities of Rotterdam and Dordrecht, located nearby. Other important landscape elements are the networks of small rivers (creeks) on the island, which originally were connected with the sea. The largest part of the Hoeksche Waard is used for agriculture.

In 2003, the Ministry of Housing, Spatial Planning and the Environment (VROM) and the Province of South Holland launched a programme to encourage public participation in environmental policy. The residents of the Hoeksche Waard were invited to formulate policies and projects to promote biodiversity in the area. The project combined two challenges: to achieve public participation and to translate a remote abstraction like biodiversity into concrete ideas relevant to residents' own lives and interests. In 2004, citizens and other stakeholders in the Hoeksche Waard enthusiastically started to work on biodiversity development. Many plans for the operationalisation of biodiversity were developed, making the initiative «*Biodiversity in the Hoeksche Waard: for and by citizens*" an example for other regions.

The project concluded with a conference on 30 June 2005 and was a great success. Creativeness and intensive co-operation between stakeholder (farmers, citizens, policy makers, and scientists) led to proposals how nature could be used to benefit both agricultural and rural development. More than thirty ideas were put forward by a host of working groups, but also strong criticism was levelled at current policies of central government and the provincial authorities and various inconsistencies were identified. The Province of South Holland and a regional consortium used the results of the project to produce a regional Biodiversity Action Plan (BAP) and to launch a number of specific projects. VROM supported the plan with additional project funding (Van Geel, 2006).

In addition, the local agricultural community produced a plan for 'Sustainable agriculture in the Hoeksche Waard' by 2015, focusing in particular on the functional aspects of agrobiodiversity.

To this end, the Alterra research institute has produced for VROM a study of the opportunities presented by the existing network of streams and dikes in the Hoeksche Waard. The main idea is:

- To create buffer strips between the arable crop and all watercourses where neither agrochemicals nor fertilisers are applied. Buffer strips act as a physical buffer to spray drift and run-off from fields into surface water. This may reduce the movement of nutrients and agrochemicals into surface waters. Thus, buffer strips improve water quality, while soil erosion of riparian field margins is also reduced.
- To sow the buffer strips with flowering plant species and realise so-called beetle banks through crop fields, to attract natural predators of agricultural pests. Perennial plants offer a hiding place and all-year habitat for predators of agricultural pests, thus contributing to inherent (natural) pest suppression.
- To realise a network of walking paths on 10% of the field margins in order to increase the area's attractiveness and accessibility for tourists. Visually attractive flowering and other species of the field margins will further enhance the touristic and recreational value of the Hoeksche Waard area.

A pilot was set up jointly with VROM, the Ministry of Agriculture, Nature Management, and Food Quality (LNV) and the Dutch Federation of Agricultural and Horticultural Organisations (LTO), to prove how effectively crops are protected by natural disease and pest prevention methods (Photo 1). This project was named Functional Agrobiodiversity (FAB, 2006; see below). Some 400 ha have been subject to sustainable planning and within those 400 ha 'beetle banks'/field margins play an important role.

## Biodiversity and natural pest control

Applied Plant Research is involved in research on functional biodiversity for the natural control of insect pests in arable farming (*e.g.* Van Alebeek *et al.*, 2002).

*Photo 1. Buffer strips with flowering plant species in the Hoeksche Waard. Some 200 kilometres of field margins have been realised in a period of one year.*

The basic idea is that field margins offer an overwintering habitat and supplemental food (pollen, nectar and additional prey) for natural occurring predators such as carabid beetles, spiders, hover flies and parasitoids. In spring time, overwintering predators enter the crop and prey upon the first pest populations (*e.g.* aphid pests) that colonize the crops (Thomas *et al.*, 1992; Marshall & Moonen, 2002; Winkler *et al.*, 2005). Pest population growth is slowed down, offering the opportunity for hover flies and parasitic wasps to further attack and reduce pest colonies (Figure 1).

Both in summer wheat and potatoes, this strategy has proven to keep aphids below the damage threshold in most years, without additional control measures (Van Alebeek *et al.*, 2002, and in prep.). However, sometimes field margins on heavy clay soils may increase slug damage in neighbouring crops.

Furthermore, research showed that specific flower mixtures can increase the fecundity and longevity of parasitic wasps and strongly improve their control of aphid and caterpillar pests. Thus, annual flower strips along and through crop fields can improve biological control (Winkler *et al.*, 2005). These principles were considered ready for on-farm testing in a pilot.

## The FAB project

In 2004, the Dutch Federation of Agricultural and Horticultural Organisations (LTO) initiated the project Functional Agrobiodiversity (FAB) (www.lto.nl/fab/). The Hoeksche Waard region was identified as an area where the agro-ecological infrastructure and the local and regional network of stakeholders were promising for the implementation of functional agrobiodiversity for sustainable agriculture. Aim of the project was to develop and improve farming systems based on FAB strategies for pest control.

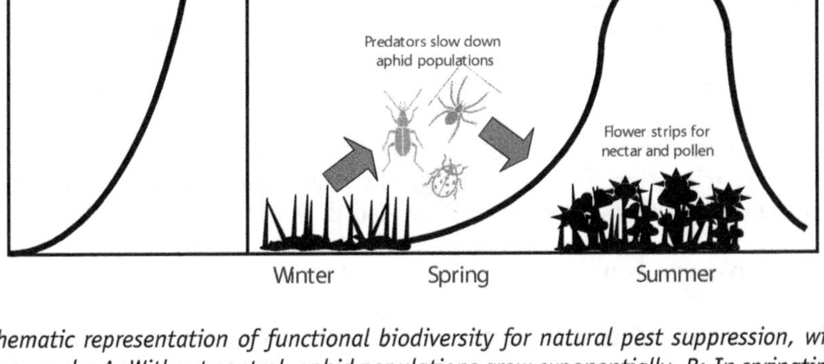


*Figure 1. Schematic representation of functional biodiversity for natural pest suppression, with aphids in cereals as an example. A: Without control, aphid populations grow exponentially. B: In springtime, predators from field margins slow down aphid population growth and in summertime, parasitoids supported by flower strips can control and suppress aphid populations below the damage threshold.*

The polder dikes, creeks and woodlots in de Hoeksche Waard were identified a well-spaced network of potential sources of hibernating natural enemies. However, current vegetation management practices are not optimal for FAB. In winter time, the vegetation of most dikes, road verges and ditches was too short for a good survival of ground dwelling predators. In summer time, most vegetations had few flowering species, so that little nectar and pollen are available for parasitic wasps, hover flies, and other natural enemies.

Thus, one of the priorities of the project is to convince the regional Water Board (Hollandse Delta) and other stakeholders to change their vegetation management into a regime of mowing and removal of cuttings. In the longer term, this will result in more flowers in summer, and more vegetation cover in winter time.

Five arable farms, adjacent to each other and with a combined surface of 300 hectares, participate in the project. Dikes, canals and ditches cross the area. Farmers, advisors and scientists jointly made a plan for FAB strategies in wheat, potato and Brussels sprouts. Perennial field margins (3 m wide) were sown alongside canals and ditches, as hibernating sites for natural enemies. Also, flower strips were sown alongside and through crop fields. Flower mixtures were designed to stimulate parasitic wasps and hover flies, without stimulating potential insect pests. Farmers were compensated financially for the crop surface lost to field margins and flower strips. Farmers were also advised on the choice of crop protection products, to minimize negative side effects on natural enemies. An elaborate monitoring programme is carried out to measure flower availability, natural enemy densities, key pest densities, crop damage and quality, the number of pesticide treatments, and costs and savings of the different measures.

Using pitfall traps for soil dwelling predators, we showed huge differences in densities of carabid beetles between farms. Densities in flower strips varied between 180 carabid beetles per square meter per 14 days trapping period on one farm, versus 20 beetles per square meter on another farm. Highest densities corresponded with an integrated farm (using selective pesticides with little side effects on natural enemies). Lowest densities were found on a farm that used broad-spectrum pesticides in an intensive spraying programme. These differences illustrate the potential for on-farm management of biodiversity. The challenge of the FAB project is to document how these predator densities contribute to the suppression of insect pest populations in field crops.

In 2005 in several fields, natural enemies were more abundant nearby flower strips than further into the crop. In some fields, aphid densities were lower close to the field margins than further away into the crop. However, in general, in the first project year the effects of field margins and flower strips were not very prominent. It was concluded that a more concentrated network of field margins and flower strips is necessary to realize significant pest suppression effects.

Therefore, in 2006, the surface of field margins and flower strips will be increased and concentrated on specific crop fields. Effects will be compared between fields without margins and fields with an intensive network of margins and flower strips.

## Towards a region-wide biodiversity experiment

Mechanised, large-scale monocultures are economically and ecologically vulnerable. For many products, farmers' prices on the world market are low and EU subsidies are declining. On the other hand (private and public) costs rise due to an increased vulnerability of crops to insect pests and diseases, the degradation of soil and water quality, increasingly stringent environmental and health requirements and a massive biodiversity and landscape decline.

The Hoeksche Waard has chosen to enhance the regional economic and ecological diversity, by restoring agricultural biodiversity at field and landscape level and by increasing the area's attractiveness and accessibility for tourists. Future plans include the realisation of an extra 2000 kilometres of field margins along all watercourses in the Hoeksche Waard. This will be the first large-scale field experiment of natural pest suppression in the world. A network of walking paths will be realised on the field margins. Colourful flowers and other species in the field margins will enhance the touristic and recreational value of the Hoeksche Waard.

Several initiatives are linked to the current project activities in the Hoeksche Waard. A scientific study will define optimal management strategies for landscape elements and vegetations as supporting structures for natural enemies of insect pests. A research programme on soil biodiversity and nutrient management will link up with the FAB project in the Hoeksche Waard region.

Project results are communicated to stakeholders in the region. Stakeholders will be challenged to participate in projects following the initiative *"Biodiversity in the Hoeksche Waard: for and by citizens"*. Regional workshops are held to formulate new initiatives that fit into the Biodiversity action plan. Local organisations are given responsibilities and budgets to realise projects in the region.

Already, the project in het Hoeksche Waard has led to similar initiatives in two other regions in the Netherlands. It is expected that more will follow in due time.

## References

FAB, 2006. http://www.lto.nl/ /fab/ (in Dutch).

Geertsema, W., E. Steingröver, W. van Wingerden, F. van Alebeek and J. Rovers, 2004. Alterra rapport 1042: Groenblauwe dooradering in de Hoeksche waard.

Marshall, E.J.P. and A.C. Moonen, 2002. Field margins in northern Europe: their functions and interactions with agriculture. Agric. Ecosyst. & Envir. 89, 5-21.

Novio Consult Van Spaendonck BV, 2005. Biodiversiteit Hoeksche Waard voor en door burgers (in Dutch).

Thomas, M.B., S.D. Wratten and N.W. Sotherton, 1992. Creation of 'island' habitats in farmland to manipulate populations of beneficial arthropods: predator densities and species composition. J. Appl. Ecol. 29, 524-531.

Van Alebeek, F.A.N, J.H. Kamstra, B. Venhorst and A.J. Visser, 2003. Manipulating biodiversity in arable farming for better pest suppression: which species and what scale? IOBC / WPRS Bulletin 26: 185–190.

Van Geel, P., 2006. Speech by State Secretary at the Conference on Biodiversity in Agriculture. Retrieved January 18, 2006 from: http://international.vrom.nl/pagina.html?id=9357

Winkler, K., F.L. Wäckers, L. Buitriago and J.C. van Lenteren, 2005. Herbivores and their parasitoids show differences in abundance on eight different nectar producing plants. Proc. Neth. Entomol. Soc. Meet. 16, 125–130.

Zuid-Holland, 2006. Retrieved January 18, 2006, from http://www.answers.com/topic/south-holland.

# Reconnecting conservation & livelihood objectives through non-timber forest product certification

*David S. Wilsey*
*University of Florida, School of Natural Resources and Environment, 7943 Southwest 47th Court, Gainesville, Florida 32608, USA*

## Abstract

Can NTFP certification provide a mechanism to integrate forest conservation and livelihood development in a particular location? My research will consider its potential role in the future viability of palm (*Chamaedorea spp*) harvest in the community forest concessions of the Maya Biosphere Reserve, where NTFP extraction generates income and, ostensibly, provides communities with an incentive to protect the forest. The current transition from extractivism to cultivated production harbingers a loss of forest product income for forest communities and a disruption of the livelihood systems that are the foundation for conservation. The issue is not whether *Chamaedorea* will remain an economically important commodity globally, as production practices shift, but whether extraction will remain a viable income-generating strategy for community forest concessions. The palm production system will be analyzed at three scales: the global commodity chain, the forest management unit, and the household. A production-to-consumption system analysis will be used to identify opportunities and constraints of the extractive production system. Ethnographic linear programming will be used to assess community and household level adaptability to a dynamic production environment. Certification as a mechanism for the integration of conservation and development objectives will depend on the viability of extractivism at all three scales, and on the identification of policy and market levers that can retain the system's comparative advantage. This multi-scale analysis represents a first step toward identifying the requisite standards of such a certification.

**Keywords:** extractivism, supply chains, green marketing

## Introduction

Sustainable non-timber forest product (NTFP) extraction was once envisioned as a catalyst for integrating sustainable forest management and economic development objectives (Nepstad and Schwartzman, 1992). Benefits generated from NTFP activities represent an important livelihood contribution and, ostensibly, provide an incentive for forest conservation. In recent years, however, it has been necessary to question the extent to which conservation and development objectives are synergistic (Wunder, 2001). Nevertheless, myriad cases exist where NTFP provide income and other benefits to forest households, and while their potential contribution to integrated conservation and development has been contested (Angelson and Wunder, 2003), it has not yet been rejected. New approaches such as certification continue to be explored. Can NTFP certification provide a mechanism to integrate forest conservation and livelihood development in a particular location? This question is being considered by palm (*Chamaedorea spp*) harvesters and stakeholders in Guatemala's Maya Biosphere Reserve (MBR).

A challenge of assessing the benefits and costs of NTFP certification, or of NTFP activities for that matter, relates to the diversity of products within the classification, as well as to their diverse and dynamic functions. For this reason my research deals with the certification of a single product and place – *Chamaedorea* palm in the MBR. Nevertheless, to ascertain whether or not NTFP certification can serve as a mechanism to promote integrated conservation and development in a particular location, it is necessary to analyze the NTFP production system at multiple scales. The extractive production system of the MBR concessions must be considered within the broader context of the global commodity chain. The long-term viability of extractivism requires that the system retain its comparative advantage relative to alternate production systems. It is also important to understand the opportunities and constraints imposed upon palm production by the community concession management unit. As comparative advantages shift over time and place, the concession arrangement will affect the dynamics of the production system. Finally, certified palm extraction must be viable at the household level. Benefits and costs of certification will impact the feasibility of palm harvest relative to other livelihood activities. Ultimately, certification will only be able to integrate conservation and development objectives in the community concessions of the MBR if it is able to ensure the continued viability of palm production at all three scales.

## Overview

The Maya Biosphere Reserve (MBR), which includes Tikal National Park, was established in 1990 to protect two million hectares of subtropical moist forest and savannah in Petén, Guatemala. A means to this end has been the integration of conservation and local economic activity; a fact reflected in the zoning of the Reserve. The MBR is comprised of three distinct zones: a core zone (36%) consisting of national parks and biotopes, a multiple-use zone (40%) that is managed as an "extractive reserve," and a fifteen kilometre-wide buffer zone (24%). Contrary to the name, the regions comprising the core zone are mainly distributed throughout the reserve's periphery, with the result that the multiple-use zones function as the *de facto* core of the reserve in terms of maintaining large-scale ecological processes. A consequence of this arrangement is that the long-term success of the MBR depends heavily on the conservation of the multiple-use zone and its constituent forest concessions.

In 1994 the Guatemalan National Council for Protected Areas (CONAP) began transferring management of the multiple-use zone of the MBR to community-managed forest concessions as a legislative and management response to the increasing rates of deforestation in the Petén (Finger-Stich, 2003). Currently, there are 12 community- and 2 industrial-concessions, ranging from approximately 25,000 ha to 83,000 ha, and covering nearly 800,000 ha of the multiple-use zone. Each concession is certified for sustainable timber management by SmartWood, an accredited certification agent of the Forest Stewardship Council (FSC). Timber, however, is but one component of a broader livelihood strategy that includes swidden agriculture and the extraction of commercial non-timber products such as *Chamaedorea* leaves (locally known as xate (sha-tay)), chicle latex (*Manilkara zapota*), and allspice (*Pimenta officinalis*) (Litow *et al.*, 2001).

Since emerging as an international commodity in the 1960s, commercial *Chamaedorea* production has relied on the harvest of leaves from wild plants. As habitat and palm abundance subsequently have been reduced by deforestation and over-harvesting, reduced availability and leaf quality have negatively impacted the comparative advantage of the wild-harvest production system. As shortages continue and the quality of leaves from wild plants decreases, stakeholders within and around the MBR are becoming concerned that the production system based on the harvest of wild *Chamaedorea* populations is approaching a threshold, described by Wiersum (1997) as the critical passage from wild-harvest to early stage cultivation outside of the forest. The observable trend toward cultivated production in Mexico and southern Petén may harbinger the loss of wild *Chamaedorea* harvest as a livelihood option and economic foundation for forest conservation in the community forest concessions. It is expected that cultivation under structurally-modified forest or artificial-shade will soon be the norm. The *Chamaedorea* production system of the forest concessions will likely have to change, or risk marginalization by intensive production of cultivated palms elsewhere.

Economist Alfredo Homma (1996) postulated that forest extraction livelihood activities emerge, boom, and gradually disappear, in both time and space, as their comparative economic advantage wanes and their livelihood contribution diminishes. Changes in comparative advantage relate to a broader economic, ecological, and socio-political context, and may be a function of changing consumer preferences for wild products as much as they are a function of the local production environment. However, it is response to change of the local production system, rather than the change itself, that will determine whether extractivism will follow the trajectory outlined by Homma. A recent comparative analysis of commercial NTFP production suggests that production systems and the associated livelihood activities for individual NTFP are determined, and perhaps bound, by local conditions (Belcher *et al.*, 2005). The unfortunate implication for integrated conservation and development initiatives is that shifts in comparative advantage resulting in diminished NTFP livelihood contributions will ultimately undermine the long-term role of NTFP as a foundation for integrated conservation and development objectives. It is within this context that the extraction of *Chamaedorea* palms from the community forest concessions of the MBR must be considered.

Almeida (1996) provided an alternate outlook for extractive livelihoods, asserting that multiple extraction models may exist. "Good extractivism" represents a paradigm that is better aligned with the integration of conservation and development objectives, one that stands in contrast to the "old extractivism" paradigm characterized by Homma. When commercialization provides perverse incentives, intervention may be necessary in order to promote "good extractivism," particularly in areas characterized by integrated conservation and development objectives. Certification has been suggested as one such intervention. Certification standards for NTFP are only recently beginning to emerge and, consequently, their efficacy in achieving conservation and/or livelihood benefits is poorly understood. Certification's role in integrating economic and conservation incentives is even less clear. The expectation that certification will be able to integrate multiple objectives in the context of a particular ecosystem may ultimately prove to be unrealistic. Nevertheless, only through defining the requisite – albeit narrow – conditions for success will we be able to truly understand the exact nature of failures to integrate conservation and development objectives.

## Methods

*Chamaedorea* palm production is inter- and multi-national. To analyze this global system I will draw upon the French *Filière* approach to commodity chains and the Global Commodity Chain analysis (Gereffi and Korzeniewicz, 1994). The strength of the former lies in its traditional application to agricultural commodity chain analysis, while the strength of the latter lies is in its coherence and focus on the full length of the commodity chain (Raikes *et al.*, 2000). The result, following Belcher (1996), will be a complete Production-to-Consumption System (PCS) analysis. A PCS analysis will highlight the diversity of palm production systems within the global production framework, and should facilitate a deeper understanding of the comparative advantages that are driving the process of change. A PCS analysis should also identify those factors that constrain production in comparatively disadvantaged systems, factors that will need to be addressed if certification is to restore some degree of comparative advantage.

Communities and their constituent households are characterized by an array of political, social, economic, and environmental factors that affect the form, function, and dynamics of extractive production systems (Belcher *et al.*, 2005). PCS analysis provides insight into the opportunities and constraints for a particular production system, but falls short of being able to assess community and household responses to interventions such as certification. Community level analysis is important for two reasons. First, the forest concession arrangements governing the stewardship and exploitation of natural resources are contracts with community-level associations. Second, the benefits and costs associated with certification will likely extend beyond the household domain. Certification may require community organization, investment in social benefits, and/or local value-added processing facilities. Household level analysis is essential for the simple reason that decisions to participate in palm harvesting, or any other activity, ultimately depend on household resources and constraints and on the array of available livelihood alternatives.

Ethnographic linear programming (ELP) is a methodology for assessing farm-household adoption of new technologies (Hildebrand *et al.*, 2003). ELP combines rapid appraisal and participatory techniques with linear programming to produce a livelihood system model. Household models can be used to evaluate adoption of new livelihood alternatives or, in this case, options for palm certification. Household models can be aggregated into a meta-model to reflect community household composition and community level constraints. Although ELP models are powerful decision-analysis tools, they cannot be used to assess comparative advantage within a given production system. Household and community level feasibility do not imply the viability of a local production model in a global production system, and vice versa. It is for this reason that the PCS analysis and ELP methodology will be undertaken together.

## Expected outcomes

A global production-to-consumption system analysis for *Chamaedorea* palms will be one product of this research, and will include an outlook for the future of extraction-based palm production. A PCS analysis will provide insight into the particular production system characteristics necessary for extractivism to remain competitive with emerging modes of production, and into the policy and market levers that can leverage such an outcome. These results are possible due to the attention given in PCS analysis to the full length and breadth of the commodity chain. Similarly, I expect to be able to describe the opportunities and constraints to palm production dynamics that are associated with the community concession arrangement. These arrangements influence harvest regimes, resource access, and distribution of benefits and costs, and consequently will affect the design and efficacy of certification standards. The production processes to which certification brings value (e.g. sustainability, social justice, etc.) will be most evident at this scale. Finally, this research will elucidate the socio-economic roles that palm harvesting plays in household livelihood strategies. The effects of certification on yield (through reduced harvest and/or increased prices) will be directly felt by households. As with new agricultural technologies, improved performance is meaningless without farmer adoption. These outcomes will be of value to the national and international agencies currently looking to palm production to help meet integrated conservation and development objectives. The results will also contribute to an ongoing discussion of the feasibility of NTFP certification and its appropriate role in integrated conservation and development.

## References

Almeida, M.W.B., 1996. Household extractive economies. In: Ruiz Pérez, M. and J.E.M. Arnold (editors), Current Issues in Non-timber Forest Products Research. CIFOR, Indonesia, pp. 119-141.

Angelsen, A. and S. Wunder, 2003. Exploring the forest poverty link: Key concepts, issues and research implications. Occasional Paper No.40, CIFOR, Indonesia.

Belcher, B.M., 1998. A production-to-consumption systems approach: Lessons from the Bamboo and Rattan sectors in Asia. In: E.Wollenberg and A.Ingles (editors), Incomes from the Forest. CIFOR, Indonesia, pp 57-84.

Belcher, B., M. Ruiz-Perez and R. Achdiawan, 2005. Global patterns and trends in the use and management of commercial NTFPs: Implications for livelihoods and conservation. World Devel. 33, 435-1452.

CEC, 2003. In Search of a Sustainable Palm Market in North America. North American Commission for Environmental Cooperation. Montreal, Quebec, Canada,

Finger-Stich, A., 2003. Community Concessions and Certification in the Maya Biosphere Reserve. In: E.Meidinger, C.Elliot, and G.Oesten (editors), Social and Political Dimensions of Forest Certification. Forstbuch, Germany.

Gereffi, G. and M. Korzeniewicz, 1994. Commodity Chains and Global Capitalism. Greenwood Press, 331 pp.

Hildebrand, P.E., N.E. Breuer, V.E. Cabrera and A.J. Sullivan, 2003. Modeling diverse livelihood strategies in rural livelihood systems using ethnographic linear programming. University of Florida, Food and Resource Economics Department, Staff Paper Series, SP 03-05. Gainesville, Florida.

Homma, A.K.O., 1996. Modernization and technological dualism in the extractive economy in Amazonia. In: M.Ruiz Pérez and J.E.M.Arnold (editors), Current Issues in Non-Timber Forest Product Research. CIFOR, Indonesia, pp. 59-81.

Litow, P., M. Baker and P.E. Hildebrand, 2001. Swidden agriculture in a forest society: livelihood strategies in the Maya Biosphere Reserve community of Uaxactún, Peten, Guatemala. J. Int. Agric. Exten. Educ., 49-54.

Nepstad, D.C. and S. Schwartzman, 1992. Non-timber products from tropical forests: Evaluation of a conservation and development strategy. Advances in Economic Botany 9. The New York Botanical Garden, New York.

Raikes, P., M.F. Jensen and S. Ponte, 2000. Global commodity chain analysis and the French *filiére* approach: comparison and critique. Econ Society 29(3), 390-417.

Wiersum, K.F., 1997. Indigenous exploitation and management of tropical forest resources: an evolutionary continuum in forest-people interactions. Agric. Ecosyt. Environ. 63:1-16.

Wunder, S., 2001. Poverty alleviation and tropical forests: What scope for synergies? World Devel. 29(11), 1817–1833.

# Socio-cultural issues of dairy production systems in the Netherlands: Results of farm visits with six citizen panels

B.K. Boogaard[1], S.J. Oosting[2] and B.B. Bock[3]
[1]Wageningen Institute of Animal Sciences, Wageningen University, The Netherlands; [2]Animal Production Systems Group, Wageningen University, The Netherlands; [3]Rural Sociology Group, Wageningen University, The Netherlands

## Conceptual framework of socio-cultural sustainability

Agricultural development in the Netherlands has reached a historical crossroad; food production is no longer the primary objective of land-use and present production systems are under discussion, consequently demand for sustainable agriculture has evolved as a general expression of collective concern. Sustainability is often assessed on basis of economic, ecological and socio-cultural aspects. Concerning livestock production systems, much emphasis has been put on economic and ecological sustainability, whereas socio-cultural sustainability is only described in a relatively general and superficial way. Therefore, the present study focuses on socio-cultural aspects of sustainability of dairy production systems in the Netherlands. Socio-cultural sustainability deals with the human affinity with their social frame of reference, with their culture (which includes agriculture) and history, and it affects the way in which their social frame of reference, nature, land and landscape, food and agricultural production systems are perceived and valued. Hence, socio-cultural sustainability is about values, subjects and processes that really matter to people. The present study had the objective to define relevant socio-cultural issues in dairy production systems essential for designing sustainable dairy production systems in the future.

## Research method: Valuable farm aspects recorded by Dutch citizens

We collected data in three areas of the Netherlands (north, south and west). In each area a citizen panel (consisting of 8 respondents) visited two dairy farms on one day. Two weeks later, the experiment was repeated in each area with a second panel that visited both farms in the opposite order as the first panel. On farm, respondents received a digital camera and recorded ten aspects they consider valuable to preserve for the future. One week after the farm visits, respondents received their individual pictures by mail and selected five pictures per farm that represented the most valuable aspects. By means of an electronic questionnaire, each respondent gave a written explanation with the five most valuable pictures, *i.e.* aspects per farm.

## Results: Socio-cultural issues of dairy production systems

Through qualitative analyses of the written explanation and the selected pictures, we derived socio-cultural issues for dairy production systems. The issues cover a range of aspects, varying from practical aspects, such as technological developments and farmers' income to more ethical aspects such as the use of cows for production (human-animal relation). Also historical and cultural values, such as identity of the Dutch and respect for farmers and animals (animal welfare) appeared to be issues for Dutch society. Values of Dutch landscape are shortly mentioned on this poster, as these are discussed more profoundly in *'Socio-cultural values of Dutch dairy systems; case study in the Krimpenerwaard and the Alblasserwaard on farm and landscape level"* by Daniël de Jong *et al.* (this publication, p. 180).

# Socio-cultural values of Dutch dairy systems: Case study in the Krimpenerwaard and the Alblasserwaard on farm and landscape level

*Daniël de Jong, S.J. Oosting and B.K. Boogaard*
*Animal Production Systems Group, Wageningen Institute of Animal Sciences, Wageningen UR, The Netherlands*

## Dairy system mapping; socio-cultural sensing

Society and agriculture lost contact with each other. A decreasing agriculture and modern lifestyles of citizens are causes for the existence of this gap. In this study values, of citizens, about the dairy production system in the Netherlands are mapped. These values are partly representing the content of the socio-cultural domain within the concept sustainability. In animal production there is much attention for the domains economy and ecology of this concept. However, for a strong future animal production system it is a necessity to balance these three aspects of sustainability.

A panel of eight citizens has visited two dairy farms. The panel members sensed valuable aspects on the farm and in the landscape around the farm. After the farm visits each panel member has given a personal semi-structured in-dept interview.

## Kaleidoscope of values

Agricultural nature conservation is valued positively by six panel members. The dairy farm landscape has a clear function in growing fodder plants for the dairy farms. The presence of cows in the dairy landscape is strongly valued by seven panel members, because this is part of the Dutch rural picture. The family farm is, for seven panel members, a valuable aspect of the dairy farm sector. Cohesion between private and business area is observed. Dairy farms with an innovative character, where production efficiency on the farm is improved, are valuable. The traditional elements of a dairy farm are valued positively. The human-animal contact on a dairy farm is valued as a positive part of dairy production system by seven panel members. It gives the profession 'farmer' an extra dimension. The presence of animal welfare measures on a dairy farm are valued positively by all the panel members.

## Conclusion

The panel members do not have any difficulty to identify valuable aspects of two dairy production systems. The aspects cover a wide range of issues. Therefore different types of dairy farming systems can improve their social sustainability from their own system framework. The identified aspects can be used within the process of defining socio-cultural indicators. Therefore further research is needed. When these indicators will be used by dairy farmers and policy makers, the societal license to produce of the dairy production system can be strengthened. This process might be one of the pillars for the connection that is necessary to overcome the gap between the dairy production system and the Dutch citizens.

# Carbon stocks in smallholder agroecosystems in Malawi: A farming systems approach

*Peter Hildebrand[1] and Paul Thangata[2]*
*[1]Food and Resource Economics, University of Florida, McCarty Hall, Gainesville, USA; [2]Social Development Consultant, Norwich, United Kingdom*

## Introduction

The adoption of agroforestry technologies in the tropics stores high biomass for soil fertility replenishment and offers potential for carbon (C) storage. Household survey and biophysical data from central Malawi are modelled to test whether farm incentives would lead to more improved fallow adoption; and therefore an increase in carbon sequestered on farms. Before maximizing C sequestration, models were run to maximize end year cash, both while satisfying household food sufficiency and other household needs.

## Methodology

In a 10-year ethnographic linear programming model, two carbon credit incentives of US$6.00 and US$12.00 are tested. With data from 40 households representing 40 models, three C sequestration scenarios were simulated:

1. Baseline scenario, no agroforestry, only annual crop C accounted for.
2. Agroforestry without a seed selling option, where seed from the improved fallows is not sold.
3. Agroforestry with a seed selling option.

## Results and conclusions

The baseline scenario (1) would sequester on average 4.0 Mg C/ha (Figure 1). The overall annual mean amount of C sequestered is 103 Mg per year, of which 12% is from soil C. In scenario 2, annual mean was 239 Mg C that translates to 4.2 Mg C/ha of which 62% is contribution from agroforestry. In scenario 3, the annual mean C was 393 Mg, translating to 4.5 Mg C/ha, of which agroforestry's contribution was 39%. Due to the windfall from seed selling and carbon credit, there is an increase in tobacco and other annual crops, resulting in the reduced contribution of agroforestry to overall C.

Our results show that agroforestry can increase C sequestration on farms. These findings are within the range of most tropical smallholder agroforestry system C sequestration rates. However, incentives are important for adoption. We conclude that smallholder farmers can benefit from the Clean Development Mechanism (CDM) of the Kyoto Protocol.

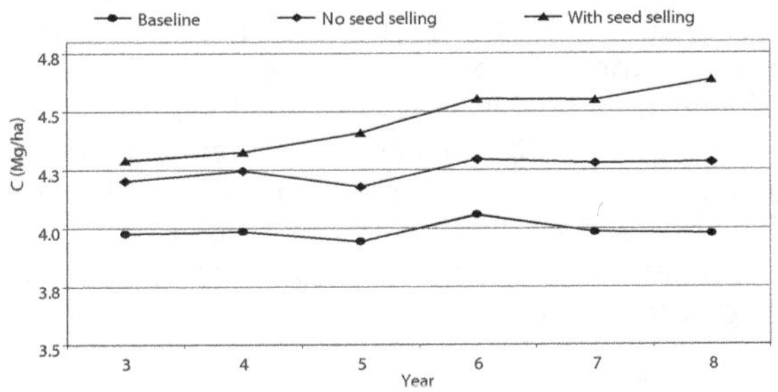


*Figure 1. Simulated annual carbon stocks (Mg C/ha) sequestered for the scenarios 1, 2 and 3.*

# Evaluation of landscape preferences of Montado in southern Portugal as a contribution for management

*Diana Ježová and Teresa Pinto*
*Correia, University of Évora, Portugal*

The Montado is an agro-silvo-pastoral system dominant in the Southern region of Portugal, Alentejo. Apart from agro-forestry production, the Montado is used by different groups of people with various interests. Even though Montado is a multifunctional system, there is lack of knowledge on its different non-productive functions and their integration. For better management of in the system it is essential to know peoples' preferences concerning this type of landscape. The Montado is an extremely flexible system in terms of many variants of management options and tree densities, resulting in a high variation of types. The objective of this study is to clarify what type of Montado is appreciated by which group of users.
Individual interviews with different users of Montado were carried out to collect data. Each interview consisted of both specific and open-ended questions. The responses to the open-ended questions provided useful information on the underlying reasons for preferences which were not obtained through the specific questions. As visual stimuli, photographs representing heterogeneous types of Montado of Cork Oak were used. The photographs differed in land cover of trees, composition of trees, presence of shrubs in understorey and presence of livestock. The respondents chose from the fourteen photographs a preferred combination of three photographs of different Montado types. Then they were asked to explain their preferences. Data from 60 interviews with four different user groups, landowners, hunters, rural inhabitants and new rural habitants, were analysed.
Results from the statistical analysis of closed-ended questions and content analysis of open-ended questions show some interesting differences between preferences of the user groups. The *landowners* chose mostly those photographs with plantation of trees in lines and without presence of shrubs in the understorey. The presence of livestock in the Montado was valued positively because landowners related this with higher income from the exploration. Contrary to the landowners, most of the *hunters* preferred Montado with shrubs in the understorey. High density of trees and shrubs gave them the sensation of mystery during the hunting, and a higher probability to find varied game. For small game, hunters preferred open Montado with a low density of trees. The livestock presence in the Montado was not appreciated by hunters for their activity. The group of *rural inhabitants* is composed by people that lived in Alentejo since their childhood. The majority stressed that they prefer the Montado types with which they are familiar. The photographs that they chose were strongly related with cultivated land, *i.e.* the Montado without shrubs and with livestock presence. The group of *new rural inhabitants* prefers to see a high variety of landscape types during their recreational activities in Montado. They chose the photographs of cultivated Montado types with a combination of less cultivated types. The photographs with composition of trees in lines were less chosen by this group of users. Clear differences in preferences on land use systems by different user-groups are shown by this study. That reveals the importance of the studies on preferences of a variety of groups of users. This study identified preference of four users groups in Montado, but actually even more users groups exist, *i.e.* tourists, beekeepers, workers, that should also be studied. The consideration of preferences of land-users in management of Montado should be a component in sustaining the multifuctionality of this traditional landscape in southern Portugal.

# Towards a dairy system that produces more than milk

*Jan J.H.M. Ketelaars, Frits K. van Evert and Ben Rutgers*
*Plant Research International, PO Box 16, 6700 AA Wageningen, The Netherlands*

Current animal production systems have low energy efficiency and for that reason use a lot of space. A substantial increase in energy efficiency can be achieved only by means of a system innovation, which we define as the simultaneous implementation of two or more technological innovations. In this work we use a simulation model to explore designs for a highly productive dairy farm that uses less space, yet generates energy and raw materials in addition to milk.
The objective of our study was to determine the potential of an increase in grassland productivity to free some of that grassland to uses other than production of roughage. In this study, that other use is as unfertilized, semi-natural pasture which is not cut until after the end of the bird breeding season in mid-June. The grass from this delayed cut is not suitable as feed; instead, it is processed in a factory for extraction of fiber and protein. Having been cut, the semi-natural pasture is grazed for the rest of the season by the young stock. In this study, we considered a dairy farm on peat soil which produces grass and buys concentrates, grass and maize silage as necessary. This farm is subject to Dutch environmental law which imposes a limit on the farm gate nutrient surplus. We used a simulation model to calculate production and environmental performance for several scenarios of stocking rate, the fraction of land managed as semi-natural pasture, and grassland productivity. In a first exercise, stocking rate was varied from 1 to 2.5 dairy cow per ha and the fraction of land managed as semi-natural pasture was varied from 0 to 0.25. Grassland productivity was set at the default value of 11016 kg/ha. This exercise showed that at a stocking rate of 1.4 cows per ha and without semi-natural pasture, the roughage needed could just be produced on-farm. At higher stocking rates, grass and/or maize silage had to be bought; at lower stocking rates, some of the highly productive grassland could be converted to semi-natural pasture. In a second exercise, we fixed the stocking rate at 1.4 cows/ha and varied the yield potential of the highly productive grassland from its default value of 11016 kg/ha to 15000 kg/ha. Delayed mowing and improvement of sod quality are measures through which this higher yield potential could be realized. The shape of the production curve used implied a higher N-use efficiency with a higher yield potential. Calculations showed that every step increase in productivity allowed an increase of the area of semi-natural pasture while maintaining self-sufficiency for roughage (Figure 1). Interestingly, the total nitrate and ammonia emissions did not increase when roughage production was concentrated on a fraction of the total area. However, $N_2O$ emission and variable costs increased slightly when more than 10% of the area was converted to semi-natural pasture. In conclusion, there seems to be scope for meeting societal demands for space (for birds, for herbs, for recreation) while maintaining milk output and farm profitability, by converting some of the highly productive grassland to semi-natural pasture.

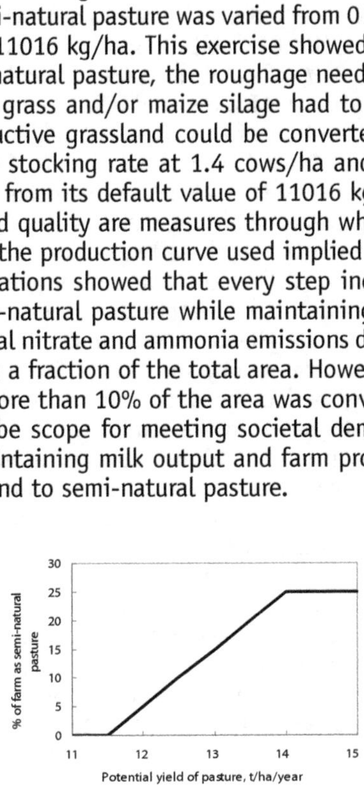


*Figure 1. Percentage of farm area that can be converted to semi-natural pasture when potential yield of pasture is increased.*

# Organic and integrated farming systems in Greece: A comparative analysis

*Eleni Oxouzi[1], Athanasios Theocharopoulos[1], Vasilis Bagiatis[2] and Evangelos Papanagiotou[1]*
*[1]Department of Agricultural Economics, School of Agriculture, Aristotle University of Thessaloniki, Greece; [2]Department of Biochemistry and Biotechnology, University of Thessaly, Greece*

Over the last two decades, attention in industrialized countries has focused on reducing pollution by fertilisers and synthetic pesticides in intensive agriculture. Especially, the increasing consumers concern about food safety and environmental pollution escalated the value of sustainable farming systems (SFS), such as Organic and Integrated Farming Systems (OFS and IFS). OFS is a sustainable way of farming without chemical inputs during the cultivation and IFS is a sustainable way of farming which falls somewhere between conventional and organic farming system.

This paper analyses and compares the development of these two alternative farming systems in Greece. The comparative analysis of the geographical development was conducted at region level (13 regions in Greece, *NUTS–2*) and region-prefecture level (52 regions, *NUTS–3*). For this purpose it was used the GIS methodology, which is designed to capture, store, manipulate, analyse and display diverse sets of spatial or geo-referenced data.

OFS has already been implemented for 10 years and IFS for 4 years. An significantly growth of area under OFS and IFS was observed in the entire implementation period. In 2003, 36,162 hectares were cultivated under the OFS and 12,556 hectares were cultivated under the IFS, which represent 1.07% and 0.37% of total utilisable agricultural area of Greece. In NUTS-2 level there is an unequal allocation of SFS, OFS and IFS. The distribution of IFS between regions is more unequal than the distribution of OFS. It is noticeable that 71.9% of total area under IFS is cultivated at the region of Central Macedonia (1.27% of total utilisable agricultural area of Central Macedonia) and only the other 28.1% at the other 12 regions. OFS has developed mainly at the regions "Peloponnesus" and "West Greece" where represent 44.2% of total area under OFS. The rate of adoption of OFS at these two regions is 2.72% and 2.29%, respectively. In NUTS-3 level, the results showed that the two alternative farming systems are applied in completely different regions. It is noticeable that the 8 regions with the major area under OFS are different from the 8 regions with the major area under IFS.

The type of crops at the regions, the policy support for Sustainable Farming Systems and the different extension – advisory services between the two systems explain the differences at the development of OFS and IFS between the regions of Greece. The intensity of main crops at each region (relative to the use of chemical inputs) is an important adoption factor of two systems. Thus, OFS is applied mainly at regions with less intensive crops and IFS is applied mainly at regions with more intensive crops (like peach crop that require many applications of chemical inputs). Also, the policy is different between two sustainable farming systems. Financial support is provided direct to organic farmers, during the last decade, through civil services that are the main advisory services in OFS (52 civil services, 1 at each region-prefecture of the country). On the other hand, IFS receives limited financial support during last 4 years through agro-environmental measures in Rural Development Plans. Extension is conducted mainly by private companies, which are the main advisory services in IFS. In contrary to OFS, these private-advisory companies are still only in a small part of the country (mainly in Central Macedonia).

# The socioeconomic integration of expatriates in the rural prefecture of Rodopi, Greece

*H. Theodoropoulou and E. Pastrapa*
*Harokopio University of Athens, Greece*

## Introduction

During the beginning of 1990s a large number of the descendants of expatriate Greeks living in the former Soviet block eastern countries as political refugees, due to the Greek civil war after the 2nd World War, were repatriated back to Greece following the opening of the borders between western and eastern Europe. The concomitant social and economic changes following the repatriation of the expatriates affected the prefecture of Rodopi significantly.

## Methodology and results

The objective of the present study was to examine the social and economic status of expatriates in the prefecture of Rodopi through a questionnaire survey by collecting data on their demographic, educational and employment characteristics. A total of 179 questionnaires were collected. The profile of the expatriates that emerged from this study was that was a man (84%) aged 26-45 years old (49%), married with two children (80%)with high educational level (32%) employed in farm and construction (37%) activities with a personal monthly income up to 500 (58%). Even though, most of the expatriates went to reside at the beginning in urban areas, where they had relatives and they relied on them for locating available jobs, later on they moved in rural areas due to governmental endowments for housing. The government aim was to move those expatriates to Rodopi in order to increase the Christian population in the area and strengthen the agricultural sector, which during the last decade was abandoned by the locals who moved to urban areas. The establishment of expatriates in the region revitalized these rural areas by their contribution to the local farming production. Cross tabulation analysis showed that the better expatriates spoke the Greek language the less they reported problems with social exclusion, unemployment, or economic exploitation ($p<0.05$). Also, immigrants' satisfaction from their income depended on how long they work in the same job. The longer they work in the same job the more satisfied they are from their income ($p<0.00$).

## Conclusions

The results of the present study indicate that two factors related to the expatriates' characteristics, the knowledge of the Greek language and the number of years that they are in the same job, affect the social and economical integration of the expatriates in the Greek way of life. Apparently, expatriates with a good knowledge of the language and a permanent position for many years in the same job achieve a better treatment from the Greek society. These two factors are ominous for the future of a stable social and economical life in Greece, since expatriates of low skills in language will not be able to adapt to the future demands of the market. Therefore, it is significant for the governmental policy to take into account those weaknesses and to help these people to adapt to their new life for the sake of a good and productive Greek socio-economic life through special vocational courses mainly to teach the Greek language. On the other hand, even the locals agree that expatriates have contributed to the increase and rejuvenation of Rodopi's population, which may be the hope for the area's future.

# Rural policy in Europe: Experiences with the LEADER approach in the Netherlands

*Anouk van Gils*
*Ministry of Agriculture, Nature and Food Quality, The Netherlands*

LEADER+, one of four community initiatives financed by EU structural funds, is designed to help rural actors improve and realise the long-term potential of their local region. LEADER+ (Liaison Entre Actions de Développement de l'Economie Rurale) is an experimental program that aims to give socio-economic impulses at local area level, through funding of local and regional initiatives. These initiatives may be developed by individuals, local associations, private companies or regional and local government bodies. A total of € 5 billion will be allocated on European level in the period 2000-2006. Main objectives of the program are: (i) to help rural stakeholders to become aware of the potential of their area and to stimulate them in their search for improvement; (ii) to encourage high quality integrated original strategies for sustainable rural development; and (iii) to create room for development and conservation of natural and cultural heritage and improvement of economic conditions. In this way LEADER+ is to contribute to new employment opportunities and improve the capacity for self management of rural communities.

Each member state has set up a National Network Unit to stimulate networking and mutual learning. The network team supports project initiatives, area councils and local communities or municipalities throughout the country and stimulates exchange of experiences and cooperation between different areas at the national and international level. The Dutch team has been doing this by organizing workshops and conferences, publishing a network bulletin and maintenance of different websites.

LEADER+ has developed a specific approach with local, trans-local and vertical features, such as area-based approach, bottom-up approach, use of local partnerships, emphasis on innovation, applying a multi-sectoral approach (local), a focus on networking and trans-national co-operation (trans-local) and decentralised management and financial organisation (vertical). In the Netherlands, four regional programs have been developed, including 28 local action groups. Each action group was requested to formulate a development strategy for their area; this strategy was used to evaluate project proposals that were to fit in the local strategy. Program implementation was done by regional administrations.

The program has been evaluated both at the national and international level. Main conclusion of the evaluation is that the approach (i) has been flexible (being applicable in different social-economic contexts, bridging gaps between local stakeholders); (ii) reduced the gap between (inter-)national and local ambitions; (iii) contributed to sustainable development at local level, building problem solving capacity and local partnerships at strategic level; (iv) contributed to diversification of rural area activities, both within as outside agriculture; (v) contributed to economic development, increasing rural employment opportunities; while it (vi) has been complementary to other programs on EU-level as well as on national level.

Problems related to implementation of the program include:

- limited time allocated in planning to the start-up phase (more time and resources should be allocated at the start of projects);
- very strict rules and regulations for project implementation;
- political and institutional restrictions, often not allowing decentralised project management, as far as financial and decision-making aspects are concerned;
- administrative restrictions, as institutions need time to organise necessary capacities for decentralised project implementation and management;

- lack of local social capital, as local stakeholders need time to build up their strategic and operational capacity required for project implementation and management.

Notwithstanding these limitations, it is felt that the LEADER+ approach has been successful. Presently, discussions have started how to integrate this approach in a new rural development program. Although, in general, it appears to be difficult to explain the LEADER+ approach to non-participants, it is hoped that successful elements of this decentralised, participative multi-stakeholder approach will be used in further EU programs and national policies.

# Sustainable development in Polish agriculture: Main streams and development directions

*Jadwiga Ziolkowska*
*Humboldt-Universität zu Berlin, Institute of Agricultural Economics and Social Sciences, Chair Agricultural Policy, Germany*

Since the year 1992, when the Earth Conference for Environment and Development of the United Nations in Rio de Janeiro took place and the Agenda 21 was adopted, "sustainability" has become an important term in many sectors of the European economy, also in agriculture.
Due to diverse natural conditions in rural areas and different farming systems the sustainable development in rural areas is a great challenge. The effects of sustainable farming depend on farmers' openness to economic risk and awareness that the innovative farming system promotes their objectives (see: Pannel, 1999).
The sustainable measures in agriculture are realised in the most countries with agri-environmental programmes. While the idea of sustainable agriculture is already common in the former EU-15 countries, it is said to be a new stream in the new EU member states. Agriculture is an important sector in Polish economy (the employment share amounts to 26%). Therefore, it can influence sustainable development to a wide extent. The implementation of regular agri-environmental measures in Poland started in 2004, however organic farming is already known since 1990.
To analyse the sustainability of agriculture in Poland in the last decade, some indicators characterising the agri-environmental state were chosen from the indicators set created by the European Environment Agency such as: number of agri-environmental contracts, agricultural intensity, agricultural eco-efficiency as well as number of organic farms (EEA, 2005).
According to statistical data from the Statistical Office in Poland, a very differentiated picture of Polish agriculture can be drawn with regard to chosen indicators of sustainable development. While the number of organic farms in Poland is increasingly growing since 1990 (even rapidly since 1999) and also the interest of farmers in the measures of the Agri-Environmental Programme seems to be relatively high in the first EU membership period, still the amount of applied fertilisers in agriculture is relatively high. Simultaneously, a drop in the agricultural area and in the number of farms as well as shrinking area of arable land can be noticed. On the other hand, however, the gross value added in agricultural sector (a measure of agricultural productivity) was consecutively growing in the period 1992-1997 and is since 1998 on a relatively stable level.
Analysing these correlations, the presented topic should encourage to undertake and discuss the subject of sustainable agriculture in Poland and in other new EU member countries during the time of changing patterns in farming systems.

## References

EAA (European Environment Agency), 2005. Indicators for evaluation of agri-environmental policy. (08.03.2005) In: http://themes.eea.eu.int/Sectors_and_activities/agriculture/indicators.

Pannel, D.J., 1999. Social and economic challenges in the development of complex farming systems. Agroforestry Systems 45, 393-409.

# Section 3 – The future of farming

# Introduction to section 3 – The future of farming

*Benoît Dedieu[1], Colin Newsham[2] and Rebecka Milestad[3]*
*[1]INRA TSE, UMR Metafort, 63122 St Genès Champanelle, France; [2]Forrest Hills, Rural Centre for Business Learning and Leisure Hazelrigg Lane Ellel, Lancaster LA2 0PL, United Kingdom; [3]Department of Urban and Rural Development and Centre for Sustainable Agriculture, SLU, Sweden*

This section is focused at the farm level. It will explore farming as being part of rural systems and food chains, and as provider of services in relation with the value systems and discourses in society (Kristensen and Halberg 1997). Discussing the future of farming leads to several questions that will be debated throughout the papers and posters in this section. We can summarize the principal axes of the debate into four points:

- *What makes farming change at the European or national level? What are the scenarios for evolution of the overall agricultural sector or of more specific sectors (livestock, crops...) at that scale.* The CAP is a well known factor of change. Theocharopoulos *et al.* analyse with an econometric model the effect of decoupled payments on the cropping area of Greece and the type of crops the farmers will develop. Environmental issues are also taken into account in the CAP, notably pollution and biodiversity. But over and above these ones, Macombe explores how European governments, firms and farmers are preparing new responses, political and practical, to the major climatic changes and the challenges with respect to energy use. Market pressures still deeply affect farming and its future (in terms of price squeeze and qualitative attributes of the goods that are requested). Bonny, with French data, shows that the power of the limited number of the downstream distributors (110) who connect the hundreds of thousands of farmers to the millions of consumers must be clearly taken into account and evaluated when discussing the future of farming. Yet, with a Finnish case study, Mikkola suggests that farmers still have ways to master their participation in the food chain by making horizontal partnerships with others producers and vertical networking with large chains. Such social connections provide resources and skills to satisfy the quality and quantity demands. Posters from developing countries (Cameroon – Djamen *et al.*, China – Fok *et al.*, Algeria – Foudil, Benziouche) pay more attention to the big changes in the popular demand for food (quantity, quality, nature: more meat and milk) and the consequences of off-farm jobs for the future of family farming. Farming here is clearly connected to the population's welfare, income and employment dynamics and to the transport facilities.
- *Farming is also connected with the local society*, notably when debating how off-farm job opportunities influence family farming (Lobley and Potter 2004) or *when defining what can be multifunctional ways of farming*. Björklund and Milestad in Sweden, Hermansen *et al.* in Denmark suggest clearly that the dialogue among farmers, local communities, or local stakeholders is a condition to specify what can be productive activities, and how farmers' productive activities and services can be combined. This dialogue could lead to a definition of concrete indicators of multifunctional farming that could guide its evolution in the local context (Halberg *et al.*,2005).
- *Farmers and farming do not represent a homogeneous world.* Learning between farmers and society is important, but learning about the diversity of farmers, their reasons, and their future potential is also a major issue (Gibon *et al.*, 1999). The local soil and climatic conditions play a basic role to explain these differences, as shown by Thenard *et al.*: livestock farmers of the Massif Central hills do not adapt their forage systems in the same ways in a granite (and drought-sensitive) region as in a volcanic one, when they adapt to

new PDO[1] cheese regulations that forbid maize silage. Farming is still dealing with natural areas, the agronomic potential of soils, saline conditions (*cf.* Brandenburg's paper), *etc.* which are not equally distributed. But farmers also do not have homogeneous attitudes when considering what style of farming they refer to, the kind of workforce (family, with paid workers or sub-contractors) their farm should be based on, or the way they will implement multifunctional farming. Ahnström and Hallgren illustrate that Swedish farmers notably differ when thinking about wild life and nature conservation programs, these programs being more or less imposed from "outside" their world. They conclude that the diversity of farmers' attitudes must be taken into account when building those conservation programs, with opened discussions between them and with other stakeholders. Even with an activity like pig production, often considered as an achievement of the industrial standardisation in agriculture, Commandeur and Casabianca show that the farming styles (Van der Ploeg 2003) are not homogeneous, neither is the idea of the future for their children. In all situations, many different attitudes of farmers can be identified, and many conceptions exist of the nature of the transformations that are needed, given the more or less uncertain times ahead. That diversity will build the future of farming.

- *Working conditions in farms are changing,* as is the workforce composition. These changes must be taken into account when considering the number of farms, the contribution of the permanent or non-permanent agricultural workforce to the rural dynamics, and the ways farmers can make choices between "get big", "get special" (Bjørkhaug) or "get diversified". It concerns particularly the future place of women as skilled workers, as active members of the decision cell with their partner for a shared activity, or as independent entrepreneurs for a separate specific activity (Branth 2002). Analysing the European replacement program in West Macedonia, Gidarakou *et al.* point at that women do have very different attitudes when entering these programs and argue their subsidised establishments with various reasons form a contribution to the growth of the farm to the development of specific activities. Bjørkhaug and Heggen recall that intensification and mechanisation have played an important role in the deskilling of women and in reducing their decisional status within the couple; qualitative interviews of Norwegian female farmers suggests that multifunctional farms with green care, tourism and small scale food production give them opportunities to valorise new skills within farms. However combining multiple activities (in and/or off farm), responding to environmental programs and market chain specifications, as well as improving work conditions: all this amounts to a big challenge for work organisation. Contributions from France (Cournut and Dedieu) and Brazil (Hostiou *et al.*) suggest that this challenge must be seen in terms of its technical content (the work to be done can be adjusted); in its social content (who is doing what and when) and in its consequences (time spent, time remaining and flexibility). Petit *et al.*, engaged in a participatory research with groups of farmers, illustrate that learning is a major issue to allow farmers themselves to design the solutions that preserve their welfare and their passions. Melberg and Berg, studying the transitions between generations in Norway underlines that if farm families still have expectations toward the young generation to take over the farm, there is a gap between the youth expectations and the life style in the contemporary farms.

All those dynamics, knowledge and learning processes lead to linked questions about how the frameworks and methods for farming system analysis nowadays could evolve. Notably farmers have to think and act in an increasingly uncertain world. The way they give consistency to their farming system must preserve its capacity of resilience. The farmers mobilize different sources of resilience (Milestad and Darnhofer, 2003) that have to be integrated in the farming system

[1] PDO: Protected Denomination of Origin.

framework of analysis. Ingrand *et al.* focus their analysis on the "relational" flexibility of French beef cattle farms: it is based on either technical or social levers: the spread of sales and the relations with cooperative or independent buyers. In a more long-term survival perspective, Darnhofer suggests to consider the farmers as adaptive managers. It notably implies to consider their ability to address changing conditions and their possibilities to initiate new development trajectories. Carlsen *et al.*, discussing the improvement of organic farming in Norway, suggest that a farm case study which involves stakeholders both within and outside the farming system could be a framework for such dialogue.

Defining the indicators of multifunctionality at the farm level, understanding the levers for work organisation, as well as the skills at stake; combining knowledge on the farmers' attitudes, as well as on the agronomical constraints and on the technical content of farming practices; connecting farm management purposes, resilience issues and horizontal and vertical networking needs as a whole: all that is important, stimulating and all that is a challenge for pluri-disciplinary research (Holling 2001), participatory learning and new design processes for multifunctional and sustainable systems. We hope the papers presented in this section will facilitate discussions on these issues and take the argument even further.

## References

Brandth B., 2002. Gender identity in European Family farming: a literature review. Sociologia Ruralis, 42 (3), 181-200.

Gibon, A., A.R. Sibbald, J.C. Flamant, P. Lhoste, R. Revilla, R. Rubino and J.T. Sørensen, 1999. Livestock farming systems research in Europe and its potential contribution for managing towards sustainability in livestock farming. Livest. Prod. Sci. 61, 121-138.

Halberg, N., H.M.G. van der Werf, C. Basset-Mens, R. Dalgaard and I.J.M. de Boer, 2005. Environmental assessment tools for the evaluation and improvement of European livestock production systems. Livest. Prod. Sci. 96, 33-50.

Holling, C., 2001. Understanding the complexity of economic, ecological and social systems. Ecosystems 4: 390-405.

Kristensen, E.S. and N. Halberg, 1997. A systems approach for assessing sustainability in livestock farms. In: Livestock farming systems – more than food production. (ed. J.T. Sørensen). EAAP publication 89: 16-29

Milestad, R. and I. Darnhofer, 2003. Building farm resilience: the prospects and challenges of organic farming. J. of Sustainable Agriculture 22: 81-97.

Lobley, M. and C. Potter, 2004. Agricultural change and restructuring : recent evidence from a survey of agricultural households in England. Journal of Rural Studies, 20, 499-510.

Van der Ploeg, 2003. The virtual farmer. Assen/Maastricht, NL. Van Gorcum.

# Farmer relationship to nature conservation and biodiversity

*Johan Ahnström[1] and Lars Hallgren[2]*
*[1]Swedish University of Agricultural Sciences, Department of Entomology, Section of Landscape Ecology, Box 7044, 75007 Uppsala, Sweden; [2]Swedish University of Agricultural Sciences, Department of Landscape Planning, Section of Environmental communication, Box 7012, 75007 Uppsala, Sweden*

## Abstract

The biodiversity decline and the habitat loss in the agricultural landscape have brought up the importance of nature conservation. In the managed landscape with many landowners and other stakeholders there is a great need for a common agreement on the aims and the means of nature conservation. Farmers' attitudes and knowledge about nature and nature conservation will affect the farms effect of the surrounding environment. Our study focuses on these attitudes and how that affects biodiversity on the farm. Sixteen Swedish farmers have been interviewed and biodiversity inventories have been conducted on their farms. The definition of nature conservation differs greatly within and between stakeholder groups and that might lead to severe communication problems. Keywords in the definition of nature conservation were to conserve, to manage, or to exploit (in a sustainable way) nature. The choice of word is reflected on the management of the farm and the conservation actions taken. The farmers have a deep and broad awareness and knowledge of their land but this is seldom recognized by nature conservation administration. The lack of respect of their knowledge makes the farmers uneasy. Farmers are not a homogenous group but unique individuals but nature conservation is often general. Can it be flexible enough to fulfil requirements of different organisms and stakeholder interests? Sustainable nature conservation in the agricultural landscape can only be achieved with a continuous dialog between stakeholders. The dialog ought to be open-minded and the participants have to respect and acknowledge each other's knowledge and interest.

**Keywords:** nature conservation, multidisciplinarity, interviews

## Introduction

Agricultural production methods have to be sustainable in an economic, ecologic and social respect, in order to provide food for the growing global population. However, the intensification and specialization in agricultural ecosystems has resulted in declining biodiversity and other environmental problems (Krebs *et al.*, 1999; Benton *et al.*, 2003). Society has recognized these problems and created nature conservation and environmental programs to counter the negative trends. The efficiency of some of these subsidies has been questioned (Kleijn *et al.*, 2001). The subsidy programs and the supporting research often seem to neglect farmers' perceptions and knowledge of nature. According to a literature search only 4% of the articles dealing with nature conservation issues had taken attitudes into account (ISI Web of Knowledge; Nature Conservation and Attitudes) (Ahnström, unpublished data). Farmer's knowledge of local growing conditions, weather changes and traditions are often at least as important to understand in terms of the unique natural and farm society context as general ecological knowledge that is context free. The farmer or manager of the land has that local knowledge and it is internalised in their decision-making. Farmers' appreciation of the natural resource base and how it can be preserved ought to be included in the design of nature conservation programs. Subsidy and conservation programmes need to be flexible to fit farms with different site-specific conditions, and in different landscape contexts.

The knowledge of biodiversity in the agricultural landscape is vast but not in any way complete. However, very few have tried to link the qualitative data from interviews with the quantitative inventory data. In other words few have tried to examine how farmers attitudes of nature and nature conservation affect biodiversity on the farm.

The aim of the author's current research project is to try to do this combination of qualitative and quantitative data to strengthen our knowledge on parameters affecting biodiversity in the agricultural landscape in Sweden. The study has also ecological aims, but they will not be dealt with in this paper.

## Materials and methods

### *Field study*

Sixteen farms in south-central Sweden were included in the study. The size of the farms range from 34 to 600 ha and the farmers age from 34 - 77. The farmers have been interviewed, semi-structured and open-ended, about the farm history, management strategies, definition of nature conservation and feeling for nature. The main author conducted all the interviews at the farms. The current paper is based on interviews conducted during spring 2004. We have also done ecological field inventories on these farms. The farmers have been revisited for further interviews, which will be presented in coming papers. The major data analyses, qualitative interview analysis, traditional ecological statistics *e.g.* multiple regression and ordination, and the efforts to combine the data, are in progress.

### *Literature study*

To learn about the current knowledge about farmers attitudes concerning nature and nature conservation we conducted a literature search and are currently writing a review article about this (Ahnström, Höckert, Bergeå, Francis, Skelton and Hallgren, unpublished). We have confined the study to information published in scientific journals. There is a broader information base available in reports (for example Toogood *et al.*, 2004) and working papers, but in order to make the selection of included studies transparent and standardized we chose to only use easily accessible and peer reviewed resources. The literature was collected until spring 2005.

## Results and discussion

### *Nature conservation – definition*

In the Convention of BioDiversity (CBD) the keywords are conservation, sustainable use and an equitable benefit sharing of the products and services that nature provide. These would be excellent words in a definition of nature conservation but most often biologists and governmental conservation personnel only use the term conservation. The question is also what should be conserved; a nature area, species or natures ability to evolve, this will affect how conservation areas and conservation subsidy programs will be designed. The farmers in our study were asked to define nature conservation. The definition differed quite substantially between the farmers. However, none of the farmers used the term to *conserve*. Instead they used *manage* the landscape and *exploit* the land in a sustainable way. It is worth noting when translated to Swedish; the English concept "nature conservation" is rather translated to "nature care". However, when the Swedish nature conservation administration uses care, in nature care, the meaning is often to conserve.

> *Interviewer "How would you define nature conservation?"*
> *Farmer "That you exploit nature. Cut some forest, produce some cereals, in a good way as we do"*
> *I. " I see. Are you a conservationists then?"*
> *F. "Yes. I show extreme care"*
> *I. "How?"*
> *F. "Well, green zones, pesticide free border zones. That is good. I am not driving and cutting them when the birds have eggs and the roe deer have youngsters and... That is my way to contribute to nature conservation."*

When the definition of nature conservation and thus the aim of nature conservation differ between and within stakeholders there is no wonder that there is some communication problems. One of the farmers in our study was very annoyed with all the talk of the need to conserve nature. *"Not even this wheat field, such a mono-culture, are the same year after year... Nature does change...No matter what we do. It is a constant movement. It does not matter, nothing will look the same ever again no matter where you are".*

***Stewardship***

The farmers of today experience growing external pressures of economic, social, and environmental nature but still few are willing to give up farming. An English farmer in Harrison and Burgess' study (2000) said, *"If it was just a matter of economics we would not be here"*. Australian farmers claim that they want to be better stewards of the land by adopting environmental strategies and taking conservation actions but these wishes is continuously challenges by immediate and long-term economic concerns (Millar and Curtis, 1999). Sustainable nature conservation needs economically viable farms that can manage the land. A farm in economic shortcomings will never be able to manage hedges, ditches or pastures. Thus it is crucial to see nature conservation as a multidisciplinary challenge or we will never succeed to make it sustainable.

The aim of farming was, of course, to earn money but as important was to have a nice place to live and the independence of the work, in a study of Willock *et al.* (1999). One of the farmers in our study said while we were talking about cereal prices over a cup of coffee in his garden early one sunny morning in July 2005; *"One should earn enough to get by but... be able to sit like this. It is wonderful".*

In our study 12 of 16 farmers live on old family farms and that is reflected by strong connections to the land. Nature conservation actions were taken to conserve the land for future generations and to enrich the connections to the land (Ryan *et al.*, 2003).

Trees and other natural objects have sentimental values. The strong history can also be a burden because the farmer feel obliged to continue the farming heritage and there is a fear of being the one who loose the farm.

The farms in our study differ in size and main occupation; organic dairy farm to conventional piglet production and from diversified organic crop production to effective large-scale conventional (IP) cereal production. Their prerequisites to contribute to nature conservation differ, as do their interest. However, the places that they showed us, when asked to show a place they liked to visit at the farm or in the vicinity, were similar. No matter what kind of farm, no matter the size and no matter the interest in nature the farmers chosen place were on a high spot with a nice view facing the own fields and the farm. The place were often species rich in terms of vascular plants but the farmers did not mention that when they said why the chose the place. The larger context (the landscape) was more important than the parts (the species). The author, as a biologist, would in most cases faced the opposite way compared to the farmers to be able to study the species and then look at the larger context via a GIS-analysis of the surrounding landscape. We are aware of that the farmers know that we are interested in biodiversity issues and that might have influenced them in picking a place to show.

### *Farmer knowledge about nature*

When we asked farmers if they knew bird species, most of them said no. However, during the interview they mentioned many bird species and their ecology. It was as if their species knowledge, to them did not fit into their self-image of being a farmer. The farmers spend much of their time outdoors and with that comes an awareness of connections and relations in nature. Thus they might know and understand that their actions might cause harm to some organism but they have to take the actions to get a good crop or an effective treatment. Farming is not just about taking management actions but to take them in right time to get good effect. The studied farmers see themselves as producers of cash crops not of biodiversity and thus it is not at all strange that they do not primarily focus on conservation actions and this has to be recognized by conservation administration. Farmers' knowledge and interests have to be acknowledged. If nature conservation could be based on dialog between biologists and farmers the likelihood of well-accepted strategies and actions would increase. The current situation where farmers are told, via laws, subsidy schemes or extension experts, what they can do, when they can do it and how they should do it, will never give a sustainable production neither of crops nor biodiversity. The farmers in our study claimed that they felt as the governmental bodies gave them a declaration of incapacity when they came to the farm and told them how to manage the land. Farmers in UK said if their previous management would have been wrong then it would not have been anything left to conserve (Harrison *et al.*, 1998).

## Conclusions

The main points of this paper are: 1) farmers' perception that nature conservation is about using not conserving the land; 2) farmers' sense of stewardship; and 3) their ecological knowledge is based on their management practice. These aspects relate to each other dialectically. These points are based on a finite analysis of the interview material and further analysis will probably provide us with a more complicated picture of farmers and nature conservation. An aspect that we will study is how management practice relates to the attitudes regarding nature and nature conservation of these 16 farmers. In the material there is also information about how the farmers relate to representatives of the formal nature conservation visiting the farm.

Farmers' knowledge about nature and ecology develops through their management practise and this is in contradiction to the knowledge of the formal nature conservation personnel, which is based on science and investigations (this generalisation is of course a simplification).

Not presented here and not yet analysed for is how farmers, based on their stewardship and knowledge of ecology, relate to biodiversity on the farm, as it is formulated/defined by nature conservationists (with tables and figures of species richness and abundances), and again how this is related to the farmers experience of nature on the farm. Ecological inventories are based on the same thinking as the formal nature conservation. To study how farmers react when meeting this thinking we showed the results of the ecological inventories performed on their farms. We will link these findings with the presented conclusions about different definitions and views on nature conservation. To even more deepen and enrich our understanding of farmers' appreciation of nature on the farm we asked them to take us to "a place on their farm they like to visit".

Based on the farmers' perception on nature conservation and stewardship is it possible to assume and understand their reactions on proposed nature conservation actions? How would nature conservation programs/politics look like if they were based on farmers' perspectives, or rather, how would nature conservation look like if the farmers were further involved in the policy and program design? This is a prerequisite to achieve legitimacy needed to implement the nature conservation programs/politics.

## Acknowledgements

This study and the PhD-project are funded by FORMAS, the Ekhaga foundation and Oscar and Lili Lamm's foundation.

## References

Benton, T.G., J.A. Vickery and J.D. Wilson, 2003. Farmland biodiversity: is habitat heterogeneity the key? Trends Ecol. Evol. 18, 182-188.

Harrison, C.M., J. Burgess and J. Clark, 1998. Discounted knowledges: farmers' and residents' understandings of nature conservation goals and policies. J. Environ. Manage. 54, 305-320.

Kleijn, D., F. Berendse, R. Smit and N. Gilissen, 2001. Agri-environment schemes do not effectively protect biodiversity in Dutch agricultural landscapes. Nature 413, 723-725.

Toogood, M., K. Gilbert and S. Rientjes, 2004. Farmers and the environment assessing the factors that affect farmers' willingness and ability to cooperate with biodiversity policies. Biofact, European Centre for Nature Conservation.

Krebs, J.R., J.D. Wilson, R.B. Bradbury and G.M. Siriwardena, 1999. The second Silent Spring? Nature 400, 611-612.

Millar, J. and A. Curtis, 1999. Challenging the boundaries of local and scientific knowledge in Australia: opportunities for social learning in managing temperate upland pastures. Agric. Human Values 16, 389-399.

Ryan, R.L., D.L. Erickson and R. De Young, 2003. Farmers' motivations for adopting conservation practices along riparian zones in a Mid-western agricultural watershed. J. Environ. Plan. Manage. 46, 19-37.

Willock, J., I.J. Deary, M.M. McGregor, A. Sutherland, G. Edwards-Jones, O. Morgan, B. Dent, R. Grieve, G. Gibson and E. Austin, 1999. Farmers' Attitudes, Objectives, Behaviors, and Personality Traits: The Edinburgh Study of Decision Making on Farms. J. Vocat. Behav. 54, 5-36.

# Future prospects of the average Norwegian farm

*Hilde Bjørkhaug*
*Centre for Rural Research, Norwegian University of Science and Technology, 7491 Trondheim, Norway*

## Abstract

As a result of agricultural restructuring, Norwegian farms are increasing in size, both in crop growing area and livestock capacity. Fewer farms can offer a liveable income for the farming couple, and as many women have left farming for off-farm income male farmers are also seeking additional income off the farm. Norwegian agriculture is highly dependent on a variety of subsidies. Present agricultural policy is leading to reductions in direct production subsidies. Future farmers are advised to either rationalize or find new ways of developing businesses or niche productions based on agricultural resources. In this paper, farmers' adaptation to agricultural policy is explored, and with that, farmers future prospects, measured through an assessment of their 'will to invest' in their properties. What is the reality of the average Norwegian farm? Who runs the average farm? How do the farmers view the future? What distinguishes an optimistic owner of a farm as compared to a pessimistic one? This paper bases it discussion on empirical survey data of which questioned Norwegian farmers on future plans in agriculture.

**Keywords:** agricultural restructuring, farm work, farm income, future farm prospects, Norway

## Introduction

Family farming is the most common way of organizing agricultural production in Norway, but the content of actual participation in agricultural production has changed. From occupying extended families in production, the majority of farms are hardly able to support one person from farm income (Almås and Haugen, 1991; Blekesaune, 1996; Bjørkhaug and Blekesaune, 2004). From the 1980s, part-time farming has become the dominant type of strategy among Norwegian farmers, a strategy where the farmer or spouse, or both, combine farming with off-farm work (Blekesaune, 1996:49). In 2003, near 70% of all farmers in Norway reported working off-farm, as did 80% of their spouses. In a European context this level of off-farm work is high. EU-statistics on employment in agriculture shows that the neighboring countries of Sweden and Finland have a high level of farmers who have also adapted to current policies "with other gainful employment", the former is close to 60%, whilst the latter near 40% (EU Commission, 2003). In Norway, part time farming is a stable strategy on farms that need off-farm income (Blekesaune, 1996:49). While pluriactivity, or part-time farming, can be seen as a strategy or movement away from farming, pluriactivity might also be a factor that keeps people on the land, reduces the decline in numbers of farms and strengthens the basis of local services (Kinsella *et al.*, 2000).

In this paper, household strategy is used as the unit of analysis to help understand the general process of agricultural change. The argument of a survival or adaptation strategy in farming is built on a model including reproduction of capital like investments in the farm, share of family income derived from the farm and household members adaptation to the labour market outside the farm (Blekesaune, 1996). Those households that may sustain in the future, are those that are able to increase production on their farms (Blekesaune, 1996:50). In this paper farmers' adaptation to agricultural policies are explored and with that the future prospects of the *average* farm; What is the reality of the average farm; Who runs the average farm; how do the *farmers* view the future? What distinguishes an optimistic owner of an average farm compared to a pessimistic one? The paper is based on analysis of empirical survey data of Norwegian farmers in 2004.

## The farming context

In Norway, 3% of the land is under agricultural cultivation. In 2005 there were about 55,000 farmers, this is half of the number it was in 1969. Norwegian agricultural policy has changed gradually in the post World War II period. From the 1950s, productivist ideals dominated agricultural policy, encouraging techno-scientific development, rationalization and the mechanization of agriculture. Protection of Norwegian agricultural interests became important (Almås, 2004). In the 1970s welfare policy became an issue also in agriculture. Social rights, income goals and gender equality entered the arena. From the 1980s a greening and a re-regulation of agriculture started and by the 1990s the policy, known as green liberalism, prevailed (Almås, 1994), increasing its attention to sustainability in agriculture, organic farming and economic support for the maintenance of cultural landscapes.

Present agricultural policy is characterized by further liberalization with support for efficient production but also to explore new ways of utilizing agricultural production. Norway faces a certain external pressure from organizations such as the WTO which questions Norwegian agricultural subsidies. One response to this on behalf of the Norwegian government has been to emphasize the non-tradable commodities through the argument of multifunctional agriculture, where qualities like food safety, biodiversity, rural communities and cultural heritage are highlighted. In this way, Norwegian agricultural authorities want to move economic support in farming from the "yellow box" in WTO-terms, where most funding is found today, to the "green box", transfers that do not disturb international production and trade (Prestegard, 2004).

## Analyzing Norwegian farmers strategies and prospects

The analysis is based on data from a survey of a representative sample of Norwegian farmers. Data were collected by the Centre for Rural Research in 2004 via a postal survey. 1712 Norwegian farmers returned completed questionnaires, a response rate of 55%. The data has been shown to form a representative sample of Norwegian farmers, with good rigour in terms of validity and reliability (Rye and Storstad, 2004).

One dimension of a strategy of the adaptation to changing policy and economy is income from off-farm work as reported earlier. The structure of strategies can be seen from the amount of income coming from farming in a household (Table 1).

Sixty seven percent of Norwegian farming households earned more than 50% of their income from sources outside the farm enterprise in 2003. This is a long and ongoing trend. Off-farm income exceeded farm income for one half of Norwegian farmers in the early Eighties (Jervell

*Table 1. Amount of income from farming in 2003, percentages (N 1681).*

| Amount of income | 0 | 1-25 | 25-50 | 50-75 | 75-99 | 100 |
|---|---|---|---|---|---|---|
| Percent | 7 | 40 | 20 | 15 | 12 | 6 |

and Løyland, 1998). The median as well as the mean on this variable, is farms that derives 25-50% of their income from farming. Out of this, three groups were constructed for the purpose of describing the structure of income on Norwegian farms; farms with low amount of income from farming (0-25%), average amount of income from farming (25-50%) and high amount of income from farming (more than 50%).

Source of income may be of importance for adaptation to the changing economic environment of the family farm and the following description gives an insight into farmer attributes and production levels, and also reveals some similarities and distinctions between farmers. The Norwegian farmer is, on average, close to 50 years of age. Thirteen percent are women, but the amount of women differs significantly in the three "income groups". While 19% are women farmers in the group of low amount of income from farming, the corresponding amount of women are 9% in each of the two other groups.

While 20% of all farmers have nine years of compulsory primary education, the level of further education differs significantly between the groups. More farmers in the *low farm income* group have education on a university level (27%). Fifty percent in the *average* group have technical/ agricultural secondary school qualifications. Nearly 60% of the farmers in the *high* group have a similar educational level. The amount of farmers with education at a university level is similarly lower.

The main agricultural production in Norway is milk and meat production. Near 20% produce grain as their main commodity. Near 90% of farmers deriving a *high* amount of their income from farming, produce milk (66%) or meat (22%). 65% of all milk producing farmers are in this group, only 7% in the *low* group produce milk. The majority of the *low* group produce meat (47%) or grain (30%). They constitute nearly three quarters of Norway's grain producers. The *average* farmers produce milk (42%), meat (29%) or grain (17%). The remaining farmers produce vegetables (7%), wood (3%) or other industries based on agricultural resources (3%).

As most farms in Norway are transferred within families on allodial rights (82%), prospects of a future farmer within the family is important for most farming families. The data reveals that 58% of the farmers believe that someone within the family will take over the farm in the future. There are no differences between the groups on this issue.

The following regression analysis is based on an additive index of farmers' prospects of investments in the farm premises, machines and tools and increasing the area of agricultural production. Prospects of possible investments were given positive values, status quo, 0, and wearing on buildings and equipment or reducing the farming area were ascribed negative values. The final index ranges from -3 to 6, finding its median in 1. The variable is constructed out of the argument that a future in farming is dependent on a certain level of investment.

What distinguishes an investment-willing farmer from a non-investing one? The variables described above are included in the analysis as independent variables in addition to a scale of the farmers' economic prospects of farm income, ranging from negative (-1), no change, to positive (1).

The regression analysis (Table 2) shows that several of the variables included in the analysis reveal significant information on farmer's differences in the 'will to invest' in the farm property. First of all, farmers with an optimistic view on prospect farm income are more likely to invest than others. Further, farmers with a low amount of income from farming are significantly less prepared for investing in their farm property than farmers in the *average* group (control

*Table 2. Regression analysis of will to invest in Norwegian farming (N=1440). Model Summary: R Square:. 289, Std. Error of the Estimate: 1.3101. Regression: Sum of Squares 1005.402, df 12, M.Sq. 83.783, F 48.811, Sig. 0.000.*

| | Coefficients | | | | |
|---|---|---|---|---|---|
| Model | Unst.Coeffi. B | Std.Er. | St.Coeff. Beta | t | Sig |
| Constant | 1.391 | 0.172 | 8.076 | 0.000 | |
| Ec optimism | 0.702 | 0.050 | 0.327 | 14.084 | 0.000 |
| Am.inc farming | | | | | |
| Low | -0.358 | 0.099 | -0.115 | -3.611 | 0.000 |
| High | 0.159 | 0.101 | 0.049 | 1.576 | 0.115 |
| Production: | | | | | |
| Plant | -0.197 | 0.153 | -0.032 | 1.284 | 0.199 |
| Animal | 4.642E-03 | 0.096 | 0.001 | 0.048 | 0.961 |
| Grain | -0.127 | 0.116 | -0.032 | 1.092 | 0.275 |
| Other | -0.550 | 0.171 | -0.080 | -3.209 | 0.001 |
| Male farmer | 4.057E-02 | 0.106 | 0.009 | 0.382 | 0.702 |
| Age | | | | | |
| Younger farmer | 0.691 | 0.087 | 0.188 | 7.947 | 0.000 |
| Older farmer | -.546 | 0.098 | -0.130 | -5.584 | 0.000 |
| Education | 7.563E-02 | 0.046 | 0.038 | 0.662 | 0.097 |
| Family success | 0.608 | 0.072 | 0.193 | 8.405 | 0.000 |

group). There are no major differences between farmers with average or high amounts of income from farming. Farmers within different main production area (grain, meat, milk) do not differ significantly from each other. Milk producers, as a distinct and large group, represents the control group in the analyses, and only farmers in "other" productions (wood *etc*) are significantly less ready to invest. When looking at farmer characteristics, some interesting distinctions are revealed.

There are for example, no differences between men and women on the issue of investing. Farmers age is an important factor explaining differences in will of investments. Young farmers (aged 18-40), are more prepared to invest and old farmers (60-85) less willing to invest than farmers in the mid age group (41-59) (control group). This pattern is quite natural as few farmers at the end of their farming career chose to invest substantially on their farm. They reserve those choices for future successors. Prospects of a future kin successor are the last variable of explanation in the model. Farmers with a kin successor in sight are also more willing to invest in the farm.

## Discussion

The aim of this paper is to show how some features of Norwegian farming households adapt to changing agricultural policies and decreasing financial output in farming and to give an insight to how different groups of Norwegian farmers view the future of farming.

The introductory analysis showed that the long ongoing trend of collecting household income from farming still applies, but an increasing number of farmers derive income from off-farm work. One half of the Norwegian farms earn 0-25% of household income from farming, a *low* amount of farm income. On such farms, the level of female farmers are double that of other farms. This corresponds to earlier studies of work and income patterns on Norwegian farms, where female farmers often report to work less on their farms in combination with a husband who also undertakes considerable off-farm work (Bjørkhaug and Blekesaune, 2004). Farmers on these properties have higher educational levels - one assumption is that a high skill base in another occupation might dissuade farmers to work more on their farms, combined with potential for earning a higher income from off-farm work. These farms are often in meat or grain production, production-types that are less time consuming and lend themselves to combining flexible on farm production with off-farm work.

The *average* group of farms supports household with 25-50% of its income. The *average* farm, when amount of household income from farming is considered, is not very different from a farm with *high* amount of the household income derived from the farm, still the difference is the adaptation to the off–farm income. In both of these groups the milk producers are found. Milk production is a time consuming operation, not allowing the farmer much time for off-farm work.

The future prospects of farming were measured through an index of 'will to invest' in farm property and land. 'Will to invest' can be seen as an optimistic view on the future in farming and those with an optimistic view on economic prospects do hold a stronger will to invest than others. The investment prospect is also much stronger among younger farmers than older farmers. A future successor in sight is also a positive factor for 'will to invest'. This might indicate that farm maintenance is connected to a farming ethic of keeping the farm for the family, not investing for an increased market value of the farm. This is interesting as Norwegian market polls state that there is a higher demand than supply of farms on the open market.

While most farmers have for a long time adapted a strategy depending on a substantial amount of off-farm income, a large group of farmers are adapting so well to an off-farm labour market that farming remains more like a hobby. Analysis presented in this paper indicates that that money is not necessarily transferred into the farm, to invest in further production. In time production on these farms will possibly end, the land sold or leased out, or in some scenarios it may remain in the family as a holiday property. Willingness to invest does not differ between the *average* farmer and farming households depending on income from farming. These two groups do still hold a willingness to invest in their farms and production. Future farming in Norway depends upon these groups ability to adapt to a further decrease in farm income and income generating work outside farming.

## Acknowledgements

The author acknowledges Arild Blekesaune, Reidun Heggem and Carol Richards for valuable comments on this paper.

## References

Almås, R., 2004. Agricultural history of Norway. From state driven modernisation to green liberalism: 1920-2000. Trondheim. Tapir Akademisk Forlag.

Almås, R., 1994. Norwegian Agricultural Policy and the Family Farm, 1945-1982. Sociologia Ruralis, XXIV-2, 120-132.

Almås R. and M.S. Haugen, 1991. Norwegian Gender Roles in transition. The Masculinization Hypothesis in the Past and in the Future. Journal of Rural Studies 7, 79-83.

Bjørkhaug, H. and A. Blekesaune, 2004. Work and income patterns of men and women of Norwegian family farms. Masculinisation, feminisation, or professionalisation of farm work? Proceedings from the 6th European Symposium on Farming and Rural Systems Research and Extension, http://home.utad.pt/~des/ifsa/ifsa_6th_eu_proceed.pdf.

Blekesaune, A., 1996. Family Faming in Norway. An analysis of structural changes within farm households between 1975 and 1990. R 6/96. Trondheim: Centre for Rural Research.

EU Commission, 2003. http://www.eu.int/comm/agriculture/agrista/2003/table_en/3516full.pdf.

Jervell, A.M. and J. Løyland, 1998. Endringer i jordbrukshusholdenes inntekt utenfor bruket 1986-96. Landbruksøkonomisk Forum (LØF) 4, 63-75.

Kinsella, J. S. Wilson, F. de Jong and H. Renting, 2000. PLuriactivity as a livelihood strategy in Irish farm households and its role in rural development. Sociologia Ruralis 40, 481-496.

Prestegard, S.S., 2005. Multifunctional agriculture and the design of policy instruments: Application to the WTO negotiations on agriculture. WP 8. Oslo: Norwegian Agriculture and Economics Research Institute.

Rye, J.F. and O. Storstad, 2004. Trender i norsk landbruk. Frekvensrapport R.no. 4/04. Trondheim: Centre for Rural Research.

# Diversity of women's role and motivation in Norwegian farms

*Hilde Bjørkhaug*[1], *Reidun Heggem*[1] *and Kjersti Melberg*[2]
[1]*Centre for Rural Research, Norwegian University of Science and Technology, 7491 Trondheim, Norway;* [2]*International Research Institute of Stavanger AS, PO Box 8046, 4068 Stavanger, Norway*

## Abstract

Women have always been an important part of Norwegian agriculture. However, their role as farmers who are equal to their male counterparts has rarely been acknowledged. Literature has pointed to the fact that women's roles have been subordinated, her work has not been recognised and her status has been as the farmer's wife rather than the farmer. The practices and tasks carried out in agriculture are gendered. In fact, the rationalisation and mechanisation of agriculture has 'pushed' women out of agriculture whilst work opportunities outside of agriculture have "pulled" them out. This paper is based on qualitative interviews of women in agriculture, both as farmers and farmers' wives, who also represent different generations. The analysis focuses on these women's stories about their attitudes toward farming and their motives and prospects for a life in farming. What is the present situation for women in Norwegian agriculture, as independent farmer, farming partner or farmer's wife?

**Keywords:** farm women, diversity, motivation, allodial law, Norway

## Gender roles and the role of women in farming

Earlier research on gender issues within agriculture sought to understand the position of women within farming. These studies attempted to explain why women have an inferior position to that of men in agriculture. In particular, studies focussed upon why women's status has been subordinate to the power of historically based discourses which assume a higher male status, or, as 'products of discrimination' where farm women tended to choose loyalty and considerable sacrifices to ensure the survival of the farm (Brandth, 2002). Farming has largely been regarded as a male occupation. Women on the farms are often presented as farmers' wives, mothers or daughters and when participating in the production on the farm, or as helpers or assistants (Alston, 1998). Different categories of work have had different connotations and values, some of which also render farm women invisible. Paid work relates to two sites: 'on' the farm or 'off' the farm. Women's unpaid farm work, as wives or equally positioned farmers, deal with both traditional tasks in the house, but also a substantial level of contribution in the farm work itself. Correspondingly, women operate in paid and unpaid off-farm work.

The presence of women in Norwegian agriculture has changed over the last sixty years. From being farms which offered extended families and other people work for a living, an average farm can hardly give sufficient income for one person today (Bjørkhaug and Blekesaune, 2004). Women have left agriculture through many different phases. Almås and Haugen (1991) argue that mechanisation of agriculture was the most important factor in pushing out superfluous labour in the first phases, while new labour market opportunities emerge as important pull factors from the seventies. The latter gives women new identities disconnected from the family farm or from a farmer's wife identity.

While many women left Norwegian agriculture, the revised Norwegian Allodial Act of 1974 opened up opportunities for women to become farm managers. It now gives the oldest child born after 1965, the right to inherit the farm. Before that, male offspring held the right to family farm succession, independent of the number of older sisters. Since the proclamation of the Act in 1974, the number of women farmers has been slowly rising (Bjørkhaug and Blekesaune, 2004). By 2004, thirteen percent of main farmers were women and about twenty percent of the working hours on the farm were carried out by women. While many of women choose a traditional farming strategy together with a partner (Bjørkhaug and Blekesaune, 2004), studies indicate that there is a group of modern female farmers who have moved away from traditional gender roles. These women have managed to construct an identity partly built on tradition and partly on a modern role for professional farmers (Haugen, 1998).

Research has further emphasised the fact that women and men have different interests and motives for farming (Bjørkhaug, 2004). While men more often express the goals of economically viable production, women's motives are more often connected to an interest of animals, the farm as a family household and the rural lifestyle.

The analysis in this paper focuses upon the stories of women in agriculture and emphasises their motives, attitudes and prospects for a life in farming. What is the present situation for women in Norwegian agriculture, as independent farmers, farming partners or farmers' wives?

## Women in Norwegian agriculture

The paper bases its discussions on qualitative data from two separate studies. One source is eleven in-depth interviews in with women with various positions, both as farmers and farmers' wives. These data were collected in 2003 -2005 in the west and the mid part of Norway. Interviews were carried out with the main farmer and sometimes the farmer's spouse. The second source of data is a case study of ten farm families. Interviews were performed with the main farmer, her or his spouse, parents (former farmers), and one child (potential farmer) and his/her partner. These data were collected during 2003-2005 in western and southern regions of Norway.

When analysing the interviews, we have been searching for parallels and commonalities between the responses. It was revealed that women's stories about their motives, attitudes and prospects for a life in farming were following similar lines of argument that made it possible to highlight five life stories as exemplars of what we believe describes the present situation of women in Norwegian agriculture.

### *Anna, the on-farm-working allodial woman*

Anna has taken over the farm from her parents and runs the farm the same way they did. Her husband is working full time off the farm. Anna distances herself from the term farmer's wife, and calls herself a farmer. She views the farmer's wife as one who is not involved in the daily farm production.

> *I am not a farmer's wife. No, I am not, I am a farmer. It is a bit mean to say, but farmers' wives, they just live on the farm, they do not participate much in the daily operations. (...) It might be that women farmers have a bit different values, views things differently. It is not a goal to own the biggest tractor.*

Anna reports that her main focus is on the work with animals and she cites the opportunity to see the grass growing as the most important motivation for being a farmer.

> *When you see that, when you have sown and you see it grows well, (...) it is nothing better than that.. to work the land and the animals.*

Anna also values living at the place she works so that she can be there when the children return from school. As a less positive aspect, she mentions the lack of social contact during the day as a growing problem in agriculture.

***Bertha, the off-farm working allodial woman***

Bertha took over the farm from her parents and works eighty percent off-farm whilst her husband manages the farm full-time. Bertha says that she would like to work more on the farm, but that is a question of economics.

> *Well, I guess I wanted to work more on the farm. We went through a process where we discussed to develop something on the farm, but it was that versus economic security, having a safe work position off-farm.*

Despite the fact that both parties are not engaged in full time farm work, she feels that husband and wife are equal because of the fact that they both have agricultural education and experience. On that basis they have discussions about agricultural issues.

> *... it is exiting to discuss and present different views. But since he sort of is the full time farmer, he needs to do it the way he feels is the right way, at least for him, since it is his job.*

At the same time, Bertha says that they have different approaches to the daily conduct, but he is the one who makes the decisions due to the fact that he is working full-time on the farm.

> *I think maybe I am more "small scale" while he maybe is more into rationalisation. (...) I believe that women can take pride in work over values and such. Maybe men count money more, they are more related to economy.*

***Cecelia, the on-farm working wife***

Cecelia is married to an 'allodial boy', but is the main operator of the farm. At first she was not prepared to move to a farm. Cecelia was scared of cows and she had no genuine interest in farm work. Since then, she has developed an interest in animals, and cites this as an important motivation of farming for her.

> *I carried with me something from my previous work, care, care for cows and especially calves. I am always there when they calve, as far as I know it is happening. I know the cows want me there. They almost never calve at night. It is just as they are waiting for me in the morning. It is quite funny (...) And I find it pleasant to be there when a new individual is born.*

Cecelia also distances herself from the term farmer's wife and feels that it is a term that describes someone who walks behind her husband and does all the things that he does not do.

> *...it is not being a farmer's wife that is my main occupation. It is being a farmer. Even though I do not often drive the tractor.*

Cecelia and her partner produce organic milk on the farm. The motivation for the re-adjustment from conventional farming was animal welfare and a strong belief in "clean" food. Despite some economic concerns, this woman farmer describes how she was the key motivational force behind the decision.

> *Yes, I guess it was me that had the motivation for.. I do not know how it started, but I have been curious about alternatives all the time, so a few years ago I took a course on 'homeopathy in the barn'.*

One of her key areas of focus is upon maintaining social networks with neighbours and others in the industry. Cecelia talks a lot about the importance of backing up each other and sees the signals when something troubles her neighbours.

***Dina, the on-farm working (niche-production) wife***

Dina grew up in a community near the farm, and married an 'allodial boy' 25 years ago. Being the daughter of a shopkeeper, Dina was used to sales and business development. After the farm take-over, Dina kept on working in her father's stores. She stayed in the clothing business until five years ago, when they started a bed-and-breakfast on the farm. Dina explains the need for starting the new business on the farm:

> *As time passed, there was only farm work for one person. This farm is so small, you need to work elsewhere (...) I attended regular paid work to earn money and finance farm improvements (...) And I found farm work boring.*

Mentally, practically and socially, Dina was not part of farm life for 20 years. This changed when they decided to expand the number and standard of the buildings. Currently, the farm is a well-renowned and popular place for team-building, meetings, weddings, *etc.* Both husband and wife share the responsibility for the enterprise, and enjoy being together running a profitable business:

> *We started to renovate existing buildings, and then built new buildings, proper kitchen facilities (...) We started with this, and then reduced the farm production. (...) The way it is today, we earn more money on the extra business than on the farm. But there is a lot of work behind it...*

In this way, this couple has created a new role for Dina and a way for her to make a living on the farm, which was absent when they took over the farm a quarter of a century ago.

***Ethel, the off-farm working wife***

A couple in their late twenties are running the farm in this case study. They took over the farm nine years ago, and have established a well-functioning family with two small children and the husband's parents live in the house next door. The husband is the main farmer and does most of the farm work, but his father also helps him out and steps in during vacations and so on. The woman, Ethel, works from her home-office on the farm as an illustrator of children's books. She describes her changing role on the farm through those nine years:

> *I didn't know what it was going to be like to live on a farm. I'm not from a farm. I had never thought I would end up on a farm, but I think it is exciting. There are many advantages, (...) to live so freely, to work with the nature. Peasant*

> *romanticism... And I like doing gardening; I saw the possibilities of combining different activities.*

Ethel seemed to have had a quite naïve view on farm life, being only 20 years old when taking over. In the start she worked quite a bit on the farm, but this changed radically when they had their first child three years ago:

> *Before we had children, I was responsible for farm work in the weekends, now it's only "now and then". (...) and I wash his clothes (laughs). We used to share work outside and indoors more equally before we had children. But I really like living like this. (...). It's a lifestyle.*

Farm life makes it possible for her to practice in a flexible occupation and still be part of farm life. Ethel has established her own, independent business as an illustrator on the farmyard:

> *I've made a choice now that the children are small, I don't want to work a 100% in the city. It would take so much time to commute, and here I can work effectively from my home office. And I'm a part of the farm family and the everyday life on the farm, staying at home during the days.*

## Discussion and conclusion

This paper is based on analysis of qualitative interviews with women in agriculture, as farmers and as married to farmers. The stories told by these women draw us to the conclusion that there is a great diversity in female farm activities and motives. The stories outlined above clearly reveal that women in Norwegian agriculture are not a homogeneous group. Still we find lines of similarities valuable for a general discussion about the position of women in farming. It is obvious that some of these women distance themselves from the term farmer's wife and the daily practices associated with such a term. Traditionally this was the role of women in farming.

Today, it seems like the farming women are engaged in defining themselves into the farmer role, namely as the ones who have an active role and the responsibility for the farm and who do not merely follow her husband and attend the work that he does not do. These women also define their role as a farmer as something else than the traditionally definition of farmers, as a person that drives the tractor and manage heavy machinery. They believe men are more into and aware of that kind of symbolism. Farming women's definition of a farmer has a content of care for animals, small scale productions and alternative modes of production. Other women use the farm as a basis for running their own businesses, either farm-related or not farm-related, thus enjoying the freedom, closeness and flexibility of farm life.

On this basis we see that some farming women want to be visible as farmers and the way to do it is to get involved in the daily farm operations. When comparing farming and non-farming women's stories, it seems like working in agricultural production is more important than allodial rights when it comes to real empowerment in the decision making process. For these women it is important to be involved in the practical work on the farm. This can be both as being the main manager of the farm and/or being responsible for a niche production on the farm. The women in our study paid attention to the rural lifestyle, the farm as a place of work and to social networks. Being accepted as members of the agricultural society is one dimension of this. Another advantage that women acknowledge is flexible working hours on the farm. Flexibility enables them to spend more time with their children and other family members, an opportunity appreciated by both active farmers and farmers' partners. Even women, who are not interested in traditional farm work, have showed interest in running a business based from the farm, as was the case of the illustrator.

The increasing numbers of girls and women taking over the farm on allodial rights does not necessarily mean that we will see more women in farming and with that changing standards. However, in those cases where a woman has either taken over the farm or married into a farm and becomes the main farmer, we can see that women have brought about a change in the farm focus. In this study some aspects of women's motivations in farming have been revealed, and through the stories analysed, the most striking discovery is that an active farming role is determining the influence on the farm practice.

## Acknowledgements

The authors acknowledge Carol Richards for valuable comments on this paper.

## References

Almås, R and M.S. Haugen, 1991. Norwegian gender roles in transition: The masculinization hypothesis in the past and in the future. J of Rural Studies 7, 79-83.

Alston, M., 1998. Farm women and their work: why is it not recognised? J of Sociology 34, 23-34.

Bjørkhaug, H., 2004. Is there a 'feminine principle' of farming – and is organic farming a way of expressing it? P 5/04. Trondheim: Centre for Rural research.

Bjørkhaug, H. and A. Blekesaune, 2004. Work and income patterns of men and women of Norwegian family farms. Masculinisation, feminisation, or professionalisation of farm work? Proceedings from the 6th European Symposium on Farming and Rural Systems Research and Extension, http://home.utad.pt/~des/ifsa/ifsa_6th_eu_proceed.pdf.

Brandth, B., 2002. Gender Identity in European Family Farming. Sociologia Ruralis 42, 181-200.

Haugen, M.S., 1998. Women Farmers in Norway. Report no. 9/98. Trondheim: Centre for Rural Research.

# Multifunctional farms and rural development: A study of four Swedish rural areas

*Johanna Björklund and Rebecka Milestad*
*Department of Urban and Rural Development and Centre for Sustainable Agriculture, SLU, Sweden*

## Abstract

This paper discusses aspects of multifunctionality emerging from a local rural perspective. The concept is discussed on the basis of interviews and discussions with farmers and rural development groups. Multifunctionality is used in the sense of social, ecological and economic functions of farms. The study is research-in-progress and focuses on four rural areas in Sweden where agriculture has low viability and where rural development groups are working for the persistence of farms. Semi-structured interviews, group interviews and collection of physical data of farms are methods used in order to explore the multifunctionality of farming and its contribution to rural development. Despite the declining economic importance on the national level, agriculture may still locally contribute as an important basis for rural development. Agriculture concerns land use and resource management, which distinguishes farming from other rural enterprises. The appearance of the landscape, the natural and cultural values existent are important for the willingness to stay or settle in an area. The mode of agricultural production is decisive for ecological functions important for other economic activities, as well as for human wellbeing.

## Introduction

Since the fourties the economic importance of agriculture in Sweden has decreased. Increasing farm size, increasing degree of mechanisation and decreased prices on agricultural commodities are some reasons for this development. As a result, agricultural enterprises have lost their unique position, *e.g.* in the policy debate, as the basis for economic rural development (Westholm and Amcoff, 2003; EuroFutures, 2005).

Multifunctionality is a feature of all forms of agriculture. OECD (2001) has proposed that multifunctionality means that an activity may have multiple outputs and can contribute to several societal goals at once. In agriculture, both market goods and non-market goods are produced at the same time. Many of the non-market goods (or externalities) are public goods, for which no market exists or functions properly (Lankoski, 2000; OECD, 2001). Different farms and different production systems can deliver different kinds of desirable non-market goods, and at differing degrees (Miettinen, 2000). Multifunctionality is an important component in traditional agriculture assuring food security, income generation and conservation of natural resources (Butler and Carkner, 2001). Such systems generate a variety of products while contributing to multiple social and ecological functions.

In the path of modernisation the multifunctionality of farms has decreased (Pretty, 1998; Westholm and Amcoff, 2003). Mechanisation and specialisation, both on farm and regional level, has led to an impoverished agricultural landscape and less support from agriculture to ecological functions (Björklund *et al.*, 1999, Donald *et al.*, 2001). Larger farms, less local markets and less employments in agriculture have reduced the amount of ecological and social functions associated with farming (Van der Ploeg and Renting, 2000; Darnhofer, 2005).

However, farmers are seeking alternative ways to survive economically without increasing in scale or specialisation. Examples include reducing the dependence on external markets by recycling and reuse, diversification of the production with the aim of local sales, adding value by further processing the food produced and getting payments for ecological and social functions produced by stacking on tourism and recreation (Pretty, 1998; Darnhofer, 2005).

## Objective

The objective of this paper is to describe and discuss critical aspects of multifunctionality emerging from local development groups and farmers. The focus is the farm level and the rural community. The paper is a report from research-in-progress. The overall research project also has the following objectives: i) to explore how local rural development groups can work for the benefit of local farms in order to maintain and develop those functions that are important for the rural area, and ii) how resource use and scale are linked to multifunctional agriculture and rural development.

## Material and methods

The research period is May 2005 to December 2006. Four out of five rural areas have been studied so far. The study areas were chosen in places where one or more rural development group is working actively with issues of farming. All study areas are rural areas where economic viability of agriculture is low and most people commute to nearby towns and cities to work.

*Bokenäset* is a coastal area on the west coast of Sweden. The number of summerhouses equals the number of year-round homes. There are farmers living solely on the income from their farms. The production ranges from dairy and meat, to eggs, cereals and potatoes. Due to the high influx of tourists and the vicinity to larger towns, there is a potential for direct marketing used by some farmers.

*Hulta* is situated in central South Sweden, just south of one of the best farm areas. In three small villages, inhabitants and one farm have developed cooperation around recycling nutrients and growing food together. The farm produces milk and meat, but also offers a number of services to the villagers such as land for potatoes, nutrients for home gardening and a farm shop.

*Tived* is a forest area in central South Sweden. Here, additional incomes have always been necessary besides farming. Farms are small and mainly produce meat. A state company owns the forest. Many farms were deserted and forested in the 1960s.

*Trönö* is a small valley in Northern Sweden, 30 kilometres from the east coast. The area consists mainly of forests and the farmland is situated along the valley bottom. The forest largely belongs to farmers. This used to be a dairy production area, but now there are almost no dairy cows or any other animals left in the valley. Farms produce cereals or grass according to the EU subsidy rules.

All areas were visited twice. During the first visit, a group interview using visual tools was conducted with members of the rural development group (*cf.* Pretty *et al.*, 1995). Typically, 5-10 persons participated. The groups made maps of their area and described the development; they listed resources available in their area and made a vision for 2025. Finally, they made a SWOT (strengths, weaknesses, opportunities, threats) analysis for the vision. The group interviews were taped and partially transcribed. One or two representatives of the rural development group were interviewed concerning all farms in the area so that we got an overview of local agriculture. In each study area, 4-7 farmers were interviewed using semi-structured interviews (*cf.* Kvale, 1996). These interviews focused on the farmers' opinion of the contribution of the farm for the local community in terms of ecological, social and economic functions. We asked what would happen and what their area would look like if their farm would be abandoned. The interview also concerned social networks and the farmers' conception of the usefulness of the local development group for the farm enterprise. Furthermore, questions about the farmer's view of the future for the farm and the farming family were raised. Farms were selected according to their different possible functions. All interviews were taped and transcribed. The transcripts were analysed on their content, focusing on the functions of the farm, the contribution of the farm to rural development and how the rural development group could work for the farmers. When possible, we also made a short farm walk together with the farmers. Data on the economic situation of the farm was collected as well as data on production and material flow (not yet analysed). 17 farms have been visited so far. In 5 of the cases, both the female and male farmer was present during the interviews. In most of the other cases, we were only able to talk to the male farmer since the woman had a job off the farm and did not take part in farming activities. The second visit consisted of a workshop where all interviewed farmers and the rural development group were invited. We made a presentation of our preliminary results and asked the participants to comment and discuss our findings. The objective was to verify our description of the area and the relations between the farmers, farming activities and the community. It was also important to discuss how the functions of the farms could be maintained and improved. Some of the results of the data analyses are presented below.

## Preliminary results and discussion

Farms have a number of functions, many of which depend on the fact that farmers manage land and forests. Table 1 displays an overview of the local social, ecological and economic functions of farms that the farmers and development groups expressed.

Maintaining the agricultural landscape was a function that was mentioned by all farmers. They held a notion that farming contributed with the maintenance of an open landscape that was more attractive for human settlement than a landscape covered with closed forests. This was also perceived as a major reason for the efforts of local development groups to keep all active agricultural enterprises in business, especially in the two most forested areas, *Tived* and *Trönö*.

Grazing animals in the landscape was perceived as crucial both by farmers and local development groups. In some areas such as, *Tived*, the tradition of harvesting leys and meadows for fodder was the rationale to keep the landscape open around the settlements, even though the economy of animal husbandry was weak.

The development groups perceived mixed farms with local sales to provide more local functions while monocultures of cereals on large fields for a national market were less desirable. Moreover, organic farms, when present in the area, were seen to more readily use local resources and sequence cropping leading to larger diversity of crops as well as wild flora and fauna.

Direct marketing of farm products has a social function in some of the areas, offering meeting places where people could share information, narratives and perspectives and where collective action could be initiated. In *Hulta* discussions in a milking room initiated a project for local

*Table 1. The social, ecological and economic functions of farms expressed by farmers and local development groups.*

| Social functions | Ecological functions | Economic functions |
|---|---|---|
| farms are meeting place for villagers | farmers maintain and enhance biological and genetic diversity | farms produce food and fibres |
| farms are places for social activities | farmers maintain the agro-ecosystem | farmers maintain roads important to all inhabitants |
| farms represent cultural heritage | farmers maintain and increase soil fertility | farms create local job opportunities |
| local farms offer knowledge on where the food comes from | farmers have local ecological knowledge | farmers have machines possible to use for other activities than farming |
| farms create feelings of safety and trust since someone is at home at most times and since the farmer is known to most people | farms are places for recycling of nutrients locally | farmers have buildings possible to use for other activities than farming |
| farmers create an aesthetic open landscape | farming provides food | farms with farm shops offer products with added value |
| farmers create the cultural landscape | | farmers are often those with available risk capital for investments |

circulation of human urine and sludge. This contributed to the ecological functions of the farm. In other areas direct sales had economic importance for a few farmers. This was the case in *Bokenäset*, an area situated near larger cities and with many tourists.

An interesting notion from the discussion in one of the areas, *Hulta*, was that efficiency might, in a rural development perspective, not be properly expressed by conventional economic measures, such as production per working hour, $m^2$ or hectare. An alternative measurement proposed was the number of people that can earn their living from a farm. Such a measurement would capture the farm's contribution to the area both in economic terms (*e.g.* more employment and money circulating locally) and in social terms (*e.g.* number of people attending schools, day care and associations). Taking one farm in *Hulta* as an example. Three persons were earning their living on 70 hectares of agricultural land, 30 hectares of forest, 18 milking cows, and a few complementary activities. This is exceptional for Swedish standards.

Farms and farmers maintain cultural and historical contexts to different degrees. The notion of this was not explicitly expressed but the farmers proudly described that they were taking care or reconstructing remnants of the old agricultural landscape. Striking examples was farmers in *Trönö* maintaining old animal sheds traditionally used for summer grazing (*fäbodar*), as well as forest grazing in *Tived* or maintenance of old species-rich meadows in *Hulta* and *Bokenäset*.

Finally, doing off farm work in the neighbourhood was a function that was mentioned by all farmers. Such work could be road maintenance in summer and winter, digging and forest work. For the farmers, such income also played a large importance for the viability of the farm.

## Conclusion

As the paper describes research-in-progress no final conclusions are developed. However, one concluding remark so far is that, on the local level, the contribution of farmers and farming is mainly the production of a landscape and the contribution to meeting places and social networks. The production of food for the local area was not perceived as a function today. Presently, there are small or non-existent economic incentives to encourage localised food production. We argue that policy could change this and that such a change would have large implications for the farms' contribution to local rural development economically as well as socially and ecologically.

## References

Björklund, J., K.E. Limburg and T. Rydberg, 1999. The impact of production intensity on the ability of agricultural landscape to generate ecosystem services: an example from Sweden. Ecol. Econ. 29, 269-291.

Darnhofer, I., 2005. Organic farming and rural development: some evidence frm Austria. Sociol. Ruralis 45, 308-323.

Donald, P.F., R.E. Green and M.F. Heath, 2001. Agriculture intensification and the collapse of Europe´s farm bird populations. In: Proceedings of the Royal Society of London 268, 25-29.

EuroFutures, 2005. Landsbygden 2020 - fyra framtidsbilder. Ministry of Agriculture, Food and Consumer Affairs, Stockholm. 84 pp.

Kvale, S., 1996. An introdruction to qualitative research interviewing. SAGE Publ. Thousand Oaks, London, 306 pp.

Lankoski, J., 2000. Multifunctional agriculture, non-trade concerns and the WTO. In: J. Lankoski (editor). Multifunctional character of agriculture. Agricultural Economics Research Institute, Helsinki, pp. 5-12.

Miettinen, A., 2000. On the effects of multifunctional agriculture on food security and viability of rural areas: review of current knowledge. In: J. Lankoski (editor). Multifunctional character of agriculture. Agricultural Economics Research Institute, Helsinki, pp. 23-38.

OECD, 2001. Towards an analystical framework. OECD, Paris.

Pretty, J.N., 1998. The living land: agriculture, food and community regeneration in rural Europe. Earthscan, London.

Pretty, J.N., I. Guijt, J. Thompson and I. Scoones, 1995. Participatory Learning and Action – a trainer's guide. IIED Participatory Methodology Series. IIED Participatory Methodology Series, London.

Westholm, E. and J. Amcoff, 2003. Mot en ny landsbygdspolitik. SOU 2003:29, Regeringskansliet, Stockholm. 180 pp.

# What is the future for farming inside the agri-food chain?

*Sylvie Bonny*
*INRA (National Institute for Agricultural Research), UMR Economie publique, 78850 Grignon, France*

## Abstract

Farming is the subject of frequent criticism as regards the environment, food quality and sustainability. Thus it seems useful to look at the factors that determine the trends of farming systems. This paper deals with factors often overlooked although important: the influences of the downstream sector on the evolution of farming. The text tackles these influences over the last decades, the economic position of farming in the agri-food chain and finally the growing role of large-scale retail.

**Keywords:** agri-food chain, large-scale retail, value chain, productivism, changes in agriculture

## Introduction

Over recent years and even decades, much criticism has been levelled at farming systems in France and the rest of Europe. Some of the issues thus called into question include the simplification of these systems, the genetic standardisation of animal breeds and plant varieties, the increasing size of animal production units with a heavy concentration of cattle or poultry, pollution and the reduction in food quality; in short their insufficient sustainability. At the same time, many demands are made upon farming in terms of fulfilling new roles, ensuring food characteristics, protecting the environment, meeting consumer demands and maintaining natural areas.

But what are the factors that determine the evolution of farming systems? The agri-food industry is often analysed only in the direction of the physical movement of goods, from upstream industries to consumers. However, farming systems are also strongly determined by their place in the agri-food chain, particularly in relation to the downstream sector and above all by their position in the value chain. This influence of downstream buyers – processing industries and large-scale retail, a group that will be referred to globally as PMR for the Processing, Marketing and Retail sectors – deserves greater consideration given its repercussions. This text thus presents the importance of the downstream sectors in the evolution of farming systems (1), the economic position of farming in the agri-food chain (2), and the growing role of the large-scale retail sector (3). The analysis is focused on the evolution of the model of agricultural production in France over recent decades as well as on its determining factors and its features. Changes in the general socio-economic context, in demand, and in technologies have an influence on the techno-economic paradigm of agriculture, which is reflected in changes in its technological, economic and environmental characteristics (Bonny, 1998; Moreira, 2004).

## The importance of downstream demand in the evolution of farming systems

### *Demand-driven production: A strategy that is already a reality*

Criticism directed at farming as regards the environment or food quality often suggests as a new watchword moving from a 'supply push' approach to a 'demand driven' approach. However, promoting demand in this manner, for which there is almost global consensus, asks few questions about its development and construction processes, and especially its impacts. Demand is seen as an objective that must only be worked towards in order to adapt supply to it, as suggested by Adam Smith in "The Wealth of Nations" as early as 1776: *"Consumption is the sole end and purpose of all production; and the interest of the producer ought to be attended to, only so far as it may be necessary for promoting that of the consumer."*

This objective of steering agriculture according to demand seems to overlook or underestimate the fact that, to a large extent, this has actually been the case over the last few decades. The demand that has played a part is principally that of FPIs (Food Processing Industries) and above all, to a greater and greater degree, that of large-scale retail, whose demands have influenced production systems (Table 1).

*Table 1. Examples of the evolution of production systems linked to downstream demand over recent decades.*

| FPI (food processing industries) and LSR (large-scale retail) factors influencing the evolution of farming systems. | Impact on the evolution of farming systems. |
|---|---|
| Significant development of FPIs and LSR dealing with a growing proportion of ag products. They capture an increa-sing share of the added value while agricultural prices fall. | Production of raw materials with a decrease in on-farm processing and direct sales. |
| FPIs and LSR look for cheap agricultural raw materials. Supplies are bought on the world market with low transport costs, hence fierce competition where agricultural commodities are concerned and great pressure to reduce their costs. | To reduce costs of production, farmers aim to produce more per worker. This leads to the use of more productive breeds and varieties and an increase in the size of farms and production units. |
| FPIs demand products suited to processing; LSR buys varieties and products with a long shelf-life and only accepts those that meet this requirement. | Choice of varieties suited to these demands, hence modification of certain qualitative characteristics of products. |
| FPIs and LSR look for relatively standardised agricultural raw materials in order to simplify processing (thanks to homogeneous raw materials) and marketing (grading, classification standards for products for trade, *etc.*). They indirectly select the varieties suited to these objectives. | Plant varieties and animal breeds become standardised and decrease in number at first, hence a reduction in biodiversity. |
| The downstream sector more recently developed a differentiation strategy for its supply to consumers and increased so the level of contractualisation with producers. | Better adaptation to different demands according to quality criteria specified by the downstream sector. |
| Any very lightly marked fruit or vegetable is refused, along with different products containing a certain level of mycotoxins. Payment according to different quality criteria (composition) becomes commonplace. | Pesticides are used to meet these standards; nitrates also aim to obtain a satisfactory volume with the required composition. |

### *The effects of downstream pressure*

One of the most striking effects of downstream pressure has been the fall in agricultural prices, enhanced by the fact that supply increasingly takes place at the global rather than the local level in order to obtain low-cost agricultural raw materials. To offset the fall in prices, farmers have been obliged to increase the volumes sold and to reduce the unit costs of the product delivered. But this has led to the specialisation and simplification of systems, with notable effects, particularly as regards the environment with the dissociation of crops and animals. The localisation of production has become more pronounced by the establishment of supply industries in places where input materials are cheapest – for example near ports – and by the location of processing industries. This has led to regional specialisation, with serious environmental consequences due to the excessive concentration of livestock in certain areas and almost total absence in others.

Many farmers react to the fall in the prices of their products by increasing the volumes delivered, and supply is therefore often high. This in turn intensifies the tendency for prices to fall, hence a spiral of generally high production leading to low prices, and so on. This has resulted in the creation of surplus stocks, especially given the fact that mobility is difficult once established in farming. In the longer term, the remaining producers must have low production costs. Consequently, the often criticised increase in the average size of production units is difficult to halt, except where processing and/or direct sales of products are possible. The low prices of agricultural products are therefore the cause of two trends that are largely disparaged, while their determining factors are frequently poorly analysed: 'productivism' and the need for income support.

Criticism is often levelled at the negative externalities of farming, in terms of the environment and quality, *etc.* The blame is repeatedly placed, whether explicitly or implicitly, (a) on the choices made by farmers themselves, (b) on pressure from the upstream and agri-supply sectors, especially the agrochemicals industry, and (c) on technical progress and consequently on the general orientation of Research and Development. It is thus often forgotten that the evolution of systems, their trends and their technical choices have also been heavily influenced by downstream pressure, especially as regards prices, standardisation and composition, *etc.* (Table 1).

## The economic position of agricultural systems in the agri-food chain

Farmers, who have become simply suppliers of low-cost raw materials, have seen their share of the agri-food chain shrink, while the downstream sectors of processing, marketing and retail (PMR) have grown. Consequently the composition of foodstuffs has radically changed: they are increasingly processed and the share of pre-prepared, ready-to-eat or cook products has considerably grown, along with consumption outside the home.

### *Agricultural prices and food prices*

While the prices of agricultural products have fallen on a constant currency basis, those of foodstuffs have slightly risen or have remained practically stable (excluding inflation) (Figure 1). The evaluation of food prices is certainly approximate because it integrates both long-term trends in product prices and changes in the composition of the 'household shopping basket', but the two phenomena are difficult to dissociate. If we divide total consumer food spending into two major categories – (1) the value of agricultural products sold by farmers and (2) all spending linked to the processing and marketing of products, including transport, packaging and marketing and advertising costs – there is clearly a reduction in the share of the former

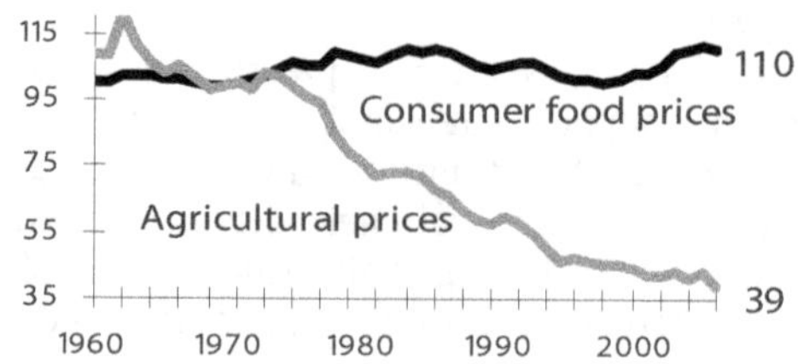


*Figure 1. The evolution of agricultural prices and food prices over the last 40 years (index 100 in constant currency). Source: APCA 2005. Data from APCA (Permanent Assembly of the Chambers of Agriculture) calculated from INSEE data (National Institute for Statistics and Economic Studies).*

in final household consumption (Figure 2). The percentage of food spending returning to the farming sector in the United States has been constantly falling over the last few decades (USDA-ERS, 2002). The same applies to France: the share of agricultural products in overall household food spending now represents only a low proportion due to changes in the composition of foodstuffs and the fall in agricultural prices.

The expansion of the PMR sector can also be linked to general changes in lifestyles: the growing proportion of women working away from home, the demand for easy-to-prepare meals, *etc.* The average daily time spent preparing meals at home has considerably dropped. In other words, jobs in the agri-food industry and retail have replaced on-farm processing/marketing work and domestic meal preparation work.

### *Effects on the environment and health*

The consequences of this evolution are often insufficiently acknowledged. In terms of environmental and health effects, farmers, technical progress and policies often bear the brunt of accusations. Comparatively, until recently, the downstream sector was very rarely called into question, especially by the media and the general public,. However, it can have a significant impact in this field. Transport, for example, which has increased at all stages of the chain with the search for low-cost raw materials throughout the world and the internationalisation of retail, has strong impacts in terms of energy and the environment, as does the increase in packaging and advertising.

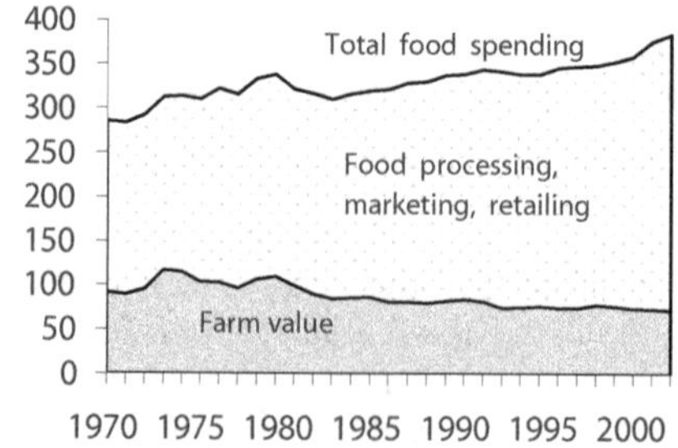


*Figure 2. Distribution of food spending between farm products and PMR over the last 30 years in the USA (in billions of constant dollars) (USDA-ERS, 2002).*

With regard to the quality of processed food products, the addition of less costly ingredients (such as salt, sugar, fat, thickening agents, sweeteners, texturing agents flavourings and emulsifiers, *etc.*), which partly replace the original raw materials, has a significant impact on the characteristics of their composition, their taste and especially their nutritional balance. Processed food products on the market often contain too much salt, simple sugar and fat, which improve their taste while reducing their cost, but which also contribute to various dietary problems, along with the increase in obesity. Another important impact of the PMR sector is the pressure it exerts, firstly to purchase low-cost supplies, and secondly to gain market share almost everywhere in the world.

## Power and dynamics of change in the agri-food chain

The agri-food chain has become increasingly complex and lengthy, with a highly heterogeneous structure: it is made up of stakeholders whose economic size and number at each level are very unequal. Grievink (2003) depicted this using an hourglass, which provides an informative illustration of the situation in Europe (Figure 3). At the top are consumers, who are very numerous but heterogeneous. At the bottom, farmers are fragmented and usually of modest economic size. The central group of stakeholders, at the level of the narrow neck, is incomparable in size and economic power: it comprises the major buying desks of the retail groups. They have significant power because a growing number of upstream stakeholders and firms have to use them as an intermediary in order to put their products on the market. As the large-scale retail sector has become highly concentrated, several large buying desks are becoming an increasingly obligatory stopping point for product marketing. They also impose on their suppliers certain conditions, standards and technical demands that must be met (known as 'private standards'), which may lead to some of them being excluded. This may concern FPIs that have to pay 'slotting fees' in order for their goods to be referenced in supermarkets, and also producers in order to meet technical standards.

The large-scale retail sector also has a considerable influence on consumers, the natural environment and more broadly speaking models of socio-economic development that are directed towards the progression of purchased consumption and the 'commodification' of a growing number of goods.

## Conclusion

There is high demand for a change in the agricultural production model. However, the frequent criticism of the current model sometimes appears rather short-sighted, as by focusing on

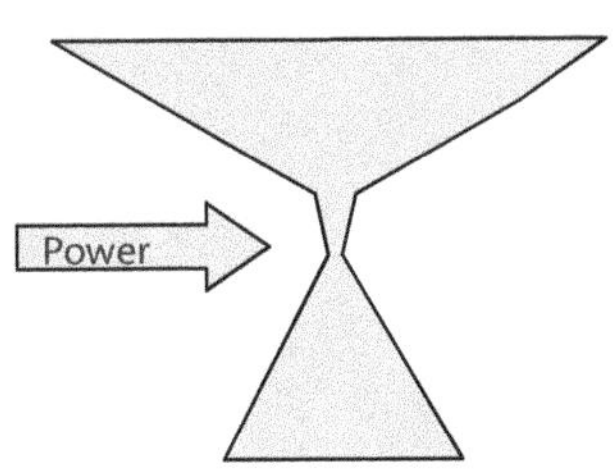


| | |
|---|---|
| Consumers | 160,000,000 |
| Customers | 89,000,000 |
| Outlets | 170,000 |
| Supermarket formats | 600 |
| Buying desks | 110 |
| Manufacturers | 8,600 |
| Semi-manufacturers | 80,000 |
| Suppliers | 160,000 |
| Farmers/producers | 3,200,000 |

*Figure 3. An illustration of the agri-food chain in Europe with the number of stakeholders at each level (Grievink, 2003).*

farming and its upstream sector, it often ignores the significant influence of the downstream sector. Consequently, questions about environmental sustainability, food quality and the impact on Southern countries often seem too exclusively addressed to farming and ag suppliers, while underestimating the dependency of farming on its downstream sector, the role of large-scale retail – which imposes prices, standards and technical specifications – and the importance of the downstream sector in terms of poor nutritional balance and pollution.

There are many threats weighing on the future of the planet, including environmental and social threats. Farming could play an important role in contributing to greater sustainability of human activities and their impacts. But it must therefore have the necessary means to achieve this and, in particular, the agri-food chain must do more to enhance the value of agricultural production.

## References

Bonny, S., 1998. Prospects for Western Agriculture During a Period of Crisis, Changing Demand, and Scientific Progress: A Case Study of France. Technol. Soc. 20, 113-130.

Grievink, J.W., 2003. The changing face of the global food supply chain. In: OECD Conference, Changing Dimensions of the Food Economy: Exploring the Policy Issues. The Hague, 6-7/02/2003. webdomino1.oecd.org/comnet/agr/foodeco.nsf.

Moreira, M.B., 2004. Globalization: the end of the social contract in agriculture? 6th IFSA European Symposium. Vila Real, Portugal, April 3-8. http://home.utad.pt/~des/ifsa/.

USDA-ERS, 2002. Briefing room: food marketing and price spreads. USDA-ERS (US Dept of Agriculture, Economic Research Service). www.ers.usda.gov/Briefing/FoodPriceSpreads/trends/.

# Farming systems under saline conditions

*Willem A. Brandenburg*
*Wageningen University and Research Centre, Plant Research International, Agrosystems Research, PO Box 16, 6700 AA Wageningen, The Netherlands*

## Abstract

As salinification is worldwide a major abiotic stress to agricultural production, it is an urgent need to design agricultural production systems that can cope with saline conditions. It is then important to develop such scenarios for both dryland and wetland conditions, each of them with their own set of physical and societal conditions.

The improvement of salt tolerance in major crops is one factor that can contribute but in order to produce under saline conditions we have to deal with all aspects of an agricultural production system such as plant soil interaction, pest and disease management, livestock under saline conditions, the introductions of new organisms to be domesticated and consequently the development of new production chains.

Salinification occurs under both dryland and wetland conditions. It has under both conditions different constraints but offers different opportunities as well. As agricultural production systems is strongly embedded in regional societal life and so in cultural patterns, a participatory approach with local stakeholders is necessary to face the constraints and to seek the opportunities.

Since there is an increasing awareness of the limits of natural resources, conservation of nature and the environment, there is also a need for designing agricultural systems that are efficient, close nutrient cycles and saves fresh water for other functions. Designing farming systems under saline conditions aims at meeting these challenges. Especially in the coastal zone as a gradient from saline to fresh water and often as a zone where many societal functions meet, there is an urgent need for a future agriculture that can cope with these challenges.

Scenarios for the design of farming systems and a methodology to get the general scenarios accepted by local communities are evaluated for both dryland and wetland conditions.

**Keywords:** farming systems, saline agriculture

## Introduction

About 10,000 years ago, agriculture was initiated simultaneously by human beings at different places. We have to realise that it was the essential step in the change from migrating (hunting and gathering human populations) towards settlement.

Among the earliest types of agriculture belongs the shifting agriculture type (After exhaustion of resources migrating to a new locality). However, when domestication took effect on both plants and animals, a mixed cultivation system was born: the animals for meat, milk and traction power, and at the same time for manure that warranted the soil not being exhausted by plant production. At certain sites permanent settlement was made possible as is obvious from archaeological research, for example in the Middle East. One may deduce from palaeobiological evidence that there are three parallel developments:

- The domestication process, the adaptation of both animals and plants to cultivation conditions, was initiated.
- Animals got an important role in plant nutrition.
- An agrosystem was set up in which plants and animals were not only producing desired agricultural products, but they also created the cultivation conditions for each other.

In essence, the above system has not been changed up to the middle of the $19^{th}$ century. At that time two scientific events give rise to new developments in agriculture: Justus von Liebich showed the principles of plant nutrition, thus leading to the high productive plant production schemes of today. In the same period, Georg Mendel carried out his first systematic hybridisation experiments and with that genetics took off. Was the domestication process until that time a process that showed a gradual progress by finding new types, from the beginning of the $20^{th}$ century, it was replaced for the major crops by intentional plant breeding schemes; and for the major domesticated animals in schemes of intentional animal breeding.

Both breeding and nutrition were main factors in the development of modern high productive agriculture. Nowadays some longterm effects of modern high productive agriculture are obvious:

- Imbalance between water consumption and nutrient uptake has caused salinification – and in some regions serious salinification.
- Water consumption has drametically increased by agriculture, thus leading to fresh water shortages and increasing salinisation.
- Not closed nutrient cycles in agriculture lead to serious environmental problems.

At the same time high productive agriculture is imperative by just maintaining food security for a double world population in about 50-75 years at only 70% of the current area of arable land. When we then also aim at a 10% improvement of the mean supply of agricultural products (both food and non-food), the implication is that the mean production/ha/yr should be increased by a factor of 3.5x! The fact that more than 50% of the world population lives in a region less than 200km from the seaside makes us realise that the application of brackish conditions (both land and water) will greatly contribute to the worlds food security and will create new opportunities for farmers (FAOSTAT, 2005).

## Saline agriculture

Two types of saline agriculture are to be distinguished (Brandenburg, 2005): dryland saline agriculture and the wetland type.

Dryland saline agriculture is mainly to be found in Australia, Middle East and Central Asean republics. Salinisation is either due to a combination of wrong irrigation and plant nutrition or by water supply from deep aquifers of brackish water. Although it is difficult to design sustainable systems for it, we are in the process of designing such systems, based on drought and salt tolerant grasses for feed, the manure to be used for growing *e.g. Jatropha* for biofuel. This crop will provide shelter for other crops thus avoiding an excess of evaporation and at the same time improving degrated soils.

In the coastal zone there are more possibilities to turn wastelands into arable land with the application of saline agriculture and to set up systems of mixed cropping cultivation. Main areas – either where saline wastelands or where already components of saline, non-sustainable production exists – are shown in Figure 1.

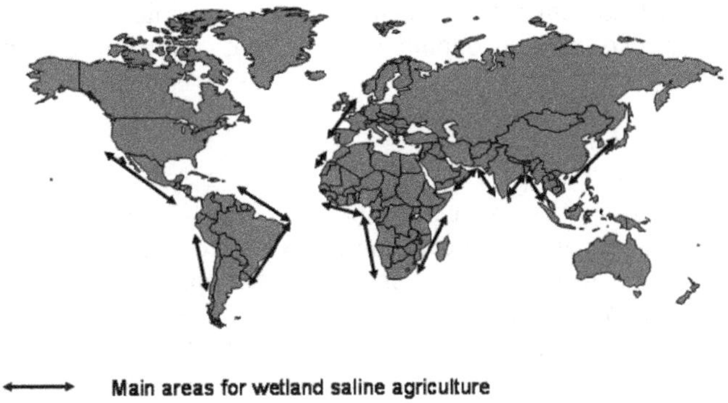


*Figure 1. Main areas for wetland saline agriculture (arrows).*

In order to establish an economically vital and ecologically sustainable wetland saline agricultural system it is essential to create a participative process with local stakeholders (Figure 2). In such a way under all climatic conditions it is possible to set up systems that are productive and at the same time are saving vulnerable coastal ecosystems.

Starting from sole production, we have designed a production system for the southwest estuarine area of the Netherlands that has a closed nutrient cycle in which the water quality of the effluent is the same as the water entering the system. It is based on a mixed cropping system in which baits are used for feed of the sole, the baits fed by unicellular algae, the water is cleaned by filter feeders, macroalgae filters, and saline vegatables, such as *Salicornia* spp. and *Aster tripolium* (Brandenburg, 2000; Brandenburg, 2005; Wagenvoort *et al.*, 1989), all of them to be used as a product. We are in the process to validate the system at an experimental site.

We are in the process of designing a comparable system under tropical conditions, starting from shrimp production, where the shrimp will be also fed by unicellular algae either as a polyculture or the algae growing in separate ponds, the effluent will be cleaned by *e.g.* wetland salt tolerant rice cultivation (CGIAR Science Council, 2003) or by macroalgae for carrageenan production or by saline vegetables such as *Atriplex*, *Salicornia* spp. thus saving vulnerable Mangrove ecosystems along the coast and preventing the collection of macroalgae in marine coral ecosystems (UNEP, 2004/2005).

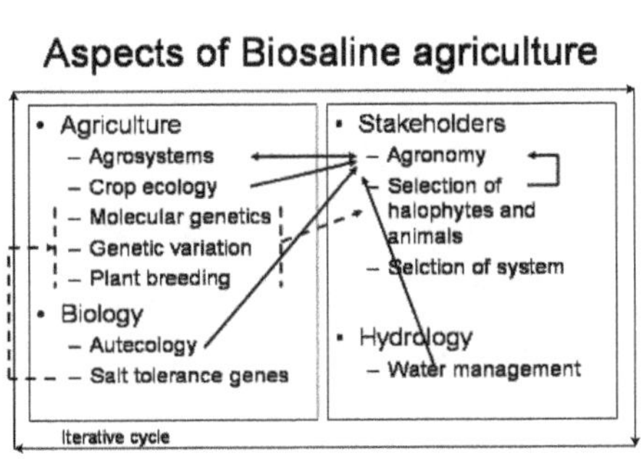


*Figure 2. Aspects of biosaline agriculture.*

The determination which components will be used in both examples is subject of the participative process and will remain so: when economic or ecological conditions are changing, the system will develop accordingly.

## Discussion

In order to get the outlined systems established, for which there is an urgent need in terms of sustainable development of rural areas, the iterative process of stakeholder involvement is essential. Therefore, the outlined systems show only examples with potential. However, when not including stakeholders' needs these systems will not be applied. Systems will be subject to further local development. In this process three aspects are vital: commitment by producers along the production chain, accompanying research and development, communication to society inclusive of education and training, thus warranting skilled human resources in the future.

## References

Brandenburg, W.A., 2000. Saline agriculture: vegetables from the sea. In: P.S. Bindraban and C.W.J. Roest (eds.) Water for food, Plant Research International, Wageningen, pp. 26-28.

Brandenburg, W.A., 2005. Salinisation and its effect on water and nutrient efficiency: opportunities for genetic improvement. The International Fertiliser Society, Proceedings 572. 22 pp.

CGIAR Science Council, 2003. Applications of Molecular Biology and Genomics to Genetic Enhancement of Crop Tolerance to Abiotic Stress - A Discussion Document. FAO. 42 pp.

FAOSTAT, 2005. www.faostat.fao.org.

Wagenvoort, W.A., J.G. van de Vooren and W.A. Brandenburg, 1989. Plant domestication and the development of sea starwort (*Aster tripolium* L.) as a new vegetable crop. Acta Horticulturae 242, 115-122.

UNEP, 2005. GEO Yearbook 2004/5. An overview of our changing environment. 96 pp.

# How do dairy farmers face work problems? Some elements about the adaptation of the technical systems in Central France

*Sylvie Cournut[1] and Benoît Dedieu[2]*
*[1]ENITAC UMR Metafort, 63370 Lempdes, France; [2]INRA TSE, UMR Metafort, 63122 St Genès Champanelle, France*

## Abstract

A 30 farms survey was made in Segala (S-W Massif Central, France), with the objective of a better understanding of what means "simplified management" as far as work and dairy herd management are concerned. Six options of simplified management were studied dealing with i) milking: temporal close of the milking parlor (TCM), suppression of the late Sunday milking (13 to 14), once a day milking (1M) and ii) feeding: grazing based management (Past), self service (Self), delegated distribution of the complete diet (CD). The options do not lead to "better working conditions" in the same ways. They target i) less daily routine work duration (Self, 1M, CD), ii) more time available for busy periods or family holidays (TCM), iii) a week-end rhythm with a real "out of milking slavery" on Sunday (13 to 14), iv) changes in the indoor - outdoor tasks equilibrium (Past). Radical innovations on milking practices such as once a day milking or suppression of the late Sunday milking are disturbing the local conceptions of what "to be a good dairy farmer" means. A multiple components factorial analysis was performed with 11 dairy herd management decisions criteria. The projection of each simplification option on the factorial plan shows that some options (TCM, Past) determine a herd management style (herd collective style for TCM and resource use style for Past). At the opposite, Self and CD options can be part of several management styles. 12 to 14 and 1M are intermediary, the first being closer to the maximisation of milk production style management, the second being closer to the resource use style.At the end, "simplifying livestock management" doen't mean much and must be detailed as far as implications on herd management rules and on working conditions are concerned.

**Keywords:** work, dairy farms, livestock management, simplification

## Introduction

The dairy farming sector has not escaped the dominant trends of increasing farm size and decreasing family labor which have characterized agricultural development for over 30 years now. It has also been severely affected by sociological changes impacting on the relationship farmers have with their work: expectations are increasingly oriented towards fully controlled work patterns that are separate from family life (Barthez,1986; Guillaumin *et al.*, 2004), to the detriment of a family farming system where work and private life are indivisible.

Generally speaking, the solutions experimented by farmers in response to their work problems fall into three categories (Dedieu and Servière, 2004): 1) reshuffling the organization of work loads (*e.g.* farm partnerships, employer groups); 2) improving facilities and machinery (*e.g.* robot milking), 3) adaptating livestock management strategies and practices. This paper focuses on options for simplified management currently being discussed in dairy farming circles in the south-west of the Massif Central (France). These options cover an extensive panel of technical improvements to milking and feeding systems. This farm survey-based study has a dual objective:

1. to determine farmers' expectations from these options in terms of work patterns.
2. to assess the impact of these options on overall herd and land management: do they involve technical improvements that can be integrated into a wide panel of livestock management strategies or do they imply a full-scale reappraisal of the strategies themselves?

## Materials and methods

The study was conducted in the Ségala region (a common part of Aveyron, Lot and Cantal departements) in partnership the local dairy professional bodies and the Institut de l'Elevage. Thirty dairy farms were surveyed. We selected 6 simplification options: temporary closure of the milking parlor (TCM), suppression of the late Sunday milking (13to14), once a day milking (1M), pasture-based management system (a longer pasture feeding phase than the local standards) (Past), self service feeding (Self) and delegated distribution of the complete diet (CD). The list of farms to be surveyed was compiled in such a way as to ensure that at least four farms were surveyed per option. The sample had average structural and productive characteristics for the region (244,000 l; 50 ha and 39 cows; 6,500 l/cow/year).

The questionnaire targeted i) the farm's development strategy; ii) expectations and conditions for successful implementation of the option; iii) management practices and their rationales for the following seven items: reproduction, replacement, feeding, milking, drying-off, batching, healthcare. Special attention was paid to the factors deciding the practices implemented according to whether they relate to the characteristics or the cumulated yield of the total dairy herd, or whether they relate to the status or production levels of each individual dairy cow. Milk controls data (herd yield and demographics) were incorporated in the analysis.

## Results

The deployment of one management option does not preclude the deployment of other options, which therefore means that the total number of cases considered for the analysis of certain options (1M, Past and Self) is significantly different from the numbers considered at the sample selection.

### *Farmers' perceptions of the management simplification options*

TCM (n=4). The main reason farmers would opt for TCM is to be able to free up time for summer holidays. This strategy results in a peak work period in response to the calving peaks in early autumn but also generates more down time in the spring (no calvings). Day-to-day herd management must constantly integrate the date factor (date of closing the milking parlor, date of the end of the calving period), which may negatively effect the optimization of milk production for some cows.

13 to 14 (n=5). More than the time gain, it is the break in the routine, the absence of a 'compulsory' task that has to be performed while others are enjoying extended downtime, that drives the choice of this option. It is implemented in farms that are relatively strong in terms of per cow milk yield. Farmers are prepared to lose a small part (-2 to -7%) of their milk production and are able to handle the increase in somatic cells that occurs at the start of each week in exchange for an enhanced quality of life.

1M (n=7). This technique has only recently been experimented in the sample population. This simplifies the day-to-day organization of workload and cuts the work hours required, but the system will need to be watched over carefully since farmers are largely unaware of its long-term effects. Discussions on this technique concern the inevitable loss of milk (which can be difficult to evaluate in late lactation), the improvement of milk composition, the risks related to changes in somatic cell count, the improvement in the cows' health status and the herd's reproductive performance, and the possibilities for reducing concentrate volumes and for using distant plots.

*Past (n=8)*. The 'maximization' of the pasture component results in a wide range of situations: only two farms run a 100% grass system, whereas the others have opted for spring-summer pasture grazing with a switch to maize silage-based winter diets as soon as there is insufficient grass at pasture. The reduction in feed costs is the crucial deciding factor, although the nature of the work involved (spending time with the cows instead of in a tractor) and its distribution over the season (no peak for silage feeding) also play a role.

*Self service feeding (n = 14)*. This option was already common practice: many farms chose to continue with this system since it requires less investment and is less labor-intensive on a day-to-day basis. The equipment upgrades required undermine the long-term implementation of this system in farms where the farm managers are reluctant to invest in covering concrete-floored areas or building large-capacity pits.

*CD (n=5)*. The delegated complete diet option is often applied in GAEC (collective farming groups) combining dairy and off-soil farming. Their prime motivation is technical: the complete ration improves milk yield. However, the consequences on work patterns are far from negligible, since a significant amount of routine animal husbandry tasks are subcontracted. Although farmers are satisfied with this option, it remains a collective organization with all the constraints attached (the round has to be both feasible and profitable).

### *Do options define technical management styles?*

In order to represent the range of management systems applied by farmers in our sample, we developed a multiple components factorial analysis (MCFA) to express the associations between livestock technical practices. The first axis of the MCFA opposes herd-based management systems (reproduction, and drying-off of individual cows have to follow the herd milk production distribution schedule) versus 'individual cow'-based management systems (the rules are essentially directed by individual data). The second factorial axis opposes "meeting production needs" to "efficient use of resources".

The projection of data related to the simplification options within the 1-2 factorial plan (28% of total contribution of inertia) provides an indication of how these options are positioned. The TCM option strongly impacts herd-based management policies due to the constraints that stem from production distribution and the phasing of calving (Figure 1). This management option guarantees a halfway-house between optimization of milk production and efficient resources use. Indeed, the peak of lactation is covered by silage in winter, with cows being put to pasture during the second phase of lactation. The Past option, quite logically, is heavily weighted towards the 'resources use' pole. Nevertheless, grouping imperatives often have relatively little impact, even though they weigh upon some farmers who choose to concentrate all or part of the calvings in a single period (Figure 1). 13 to 14 clustered more closely towards the 'individual' and 'production' poles, confirming that this option is part of a process that introduces the work issue in relatively intensive herd management strategies.1M is closer to the 'resources use' pole than the 'production' pole. This corroborates what the surveyed farmers expressed, since the 1M option is easily adaptable to more autonomous systems that can make savings on milk production costs. The Self option is a highly interesting case (Figure 2), since farms electing to

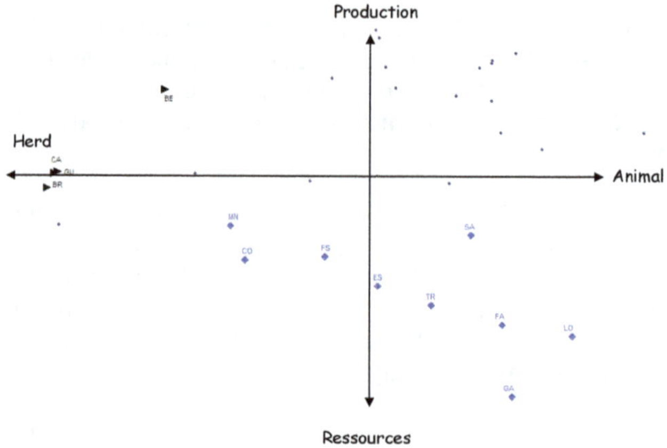


*Figure 1: Projection of TCM option (▸) and Past option (♦) on the MCFA model of livestock management diversity (each point is a farm).*

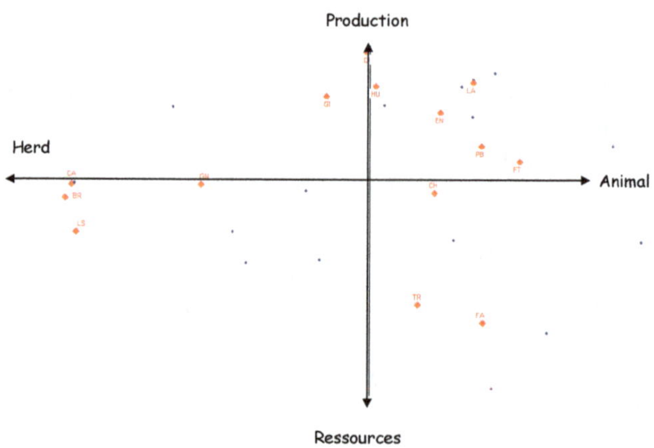


*Figure 2. Projection of the Self option (♦) on the AFCM model of livestock management diversity (each point is a farm).*

implement this strategy are positioned across the whole factorial design, thus indicating that this option has little impact on the organization of management policy: it thus appears as an ad hoc technique, in a similar but less clear-cut way than CD.

## Discussion and conclusion

The family-based farm sample used in this study cannot be considered as representative of the dairy farmer population in the Ségala region. In the largest farms, particularly those involved in collective farming groups or with salaried workers, the 'work 'issue is more centered on work efficiency than on free time (Rault 2005). However, the sample provides a good illustration of the diversity of work-related components that have to be taken into account: i) to reduce the work hours required for daily routine animal husbandry and generate flexibility in terms of workload (Self, 1M, CD), ii) to generate free time at certain periods of the year where interactions between downtime and other farming tasks is a problem (periodic 1M, TCM), iii), to introduce variation in the weekly routine (13 to 14), iv) to adjust the balance between different types of tasks (indoor/outdoor) and how they are distributed over the season (Past). The projection of the options in a factorial design also underlines the wide range of technical implications involved in the choice of management simplification options, some of which can be integrated into a range of management strategies while others have a huge impact on management patterns. The rather vague term "simplification" evidently covers a wide variety of issues, as much in terms of impact on work patterns and impact on technical implications.

The surveys highlight the fact that adopting pioneering techniques, particularly the suppression of milkings, brings into play "reactions from neighbors" (13 to 14) or "pride in showing that [the 1M system] works". This is indicative of the socio-technical dimension involved in changing farm practices (Darré *et al.*, 2004). The 'quality of life' model for a more attractive profession, even going as far as clearly under-performing in terms of milk production appears at the same time as another dialogue on the need to cut social costs and increase labor productivity due to the expected drop in milk prices. The challenge lies in identifying the synergies between these two models and ensuring ways for their driving factors, *i.e.* competitivity and controlled work patterns, to co-exist. Work-related problems can be resolved through a combination of solutions affecting one or more farming techniques and (or) the relevant equipment, and or even labor management (Malderieux *et al.*, 2004). Thus, even though studies on the simplification of farm management policy and on the reference solutions have mainly focused on investigating options taken one-by-one (*e.g.* robot milking or once a day milking), farming consultants need to be able to formalize and evaluate systems providing consistent and compatible technical and social solutions.

## Acknowledgements

Thanks to J.C. Moreau, J. Seegers, J. Foucras, J. Brial, C. Antheaume, J. Weber and P. Riol.

## References

Barthez, A., 1986. Du labeur paysan au métier d'agriculteur. L'élaboration statistique en agriculture. Cah. Eco Socio Rur. 3, 45-72.

Darré, J.P., A. Mathieu and J. Lasseur (Eds), 2004. Le sens des pratiques. Conceptions d'éleveurs, modèles d'agronomes. Science Update INRA Paris. 320 pp.

Dedieu B. and G. Servière, 2004. Des pistes pour améliorer le travail des éleveurs. POUR 182, 32–38.

Guillaumin A. and F. Kling-Eveillard, 2004. Marty M. Attentes des éleveurs laitiers d'Aquitaine en matière de qualité de vie et de conditions de travail. Renc. Rech. Rum. 11, 208.

Madelrieux S., N. Girard, B. Dedieu and L. Dobremez, 2004. Modéliser une activité collective combinant différentes échelles de temps: l'organisation du travail en élevage. IC2004, 5ème journées francophones d'Ingénierie des Connaissances, Lyon, 5 – 7 mai 2004, pp. 115–126.

Rault, G., 2005 Approche sociologique du travail en élevage et questions sur le conseil agricole. Actes du Séminaire de l'Association Française de Production Fourragère «Elevage, prairies, travail», Paris, 20 octobre 2005, pp. 29-35.

# Can family farmers be understood as adaptive managers?

*Ika Darnhofer*
*University of Natural Resources and Applied Life Sciences, Feistmantelstr. 4, 1180 Vienna, Austria*

## Abstract

Although farmers are often seen as the (passive) recipients of research results, this may well be a skewed 'modernist' perception underestimating the farmer's learning processes based on own experimentation. It also seems to disregard the farmer's awareness of the world as characterised by rapid and unforeseen change. Family farmers may thus focus on maintaining the adaptability of their farms through a shifting mix of on- and off-farm enterprises as well as constant experimentation and innovation so as to strengthen the adaptive flexibility and thus the resilience of their farms.

**Keywords:** adaptive management, resilience, evolutionary approach

## Efficiency-lead approach to farm management

Farm management is a research topic often studied by agricultural economists. Within the modernisation paradigm, efficient farm management came to be characterised by the continuous application of innovatory technologies aimed at increasing output and productivity. Agricultural scientists thus continuously seek more refined agricultural 'packages' which combine fertilisers, pesticides, herbicides and crop varieties; as well as more powerful (and expensive) machinery (see Morgan and Murdoch, 2000; Van der Ploeg *et al.*, 2000). The farm development trajectory inherent in this modernisation model requires the farmer to specialise, to increase the scale of her operation, and to combine this with intensified production (Van der Ploeg *et al.*, 2000). This can lead to a high level of efficiency, due to an optimal adaptation to the current social, political and economic environment. Some – mostly large-scale – farms pursue this development model (Walford, 2003; Burton, 2004).

However, should prices become unstable or the socio-political environment change in an unexpected way, the dependence on external inputs and use of expensive technologies can lead to entrapment or 'lock-in' (Van der Ploeg and Frouws, 1999). Indeed, this efficiency-lead farm management approach is best adapted to a setting where the goals are not contested, where the environment characterised by gradual and predictable changes, and where there is a high level of control over implementation.

## Long-term perspective of farm management

The assumptions underlying the approach to farm management as characterised by the modernisation paradigm differ from the evolutionary perspective with its emphasis on the fact that uncertainty about future developments cannot be reduced. An evolutionary approach takes a longer-term view of the farm and points out that making and taking decisions, solving problems, designing and re-designing systems all have to take place under a high level of complexity and uncertainty (Rosenhead and Mingers, 2001). There is no such thing as an ever stable system and farmers have always lived in changing environments – politically, socially, economically and ecologically – where surprise and structural change are inevitable. Indeed, there is an immense uncertainty surrounding any decision about long term growth and development: there are too many factors and many of the relationships between them are unclear or in dispute (Rosenhead and Mingers, 2001).

The emphasis on the need to consider not only the complexity of the system, but also its dynamic aspects is at the heart of the concept of resilience (Holling, 2001). When considering temporal dynamics, a farm's sustainability thus depends on its ability to cope with changes in both external and internal conditions and implies the capacity to create, test and maintain this adaptive capacity (Milestad and Darnhofer, 2003). In particular, to manage a farm for long-term development requires the ability to cope with unexpected events and to juggle shifting objectives (Holling, 2001).

From an evolutionary view point, continual development and change is needed to maintain a farm's 'fitness' relative to the systems it is coevolving with. As Rammel (2003) points out, this imperative to change highlights that there can neither be any best state, nor a stable equilibrium nor an optimal path of development. Evolutionary perspectives do not relate to stability in a static sense as systems are faced with moving equilibria and the dynamics of coevolutionary interactions which cannot be foreseen ex-ante. As Rammel and Van den Bergh (2003:127) point out, given this permanent process of unpredictable change any kind of optimising must be understood as local and myopic. If optimality exists, it will be temporary, because through evolution, selection, innovation and environmental change it is easily transformed into a maladaptive trait. There thus seems to be a trade-off between (economic) efficiency (*i.e.* taking advantage of existing favourable conditions) and adaptability (*i.e.* the ability to sustain long-term stability by maintaining high compatibility in the face of a changing environment) (Giampietro, 1997:85).

## Adaptability-lead approach to farm management

As the research on farming styles led by the Wageningen School (Van der Ploeg, 2000; Commandeur, 2003) has shown, farmers have tended to be careful when faced with a 'package solution', carefully selecting the practices and technologies offered to them. Indeed, if the core objective of many family farmers is less the short-term profit maximisation than the long-term survival of their farm, risk minimizing may be a key consideration. Risk minimising, as understood here, is not based on transforming uncertainties into calculated or insured risks, but on the awareness that evolving systems are complex and can neither be understood nor controlled completely (Rammel and Van den Bergh, 2003). This understanding leads to an alternative orientation of farming strategy which shifts the weight from economic efficiency and short-term optimality to conditions fostering adaptive flexibility and long-term stability.

Although the link between diversity and adaptability is still being debated, it seems likely that diversity can be a key element of long-term stability and even farm survival. Indeed, maintained diversity (on- and off-farm) represents a repertoire of alternative options and increases the possibility that altered conditions or new socio-economical objectives can be successfully met through pre-adaptations and further changes.

This change in perspective emphasises the importance of analysing the family farm as a whole and not separating the agricultural production from the farm family's other activities. This is all the more relevant for family farms orientated towards multifunctionality, *i.e.* engaged in a variety of para-agricultural activities. For example in Austria some 70% of farms are managed by part-time farmers (Darnhofer and Schneeberger, 2006). As the family income depends only in part on the income generated from agricultural production, farmer choices are guided not only by the optimal allocation of factors of production, but also by the values and preferences of various family members, the need to accommodate off-farm activities as well as the constraints and opportunities arising from the farm family life cycle. Family farmers thus seem to take an integrative, holistic look at their farm, and consider the effect of strategic choices on family cohesion and thus long-term farm survival. In Austria, this is not least linked to the fact that farms are typically inherited from forefathers, resulting in a strong bond to the land and a deeply felt responsibility to keep the farm going.

The situated change perspective (see Orlikowski, 1996) is particularly useful when turning away from the assumption of stability towards flexibility, self-organizing and learning. This perspective conceives change as endemic to practice and hence as enacted through the situated practices of farmers, as they improvise, innovate and adjust their work routines over time. In this view series of subtle but nonetheless significant changes are enacted over time as farmers appropriate a new technology into their practices and then experiment with local innovations, respond to unanticipated contingencies, initiate opportunistic shifts in structure, improvise various procedural and cognitive variations to accommodate their evolving use of the technology. Each variation of a given practice is not an abrupt or discrete event. Rather, modifications are enacted through a series of ongoing and situated accommodations, adaptations and alterations that draw on previous variations and mediate future ones. There is no deliberate orchestration of change, no technological inevitability, no dramatic discontinuity, just recurrent and reciprocal variations in practice over time. Each shift in practice creates the conditions for further improvisations, unanticipated outcomes and innovations which in their turn are responded to with more variations. And such variations are ongoing; there is no beginning or end point in this change process. Thus the transformation of farming practices emerges out of farmers' (tacit or not so tacit) accommodations to and experiments with the everyday contingencies, challenges, exceptions, opportunities and unintended consequences that they encounter.

The situated change informs the view of farmers as adaptive managers. Indeed, family farmers seem to be aware of the trade-off between efficiency and adaptability (Darnhofer, 2005). They are aware that the drive to specialisation is based on a false sense of stability and that decreasing the number of on- and off-farm activities may lock their farm into a narrow development trajectory in which the lack of alternative options goes hand in hand with increased vulnerability.

This approach to farm management aiming at maintaining diversity and fostering adaptive flexibility can be illustrated using the results of a case study on organic farmers' approach to pig husbandry, a fairly new farm enterprise serving a young market (see Darnhofer, 2005). For a growing number of farmers in the study area (the Horn district in Lower Austria) conversion to organic farming is becoming attractive, not least due to its potential to support a reorientation of farm activities from a focus on the production of agricultural commodities towards rural development activities. These cover a wide range of potential avenues, including innovations related to food quality and repositioning agriculture within the food supply chain (*e.g.* on-farm processing and direct marketing); non-agricultural activities located at the interface between society, community, landscape and biodiversity (*e.g.* agro-tourism, management of landscape and nature, energy production, provision of communal services); as well as engagement in off-farm income-generating activities (*e.g.* work in a non-farm related field, contract harvesting, organic farming inspector).

The approach of these farmers to rural development activities shows a step wise process, where long-term commitments are avoided due to the awareness of the uncertainty of future developments. For example one of the features of organic pig farming that is particularly attractive to interviewed farmers is the fact that they do not need to keep a large number of pigs and thus can venture into this enterprise without major investment in animal housing. They can adapt the available buildings using resources available on-farm (*e.g.* wood from own forests, family labour). They may thus start organic pig farming with 10 sows or 100 pigs; a number that would be considered uneconomical in conventional pig farming as it does not allow to benefit from the economies of scale. Farmers can thus venture into a new enterprise with limited commitment of financial capital and see how animal husbandry it fits into their farm (*e.g.* use of manure for crop production, labour requirements), how the market develops while at the same time building their experience-based skills. Should the enterprise develop satisfactorily, they may expand it, unless a more attractive alternative use of their resources has developed. For example in the study area organic farmers have joined to invest into a biogas plant which provides electricity to the grid and digested material as fertiliser for their fields. The biogas plant project developed during a short widow of opportunity opened by a law that guarantees farmers a price per kilowatt produced from renewable resources.

This case study shows the farmers' innovation potential and their capacity to recombine and reconfigure the resources at their disposal. Indeed, farmers are well aware of the dynamic nature of the very specific resources available on their farm, *i.e.* family labour, off-farm opportunities, policy incentives, market situations, cooperative networks, climate, financial resources, human capabilities and social networks. Flexibility and adaptability are seen as important attributes to enable the farm to co-evolve with its environment.

## Summary and outlook

There seems to be a divergence between the myopic optimisation strategy focusing on economic efficiency and the long-term farm sustainability strategy followed by many family farmers. Their goal is often to find a balance between securing current income and securing the farm as a viable entity, so as to be able to pass it on to the next generation. Family farmers may thus seek to foster a sound balance between short-term efficiency goals and long-term stability. Farm sustainability can be enhanced by maintaining the evolutionary potential and adaptive flexibility, *i.e.* a continuous process of adaptive learning and the possibility to initiate new development trajectories. It may be helpful to better understand to what extend and under which conditions family farmers use an adaptive approach to farm management. Also, a better understanding of the perceived advantages of a situated change approach as a strategy to secure the long-term survival of their farm might help to highlight the (perceived) shortcomings of normative, optimum-oriented management recommendations.

## References

Burton, R., 2004. Seeing through the ‚good farmer's' eyes: Towards developing an understanding of the social symbolic value of ‚productivistic' behaviour. Sociologia Ruralis 44, 195-215.

Commandeur, M., 2003. Styles of pig farming. A techno-sociological inquiry of the processes and constructions in Twente and The Achterhoek. PhD Thesis, Wageningen University.

Darnhofer, I., 2005. Organic farming and rural development: Some evidence from Austria. Sociologia Ruralis 45, 308-323.

Darnhofer, I. and W. Schneeberger, 2006. Impacts of voluntary agri-environmental measures on Austria's agriculture. Int. J. of Agricultural Resources, Governance and Ecology (in press).

Giampietro, M., 1997. Linking technology, natural resources and the socioeconomic structure of human society: A theoretical model. Advances in Human Ecology, 6, 75-130.

Holling, C., 2001. Understanding the complexity of economic, ecological and social systems. Ecosystems 4, 390-405.

Milestad, R. and I. Darnhofer, 2003. Building farm resilience: the prospects and challenges of organic farming. J. of Sustainable Agriculture 22, 81-97.

Morgan, K. and J. Murdoch, 2000. Organic vs. conventional agriculture: knowledge, power and innovation in the food chain. Geoforum 31, 159-173.

Orlikowski, W., 1996. Improvising organisational transformation over time: A situated change perspective. Information Systems Research 7: 63-92.

Rammel, C., 2003. Sustainable development and innovations: Lessons from the Red Queen. Int. J. of Sustainable Development 6, 395-416.

Rammel, C. and J. van den Bergh, 2003. Evolutionary policies for sustainable development: adaptive flexibilities and risk minimising. Ecol. Econ. 47, 121-133.

Rosenhead, J. and J. Mingers, 2001. A new paradigm of analysis. In: J. Rosenhead and J. Mingers (eds) Rational analysis for a problematic world revisited. 2nd Ed. Chichester: J. Wiley & Sons, pp. 1-19.

Van der Ploeg, J.D., 2000. Revitalizing agriculture: Farming economically as a starting ground for rural development. Sociologia Ruralis 40, 497-511.

Van der Ploeg, J.D. and J. Frouws, 1999. On power and weakness, capacity and impotence: Rigidity and flexibility in food chains. Int. Planning Studies 4, 333-347.

Walford, N., 2003. Productivism is allegedly dead, long live productivism. Evidence of continued productivistic attitudes and decision-making in South-East England. J. of Rural Studies 19, 491-502.

# Interests and policies for becoming farmers: The case of young women farmers in Greece

*I. Gidarakou, L. Kazakopoulos and A. Koutsouris*
*Dept of Agricultural Economics & Rural Development, Agricultural University of Athens, 75 Iera Odos, 11855 Athens, Greece*

## Abstract

EU's setting up (young farmers) incentives provide a good opportunity for young people up to 40 years old to acquire private property and thus establish and operate their own farms. For young women property ownership and becoming a farmer is, in addition, an opportunity to strengthen their position in the farming household and their image as a farmer. The present paper uses data collected in the Western Macedonia region, Greece, in the framework of a wider study on "Employment and gender equality in the primary sector" in the region. The reasons for entering the programme, their managerial role, the production orientation and evolution of their farms as well as their views towards agriculture are examined. Research shows the persistence of the gendered management of farms along with trends for the further development of women's farms as well as the development of an ex-post positive attitude towards the programme and farming on the part of quite a number of women who entered the programme. A number of policy implications concerning the provision of appropriate information and advice as well as the training of young women in the Greek countryside aiming at strengthening their professionalisation are thus indicated.

**Keywords:** young farmers, gender, setting-up incentives

## Introduction

The aging of the agricultural labour force and the difficulties faced in the farming succession process are major problems of the European agriculture. Thus, since the early nineties, the CAP put special emphasis on the provision of incentives through the "young farmers" programme, thus assisting the establishment of young people (under 40 years old) in agriculture. Main instruments are the provision of financial assistance for setting up and the subsidy upon the rate of interest for loans aiming at covering relevant expenses. New entrants also enjoy a number of advantages as for example participation in modernisation schemes and access to quotas as well as to the land made available through the early retirement scheme.

The gender dimension is a basic variable as far as the renewal of farming is concerned. The stronger rejection of farming as an occupation on the part of women in the industrialised world is widely documented, owed to the male inheritance of the farm assets and management and the modernisation of farming which, in turn, deskilled women into the role of farmer wives (Shortall, 1992; Gasson and Errington, 1993; Fonte *et al.*, 1994; *etc.*). The facilitation of entering farming in the role of a farm manager is a substantial challenge and, potentially, a chance for reconsideration of their attitude towards agriculture given the unemployment in off-farm economic activities. It is also a means of empowerment of rural women within the household and the local societies since the role of the manager implies the possibility of occupational training, the development of entrepreneurial identity, access to resources, financial and farming institutions and finally to the public image of farming. Such elements are considered to be of key importance in the empowerment process for rural women (Pettersen and Solbakken, 1998; Alston, 2003). How, then, are such incentives utilised by women and what is their role in the restructuring of labour relations in agriculture? It is known that the role of incentives has marginally occupied research, while generally the gender dimension in the utilisation of the instruments for rural development, such as the subsidy schemes, is restricted and, thus, it is not known to what degree and under which circumstances the policies are transformed into practice (Bock, 1999).

In Greece, such measures are implemented since the beginning of the nineties within the operational programmes of the Ministry of Rural Development and Food. From the few research projects referring to gender it is known that the measures attracted quite a substantial number of young women which, however, did not change the labour relations within the farming family (Kazakopoulos and Gidarakou, 2003; Gidarakou *et al.*, 2005).

Based on such a problematique, the present work intents to explore aspects of the profile of young women and their farms. More specifically, the framework within which their decision to utilise the measures in order to set up into agriculture will be explored as well as some aspects of their managerial role in the farm and the evolution of their farm. Finally, their perceptions on their choice to enter farming will be presented.

## Methodology

The data refer to a sample of 126 young women farmers coming from three Prefectures (Kastoria, Kozani and Grevena) of the West Macedonia region, Greece. The sample was stratified in order to take into account the differences of the rural communities as far as their altitude is concerned (plain, semi-mountainous and mountainous) and accounts for 43% of the total young women farmers in the research area The data were collected through personal interviews based on a structured questionnaire as well as through in-depth interviews addressing 19 young women identified as being actively involved in farming.

## Basic characteristics of the young women farmers and their entrance into farming

The percentage of the young women, entering farming through the programme, within the totality of young farmers in the research area is considered very satisfactory (38%). Similar were the findings of research in Central Greece (Kazakopoulos and Gidarakou, 2003). The young women are mostly married (88%) and have a quite high educational level and in any case better than that of their husbands. Their level of occupational training in agriculture is also considerably higher as compared with their husbands or fathers since such an attendance is a prerequisite for enjoying the financial assistance provided by the programme. Theoretically, thus, young women appear to be more qualified than their husbands in being involved with agriculture. Nevertheless, their significant, in numbers, engagement in the programme does not relate to the redefinition of their attitude towards agriculture. Out of the 126 young women only 4 wished to become engaged in farming after their schooling. Their number increased to 6 at the time they entered the programme. It then becomes obvious that their decision to join the programme does not owe to their need for financial assistance in order to initiate their farming business Agriculture was not a preferred job for them; 21% held a job while entering the programme while, overall, 48% had working experience mainly non-formal ones (*i.e.* without social security). Exploring the reasons for their engagement in the "young farmers" programme two main axes of reasoning emerge: the first concerns the rising unemployment in off-farm economic activities and the second, through a number of arguments, consists in family strategies for the utilisation of the financial incentives. Personal reasons (*i.e.* willingness to develop their own business or to utilise their qualifications) were marginal. The attraction of quite a number of young women in the programme owed to 'collective' family decisions and not to either a redefinition of agriculture or the seizing of an occupational opportunity provided by the programme on the women's part.

## Young women's managerial role

Despite the fact that young women farmers received occupational agricultural training, their engagement in agriculture does not, in general, reflect the image of the farm manager. When the new farms were established 54% of them did not work in the farm and only 7% carried out the farming work by themselves. Still today, 24 women are not engaged in any way in farming. The latter present some distinct characteristics such as that they were never occupied in agriculture, they come form urban areas, the main income source of their household comes from off-farm employment or the size of the farm is big (thus most of the farming work is mechanized and there is no need for them to get involved).

Nowadays, the examination of the labour and the decision-making division in the new farms shows that in the majority of the cases someone but the women are responsible (none of them ploughs alone, only 9% irrigate, 4% do pruning, 1% apply fertilizers but 40% collect tobacco and 25% hoe). Mechanisation is shown by research to be still the basis of labour division (*e.g.* Brandth, 2002). On parallel, there are very few cases in which women are the main decision-makers (17% in regard to the cultivation system, 13% for the purchase of tools and 14% for products' sales and prices). Common decision-making is the usual practice though in quite a number of cases a third person (husband, father, brother) plays a decisive role (57% for the purchase of agrochemicals and 41% for tools, 42% for animal sales, 37% for investments in the farm).

The hypothesis of the women's enhanced access to assets and especially land, through the transfer of land titles from husbands and/or fathers as well as the purchase of land for the establishment of the new farm, is verified albeit in a limited way. Compromises by the Ministry (possibility of either 10-year long lease of land or transfer – even quasi-transfer – of already existing family land to young women) in order to facilitate the access of as many young people as possible to the "young farmers" programme undermined the possibilities for women's access to the land over which they are presented as heads. A more detailed analysis shows that 64.6% of the women did not own land when entering the programme while nowadays 50% still do not. On the contrary, evidence from similar research concerning young men farmers shows that for them land transfer is more common.

## Production orientation and investments

In the majority of the cases (87%) the farms of the young women farmers were oriented upon their establishment towards plant production. Around 44% of them had less than 2 ha. with only 27% being over 5 ha. The small size was essential for the orientation of the farms towards labour intensive cultivations (tree orchards and tobacco) which contributed to the minimum annual working units required by the programme and in which women contribute more in terms of labour. Thus, the new farms diverted from the cultivation pattern of the area under consideration where cereals dominate. A much lesser number (16 farms) was oriented towards animal production including apiculture (5 farms).

The evolution of the farms is substantial in terms of either land size for plant production oriented farms or the number of animals for animal oriented ones. Such an evolution is depicted in the evolution of European Size Units (ESUs). While upon entrance in the programme 74% of the farms were less than 7 ESUs, nowadays such a size accounts for 53% of them. Similar are the changes in specific farm types as well.

Despite the fact that the participation of women in farm decision-making is as already mentioned limited there is an encouraging 41% of them who at least are aware of investment plans in the farm. Some of women (33) have successfully submitted investment plans (within the modernisation schemes of the Ministry of Rural Development and Food). In generals, the objective of smaller farms is enlargement while of the bigger ones restructuring of their production infrastructure. In only 19% of the cases no plans were mentioned either due to the fact that the topic has not been considered or because women are totally disengaged from such decisions.

## Women's attitudes towards their setting up in agriculture

The majority of the women (65%) in their ex- post assessment of their entrance into the programme claim that they took the right decision. The justification of such a declaration refers to the issue of unemployment outside farming as well as to the benefit accrued for the family through the financial aid and thus the contribution towards the fulfilment of family needs and targets in agriculture. Nowadays, one third of the aforementioned women also believe that such a decision gave them the opportunity to own some assets and/or to have some personal income; these are personal reasons which upon their entrance had a marginal role and thus relate to a notion of women's empowerment within the household. Women's viewpoints are quite homogeneous with regard to community type or farm size.

The 23% of the women expressed the view that they took the wrong decision. Such a view is based on the perception that either agriculture does not provide them with a satisfactory income or the commitments of the programme (staying for 10 years in farming) are too heavy for them. The fact that the latter is the main reason of disappointment for 45% of the total of the young women shows that they are looking for off-farm jobs.

Indeed, 57% of the women are willing to abandon agriculture for an equally profitable off-farm job. Nevertheless, the fact that 41% of the women would not is a significant fact vis-à-vis the marginal number of women who wished to engage in agriculture upon entrance.

## Conclusions

Despite the fact that, at the outset, the number of women who were attracted to the "young farmers" programme is promising and shows a slow down of the masculinisation of farming, such numbers were found, by research, to be misleading. On the one hand, the number of young women who wished to become engaged in farming and had personal choices and objectives was negligible. On the other, their involvement in managerial tasks as well as the acquisition of land of their own were not fully accomplished. Therefore, power was not devolved and a gendered relation between farming and sex is, despite numbers, still prevalent.

The majority of the new (women's) farms were found to be small and thus labour intensive (deviating from the cultivation pattern of the research area) due to the requirements of the programme as far as the minimum annual working units allowing for entrance are concerned. Nevertheless, the farms were enlarged in terms of ESUs thus showing an interest in either land expansion for smaller farms and structural improvement for the bigger ones.

The majority of the women assess their decision to enter the programme as being right. Despite the fact that their image does not correspond to that of the farm head, a significant number of women believe that such a decision empowered their identity and gave them the opportunity to contribute to the household income. Nowadays, a substantial number of women, despite the fact that the great majority of the young women were not interested in agriculture while finishing their schooling as well as upon entrance in the "young farmers" programme, have re-oriented their attitude towards farming and they would not abandon agriculture for an equally profitable off-farm job.

The latter also point to some policy implications within the Greek setting. On the one hand, a more intensive communication campaign, tailor-made for young rural women which would inform them about opportunities/programmes and potential benefits should be pursued. This is so since the structure of roles in the farming household and the gendered structure of institutions makes it very unlike that young women are the prime receivers of information on development programmes (young farmers, modernisation schemes, LEADER Initiative *etc.*)[1]. On the other, within a concrete strategy of enhancing the occupational identity/status of young women farmers, the provision of targeted advisory services and training should become among the priorities of the Ministry of Rural Development and Food since such women may well play the role of attractors for the sensitisation and involvement of other young women in farming. Advisory and training schemes should thus aim at actively supporting the viability of young women's farms identified as having a long term perspective in farming as well as facilitating them to enter various programmes especially vis-à-vis the reformed CAP (sustainable rural development focus) and the emerging needs for inter-generational continuity.

## Acknowledgements

The data used in the paper are based on a research co-funded by the European Commission and National resources – EPEAEK II.

## References

Alston, M., 2003. Women's representation in an Australian rural context. Soc. Rur. 43, 474-487.

Bock, B., 1999. Women and rural development in Europe: Appreciated but undervalued. In: Gender and Rural Transformation in Europe. (Compiled Conference Papers). Wageningen University, 14-17 October 1999, The Netherlands, pp. 84-88.

Branth, B., 2002. Gender Identity in European family farming: A literature review. Soc. Rur. 42, 181-200.

Fonte, M., M. Minderhood-Jones, L. van der Plas and J.D. van der Ploeg, 1994. The menial and the sublipe. In: L. van der Plas and M. Fonte (editors). Rural Gender Studies in Europe. Van Gorcum, Assen, The Netherlands, pp. 1-13.

Gasson, R and A. Errington, 1993. The farm family business. CAB International, 290 pp.

Gidarakou, I., L. Kazakopoulos and A. Koutsouris, 2005. Patterns of young women's setting up in agriculture. Paper in XXI ESRS Congress, Keszthely, Hungary, 22-26 August, 2005.

Kazakopoulos, L. and I. Gidarakou, 2003. Young women farm heads in Greek agriculture: entering farming through policy incentives, J. Rur. St. 19, 397-410.

Shortall, S., 1992. Power analysis and farm wives: An empirical study of the power relationships affecting women on Irish farms. Soc Rur 32, 431-451.

Pettersen L.T. and H. Solbakken, 1998. Empowerment as a strategy for change for farm women in Western industrialized countries. Soc. Rur. 38, 318-330.

[1] From the research undertaken in the West Macedonia region it becomes obvious that only one in five young women was provided with information on the young farmers programme by the Extension Service; the rest of them were informed by family members, neighbours, friends *etc.* Additionally, 80% of them did not take advantage of any of the additional incentives provided to young farmers with half of the young women hardly knowing any of those incentives.

# Exploring the multifunctional role of farming systems

*John E. Hermansen, Egon Noe and Niels Halberg*
*Dept. of Agroecology, Danish Institute of Agricultural Sciences, Research Centre Foulum, PO Box 50, DK-8830 Tjele, Denmark*

## Abstract

Public expectations of farming practices are changing from a demand for environmentally "sustainable farming practices" to farming making an "enhanced contribution to the development of the rural areas", the so-called multifunctionality. Based on our research model of including farmers in the development of eco-friendly farming systems, we propose that the achievement of these changed expectations could be facilitated through an appropriate research and development initiative in several European regions. Key elements in such a project should include: (i) the establishment of platforms for dialogue and development of relevant indicators of multifunctional effects (ii) documentation of results obtained from a number of farms and (iii) analysis of development opportunities in the light of the changes "requested". Participants in the regional platform for communication will include farmers, regional policy makers and administrators, grassroots movements and research staff. It is expected that such a coordinated research initiative can revitalize the contribution of farming to rural development and yield important insight to be used by the individual farmer in coping with future challenges.

**Keywords:** indicators, platform for dialogue, farm performance, systems research

## Introduction

Like any other business, farming is constantly challenged by the changing demands and expectations of the surroundings. The development of sustainable farming systems has until now mainly focussed on how to farm without unacceptable environmental impacts. Research and development efforts include decision support tools for farm management in relation to environmental impact (*cf.* Halberg *et al.*, 2005), and the interaction between land use, landscape and ecological services (*cf.* Gibon, 2005).

The public expectations of farming are now moving from a demand for "environmentally sustainable farming practices" to an expectation that farming should be able to contribute significantly (more) to the development of rural areas. This can be translated into an expectation of multifunctionality of farming. At the same time, European food production in its present state is under heavy competition from food production in other parts of the world. Therefore, it is also in the interest of many farmers to look for other ways of generating income. So, for several reasons it will be important for many farmers to adapt to such expectations and opportunities. This calls for a better understanding of the development opportunities of farms in relation to multifunctionality and to explore the possible role of different farming practises in relation to rural development.

The purpose of this paper is to present our experiences in researching farmers' options and motivations for adapting to more environmentally sustainable systems and – based on these experiences – to suggest how an intensified effort in researching multifunctional farming systems could appropriately take place.

## Theoretical approach

It is well accepted that the term sustainability has a major normative component and that sustainable development includes ecological, economic and social issues. A group of European scientists engaged in livestock systems research jointly developed a model based on systems theory and systems thinking (Gibon *et al.*, 1999). This model is explicit in terms of the normative dimension. It represents the duality of the farming system seen as a human activity system as well as a production process.

While the view of a farm as a production process is appropriate to operationalize parameters for production and farm economy, it is less suited to operationalize externalities that have a major normative component. In considering the farm as a human activity system, the farmer (family) is having his goals fulfilled through farming activities. The farmer uses information from the surroundings as well as from the production system to adapt the activities to the family's goals and to respond to the pressure from the surroundings. Therefore, this model gives a good framework for describing and analysing social values in terms of sustainable development as well as the communication between the farmer and the surroundings related to the adaptation and development of the farm.

The research model in Figure 1 exemplifies the use in studying organic farming. It illustrates, that the farmer probably does not develop his norms and agricultural practise in isolation from others. No doubt, the individual farmer may put very different emphasis on to what degree other actors impact on his views and norms. However, the farmer will be influenced by the dominating views in his environment in an effort to ensure the survival of the farm in the longer term.

## Farmer's possibilities and motivation for adapting to a more environmentally sustainable production

During a number of years our research group has investigated how farmers could include environmental issues in their management practices with the aim of fulfilling, on the one

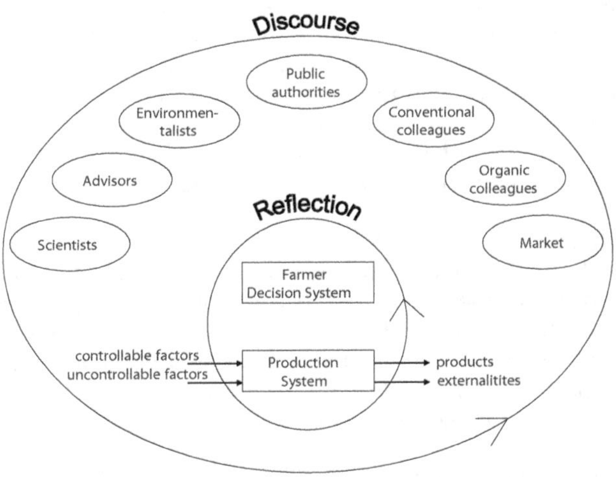


*Figure 1. A model of the farmer's reflection on farm sustainability in light of the value systems and discourses in society (from Kristensen and Halberg, 1997).*

hand, their own goals and, on the other, the social expectations regarding environmental impact. The work was based on the aforementioned duality model. A major project was named "Ethical accounting in livestock farming". The idea was to investigate means of decision support for farmers in order to respond to the increasing public criticism of animal welfare and environmental impact.

The project, which was carried out over a 3-year period in cooperation with 20 farmers, included the development of environmental indicators based on the actual farm performance. The indicators included aspects of ground water pollution, use of limited resources including fossil energy, and impact on wild flora (Halberg, 1999). The indicators were all intended to be scientifically sound, to reflect a concern shared by the community, to make sense to the farmer, and to be adaptable. After finishing the project an independent analysis of farmers' perception of the ideas in the project was performed (Michelsen and Al-Seadi, 1998).

The overall experience was that the majority of the farmers found the information presented to them in the form of indicators valuable and felt motivated to include these aspects in their farm management, and also did so to varying degrees. A drawback mentioned was the lack of reference information for benchmarking at that time, which made it difficult for the individual farmer to assess their own performance and communicate the results.

The project illustrated clearly, that farmers basically do have an interest in how the farm performs in aspects beyond direct regulation and that are of interest to others, and that it is possible to develop indicators to address such issues. Research also shows that farmers and others may meet in dialogue on issues such as landscape changes and aesthetics if the ´right´ platform for dialogue is established (Noe and Halberg, 2002; Højring *et al.*, 2005).

## Researching multifunctional farming

The term "Multifunctional agriculture" is not a well-defined term with a well-accepted meaning among different groups in society. OECD (2005) focuses on the added value potential in an economic sense of improved or diversified production processes, whereas FAO (2005) views the term in particular from a biodiversity perspective. Among many researchers landscape quality is a core element in multifunctional agriculture (Brandt and Vejre, 2004). Without discriminating among the meanings mentioned, we suggest that multifunctional agriculture is the attribute of (the degree of) how different important local aspects of ecological, economic, and societal character are been included in the actual farming management practices. The multifunctionality in that sense is then a matter of the relation between the farm (farmer family) and other interest groups and stakeholders in the local environment.

In the reflection by farmers on the development of the farm under stimuli from the surroundings it was shown earlier that indicators for environmental issues could be developed whereby such reflection was supported and operationalized. It is reasonable to expect that locally important impacts of the production system will have an even greater influence on direct management initiatives. This underpins the concept of supporting the farmer's opportunities to understand and reflect on the impact of the production on the local/regional scale *i.e.* the performance in relation to the site-specific rural development.

If we consider farming to have a potential for (improved) contribution to rural development, significant research and development efforts will be needed to meet the challenges in this area. The analytical model described earlier can form a good starting point for such an effort. However, based on our experience a number of challenges still remain.

First of all, the issues that need to be addressed in such a context include a wide variety of research disciplines. Secondly, the indicators that are relevant in a local context are probably of a less universal character, why it may be difficult to communicate and transfer research results. In order to determine relevant research methods for such complex aspects and also create knowledge about the potential of different farming practices for rural development, we suggest that research and development initiatives should include:

1. The establishment of local platforms for dialogue and development of relevant indicators of multifunctional performance adapted to the local environment. The expectations of and opportunities for farming to contribute to local development are defined among a range of stakeholders (regional policy makers and authorities, professional associations and councils, farmers, agricultural advisors, grassroots movements and researchers of different backgrounds). Through dialogue, the most promising areas are identified and research staff proposes relevant indicators for farm performance.
2. Documentation of results obtained from a number of farms. The indicators of farm "performance" will be evaluated at farm meetings to examine the farmer's affiliation to the indicators and the farmer's views on possibilities and constraints to improve farm performance. These analyses will form new inputs to the platform for dialogue.
3. Analysis of development possibilities in the light of changes "requested'. Using agricultural advisors and researchers, a range of options can be considered for use on-farm and for policy-making at regional or central level.

We conclude that there is a need for co-ordinated, multidisciplinary initiatives in researching the multifunctional role of farming. Although the proposed plan is seen from the local perspective, it is obvious that the establishment of a number of coordinated local projects in different countries represents a particular perspective in creating knowledge; not only on methodology, but also - through knowledge and knowledge transfer - on new development opportunities for the individual farmer. The contribution of farming to rural development can thus be revitalized.

## References

Brandt, J. and H. Vejre, 2004. Multifunctional Landscapes –Theory, Values and History, WIT press, Southampton. 276 pp.

FAO, 2005. www.fao.org/mfcal/i notrad.htm

Gibon, A., 2005. Managing grassland for production, the environment and the landscape. Challenges at the farm and the landscape level. Livest. Prod. Sci. 96, 11-31.

Gibon, A., A.R. Sibbald, J.C. Flamant, P. Lhoste, R. Revilla, R. Rubino and J.T. Sørensen, 1999. Livestock farming systems research in Europe and its potential contribution for managing towards sustainability in livestock farming. Livest. Prod. Sci. 61, 121-138.

Halberg, N., 1999. Indicators of resource use and environmental impact for use in decision aids for Danish livestock farmers. Agric. Ecosyst. Environ. 76, 17-30.

Halberg, N., H.M.G. van der Werf, C. Basset-Mens, R. Dalgaard and I.J.M. de Boer, 2005. Environmental assessment tools for the evaluation and improvement of European livestock production systems. Livest. Prod. Sci. 96, 33-50.

Halberg, N., G. Verschuur and G. Goodlass, 2005. Farm level environmental indicators: are they useful? An overview of green accounting systems for European farmers. Agric. Ecosyst. Eviron. 105, 192-212.

Højring, K., E. Noe, A.G. Busck and E.H. Erichsen, 2005. Landbrugslandet – skabelse og iagttagelse. Syddansk Universitetsforlag. 220 pp.

Kristensen, E.S. and N. Halberg, 1997. A systems approach for assessing sustainability in livestock farms. In: Livestock farming systems – more than food production. J.T. Sørensen (ed.). EAAP publication 89, pp. 16-29.

Michelsen, J. and T. Al-Saedi, 1998. "Hvad synes husdyrbrugerne om etisk regnskab?". I: Etisk regnskab for husdyrbrug. Sørensen, Sandøe and Halberg (eds.), DSR forlag, pp. 113-133.

Noe, E. and N. Halberg, 2002. Research Experience with Tools to Involve Farmers and Local Institutions in Developing More Environmentally Friendly Practices. I: Environmental Co-operation and Institutional Change. Konrad Hagedorn (ed.), pp. 143-161.

OECD, 2005. Building Partnerships for Progress: Multifunctionality. www.oecd.org/department/.

# The labour organization of small-scale breeders in the Brazilian Amazon is a key point for sustainable development

*Nathalie Hostiou[1], Jean-François Tourrand[2], Jonas Bastos da Veiga[3], Ailce Margarida Negreiros Alves[4] and Tienne Barbosa[5]*
*[1]Equipe TSE, INRA, 63122 Saint Genès Champanelle, France; [2]Cirad-Tera, Avenue Breton, 34398 Montpellier, France; [3]Embrapa Amazônia oriental, Tv. Enéas Pinheiro, s/n, CEP 66160-100, Belém-PA, Brasil; [4]UFPA, Campus de Marabá, Marabá-PA, Brasil; [5]UFPA-CA, Campus do Guamá, Belém-PA, Brasil*

## Abstract

To promote the sustainable development of the Amazon region, we are confronted with a lack of knowledge concerning small-scale farming on pioneer fronts. Current knowledge leads us to believe that work organization is relevant for the sustainability and transformation of these farms. Taking as an example dairy-beef cattle farms in Uruará (Pará), this paper aims to produce knowledge on farmers' labour with the view of guiding future research and development actions. The farms are extensive grassland systems, with low levels of inputs, equipment and technology. Most of the tasks (milking, grazing management) are performed manually. A study, based on the quantification of working time shows that most of the activities (milking, pasture management) are carried out exclusively by the family manpower, helped sometimes by wage earners, volunteer workers and farmers' mutual aid. The study reveals the high variability of routine work (from 913 to 3,955 hours per year for milk production) and seasonal work (from 17 to 328 days per year). In order to manage their cattle and pastures, farmers' technical strategies are also very different (time reduction of labour, control and reduction of the peak work period, modification of manpower). The analysis of their trajectories clearly proves the alterations of the interactions between agricultural and non-agricultural activities, cattle production and working groups. Our contribution confirms that labour organization must be taken into account to support sustainable changes of these dairy-beef livestock farms. A better organization of work would facilitate the adoption of new techniques and practices. An aim is to investigate ways to rationalize these farmers' labour, which is characterized by diversified situations. Actions of research should be approached through multidisciplinary point of views.

**Keywords:** livestock farming, work organization, Amazonia, Brazil

## Introduction

Sustainability of small-scale cattle farms is a crucial issue on the pioneer fronts of the Brazilian Amazon region. On dairy-beef cattle farms, livestock is put out to graze on pastures developed from deforestation (Faminow, 1997). Land use is extensive, levels of equipment and input are minimal; yields are low (4-5 litres milk/cow/day) and pastures are invaded by weeds. Working groups are largely based on family labour with few wage earners (Vosti *et al.*, 2001). Tasks are done by hand (milking, weeding and establishing pastures) and are demanding and tiring in terms of manpower (Norgaard-Richard *et al.*, 1988). The work put into these grassland farms could be a problem in relation to size (number of cattle, amount of grazing land) and the constraints on manpower are considerable. While few studies have focussed on work in these dairy-beef cattle farms, in order to ensure their sustainability and intensification (Tourrand *et al.*, 1998; Muchagata *et al.*, 2003). This study is exploratory because it takes a technical approach to analyze work organization in dairy-beef cattle farms. The aim is to increase knowledge as regards working time, workloads and breeders' strategies focussing from the example on dairy-cattle farms in the Uruará region (Eastern Amazon), and with the view of orienting future research and development actions.

## Materials and methods

This study was realized in the district of Uruará on the Transamazon road, in the east of the Pará state in Brazil. As there is no milk processing factories, the farmers sell directly their milk to the city consumers (Poccard-Chapuis *et al.*, 2002). They produce between 5,800 and 51,700 litres of milk per year (Hostiou *et al.*, 2003). The research for this paper was based on following seven dairy farmers over a period of one year (Table 1). To take account work organization, data was collected during interviews on: (i) the farmer' practises with herd management and land use; (ii) the manpower organisation and time devoted to various tasks (milking, livestock and pasture management). The "Work Assessment" method was used for the analysis to quantify work in relation to livestock management and surface areas during one year (Dedieu *et al.*, 1999) (Box 1); (iii) the trajectories of the farms to identify the history of the farm and the transformations of farming activities and labour across time.

Box 1. The "Work Assessment" method makes a distinction between the basic farm units (permanent workers for whom livestock farming is predominant in time and income) and the labour outside the basic group (volunteer worker, mutual help and the intervention of sub-contracting companies and temporary workers, *etc.*) Tasks are differentiated according to the rate at which they are completed: routine work (TA), expressed in hours and performed daily, is difficult to concentrate and defer; seasonal work (TS) includes tasks that can easily be deferred or concentrated, be it cultures, fodder or herd management and is quantified in numbers of days per year. Calculated available time (TDC) represents the time remaining for the basic farm workers to complete un-quantified tasks, *i.e.*, leeway (expressed in hours per year).

## Results

### *Working times*

Three basic farm units were identified: (i) sole farmer; (ii) farmer with a member of his family (wife or son); (iii) farmer with wage earner(s) (Table 1). Routine work corresponds to the daily tasks with producing cows and calves: milking, sale of milk, care and herd gathering. Over one year, it varies from 1,280 to 3,955 hours for the basic farm units. Considerable differences are noted for daily routine work: from 2 hours to 6h30 per person in the basic farm unit. Several factors would appear to account for these differences: the number of members within the basic family unit, the number of cows being milked and the time consuming by sale, these two latter points being the main tasks as regards time. Seasonal work includes tasks with the herd, the pastures and land management. Over one year, this work varied from 17 to 328 days and 17 to 124 days per person. Seasonal work with pastures includes cutting the weeds by sickle, uprooting the weeds' roots and applying weedkiller, repairing and placing fencing, and sewing of fodder grasses when restoring the pastures. All these tasks are done by hand. In these grassland farms, where the herds feed on grass all the year round, significant differences were noted (17 to 176 days per year). The grass surface area would appear not to be a differentiation factor between farms. The differences were seemingly more in connection with the farmers' activities and decisions. Various activities were partly delegated to wage earning workers or volunteer workers as cutting, sewing, fencing. Available time, ranging from 444 to 1,529 hours

*Table 1. Production system type, working time and work organization strategy.*

| Farm | Production system type | Grass surface area (ha) | Number of dairy cattle | Basic farm unit | TA/basic farm unit (h/y) | TS total (d/y) | TS pastures (d/y) | TDC/basic farm unit (h/y) | Work organization strategy |
|---|---|---|---|---|---|---|---|---|---|
| F1 | Milk and | 53 | 11 | 1 | 2250 | 17 | 17 | 444 | A |
| F2 | calves | 70 | 33 | 2 | 1065 | 117 | 96 | 1100 | B |
| F3 | | 75 | 45 | 2 | 1035 | 328 | 176 | 1265 | B |
| F4 | Intensive dairy production | 16 | 17 | 2 | 640 | 38 | 29 | 1848 | C |
| F5 | Milk, calves | 32 | 13 | 2 | 807 | 275 | 144 | 1128 | B |
| F6 | and cultures | 90 | 14 | 1 | 1590 | 93 | 37 | 920 | A |
| F7 | Milk, calves and fattening | 80 | 70 | 3 | 1318 | 252 | 154 | 1160 | C |

TA: routine work; TS: seasonal work; TDC: calculated available time.

per year and per person from the basic farm unit, shows the considerable differences in terms of the time remaining to other activities during the year.

***Technical management and work organisation over one year***

Three strategies, linking work organisation and the technical management of the livestock were identified (Table 1).

*Modify the routine work in order to perform all tasks single-handed (type A):* The basic farm unit, consisting only by the farmer, does many hours of routine work (between 1,580 et 2,250 hours/year). But the routine work varies considerably according to season (end of milking for example): the basic farm unit wants to organise some free time for farming activities other than livestock. The technical practices aim to simplify the herd management (allocation of paddocks near the cowshed, continuous grazing of a paddock). Seasonal activity with the pastures is less time consuming and spread over a period of several months in order to optimise the distribution of family manpower without wage earners.

*Adapt management and manpower to the work periods (type B):* Routine work in these farms can be considered as average (800 to 1,050 hours/person) but the number of days of seasonal work with pastures is high (96 to 176 days/year). Characterised by alternating periods of intense and less intense work during the year, livestock farmers adapt their management technique and manpower to requirements. During the periods when high demand for routine work (calving peaks, road transport difficulties during the rainy season to sell the milk), the basic farm unit uses volunteer family help. Seasonal work is spread out over the year in order to optimise workloads (manual cutting of pastures during the rainy season, establishing new pastures from forested land and fencing at the end of the dry season). Extra family labour or wage earners are used. During the peak of work periods, practices are modified in order to reduce time devoted to leading the herds (gathering or separation into groups, continuous grazing of paddocks for example, *etc.*).

*Employ labour and distribute activities among the workers (Type C):* In these farms the aim is to maintain high dairy productivity and pastures all the year (fast rotation of cows in the plots, manual cutting of pastures). Routine work and available time are noted for their stability throughout the year. Several factors account for this organisation: use of full-time wage earners and temporary workers, spreading of seasonal activities over the farming season in order to reduce peak workloads.

***Work in the livestock farms' trajectories***

In the Uruará region, livestock farming is based on grass with few input and mechanisation and most of the tasks are performed manually. In order to organise their work, farmers juggle with the size of the herd, the size of the working group, and the interactions between the various agricultural activities. The elements presented in this paper show a synthesis of results concerning the trajectories of dairy-beef livestock farms (Hostiou *et al.*, 2003). In these farms based on family manpower, working groups change around the years, with increasing help from teenage children and then the departure once they have grown up (Alves, 2001). In some cases, the loss of family manpower is not compensated by other labour. Indeed, only the wealthier farmers can afford to employ full-time farm labourers. The others hire temporary workers during peak periods (especially for the tasks with the pastures). Activities within the family working group are distributed between cultures and breeding tasks. However, as and when the family manpower decreases, the farmers tend to specialise in breeding. Family manpower also affects the size of the farm in that the available manpower has a stabilising effect on the number of cattle and the amount of milk collected on the farm.

## Discussion and conclusions

Based on the Work Assessment method, this exploratory study is a first analysis as regards the technical organisation of work on small-scale dairy-beef livestock farms in an Amazon frontier region of Brazil. Due to the small size of the sample, the results should not be generalised. However, the study reports a number of factors that are typical of farm working groups in the region, including daily and seasonal work times, the distribution of tasks between the workers of the basic farm unit and other labour. The study also provides a number of explanations. Work would appear to be one of the factors to be taken into consideration to support the transformation of farming practices and the technical organisation of livestock farming methods. Future research will aim at increasing our understanding of work organisation with the view of identifying possible improvements and solutions. Several suggestions have been put forward. The idea is to continue studying work times within a larger sampling of livestock farms in order to establish references for each type of farm (Chauvat *et al.*, 2003). Criteria retained for selecting the farms seemed pertinent: composition of the work groups, agricultural (livestock breeding, cultures) and non-agricultural activity combinations, the milk marketing chain. This study reveales problems of work time, management of peak periods and the strenuousness of various tasks. Farmers' expectations as regards their work should also be taken into consideration for a better knowledge of work organisation (Guillaumin *et al.*, 2005). Research and development actions are therefore necessary. They should be planned in partnership with the farmers themselves in order to avoid any hindrance or reluctance to adopt new and more rational techniques and encourage the transformation of farmers' practices. The aim being to improve farmers' work conditions and encourage the livelihood and the transmissibility of the farms from one generation to the next.

## References

Alves, A.M.N., 2004. Dynamique et reproduction sociale de l'agriculture familiale des fronts pionniers d'Amazonie brésilienne. Exemple du paysannat de la Transamazonienne associant élevage laitier et agriculture. Master, EHESS, Paris, France.

Chauvat, S., J. Seegers, B. N'Guyen, and B. Clement, 2003. Le travail d'astreinte en élevage bovin laitier. Synthèse nationale d'enquêtes bilan-travail. Analyse des aspirations et des solutions envisagées par les éleveurs du dispositif Réseaux d'élevage. Paris, Institut de l'Elevage. 51 pp.

Dedieu, B. and G. Servière, 1999. Caractériser et évaluer l'organisation du travail en élevage. La méthode Bilan Travail. Fascade 1. 4 pp.

Faminow, M., 1998. Cattle ranching systems in the Amazon. UK, CAB international. 253 pp.

Guillaumin A., F. Kling-Eveillard, J.C. Moreau and J. Seegers, 2005. Résultats d'enquêtes en Aquitaine. Quand les éleveurs laitiers parlent de leurs conditions de travail. Travaux et Innovations 115, 30-35.

Hostiou N., N. Girard and P. Lecomte, 2003. Gestion de prairies cultivées par des éleveurs laitiers sur un front pionnier en Amazonie brésilienne. In: Institut de l'Elevage-INRA (editors), Dixièmes Rencontres Recherches Ruminants. Paris: pp 101.

Muchagata, M. and K. Brown, 2003. Cows, colonists and trees: rethinking cattle and environmental degradation. Agricultural systems 76, 797-816.

Norgaard-Richard, B., G. Possio, and S.B. Hecht. 1988. The economics of cattle ranching in eastern Amazonia. Interciencia 13, 233-239.

Poccard-Chapuis, R., M.G. Piketty, J.B. Veiga, N. Hostiou, and J.F. Tourrand, 2002. Milk production, regional development and sustainability in the Eastern brazilian amazon. In: Cirad (editors), Les systèmes agroalimentaires localisés: produits, entreprises et dynamiques locales. Montpellier, Cirad. 12 pp. [CD-Rom].

Tourrand J.F., J.B. Veiga, D. Quanz, L.A. Ferreira and M. Simão Neto, 1998. Produção leiteira em área de fronteira agrícola na Amazônia - o caso do município de Uruará. In: A.K.O Homma (editor), Meio Ambiente e desenvolvimento agrícola. Amazônia, Belém, Brazil, pp. 345-365.

Vosti, A., C.L. Carpentier, J. Witover and J.F. Valentim, 2001. Intensified small-scale livestock systems in the Western Brazilian Amazon. In: A. Angelsen and D. Kaimowitz (editors); Agricultural Technologies and Tropical Deforestation. CABI Publishing-CIFOR, UK, pp. 113-134.

# Commercial flexibility of beef cattle farms

*Stéphane Ingrand[1], Eduardo Chia[2], Charles-Henri Moulin[3] and Julie Cessieux[4]*
*[1]UMR Métafort, Inra Sad, équipe TSE, 63122 Saint-Genès Champanelle, France; [2]Inra/Cirad Montpellier, France; [3]Agro.M, UMR Elevage des Ruminants en Régions Chaudes, 2 Place Viala, 34060 Montpellier cedex 1, France; [4]13, rue de Moulin, 38160 Saint-Marcellin, France*

## Abstract

This study falls within the continuity of a project dealing with the evolutions of livestock farming systems. We introduce the concept of relational flexibility, a concept resulting from management sciences, and try to find out in what way it could help to produce new ways of analysing farming systems. Some surveys were carried out in five commercial companies and then in 15 beef cattle farms. We showed the reality of a relational flexibility of farming systems, based on the relationships between the farmers and the purchasers of the animals. It combines internal flexibility (playing on the product type and the production process) and external flexibility (the type and the number of purchasers, as well as the relation established with them). The stockbreeders have different strategies to face up to risks, either preferring to work with a single purchaser, or multiplying the number of purchasers. Anyway, whatever the farmers' strategy, the level of confidence with the trading partner seems to be the driving force of relational flexibility.

**Keywords:** flexibility, resilience, beef cattle, commercial, strategy

## Introduction

Trying to understand the future of farming systems requires more precise knowledge of the ability and properties of the different existing systems to initiate changes and adaptations in face of uncertainties and evolutions of the context. Such properties of farming systems have to be analysed in the same way as technical and economic productivity, *i.e.* we have to produce concepts, analysis frameworks and evaluation parameters. The question is: what are the conditions for perennial systems together with their capacity to produce sufficient (*i.e.* from the farmer's point of view) outputs. Milestad (2003) speaks about a "prerequisite for survival" and proposes transposing the concept of resilience from ecological sciences to farming systems (ecological resilience becoming socio-ecological resilience), "farms being considered as learning systems in constant evolution with their environment". To better understand the mechanisms of adaptations to the market and to variations in the environment, we chose in a previous research program to mobilise the concept of flexibility, taken from management sciences to evaluate the ability of beef cattle systems to adapt to opportunities according to different events (Lémery *et al.*, 2005). We considered flexibility to be the capacity of the livestock system to adjust quickly to a wide range of economic, technical, marketing and climatic constraints, whilst allowing the livestock farmer to cope with the production plan in the medium term, or even the long term. The present study focuses on the relational aspects of flexibility, that is to say, the social frameworks in which farmers are engaged. We tried to connect these frameworks to the internal flexibility of livestock farming systems.

After defining a few terms, we will present the methodology used to assess commercial flexibility and the main results; then we will discuss some perspectives of this work.

## Flexibility/resilience of farming systems

The concept of resilience implies some kind of equilibrium for a system and either the ability of the system to face up to a disturbance, *i.e.* to recover initial characteristics after the disruption, or the amount of disturbance needed to flip the system toward another point of equilibrium (Milestad, 2003). In management sciences, the flexibility of a company is measured by the number of stable states it is able to reach and by its ability (cost and delay) to flip from one state to another (Tarondeau, 1999). According to the degree of anticipation of the management team, strategic (long term) flexibility can be distinguished from tactical (short term) flexibility, static (over-capacity in order to cope with predictable events) *vs* dynamic (capacity to face unpredictable events) flexibility or proactive (anticipation) *vs* reactive (real time) flexibility.

Farming systems, like all systems, can be defined theoretically by their outer limits, and then by what is inside and what is outside. Some major trends or events which have occurred during the past 15 years (EU policy, social demand, health crisis...) no longer make it possible to assess farming systems functioning independently from the context. Chia (2004) proposed distinguishing the internal and the external flexibility of farming systems, *i.e.* which source is internal or external. In our previous program cited above (Lémery *et al.*, 2005), we focused on the former case, using different disciplines as different points of view (economy, sociology, workforce organisation, management science and animal science, *i.e.* herd management) on the strategy of farmers to face a crisis or any other unpredicted disturbance (Table 1). In the present study, we focused on the external source of flexibility and as a first step, on the commercial links between farmers and the purchasers of animals. The hypothesis was that the number of purchasers (together with the type of contracts), combined with the range of

*Table 1. Strategies of livestock breeders faced with a crisis (Lémery* et al., *2005).*

| Sociological type | "Act upon" | | "Go with the flow" | |
|---|---|---|---|---|
| Farm functioning | high technicity<br>quality signs<br>collective initiatives | herd size increase<br>not enough grass, high stocking rate | experience and tradition based<br>autonomy (food commercial...) | diversification of the beef cattle production<br>other productions than beef cattle |
| Main source of flexibility | collective support | farm size | robustness | diversity |
| Properties of the system that farmers want to achieve | a steady system, non sensible to unpredictable events | | a system able to change and then return to the initial state | a system able to flip from one state to another |
| Strategy, behaviour | external frames<br>technical skills | internal organisation (size, forage, storage) | no change | to grasp opportunities |

Scheme:

categories of animals sold by each farmer is a major source of flexibility for the farming system (see the analogy with the concepts produced by management sciences in Table 2).

## Material and methods

We then defined the range (number and categories of animals sold in a year) as intermediate between internal and external flexibility (Cessieux, 2005). It is the result of a combination between the production context within the farm and the commercial context outside the farm. Both of these elements make it possible to define a potential range for the farmer which can differ from the real range (Figure 1), according to other considerations (farmer preferences, local practices, previous experience or background...). The range is defined by the number of animal categories, the number of animals per category, the fattening status of animals in each category and the inter-annual stability of this pattern.

Data were collected through surveys based on two questionnaires: one addressed to 5 commercial structures purchasing animals in the same area (*i.e.* administrative French department of Allier, to the north of Clermont-Ferrand, with specialised beef cattle farms and a great number of different types of animal purchasers: markets, large and small companies, producer groups) and the other one addressed to 15 farmers chosen by the previous structures, according to two criteria: large or small range of animals sold, high or low involvement with the commercial structure, *i.e.* the proportion of animals sold.

*Table 2. Analogy between industry and farming systems; flexibility according to the range of products.*

| Specific variable according to the products | Theorical definition (industry) | Adaptation for livestock farming systems |
|---|---|---|
| Diversity | Capacity of the system to manage different kinds of products simultaneously | Large number of animal categories sold |
| Renewal | Capacity of the system to substitute existing products by other new ones | Adhesion to quality market signs, fattening or not |
| Modifications of the specifications | Capacity of the system to modify existing products | Quality of feeding, number of heifers kept for breeding, genetics |

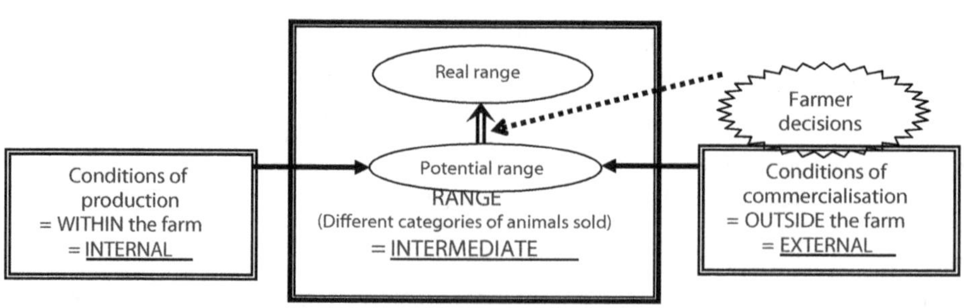


*Figure 1. The range of animals sold in a year: intermediate between production and commercialisation.*

## Results: flexibility for some means constraints for others

### *Commercial structures: flexibility is either quality or quantity*

We chose structures that were very different in size (107 to 900 members) to connect different strategies to different points of view about farm flexibility. The smallest structure does not try to increase the numbers of farmers (chosen according to their ability to sell fattened animals all year long), but tries to increase the number of animals sold in quality market chains for each of them (4200 head sold for 107 farms and about half of them sold in quality market chains). The consequences of the crisis were a decrease in steer production and a development of beef from organic farming. In the 2 biggest structures (900 members/1450 "animal suppliers" for one; 286 members, 901 suppliers for the other), the commercial target is to increase their size (animals and farms). There is no specific criterion to become a member.

In the former case, farmer flexibility is seen as the ability to sell fattened animals any time during the year and to respect technical schedules (organic farming or other). For the latter, the proportion of fattened animals is a criterion measuring the farmer's capacity to delay sales when prices are low or when the market is saturated. They consider that stability (of farming systems) is the greatest insurance of their flexibility (but they have few possibilities to influence it and they need many non-member suppliers to regulate flow according to market demand).

### *Production pattern in farms: flexibility is even unicity or diversity of purchasers*

The range of animals sold is mostly diversified (10 out of 15 farmers surveyed). Within the 5 cases with a low range pattern, only one does not produce any fattened animals (F11). This situation (fattening) corresponds to the traditional production system in this area, where crop production is possible to feed and fatten animals, as well as grass quality allowing steers to be fattened at pasture. For example, one farmer (F8) has no cows. All animals (cows, steers, heifers) are bought from other farmers to be fattened. Depending on the farmer, the diversity of categories sold is a way of ensuring security for the system. More interesting is that there seems to be no relationship between production pattern and commercial circuits, as the combination between a low range of animal sold and many purchasers exists (F8) as well as the combination between a wide range and only one purchaser (F5, F10). In fact, the number of purchasers is not a sufficient criterion to describe the commercial strategy of farmers. The other criterion is the degree of confidence between farmer and purchaser: length of membership, types of contracts, capacity of the purchaser to buy all kinds of animals (*i.e.* lower quality as well as higher quality) or on the other hand the capacity of the purchaser to valorise the best individuals in a specific market chain. This degree of confidence most often implies for the farmer a rather strict schedule allowing the purchaser to anticipate and plan his activity. Not all farmers are ready to undertake such a process, especially those who sell their animals to several non specialised purchasers.

## Conclusion and perspectives

The number of purchasers, combined with the degree of confidence between farmers and purchasers, constitute the main divergence of opinion as to what generates flexibility within the system. For some farmers, a kind of freedom is crucial to be able to cope with any unpredictable event (including the bankruptcy of the purchaser...) and they consider the global flexibility of their system as depending mainly on its internal capacity to change. On the other hand, other farmers, when thinking about the flexibility of their system, include the overall market chain, or at least a part of it (the purchasers). As shown in a previous work dealing with farmer practices involved in quality market chains (Ingrand *et al.*, 2001), some farmers consider the commercial act as a part of their activity, whilst the others delegate this task.

The analysis of flexibility as a process and/or mechanisms to adapt and anticipate either predictable or unpredictable events illustrates i) the diversity of farmer strategies ii) the key role of organisational and commercial networks. The continuation of this work will consist in a "long-term approach" of flexibility (Moulin *et al.*, 2004). Thus, the trajectories of some farms, very well described 15 years ago, will be built up and analysed under different aspects: production pattern, technical practices on herd (feeding, breeding, culling), workforce organisation and commercial circuits.

## References

Chia, E., 2004. La flexibilité relationnelle: le cas des éleveurs. Séminaire Transformations des pratiques des éleveurs et flexibilité de systèmes d'élevage. Montpellier 15 & 16 mars. 8 pp.

Cessieux, J., 2005. La flexibilité commerciale des élevages de bovins allaitants exemple de l'Allier. Mémoire de fin d'étude (DAA). INRA, Agrocampus Rennes. 20 pp. + annexes.

Ingrand, S., I. Carrasco, J. Devun, J.M. Laroche and B. Dedieu, 2001. L'implication des éleveurs de bovins viande dans les filières de qualité correspond-elle à des conduites d'élevage spécifiques? INRA Prod. Anim. 14, 105-118.

Lémery, B., S. Ingrand, B. Dégrange, B. Dedieu, H. Bardey, J. Brossier, P. Pasdermadjian and M. Séréna, 2005. Les exploitations d'élevage bovin allaitant face aux aléas de la filière: une approche en termes de flexibilité. Economie Rurale (in press).

Milestad, R., 2003. Building farm resilience. Doctoral diss. Dept. of Rural Development Studies, SLU. Acta Universitatis agriculturae Sueciae. 375 pp.

Moulin, C.H., S. Ingrand, J. Lasseur, S. Madelrieux, M. Napoléone, J. Pluvinage and V. Thénard, 2004. Comprendre et analyser les changements d'organisation et de conduite de l'élevage dans un ensemble d'exploitations: propositions méthodologiques. Séminaire transformation des pratiques techniques et flexibilité des systèmes d'élevage, Montpellier, 15-16 mars 2004, INRA. 12 pp.

Tarondeau, 1999. Approches et formes de flexibilité. Revue Française de Gestion, mars-avril-mai, pp. 66-71.

# How are European farmers dealing with global environmental changes?

*C. Macombe*
*CEMAGREF, UMR Métafort, France*

## Abstract

Is European agriculture adapting to the changes that the natural environment has in store for us? What are stakeholders in the agri-food and agro-industrial chains doing? The question is of interest to public policies because if players make choices that involve the future, the public authorities have to be able to direct them so they benefit everybody. This paper is a research proposal. We try to make an inventory of the already actual responses from states, firms and farmers. European farmers are already beginning to adapt about global warming changes. Other clues show that some farmers are anticipating scarcity of fossil energy. The research will focus on identification, for each main kind of agricultural industry, of the actor who is known as legitimate and confident speaker. Thus, public power can act upon this actor, to help adaptations at stakes. First outputs, from French case, will be presented at the symposium. To go further, we need help from colleagues, to locate the same kind of information from their own country. This question corresponds to what is happening in reality, at least in France, as we are about to show.

**Keywords:** global warming, adaptation, farmers, organisations, legitimacy

## Global change

### *Climate change and fuels depletion*

Since 1995, the green revolution has been tailing off. World cereal production started to decrease in 1984, and we are losing 60,000 km$^2$ of arable land every year (Reeves and Lenoir, 2005). Further to work by the Intergovernmental Panel on Climate Change (IPCC), set up by the United Nations Organization, there is now a fairly broad scientific consensus as to the scenarios presented by this panel of experts in their third report published in 2001 coming true. That is to say a rise in the globe's temperature of between 2° and 6°, with more rainfall in already wet areas and even less in dry ones. Variability and extreme events should also increase in number. It is an accepted fact that these phenomena are gradually taking hold, but sudden changes cannot be ruled out, with potentially shorter timescales (Schwartz and Randall, 2003). The inevitable growing scarcity of available fossil fuels, because of depletion, adds further to this questioning about climate change. Reliable sources agree in their estimations that world oil production will reach a peak between 2010 and 2020 (Wingert, 2005). After this peak, neither new extraction techniques nor new oilfield finds will manage to offset the depletion of old oilfields.

### *States awareness and research*

The joint phenomena of climate change and depletion, "global change", concern the whole world and every branch of industry, but this paper focuses on agri-food chains in the European Union, illustrated by the French case. The levels of awareness and action vary depending on the country, according to its own environmental history (Acot, 1998). The implications of scientific studies on climate warming are given more and more media coverage but, apart from England and Germany, they are not really managing to influence government thinking or energy consumption habits (Valantin, 2005). In France, public research into climate change is mainly carried out by INRA and particularly involves global warming. Researchers have been working for about fifteen years on two main points: evaluating the impact of carbon dioxide doubling and global warming on agricultural and forestry production; and quantifying the exchanges of agricultural and forest areas for all greenhouse gases and appreciating the efficiency of practices to reduce emissions or increase absorption (Seguin, 2002). These changes represent hazards but can also be tremendous opportunities for agriculture (Chopplet, 2004). In any case, we are going to have to change paradigm.

## Practices are changing

What are the people involved in agricultural-related sectors currently doing to cope with these changes? There are some signs to show that practices are moving in France among farmers and certain interprofessional sectors. Contacts with a very small number of experts (6) show a change in behaviour, crop rotation or practices in many regions.

### *Changes in farming practices*

The northern part of the Lot is a cattle-breeding area (brood and dairy cows) combined with cash crops. The farmers' average rotation of crops is changing; many farmers are introducing alfalfa since they noticed that it was the only prairie grass to grow robustly again at the end of the heat wave in 2003. Many are also planning to introduce one hectare of rape so as to use the oil as fuel for farm engines. Those who have no access to irrigation have given up on maize because of its regularly disastrous yields in dry weather. Lastly, one or two extra hectares of wheat or barley might end up in the home boiler (multi-fuel burners). New practices are become widespread; in the Lot, farmers are no longer sowing new meadows with pure grasses, which are an easy target for drought. Farmers in Limousin are storing two-year-old hay that in the past they would have sold immediately. Many farmers now favour the spontaneous re-growth of bushes when they clear riverbanks, since they have seen the erosion that bad weather can cause on bare soil. Drip irrigation techniques are expanding to the detriment of sprinkler irrigation (Agreste Cahier – an Enquête Structure 2003 survey, April 2005), which is costly in water and promotes the spread of disease. In the Gâtinais region south of Paris, selecting later varieties of rape spares farmers from using growth regulators, use of which had spread since customary varieties blossomed before the winter! New prevention strategies are being developed to combat apple scab by windrowing the leaves on the ground and pulverizing them. Systematic monitoring of new pests is in place (Heliothis moths on vegetables throughout the Centre region since 2003). Some Breton farmers want to sow their fallow land with rape and are planning factories to press it.

### *Changes drivers*

According to our contacts, the drivers of these changes boil down to four or five major influences: two for awareness and two or three for reactivity itself. All the farmers suffer the effects of the concerns circulated by the specialist and general media, who show world-wide phenomena that all contribute to the same message: the global change heralded is taking place. The media offer an explanation and establish the phenomena (storms, heat waves, growing scarcity of fossil fuel, *etc.*) as regular or even permanent events. Some measures to protect ecosystems are starting to sink in. For instance, a farmer who used to let his cows drink from the river, so long as there was enough water, is required by fishermen to apply the law as soon as a shortage threatens young fish stocks. A second driver interacts strongly with the first; this involves what the farmers see and experience in person. We can mention the effect of the 2003 heat wave on grass, the effects of drought on crops and animals, of heavy rain on soil, or of repeated fierce winds in Brittany. The spectacular appearances of pests, parasites and fungi are not to be left out. An extra generation can be formed on the spot, thus multiplying eight- or tenfold the infestation burden (in 2003, a 2nd full generation of codling moth on Val de Loire apple trees and of corn borer). Some "pests" confined to more southern climes are conquering new lands (this is the case of powdery mildew, rife in Languedoc, which went right up to Champagne in 2003). Hordes of locusts have "devoured" grass in the Lot in 2004 and in Aveyron in 2005. The information is not necessarily interpreted on a scientific basis. Some phenomena are perhaps wrongly interpreted as being the effects of global change. Conversely, probable signs of climate change (such as the presence in Creuse, witnessed in 2005, of the Wolfarthia fly, a common sheep ectoparasite in the Mediterranean area) are not identified as such by farmers. Be that as it may, a growing number of farmers are behaving as if global change were in progress. The third driver of change would seem to be the ability to access technology. A wide variety of formerly uncommon techniques are disseminated today: sexual disruption of parasites, grinding parasite-infested leaves, *etc.* Techniques available "elsewhere" have been adopted. Farm mechanization has also evolved, offering new machines for new needs. When a technique is no longer available, its role is particularly noticeable; for instance, the desire to redraw the channels that used to irrigate Limousin fields by gravity is hampered because the traditional hand tool that allows this to be done has practically disappeared. Similarly, even though the buildings have not stood the test of time well, Breton pig breeders rarely take account of the repeated fierceness of winds when designing new buildings for livestock because there is no readily available, inexpensive technology. Adapting regulations is a fourth important driver and one on which the State can have a direct effect. For example, sowing new fields with a combination of one or more grasses and one or more legumes was facilitated at the end of 2004 by a relaxing of the law that prohibited sales of "mixes" in France. Likewise, making irrigation users pay for water, even if inexpensive (in mainland France, the cost varied between 1.8 and 3.8 euros per $m^3$ in 2001 according to Agreste primeur, no.140, published on 13/02/2004), made people aware that it was not a free resource.

***The fifth driver of change is legitimacy***

We finally theorize that there is a fifth driver of change, which is the subject of the research proposal put forward here. This probably has the most influence on crop rotation and practices, provided that it makes itself felt! It involves the position of partners in the agricultural sectors, the moment that they want and have the legitimacy to act in order to adapt to global change. Looking at the sectors, we explored (week of 12 to 16 December 2005) the sites of the various French Interprofessional Offices by product and found the three latest press releases or documents aimed at the public and published on these sites. Two contrasting attitudes emerge with regard to global change. Some Offices have anticipated the changes and give information about their strategy, on at least one of the aspects of global change (French Cereals Office ONIC with transportation of grain by river and ONIOL, the French Interprofessional Office of Oleaginous, Proteaginous Plants and Textile Crops, with the development of a diester). Others are only publicising their technical missions for the time being, without referring to global change (OFIVAL, the French Office for Meat, Livestock and Poultry Farming, ONILAIT, the French Dairy Board, among others), or indeed other concerns (ONIFLHOR, the Interprofessional Office of Fruits, Vegetables and Horticulture). The first two seem to unite the sector's energies and establish themselves clearly as legitimate leaders. We are unaware of the real positioning of the others, either they are carrying out research and forming specific strategies, but without wanting to make them known to the public, or they are not in a position to deal with these issues. Initial contacts indicate that adjustments are in hand. How can these be assessed and what suggestions for action?

## Research proposal

The research question is as follows: how is European agriculture managing global change? The aim is to identify the relevant levels for public authority action. We want to study a "sector". This belongs to the agri-food arena and there are several ways of interpreting it depending on what priority we choose. Interpretation can be done by major types of foodstuff, confirming that we want to above all protect coverage of food requirements. On the other hand, if the aim is to keep the agro-industrial network happy, we need to switch to a finer-tuned division by agri-food industry sector, which depends on the official nomenclature used. Lastly, if we want to protect human food and the industrial fabric, and we acknowledge that deciding practices hinge on farmers, we have to make a division using the major types of farm activities. We will call it division into agricultural sectors to identify the prominence of agricultural activity, of which fruit becomes the borderline resource. In France, each sector has farm producers, organisations for market authorisation, primary processing, marketing and, more often than not, organisations for secondary and even tertiary processing, a technical institute, consultancy firms and associations per product or of more general nature.

We posit that most of the organisations that make up these sectors have the necessary capabilities to adapt. Most of the individuals and organisations have static routines (that provide the means to carry out a usual task reliably and efficiently) and dynamic routines (Teece, 1986): these latter ones form a body of knowledge and expertise built up through experience and practice, and links outside the organisation. The tools created as part of a competitive struggle between companies (see for example the article on the agri-food industries in Gestion 2000, January-February 2001) appear outdated when they fail to include the assumption that basic foods might run short. The matter of agricultural production has long been considered as being "self-evident" (Debair *et al.*, 1986; Ravignan, 1988).

We make the assumption that, generally speaking, there is a particularly influential player in each sector, who is acknowledged as being legitimate, to the extent that its stands make the other players in the sector act, and in particular the farmers. Our aim is to establish who this key player is for the main sectors. On this subject we are calling upon our European colleagues for their advice. We hope that by comparing the way in which the various sectors are organised in Europe, a simple shared methodology may emerge. Identifying the legitimate player would help pinpoint the most relevant leverage for action for the State. Different practical solutions need to be thought up depending on the position held by the key player. If this player is a technical institute or a group of producers, it is not hard to imagine what actions would be appropriate. The situation is slightly more complicated if it involves a primary processing company (Lambert, 2001), and especially if this firm is "cornered" between sensitive, perishable supplies (such as fruit and vegetables) and secondary processing industries or distributors, with little room to manoeuvre, *i.e.* with very little organisational slack (March and Simon, 1999). Public support may occur by proposing a survey with a view to examining scope for adaptation, but which should, if possible, be carried out at industry cluster level (Porter, 1999), so as to promote synergism and mutual aid, and in any case by considering several levels of the value-adding chain (Johnston and Lawrence, 1988). Finally, it is important that the public authority also be aware of and supervise emerging projects. It cannot be indifferent to the establishment of a rapeseed oil extraction plant in Brittany, which would press the harvest from fallow land transformed into industrial set-aside.

To venture some figures, the 60,000 farms that devote more than 90% of their acreage to field crops (cereals, oleaginous/proteaginous plants, beet, potatoes) will, in principle, be very much alive to the directions given by ONIC and ONIOL. On the other hand, we think that almost all the so-called "non-professional" farms, as well as the some 200,000 that mix livestock breeding and crops, have considerable scope for action, which allows them to do without a mobilised sector. On the other hand, the future of the 30,000 vineyards, the 5,000 farms devoted to vegetables, the 3,000 horticultural concerns and the 3,000 orchards, not to mention the 75,000 specialist breed farms with grass, will probably be closely dependent on the directions given by the reputedly legitimate players in the sector (all the figures are taken from Agreste Cahiers – Enquête structure 2003 survey, published in April 2005 and therefore concern 2003). On top of this comes the concern of what is to become of the industrial fabric that processes and turns agricultural produce into foodstuffs. We think that these developments must be steered and regulated right now by the public authority. Let us not forget what Talleyrand once wrote: "If something is urgent, it's already too late".

## References


Acot, P., 1998. The European Origins of Scientific Ecology, The history of science, technology and medicine. 2, Editions des archives contemporaines, Gordon and Breach publishers, Amsterdam, 932 pp.

Chopplet, M. and D. Thomas, 2004. Towards alternatives for vegetable crops, emergence of new bioeconomics. Futuribles 295, 5-18.

Debeir, J.C., J.P. Deléage and D. Hémery, 1986. The servitudes of power. Flammarion, Paris, 428 pp.

De Ravignan, F., 1988. Material support will not be provided. Essay on French agriculture. Cahiers libres, Editions La Découverte, 166 pp.

Johnston, R. and P.R. Lawrence, 1988. Beyond vertical integration-the rise of the value-adding partnership. Harvard Business Review, July-August, 94-101.

Lambert, A., 2001. Vertical splitting-up, an answer to flexibility requirements in food-processing firms. Gestion 2000, 1, 59-78.

March, J.G. and H.A. Simon, 1993. Organizations, 2nd ed., Cambridge, Oxford, Blackwell, 287 pp.

Porter, M., 1998. On competition. Boston, Harvard business school. 485 pp.

Reeves, H. and F. Lenoir, 2005. Pain at the Earth. Sciences, 2, Seuil, Paris. 273 pp.

Schwartz, P. and D. Randall, 2003. An abrupt Climate Change Scenario and its implications for United States National Security. Report by the Office of Net Assessment, the Pentagon's internal think-tank.

Seguin, B., 2002. The agronomic research faced with greenhouse gas. Courrier de l'Environnement de l'INRA, juin, 46, 5-20.

Teece, D., 1986. Profiting from technological innovation: implications for integration, collaboration, licensing and public policy. Research Policy 15, 285-305.

Valantin, J.M., 2005. Climatic threats upon world order. Lignes de Repères, Paris. 158 pp.

Wingert, J.L., 2005. Life after oil. Frontières, Autrement, Paris. 238 pp.

# The future of farming in Norway: Processes of recruitment and intergenerational transitions

*Kjersti Melberg and Christin Berg*
*International Research Institute of Stavanger, PO Box 8046, 4068 Stavanger, Norway*

## Abstract

This article focuses on patterns of recruitment and transitions across generations in Norwegian farm families. The main research questions are: What vision do different generations of farmers have for the future? What motivates young farmers to take over the farm? What are their driving forces, view on modernization and recruitment? Analyses are based on two data sources, the Living Conditions Survey of the Farming Population (2002) and life history data from 3(4) - generational farm families, drawing on the growing sociological tradition that examines the relationships among multiple generations of family members. One conclusion of this study is that the expectations of young people and successors often do not correspond with a life in the contemporary Norwegian farm life, regardless of the continuous modernization in the sector and rural areas.

## Introduction

This paper focuses on patterns of recruitment and transitions across generations in Norwegian farm families. The research questions on which we focus are: What vision do different generations of farmers have for the future? What motivates young farmers to take over the farm? What are their driving forces, view on modernization and recruitment? The intergenerational processes of recruitment and transitions will influence the future of farming in Norway, and calls for a study. Here, farmers represent a long-lasting, essential part of the national identity, and play an important role in comprehensive Norwegian policies directed towards stabilising settlement in rural areas. Concurrently, this is an occupational group under pressure. During the last 40 years, close to 50% of the farms have been closed down, resulting in less than 60 000 running farms (see www.ssb.no.).

One description of Norwegian agriculture is that it shows general features of modernization and individualization (Giddens, 1991; Inglehart, 1990), as well as rapid, sector-structural challenges (Melberg, 2005). A combination of financial limitations, conflicting career demands, and obligations in multiple arenas is said to threaten faming household economies and individual living standards (Løwe, 1998; Rye, 2000). A major factor which influences current transitions in and recruitment to the farming sector, is that farming has changed from being a desirable profession, to become a characteristic and less attractive life-style for young inheritors. Urbanization, educational opportunities, and a tight labour marked pull young people to the cities.

Among others, this is one cause of a marked decrease in the number of Norwegian farms the last decades. Family-owned and -operated farms do however continue to dominate in Norway despite changes in their size and dependence upon non-family resources. Norwegian farmers generally possess a strong work ethic, striving to adapt to external conditions, to maintain the self-sufficiency of the family unit, and to uphold the traditions of the intergenerational family farm (Melberg, 2003, 2005; Elder *et al.*, 1998). The vital question of inheritance, land and farm production still engage farm families. The paper set out to explore change and continuity in transitions and recruitment between generations within families. The analysis is based on a case study of Norwegian 3(4) generational farm families. These data were collected during 2003-2005. Spouses and other family members in eight farm families were. The interviews centred on individual life history (see Hareven, 1996).

## Literature review

There is a growing sociological tradition that examines the relationships among multiple generations of family members (Bengtson, 2001; Hareven, 1996; Willson *et al.*, 2003). One influence is the changing demography of the Western European countries, *i.e.* steadily increased longevity and decreasing fertility. Consequently, intergenerational dynamics have become a major force in family life research. One major finding is that the lives of grandparents and grandchildren, like those of parents and children, are linked (Crosnoe and Elder, 2002). This is in particular the case in farm families which are more likely to have parents and (adult) children working and living together. The term "intergenerational ambivalence" describes the inherent contradictions in such relationships (Lüscher and Pillemer, 1998).

Farm families likely have even more opportunities for both an affectionate and a distressing relationship. Family stress research show how life course transitions, such as births, deaths, and other changes of life, can lead to stress in the family (Boss, 2002, 2003). In a farm family, kinship is connected to the necessity of family members to work together and a mutual involvement in an economic as well as domestic enterprise. Evidently, this can create tensions and exacerbate interpersonal conflicts among family members (Lee and Cassidy, 1985; Rosenblatt and Anderson, 1981). When a farm family experiences transitions such as deaths and succession, we hypothesize that *members of all generations are likely to be involved in the process* ($H_1$).

Within a familial farming context, the life transitions of the generations are intertwined. For example, the timing of leaving or taking over the family farm is interrelated with the timing of the older generation's transition into retirement or inheritance. Retirement could be more stressful for many farm families than for many other families. In some farm families retirement is not an individual act, but rather an extended sequence of transitions in intergenerational relationships. The parallel processes of retirement and the next generation taking over the farm symbolize the family entering a new life stage and new social roles. In many farm families, however, retirement tends to be delayed and taken on more gradually (Rosenblatt and Anderson, 1981). Mutual dependence is inherent in the farm family's life, which makes it hard for people to maintain a professional relationship (Rosenblatt and Anderson, 1981). Our second hypothesis proposes that the *experienced processes of transitions and the view on recruitment will differ between the generations* ($H_2$).

Our third assumption is that there *in each farm family exists a contract across generations and age groups* ($H_3$), *i.e.* a set of shared expectations and obligations regarding the succession of generations, through time and within social structures, following Bengtson and Achenbaum (1993). This assumption implies that there is an informal, unwritten and often implicit set of expectations and obligations underlying the relations between farm generations, in terms of expectations and obligations that create solidarity or conflict, and the negotiations of these in everyday life. For example, in most farm families retirement is not an individual act, but rather an extended sequence of transitions in intergenerational relationships. The parallel processes of retirement and the next generation taking over the farm symbolize the family entering a new life stage and new social roles (Rosenblatt and Anderson, 1981). Such processes are the focus of this study.

## Results

Some expressions from two extended families illustrate what motivate new generations to take over the farm. In the first case, the farmer and his wife chose to take over the Andersen farm in their mid thirties. At the time they were well settled in an urban area, and although she grew up on a farm, they had never considered farming as a place to work or live. After having had the farm for ten years her brother declared that he was tired of farming and considered other ways of living. His sister took over, and brother-in-law then became the main farmer. *"We felt that we had to think it over. After discussing it, we still doubted, but jumped into it. I don't think we have ever regretted that"* (Farmer Andersen). The two siblings agreed upon a suitable purchase price for the farm, and the current farmer concludes: *"As far as I can se he is satisfied, and we are satisfied. Then one cannot complain."*

In the second case, the current farm couple on the Pettersen farm decided to take over the farm being in their early twenties. They were students at the time. The first heir (his brother) is portrayed by the informants as not so fond of animals and farm life. He still built a house and settled down on the farm with his family and had a full-time off-farm job. His father ran the farm. *"He had his duties and helped out when help was needed and had his mind on taking over the farm"* (Mother Pettersen). The oldest son was however clear about the fact that he would never become a full-time farmer. He enjoyed his off-farm job. He therefore offered his brothers to take over the farm in the case they wanted to run the farm full-time. *"My husband was always very clear about that they were not to take over the farm if they didn't want to"* (Mother Pettersen). The youngest brother saw this as an opportunity: *"I guess I like practical work. I can combine that with being engaged in other activities. It is a profession which demands a great deal of knowledge and you have to know what you are dealing with. And then I guess I like to decide over myself. It is also a very result-oriented profession; you often get results according to the labour you put in".*

The first heirs to both farms may be categorised as traditionally influenced first born, feeling the responsibility and duty to uphold the farm. But the stories are slightly different. The father in the Andersen family was a devoted farmer and probably firm about handing over his knowledge and preparing his oldest son to be a decent farmer. The daughter who now owns the farm with her husband says it like this:*"We all helped out at home when I grew up. Often it was my brother that helped outside and in the barn, he was the heir. I never thought of an opportunity to later take over the farm. I was only happy that I wasn't the oldest one. I never wanted to become a farmwife* [laughter]" (wife Andersen). The father in the Pettersen family is described in the interviews as more reluctant in forcing any of the children to take over the farm. When he describes him taking over the farm, he says it felt natural at that time. However if he had not felt the duty as the oldest son to take over, he most likely would have had another profession *"...but I wouldn't say I have had a bad profession"* (father Pettersen).

What are the driving forces and the view on modernization and recruitment? There is a striking similarity between the two farm cases we are referring to and several other cases. Even though farm life may be considered to be inflexible with set feeding routines, seasonal dependencies, *etc.*, the life is also described as flexible. It could be having someone to look after the children, someone stepping in doing part of the farm work, flexible work periods during the day and so on. The Anderson family emphasize their life today as a lot calmer than before. This is much due to him not having to commute to work, but instead being at home when the children come from school. The flexibility that the informants refer to may be seen as one of driving force behind choosing farm life: *"Everything changed when we actually had a choice [between living in the city and moving to the family farm]. And we thought of it as healthy to get away from "the streets" too (...) there were so many people and circumstances in the neighbourhood to take into consideration*." (wife Andersen).

The two case families have had the farm about the same number of years, but they portray a slightly different view of the future. The Pettersen farmer and his wife are several years younger than the Andersens, and still have very small children. The Pettersen farmer has a positive attitude towards the future: "*I expect and hope to manage keeping up with the development* (...) *and with a fairly good economy*." (farmer Pettersen). But he puts a strong weight on having an economic base to spend time with his family and doing leisure activities. He therefore stresses increase in production and efficiency since he soon has reached his set work load limit. He also needs to plan for not having his father as part-time employed on the farm.

On the Andersen farm, the oldest children have reached adolescence. The family seems satisfied with the decision they made taking over the farm, but are a bit more reluctant towards making possible major investments for the future. "*Off course this attitude is a bit passive; especially towards investments it's passive. Because, you can say I'm a bit scared of having too much debt. Then you have to run the farm no matter how the prospects are, only to keep it going. Now I can get rid of these tings [land and production facilities] and only live here if it should come to that*." (farmer Anderson). This view is probably rooted in that the children so far show a bit reluctance taking over the farm. "*The oldest helps a bit in the barn, but is not so fond of it (...) It is probably his need for pocket money that makes him help out (...) Next generation, no we don't know anything about that yet (...) I feel I have to know if the kids want to take over before they are 25 (...) If they don't want to take it, one can consider cousins? It's not very interesting to build a barn for several million if no one wants to take it over*." (Farmer Andersen). His wife puts it this way: "*We have always said that if we don't feel happy doing this, we have to think differently, maybe put it for lease. I have never had very strong ties to the farm. (...) We enjoy the life we have now, but it depends on the conditions running a farm in the future. If we don't have an income it's impossible to run a farm*." (Wife Andersen).

## Conclusion

The analysis above demonstrates that members of all generations are involved in the recruitment process. Decisions regarding investment, retirement *etc.* taken by the present farm generation are however closely tied to life-choices of the next generation. Thus recruitment to the sector is to an increasing degree unpredictable, as the expectations of young people and successors often do not correspond with a life in the contemporary, rural Norwegian farm life. The processes of transitions and the view on recruitment do also vary between the generations: From being in the position to expect someone among the siblings to take over, the current farm generation can't rely on the next generation to carry on. The results do however indicate that many farm families still have expectations towards *someone* in the young generations to take over the farm, but that the process is more unstructured than in times past. In both cases presented here, the first heir to take over was not the one who finally settled down on the farm. There generally seems to be a set of shared expectations and obligations regarding the succession of generations, which changes through time and within social structures. One conclusion of this study is that the future of the farming profession in Norway is increasingly open and uncertain.

## References

Bengtson, V.L., 2001. Beyond the nuclear family: The increasing importance of multigenerational bonds. Journal of Marriage and Family 63, 1–16.

Bengtson, V.L. and A. Achenbaum (editors), 1993. The Changing Contract across Generations. New York, Aldine de Gruyter.

Boss, P. (editor), 2003. Family Stress: Classic and Contemporary Readings. California/London, Sage Publications.

Boss, P., 2002. Family Stress Management. California/London: Sage Publications.

Brannen, J. (editor), 1995. Mixing Methods: qualitative and quantitative research. Aldershot, Avebury.

Crosnoe and Elder, 2002. Life Course Transitions, the Generational Stake, and Grndparent-Granchild Relationships. Journal of Marriage and Family 64, 1089-1096.

Elder, G.H., R.D. Conger, R.R. Swisher and F.O. Lorenz, 1998. The long arm of the farm: How an occupation structures exposure and vulnerability to stressors across role domains. Journal of Health and Social Beh. 39, 72–89.

Giddens, A, 1991. Modernity and Self-Identity. Self and Society in the Late Modern Age. Stanford, CA, Stanford University Press.

Hareven, T.K., 1996. Introduction: Aging and generational relations over the life course. In: T.K. Hareven (editor), Aging and Generational Relations over the Life Course. New York, Walter de Gruyter, pp. 1–12.

Inglehart, R., 1990. Culture Shift in Advanced Industrial Society. Princeton, NJ, Princeton University Press.

Lee, G.R., and M.L. Cassidy, 1981. Kinship systems and extended family ties. In R.T. Coward and W.M. Smith, Jr. (eds.), The Family in Rural Society. Boulder, CO, Westview, pp. 57–71.

Løwe, T. 1998. Levekår i landbruket. En studie av landbruksbefolkningens levekår. Report 98/25. Oslo, Statistics Norway.

Lüscher, K., and K. Pillemer, 1998. Intergenerational ambivalence: Further steps in theory and research. Journal of Marriage and the Family 60, 413–425.

Melberg, K., 2003. Landbruket som livsform. Gårdbrukerparets levekår og livskvalitet. Report RF-2003/018. Stavanger: Rogaland Research.

Melberg, K., 2005. Farm Family Transactions in Norway: Unpaid Care across Three Farm Generations. Journal of Comparative Family Studies. 36, 419-441.

Rye, J.F., 2000. Hvorfor bråker bøndene når de har det så bra? En studie av sammenhengene mellom levekår og livskvalitet i landbruket. Report 3/00. Center for Rural Research, Trondheim.

Willson, A.E., K.M. Shuey and G.H. Elder, 2003. Ambivalence in the relationship of adult children to aging parents and in-laws. Journal of Marriage and Family 65, 1055–1072.

# Farmers' new participation in food supply chains: Making horizontal and vertical progress by networking

*Minna Mikkola and Laura Seppänen*
*University of Helsinki, Ruralia Institute, Lönnrotinkatu 3-5, FI-50100 Mikkeli, Finland*

## Abstract

Market access is crucial for the food supply chain. The continuing concentration of the retail industry in northern Europe has strengthened the retailer's position. The relation between supplier and retailer has been described as "fight" over the control of the vegetable supply chain, and farmer's position in the chain has been seen as "weak". Farmers have made efforts to strengthen their position by networking, in a way comparable with supply chain management (SCM) paradigm. In this study, the position of a farmer vis-a-vis chain development is examined, basing on actors' relations within an enlarging supply chain. The relations analysed in this study are market, hierarchic, partnership and other social relations. This case study illustrates the pattern of relations active in the enlargement of a conventional vegetable chain. A socially skilful farmer strengthened his position by networking step by step both horizontally and vertically. This case study suggests that strengthening the partnership relations may induce chain development for the benefit of farmers, while also market and hierarchic relations contain supportive elements. The role of social and partnership relations in the development of vegetable supply chains needs further examination for validation of the relational effects.

**Keywords:** supply chain, horizontal, vertical, network

## Farmers' relational position in concentrating vegetable supply chains

Market access is crucial for the food supply chain. The continuing concentration of the retail industry in northern Europe has strengthened the retailer's position (Hollingsworth, 2004; Duffy *et al.*, 2003; Atkins and Bowler, 2001). The supermarkets are seen to exercise control over the supply chains (Hollingsworth, 2004) by mastering the distribution for time constrained consumers, who do one-stop-shopping while the volume of direct distribution remains limited in practice (Henchion and McIntyre, 2005; Jones and Comfort, 2004). The state of overproduction in northern European vegetable market, due to increasing supply from southern producers, is further strengthening the retail industry (Wilson, 1996). The price-sensitive major retail chains also turn to global buyer alliances (Hollingsworth, 2004).

The relation between supplier and retailer has been described as "critical" (Hollingsworth, 2004) and as "fight" over the control of the vegetable supply chain (Wilson, 1996). Farmer's position in the chain has been considered as "weak" (Henchion and McIntyre 2005) and as "adversarial" for the farmer-of-the-Middle in the US (Stevenson, 2005). The difficult position of European farmers is implicated by the unwillingness of processors to integrate backwards in the chain, while the primary producers are interested in forward integration (Henchion and McIntyre, 2005). According to these authors, a majority of primary producers are willing to use pre-agreed contracts with processors and tend to cooperate with other primary producers to form producer groups. This signals efforts to strengthen the position of farmers by networking, in accordance with supply chain management (SCM) paradigm (Porter, 1985). The integration of the supply chain by cooperation comes close to network forms of organization (Powell, 1990). Other integrative relations are internal hierarchy or external power; also market relations can be useful in sourcing supply chains (Powell, 1990; Powell and Smith-Doerr, 1994). There is also social character in economic exchange (Granovetter, 1985).

How can farmers 'stick' to the supply chain, if their position in the chain is difficult? In this study, the position of a farmer vis-a-vis chain development is examined by analysing the actors' relations within an enlarging supply chain. What are the relations like within the supply chain and how do they support farmer's activities? This study presents a Finnish exemplar of a supply chain, illustrating the pattern of relations active in the enlargement of a conventional vegetable chain. Here a socially skilful farmer strengthened his position within the chain by networking step by step both horizontally and vertically.

## Theoretical framework for economic exchange relations

Actor's position in the supply chains can be analytically approached by market, hierarchic and network relations (Powell, 1990). The Weberian market relation has been historically described as a "peaceful conflict", as a "battle of man against man", including an "abomination to every system of fraternal ethics". The market relation is characterised by independent choices of self-interested actors, entailing limited personal involvement and freedom of future commitments (Powell, 1990). Instead of the invisible hand of market, coordination can also proceed by the visible hand of management. This coordination takes place within hierarchical relations of a vertically integrated organisation. Hierarchies have clean lines of authority and decision making procedures (Powell, 1990). The authoritative power translates into orders and instructions; but between organisations, there exists also power based on overt domination, entailing possible sanctions (Powell and Smith-Doerr, 1994). Networks have been identified as *e.g.* inter-firm agreements, strategic alliances, "quasi-integration", stable relationships and partnerships (Powell, 1990). Common for these forms are long-term relations, shared information, reciprocal adjustment and mutual benefits and burdens. In this paper the concept of 'partnership' is used for long term economic collaboration relation. The 'social' was taken up as other social relations if these were involved in the chain development without direct economic exchange.

## Empirical material and methods

Among judgementally chosen chains of different scales, this large scale conventional vegetable chain represented an interesting case of chain enlargement. The chain was analysed in terms of chain structure, actors' relations and chain development. The four chain actors participated in an open-ended focussed interview about their activities. One interview took about 1.5 hours and was recorded and transcribed verbatim.

First, the relations of the actors were identified as market, hierarchic, partnership and other social relations by comparing the relevant segments of speech with relational descriptions of the literature. The dyadic relations formed 'ego networks' (Powell and Smith-Doerr, 1994), which were combined on the chain level and visualised as a network view (Figure 1). The visualisation is based on actors' interviews (marked by a letter and a number) making the actors' social reality visible, which is not, however, presented as an 'objective, true' pattern of chain relations.

Second, to grasp the dynamics of chain relations vis-a-vis development of economic exchange, the formation of relations and vegetable production were analysed as a narrative. The narrative is here understood as produced by the actors in order to create coherence and causality into the developmental history of their trade (Linde, 1993). For the interviewees, it presents a crystallization of meaningful developments and events in their professional lives.

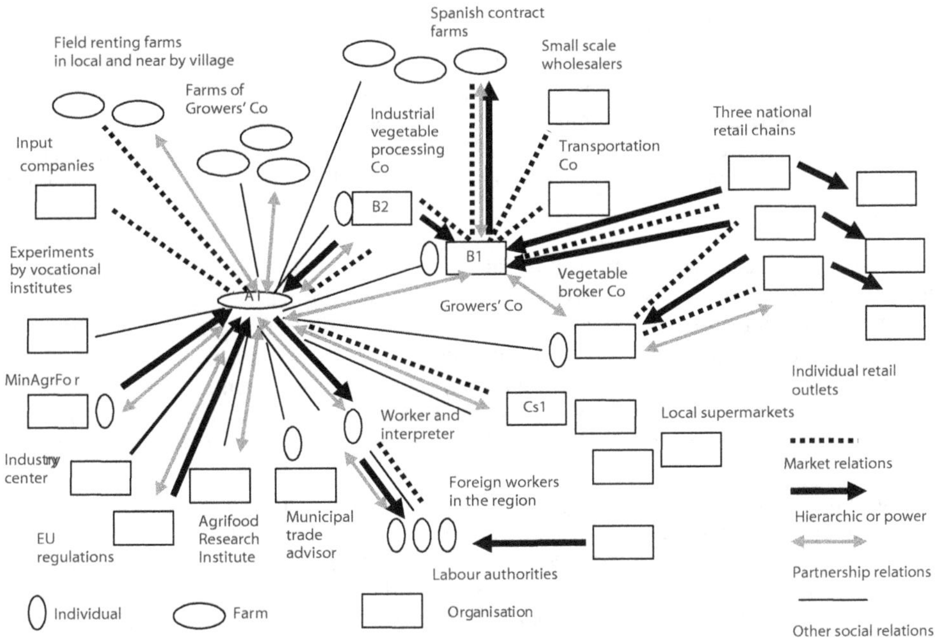


*Figure 1. View of chain relations within a large scale conventional vegetable chain.*

## Chain relations within a large scale conventional vegetable supply chain

The visualisation of the chain relations discloses the complex relational pattern on the chain level, and the farmer's (A1) relational position within the chain can be studied. Only a minority of the chain relations exhibited 'simple' market, hierarchic, power, partnership or other social modes. Many of the relations consisted of strands of two, three or even four different relational modes. In primary production there were mainly social, partnership or hierarchic relations, except for market relation in input buying. There were strong social and partnership relations with other farmers, growers' company (B1) and the vegetable broker company. The connections with industrial vegetable processing company (B2), foreign workers, contract farms and local supermarket (Cs1) were labelled by an extensive relational mix. The relations at the commercial customer end of the chain were typically market and hierarchic relations; however, there were indications of emerging partnership with one of the major national retail chains.

## Development of a large scale conventional vegetable supply chain

In the large scale conventional vegetable supply chain, the farmer (A1) has had considerably strong social relations with other farmers, university researchers and a trade advisor, a 'vegetable prophet'. These relations supported the enlargement of the farmer's (A1) production by swapping fields with his neighbours; fields were additionally rented at high market prices. The relations with the authorities were necessarily hierarchic, but this was mitigated by the subsidies, which were understood to refer to partnership relations.

Out of the social circuit of local vegetable farmers the group of eleven farmers was crystallised; they started a common growers' company (B1) to produce, package and initially also market their produce to wholesalers and major retail chains. The farmers' partnership relations were 'deep', based on trust, and including big investments. The farmers also shared on their fields the same group of foreign workers. The relations with the foreign worker group were simultaneously based on social bonds, market prices, hierarchic orders and partnership, offering benefits for both groups.

A small flow of produce was transported by the farmer himself to the local supermarket (Cs1), on the basis of social, market and partnership relations. The farmer (A1) also initially delivered to industrial vegetable processing company (B2), in a simultaneously hierarchic, partnership and market mode of exchange. The deliveries included clear specifications, and some foreseeable income, based on market price; the farmer (A1) saw the price competition as a 'fight' for markets. The farmer (A1) learned a lot in the hierarchic relation, transferring what he learned into his own activities. The industrial actor (B2) described they developed the local farming.

New developments started when the growers' company (B1) found a connection to a broker and partly financed the vegetable broker company. The farmer (A1) and the broker had a strong social relation, travelling together, and by ownership they had a partnership relation. The growers' company (B1) wanted to sell vegetables during the wintertime, too, through the broker company. Connections were made with Spanish farmers, who became contract farmers for the growers' company (B1). The relation operated in social, hierarchic and partnership modes, including market orientation as well. This connection strengthened the broker company which was able to meet volumes and quality needed by the major retail chains. The relation between suppliers and retail industry has been earlier labelled by strong market and power relations: "the chains have always determined the price". However, emerging partnership relations may be underway, allowing more room to negotiate about quantities, timing and price of deliveries. The farmer (A1) was cautious with future turns. If realised, a more negotiable access to the markets could open for the farmer (A1), the growers' company (B1) and other farms connected to the broker company.

## Horizontal and vertical progress by networking

The social core can be identified as the root of the development of the supply chain. Social skills supported the growth of the trust and partnership, developed during years of mutual cooperation and evaluation. This trust comes close to the process-based trust, identified by Lindgreen (2003) in food chains, and the reflective trust described by Adler (2001). The ability to master a great number of relations, even consisting of different, dynamic strands, calls for social skilfulness (Argyle, 1991: 115-131). The farmer of the case study appeared as a skilful actor, mastering relations up- and downstream within the chain, gaining new opportunities for learning and enlargement of his activities. The crucial social and partnership relations acted in interplay with market and hierarchic elements, supporting the chain. The progress has been stepwise and alternately horizontal (the local farmers), vertical (the growers' company), horizontal (foreign workers), vertical (broker company), horizontal (foreign contract farms) and finally possibly vertical (a major retail chain). Through his activities he also had simultaneously a different position on several levels; he was a farmer, a 'leading' farmer, a broker, and even reaching the level of major retail chain buyers. The relational development seems to imitate the ideal supply chain management (SCM) of Porter (1985).

## Concluding remarks

The chain relations are suggested to be essential in the enlargement of the studied chain. The social and partnership relations seemed to be the core of the enlargement with at least partly supportive market and hierarchic relations within the supply chain. This case study shows that it is possible for farmers to strengthen their position and increase control within supply chains. However, the role of social and partnership relations in the development of vegetable supply chains needs further examination for validation of the relational effects.

## References

Adler, P.S., 2001. Market, Hierarchy, and Trust: The Knowledge Economy and the Future of Capitalism. Organization Science 12, 215-234.

Argyle, M., 1991. Cooperation. The basis for sociability. Routledge, London and New York. 276 pp.

Atkins, P. and D. Bowler, 2001. Food in Society. Economy, Culture, Geography. Arnold, London. 328 pp.

Duffy, R., A. Fearne and S. Hornibrook, 2003. Measuring distributive and procedural justice. An exploratory investigation of the fairness of retailer-supplier relationships in the UK food industry. British Food Journal 105, 682-694.

Granovetter, M., 1985. Economic Action and Social Structure: The Problem of Embeddedness. American Journal of Sociology 91, 481-510.

Henchion, M. and B. McIntyre, 2005. Market access and competitiveness issues for food SMEs in Europe's lagging rural regions (LRRs). British Food Journal 107, 405-422.

Hollingsworth, A., 2004. Increasing retail concentration. Evidence from the UK food retail sector. British Food Journal 106, 629-638.

Jones, P., D. Comfort, and D. Hillier, 2004. A case study of local food and its routes to market in the UK. British Food Journal 106, 328-335.

Linde, C., 1993. Life Stories. The Creation of Coherence. Oxford University Press, Oxford. 242 pp.

Lindgreen, A., 2003. Trust as a valuable strategic variable in the food industry. British Food Journal 105, 310-327.

Porter, M., 1985. Competitive Advantage Creating and Sustaining Superior Performance. The Free Press, New York. 557 pp.

Powell, W.W., 1990. Neither market nor hierarchy: Network forms of organisation. In: B.M. Staw and L.L. Cummings (eds.), Research in organizational behaviour. An annual series of analytical essays and critical reviews. Vol. 12. Jai Press Inc. London, pp. 295-336.

Powell, W.W. and L. Smith-Doerr, 1994. Networks and Economic Life. In: N.J. Smelser and R. Swedberg (eds.), The Handbook of Economic Sociology. Princeton University Press, New Jersey, pp. 368 – 402.

Stevenson, S., 2005. Strategies for Agri-Food Enterprises-of-the-Middle: Values-Based Supply Chains in the Pacific Northwest. Panel Session 58., at 2005 Joint Meetings of Agriculture, Food, and Human Values Society (AFHVS) and Association for the Study of Food and Society (ASFS) 9.-12.6.2005 Portland, Oregon, USA.

Wilson, N., 1996. The supply chains of perishable products in northern Europe. British Food Journal 98, 9-15.

# Working conditions: A challenge for sustainability of mountain family farming systems diagnosis at farm level and responses designed by farmers in the Alps

*Sandrine Petit[1], Laurent Dobremez[2] and Philippe Fleury[1]*
*[1]Service d'Utilité Agricole à Compétence Interdépartementale, Groupement d'Intérêt Scientifique Alpes du Nord, SUACI, GIS Alpes du Nord, 11 rue Métropole, F-73000 Chambéry, France; [2]Cemagref, GIS Alpes du Nord, Unité de Recherche Développement des territoires montagnards, Domaine Universitaire, BP 76, F-38402 Saint-Martin d'Hères cedex, France*

## Abstract

Our paper discusses one key component of family farms sustainability in the Alps that is the working conditions. Mountain livestock systems remain labour intensive. Farmers feel the gap between them and the rest of the society deepening regarding their workload. They express a demand for more spare time. We address this issue from two perspectives. First, we analyse how a group of farmers has constructed their own solutions to address the question of working conditions. We refer to the sociology of translation to understand this collective process. It shows that over a period of 3 years, farmers have refined the problem and tested different solutions. The second approach lies in a systemic approach at farm level assessing the work organisation and the farmer's self-perception regarding his/her workloads. Crossing those two approaches, we discuss why successive phases of *re-problematisation* occurred and why the solutions imagined are not so straightforward. The single object of work organisation turns out to encompass all the complexity of small-scale farming systems in mountains areas.

**Keywords:** mountain livestock farms, northern Alps, work organisation, sustainability

## Introduction

In mountains areas, farming systems based on livestock breeding remain of small-scale and their number is steadily decreasing as anywhere in Europe. In Moyenne Tarentaise (Savoie, France), the number of farms has drastically decreased, about 70% over the last twenty years. This decrease leads to a worrying situation about the maintaining of a high nature and landscape value farmland. The maintaining of the existing small-scale farms depends on many factors (economic, social). The social component mainly expressed in terms of workloads has been identified as a major factor of farms sustainability and of taking over. Mountain livestock systems remain labour intensive and farmers are facing difficulties in articulating different activities along the year: seasonal works (hay making, manure spreading,...), daily tasks (milking and feeding animals) and off-farm activities (Madelrieux *et al.*, 2004). On the basis of an on-going European project, we propose to discuss the farmers' concern on workloads. The hypothesis of our paper is that work organisation and the way farmers address this problem reflect the coping strategies of small-scale farmers. We examine this hypothesis from two perspectives. First, we describe and analyse how a group of farmers addressed the question of working conditions in farms and constructed their own solutions. We refer to the sociology of translation to understand this collective process. From the collective search of solutions and constant adaptations of those measures by farmers, we take out hypothetical reasons why farmers refined the problem and designed successively different tries. Secondly, we discuss these hypotheses crossing a sociological analysis and a systemic analysis of farming systems based on a field-survey.

## The context: Farming systems in Moyenne Tarentaise and the Imalp project

The Moyenne Tarentaise (districts of Bozel and Moûtiers), is located in the French northern Alps.191 livestock farms were inventoried in 2000. 58% of stock farmers' couples have an off-farm activity, generally in ski resort during winter. The dairy cows systems are preponderant: 106 on 191 including 29 farms managing alpine pastures. In average, they remain of small-scale: 20 dairy cows and 1.5 annual agricultural work units. The very small farms (less than 5 livestock units) are often managed by retired people. Farmers produce milk for dairy cooperatives or ensure themselves the milk processing into cheese. In summer, they utilize alpine pastures or cattle are tended by pastoral associations while in the valley farmers cut hay. The Moyenne Tarentaise is included in the production area of the labelled Beaufort cheese.

The Imalp project, our framework, is a combined research and demonstration project (Fleury *et al.*, 2006). The demonstration part consists in the design and implementation of a consistent set of actions towards sustainable agriculture by a group of local stakeholders (farmers, elected officials and NGO members). Each action is managed by an action group. Among the different actions implemented, in Moyenne Tarentaise, solutions to limit the work peaks and improve work organisation were discussed and implemented by a group of 10 farmers. Our paper is focused on this action group. The scientific evaluation of Imalp consists in two parts as described below.

## The methodology

The methodology is an in depth on-field approach based on local action. To understand the local processes initiated by the Imalp groups, we cross two approaches. On one hand, we analyse the collective dynamics with the concepts of the sociology of translation (Fleury *et al.*, 2006). To analyse the different phases, we focused mainly on the concepts of *problematisation* (to formulate the problem), *interessement* (how to seal partnerships), *enrolment* (how to define and coordinate actors' role) (Callon, 1986). To collect the data, we attended the meetings of the action group, reported the debate and each phase over a period of 3 years (meeting content, analysis of the documents produced, reporting of all services done between farmers). On the other hand, we resorted to a comprehensive approach of farm functioning. We surveyed a sample of 16 farms including 26 farmers twice in 2002/03 and in 2005. Different sustainability indicators were elaborated. For work organisation analysis, we adapted the "Bilan Travail" method (Dedieu *et al.*, 1998). The main concepts of basic group (the workers who organize the work on the farm), daily obligatory agricultural routine work (ORW) and seasonal tasks were used. The year was divided in sequences according to the routine work and off-farm activities. We described typical days for each sequence. The workload is expressed by the annual number of work peak days for each person of the basic group. A day is regarded as work peak if (i) the ORW is more than 6 working hours, or (ii) the ORW summed with off-farm activities is more than 8 hours, or (iii) the day is dedicated to agricultural seasonal work. Another indicator is the farmer's perceptions about their workload and spare time.

## The collective process from shaping a problem to testing solutions

We use the sociology of translation concepts to describe and analyse the group process. An action group was created in 2003 to specify and implement concrete solutions to cope with farmers' workloads. We closely detail how the group tried to establish a collective set-up as a response to shortage in labour.

Two types of solutions were first considered by the group of farmers: individual solutions (each farmer attempting to improve their own work organisation) and a collective response (search for mutual help or exchange of services). In June 2003, the action group expressed that the problem of seasonal workloads could not be solved by mutual help between farmers as everyone was facing shortage in labour at the same period (hay making, manure spreading periods). In a second meeting, this assumption was questioned and farmers discussed if the seasonal tasks vary according to the farms location at different altitudinal levels. The shared consensus on the overloads of work burst out and the farmers further examined the differences between farms. From the controversy, the different benefits of a collective structure of services and mutual help were highlighted: having leisure time, working with a colleague as more pleasant than alone, diversifying the current production activities, developing opportunities for young people to come into agriculture. The group decided to assess the feasibility of a services cooperative. The feasibility study confirmed the variation of seasonal tasks and peak period according to the farms. Then the farmers' group visited a services cooperative in Jura. The visit triggered the attempt to replicate the services cooperative in Moyenne Tarentaise. After this decision, action group members launched an appeal for volunteers and elaborated a price list for each service that could be demanded or offered by farmers. We consider that they started the *interessement* phase, searching for actors to join the cooperative. The price list was an object having an active role to encourage the *interessement*. In 2004 and 2005, several exchanges were done. These exchanges consisted in: cattle transportation, land upkeep, emptying a manure tank and pressing hay with a round baler. The exchanges were often performed in situations of emergency when farmers need to be helped out because of a machinery breakdown. Some exchanges between farmers occurred in an area of vicinity on the basis of the price list but without any invoicing and without formal information given to the action group. The action group disseminated information towards farmers and organised a meeting with two farmers' associations from the upper and lower valley to enlarge the offer and demand of services. Despite these initiatives, no exchanges were done. The *interessement* of actors encountered a limited success and the *enrolment* of farmers in the cooperative was not yet reached.

At this stage, an argument emerged and became recurrent in the discourse of the farmers' group. This argument lied in the question of equipment and investment. The cooperative was at this stage seen as a mean to avoid or postpone a costly investment in equipment by demanding a service or conversely a mean to obtain a return on a machinery investment by offering a service. This last *re-problematisation* on equipment and investment seems to currently prevail. It refers to the economic component of farms. Two main hypotheses could be formulated and tested to analyse this process of successive *re-problematisations*.

- The consensus on the problem of work overload and the limited demand of services hide a diversity of farms situations and of farmer's perceptions regarding work loads.
- Investment and equipment are major issues, emerging because of an evolution of the economic and social context (CAP reform, decrease of milk price in Europe, *etc.*). It reveals a crisis of small-scale farming systems and changes in farming coping strategies.

## Discussion: Technical, social or economic factors of coping strategies of small-scale farms

### *The diversity of farmers' situations regarding workloads*

To examine the farmers' situation regarding their workloads, we break down this issue in two aspects: (i) the daily tasks; (ii) the peak time for work related to seasonal tasks (hay making, manure spreading, land up-keep). Most of farms are autonomous for daily tasks however 5 of them resort to un-paid person (*e.g.* retired parents). For 16 farmers out the 26 surveyed, the basic group has a number of peak days superior to 120 per year. In 9 farms out of 16, the farmers reported to be unsatisfied with his/her workload and free time, the spouses surveyed insisted on the lack of spare time. The causes of the workloads and peaks are mainly related to daily tasks for three profiles of farms: (i) small-scale dairy farms managed by young farmers combining the routine work with an off-farm activity in ski resort: the work peaks are in winter; (ii) farms managing an alpine pasture and producing milk in alpine pastures: the work peaks are in summer, (iii) dairy goats and ewes farms with cheese processing the important work loads occurred all over the year. For a fourth farm profile, (iv) dairy farms with more than 20 cows, which are herded during summer in alpine pastures by a pastoral association or a private farmer, the major problem of workloads lies in seasonal tasks. The farming system diagnosis has confirmed that some farms are facing difficulties for their daily tasks most of the year or during the period when they have an off-farm job whereas for others the problem lies in seasonal works.

The services cooperative as conceived mainly responds to the problem of seasonal tasks. It is not a response for the constraints of daily tasks such as milking that is a problem for farms of the first three profiles mentioned above. Regarding daily tasks, farmers have searched for individual solutions based on machinery (*e.g.* a milking pipeline, automatic cluster removal, automatic calf feeder) or have increased the workforce (cessation or decreasing off-farm activity, new salaried workers). To be efficient, the services cooperative requires organisational changes in the farm systems and changes in the farmers' representations. To offer or demand a service implies that the farmers have to plan their tasks in advance and to manage a planning. Besides, they are usually reluctant to hand on seasonal tasks such as hay cutting, as in mountain slopes it requires special skills. To join the cooperative is a new way of conceiving farmer job not as a family activity but as an enterprise resorting to sub-contracting. Making time available through a cooperative is also a challenge for farming tasks diversification to respond to social expectations in terms of landscape upkeep, manure management and agri-tourism activities.

***Uncertainties in the farmers' environment and coping strategies***

Why have farmers finally seen the problem as an investment matter? This last shift in *problematisation* was in particular striking for dairy farmers (4$^{th}$ profile). To examine possible explanations, we consider the changes in the environment of farmers from 2003 to 2005. We identify many uncertainties and shocks that emerged in 2005: decrease of milk price, decrease of Beaufort cheese sales, the CAP reform and decoupling subsidies, the implementation of new environmental regulations. Those changes represent major economic driving forces. In this uncertain situation, farmers have shifted from a social concern on workloads to an overall view of farming systems situation including also economic aspects such as investment. In this context, they seem to go back to standard strategies such as: increasing work efficiency with equipment, scaling-up the farm. These strategies are more individual and this might be an additional explanation to the difficult interessement and enrolment of farmers in the cooperative.

## Conclusion

On the same object (workload), we used the sociology of translation and the farming systems diagnosis. This crossed discipline perspective demonstrates the complexity of a stakeholders' group process. This process included different *problematisations,* difficulties in the actors' *interessement,* major changes in the overall context. We also refine the workload problem with an analysis of the farmers' seasonal and daily tasks. It was a key distinction to understand how the services cooperative can fit to the farmers' needs. We could highlight the different levels of solutions designed by farmers: the individual level and the collective level. This analysis shows that the complexity lies in the difficult combination of innovation at individual and collective level, and both in technical and organisational fields.

## Acknowledgements

This paper is based on the results obtained in Savoie by the Groupement d'Intérêt Scientifique des Alpes du Nord in the framework of an going European project entitled *"Implementation of sustainable agriculture and rural development in alpine mountains"* (IMALP, Contract QLKT5-CT-2002-01099, www.alpes-du-nord.com/imalp ). In Savoie, the local activation is assumed by the Groupement Intercantonal de Développement Agricole Moutiers-Bozel, the Chamber of agriculture of Savoie and Albtertville Tarentaise Expansion. The partners of the project are the Istituto Agrario di San Michele all' Adige (Italy), the Institut d'Economie Rurale de l'Ecole polytechnique fédérale (Switzerland), and the Universität Innsbruck (Austria). This project is supported by the European Community and the Confédération Helvétique.

## References

Callon, M., 1986. Éléments pour une sociologie de la traduction. La domestication des coquilles Saint-Jacques et des marins-pêcheurs dans la baie de Saint-Brieuc. L'Année Sociologique 36, 169-208.

Dedieu, B., J.M. Chabosseau, J. Willaert, M. Benoit and G. Laignel, 1998. L'organisation du travail dans les exploitations d'élevage: une méthode de caractérisation en élevage ovin du Centre-Ouest. Étud. Rech. Syst. Agraires Dév. 31, 63-80.

Fleury, P., C. Kirchengast, V. Mieville-Ott, N. Magnani and S. Petit, 2006. Participative approaches and extension practices towards sustainable agriculture and rural development. A methodological approach based on sociology of translation and network theory. In: Langeveld, H. and N. Röling (eds.) Changing European farmingsystems for a better future, Wageningen Academic Publishers, Wageningen, The Netherlands, pp. 314-319.

Madelrieux, S., B. Dedieu and L. Dobremez, 2004. How to model work organisation. In: livestock farms implementing a combination of economic activities? 6th European Symposium on Farming and Rural Systems Research and Extension, IFSA, Vila Real (Portugal), 3rd-8th April 2004, pp. 269-279.

# What technical changes are needed for traditional dairy farming systems in less favoured regions?

*Vincent Thénard, Jean-Pierre Theau, Olivier Thérond and Michel Duru*
*Institut National de la Recherche Agronomique, UMR 1248, BP 52627, 31326 Castanet Tolosan Cedex, France*

## Abstract

To maintain milk production in European less favoured regions, farmers must develop high quality cheese production. Although traditional milk production is based on grazing and milk production targeted at spring and summer, many changes have lead to wide diversity in dairy farming systems. To contribute to the design and assessment of scenarios for the evolution of farming systems, we study how farmers change nowadays their grassland management. In this paper, we show how changes in grassland management must be consistent with farmers' strategies of milk production, and how local milk production context could contribute to the development of these changes. Two studies were made in 3 areas in the southern "Massif Central" (France). The first study consists in 24 farmers' interviews. It describes how farmers have developed three different management systems of grassland and livestock productions to adapt the traditional dairy farming system and to maintain the consistency of farming system functioning. Based on the Laguiole cheese production, the 20 interviews of the second study have shown that the new requirements of the PDO cheese specifications (abandonment of maize silage and inducements to feed just hay) has lead to three main technical changes (a gradual or an instant change-over to using just hay). In this context, the milk price and the local dairy cooperative agreement have favoured the rapid change-over to using just hay.

**Keywords:** dairy farming systems, farm management, PDO cheese production

## Aims and approach

The adoption of European PDO (Protected Designation of Origin) quality labels is generally essential to the local development (Barjolle *et al.*, 2002; Barham, 2003). They can be one way to sustain dairy production in less favoured regions in spite of CAP changes. But to achieve this goal, the labels must make it possible to maintain or to increase the price of the milk through the recognition of a quality cheese production (Brunschwig, 2000; Chatellier and Delattre, 2003; Delattre *et al.*, 2005; Espinasse *et al.*, 2005; Barjolle *et al.*, 2002. Also the specifications of the "Geographic Indications" require many changes, particularly in the feeding system, to develop high quality cheese production and to give more guarantees to the consumers. In many cases, the new specifications require the limitation or the removal of maize and grass silage from dairy cows' diets, and a lengthening of the grazing period. In the "Massif Central" (Central France), a low mountain region, cheese production accounts for most of the milk production (8 cows'-milk cheeses are produced with PDO labels). The main challenges for the future concern not only quality cheese production but also the farms' sustainability. We need scenarios for the evolution of the livestock farming systems. The design and assessment of such scenarios must determine appropriate changes to specific contexts. To assess the evolution of livestock farming systems, we analysed them on the basis of farmers' practices (Landais and Deffontaines, 1988). Our hypotheses are: (i) changes must be compatible with farmers' strategies of milk production to preserve the consistency of the livestock farming system; (ii) socio-economic conditions could contribute to the development of these changes. We have shown previously the importance of links between management practices of grassland and livestock productions (Thénard *et al.*, 2004). Hence we defined the consistency of the livestock farming system in terms of the interactions between the forage system and the herd management. This paper presents the first results of two studies carried out in the southern "Massif Central" mountains. Firstly, we describe different dairy farming systems instead of traditional milk production. Secondly, we identify how some technical changes have occurred in response to the soil and climate conditions of the area and the local cheese production context. Thirdly, we highlight the changes that have taken place in the "Laguiole" production because in this area the current changes are linked with the new requirements of the PDO specifications.

## Case studies

This project was conducted in 3 natural areas. Each of the three studied areas is defined in terms of its natural conditions (soil, climate, altitude *etc.*). In each area, a local dairy cooperative collects milk to produce a specific PDO cheese with different specifications: Aubrac is a wet upland area (1000-1600mm/y). Since 2004, the "Laguiole" cheese is produced with new detailed specifications (removal of maize silage from dairy cow, unpasteurized milk, limitation of the milk yield per cow...). Just one local dairy cooperative can produce this cheese. The western side of the Cantal Mountains is a wet hilly area (rainfall: 1200-2400mm/year). The "Cantal" cheese (similar to "Laguiole") is produced without any such specification. Several dairy factories can produce it. One local dairy cooperative has developed a specific production based on unpasteurized milk from "Salers" cows, a traditional breed. Margeride is a dryer upland area (800-1200mm/y). The "Bleu de Causses" cheese is produced without specifications. It is a small part in the cheese production of the local dairy cooperative. The farmers do not know whether or not their milk is used for this cheese, but the cooperative and a few farmers are now trying to develop a "traditional" cheese to maintain the milk price.

Two studies were based on the descriptive and comparative analysis of management practices of grassland and livestock productions. Data were collected during interviews seeking to describe farmers' practices.

The first study was conducted in the 3 natural areas. To investigate some of the diversity in dairy production, 24 dairy farmers were interviewed chosen with the advisors of the dairy cooperative to have a sampling of the diversity of farming systems. Firstly, livestock data were analysed to identify synthetic criteria of livestock practices. These criteria were computed using Multiple Correspondence Analysis (MCA of ADE4 software) to classify farms according to their management practices of livestock production (Girard, 2005). Forage management practices (date of cutting, interval between two grazing events) were expressed in degree-days (Theau *et al.*, 2004). These data were analysed using Principal Component Analysis (PCA of ADE4 software) to describe the grassland management practices. Next, these management practices and those for livestock productions were combined to describe different types of dairy farming system (see Thénard *et al.*, 2006).

The second study was made to highlight the recent changes which have taken place in Aubrac with the abandonment of maize silage and inducements to feed just hay. Data recorded on 20 dairy farms were on: (i) the farm structures in 2004 and their evolution (2000 to 2004); (ii) the forage systems in 2000 and 2004, (iii) the management practices of livestock production in 2004. We analysed the data by Multiple Correspondence Analysis (MCA from ADE-4 software) to describe the technical changes. In this paper only the main results on the technical changes have been presented (see Thénard *et al.*, 2005). Next we analysed the role of the local dairy cooperative to these changes.

## Switch from traditional milk production to "new" dairy farming systems

The traditional farming system is based on grass (grazing, silage, hay) as well as a production of milk in spring and summer seasons. But many technical changes have produced a great diversity in the current dairy farming systems. The average farm in our sample has 92 ha of "Agricultural Area" and 37 dairy cows which produce 177 000 l of milk. The altitude is around 890m (650-1080); 10 farms have a suckler herd. The MCA of the livestock practices has shown 5 management practices of livestock production. They have been described mainly for the milk production period, the winter ration and the herd replacement rate. The PCA of the forage practices has described 4 management practices of grasslands. They have been characterized mainly by the earliness of grazing in spring (palatable grass for early grazing or cutting, tough or unpalatable grass for late grazing or cutting). The combination of these results has enabled 3 different types of dairy farming system to be defined. However some practices (late cutting, grass silage use, sown pasture use) are linked to the soil and climatic conditions:

*Type 1 - Renewal of the traditional milk production system (9 farms):* farmers produce milk mainly during spring and summer, usually with a low herd replacement rate. Seven farms improved their spring grazing by topping and one made grass silage. The aim of both practices was to produce more and better hay later on. Soil and climatic conditions permit regular grass growth in spring and summer.
*Type 2 - Intensification of the grassland and milk production system (8 farms):* farmers produce a significant amount of milk during winter that needs ample forage stocks and a large quantity of concentrates. Sown grasslands and grass silage are used to favour early date for grazing or cutting. In case of soil having low water reserve and dry summer, sowing of grasslands is viewed as the only way of forage intensification.
*Type 3 - Increasing and diversification of animal production (5 farms):* farmers produce both meat and milk. The increase in the agricultural area has lead to suckler herd development and usually to an intensification of milk production. Grassland management is based on topping in spring, and complementarity between the dietary requirements of the dairy and suckler cows. Characteristics of soil and climate are not very limiting.

## Influence of area and local cheese production on changes

The types of dairy farming system described above have been specified for each area. We will now explain how these types adapted to the local context. We can distinguish the wet and the dry areas.

In the wet areas (Aubrac and Cantal), the first and third types of dairy farming system are present:

*Type 1 - Renewal of the traditional milk production system:* because soil and climatic conditions are favourable to forage production and the milk price is higher, farmers can maintain the traditional system *"to produce quality cheeses"*.
*Type 3 - Increasing and diversification of animal production:* because the development of the specifications is recent, farmers had to develop new livestock activities and had to do a limited-intensification of dairy production *"to preserve the income"*. In these areas where cheeses are

renowned, the increased milk price is recent and some farms had previously developed an other dairy farming system than the traditional one.

In the dry area (Margeride) without a renowned cheese, only the second type of dairy farming system is present:

*Type 2 - Intensification of the grassland and milk production system:* because on the one hand summer is a dry season and the soils are drying, and the other hand the milk price is without bonus, sown meadows are used for increasing forage production earlier in the season *"to maintain milk production in the region with grassland intensification".*

## A favourable social and economic context to the change in Aubrac

The new requirements of the "Laguiole" cheese specifications, farmers had to change their livestock farming systems. An average farm from our sample has 37 dairy cows producing 174 000 l of milk on 59 ha of Agricultural Area; the altitude is around 750m (650-850). The MCA explains four different technical changes in response to the need to abandon maize silage:

*Change 1 - Maize to grass silage (3 farms):* Specialist dairy farms.
*Change 2 - Maize to grass silage with more concentrate (6 farms):* Most farms have decreased the number of suckler cows of the farm.
*Change 3 - Progressive abandonment of grass silage (5 farms):* Specialist dairy farmers. They intend to stop using grass silage in the next few years.
*Change 4 - Barn-dried-hay use (7 farms):* The smallest and largest farms in the interview sample. Maize silage has been replaced by barn-dried hay.

All the farmers in the area (26/27) have decided to change their forage systems because the social and economic context was favourable. Today, 80% of the farmers interviewed think that the new requirements have had a positive effect on the cheese quality and on the cheese image which is "terroir"-dependent. To train farmers in the technical change, the dairy cooperative has developed incentive policies to eliminate maize silage from the cows' diet. The barn-dried hay system produces high quality hay. In the framework of the regional development, farmers can obtain aid for building these systems and the dairy cooperative will lend money interest-free to begin the building. But the main incentive policy is more profitable use of the milk used in the "Laguiole" production. All farmers who produce milk with the requirements of the specifications obtain a high baseline price: 0.374 €/l whereas the price of the region is about 0.320 €/l. A high milk quality and a good bacteriological content permit bonuses (at the most 0.0305 €/l). Other bonus is given (0.023 €/l) when cows are fed with hay alone. The price can raise 0.4355 €/l (+36% of the main price of the region) and these bonuses are considered effective to favour technical changes.

## Discussion and conclusion

These studies based on the characterisation of management practices, show that there are several consistent arrangements between the forage and the herd management leading to a large diversity in dairy farming systems. This diversity appears to depend partly on soil and climatic conditions for the grassland production. However the knowledge of the cheese production context is very important to understand how the feeding systems could be change. For instance, to have good milking performance, the herds need high quality of forage. Barn dried hay respond to these requirements but it is very expensive. So the economic decision taken into a local context must favour it. For another area, Josien *et al.* (2003) have shown that the current state of the dairy systems is very important for the acceptance of grass silage abandonment in the Massif Central. However it needs a clearer focus on the organisation of these farming systems (Coleno *et al.*, 2002) in particular on the current feeding systems. On the other hand, we have seen that the dairy cooperatives' training can encourage farmers to accept some technical changes and Quetier *et al.* (2005) have shown that changes in production rules can produce changes in the farmers' land use strategy. In conclusion, this study is a relevant springboard for predicting the evolution of dairy farming in this region and to design and to assess scenarios for their evolutions we must develop.

## Acknowledgements

Many thanks to the farmers and the cooperatives involved for their contribution to this work.

## References

Barham, E., 2003. Translating terroir: the global challenge of French AOC labelling. J. Rur. Stud. 19, 127-138.

Barjolle, D. and B. Sylvander, 2002. Facteurs de succès des «produits d'origine» dans les filières agroalimentaires européennes Economie et Société 36, 1441-1462.

Brunschwig, G., 2000. Terroirs d'élevage laitier du Massif central. Ed. ENITAC, 223pp.

Chatellier, V. and F. Delattre, 2003. La production laitière dans les montagnes françaises. INRA Prod. Anim. 1, 61-76.

Coleno, F.C., M. Duru and L.G. Soler, 2002. A simulation model of a dairy forage system to evaluate feeding management strategies with spring rotational grazing. Grass Forage Sci. 57, 312-321.

Delattre, F., J.L. Reuillon and A. Farruggia, 2005. Incidences et enjeux de la réforme de la PAC pour les exploitations laitières de montagne. Four. 181, 143-159.

Espinasse, R., B. Posté and M. Grasset, 2005. Quelle production sur les territoires face aux besoins des marchés. In: Proc of 2nd Journée CEREL, Rennes, pp. 65-77.

Girard, N., 2005. Catégoriser les pratiques d'agriculteurs pour formuler en partenariat un «problème». Cah. Etud. Rech. Francophones. Agricultures (in press).

Josien, E., P. Chassard, N. Ballot, J.L. Reuillon and J.P. Garel, 2003. Etude de la faisabilité de l'évolution du cahier des charges en AOC fromagères. In: Renc. Rech. Ruminants, INRA eds. 10, pp. 207-211.

Landais, E. and J.P. Desfontaines, 1988. Les pratiques des agriculteurs. Point de vue sur un courant nouveau de la recherche agronomique. Etud. Rur. 109, 125-158.

Quetier, F., P. Marty and J. Lepart, 2005. Farmers' management strategies and land use in an agropastoral landscape. Agr. Syst. 84, 171-193.

Theau, J.P., P. Cruz, P. Anquer, V. Thénard, F. Coléno and M. Duru, 2004. Improving the management of native pastures: a phenoloy - based tool. In: Proc of 20th General Meeting of the EGF, Luzern, 9, pp. 796-778.

Thénard, V., O. Théron, J.P. Theau, D. Verger and M. Duru, 2006. Knowledge of the dairy farms functioning to assess their adaptation to the CAP changes. in Proc of 21th General Meeting of the EGF, Badajoz (Spain) (in press).

Thénard, V., C. Miquel, A. Miquel and M. Duru, 2005. Grasslands management changes in dairy farms: Case of the «Laguiole» cheese production. In: Proc of 13th Meeting of the FAO-CIHEAM Moutain Pastures Networks, Udine, Ed. ERSA (in press).

Thénard, V., F. Coléno, J.P. Theau, L. Marey and M. Duru, 2004. How to analyse technical adaptability of dairy farms involved in quality cheese production? In: Proc of the 6th European IFSA Symposium, Vila Real, pp 97.

# Structure and operation of small scale farming in the oasis in southern Algeria

*Salah Eddine BenZiouche*
*University of Biskra, CREAD Algeria, BP 904, RP Biskra 07000, Algeria*

## Introduction

Many policies have been implemented to develop palm date cropping. This poster presents research aiming at analysing the structure and distribution of farms growing date palms. One of the main issues addressed in that setting is the strategy adopted by these farms in response to occurred changes.

## Farms structure

35% of the farms growing date palms have very limited land (less than one hectare), 45% owning one to three hectares and only 20% owning more than three hectares (BenZiouche, 2000).

## Cropping systems and marketing

Only 9.5% of small scale farmers use pesticides. Most of the farmers still use local indigenous techniques. 95% of them weed manually, use manure instead of chemical fertilizers, and prune date palm. In order to improve the performance of their cropping system, they should combine different types of production in order to increase their income and minimize risks (Messar, 1996).

There are in terms of dates marketing and farmers should take advantage of them. They should also consider the way consumers are behaving in each region. They should adopt different ways of marketing. They can, for example, try to market their products on farm, especially for varieties for which the demand is important. Simultaneously, they can sell the varieties for which the demand is weak on the market. From time to time, farmers sell the whole production to a middle man who collects the dates, transports them to the nearest market: farmers do not have enough cash in order to hire labor for date collection, to transport their production to the nearest market. It is a way for them to reduce risks (risks related to markets or climate risks). Some of them use indigenous techniques for date processing and storage: storage of 50% dried dates in small wooden boxes, date jam processing, dates storage in a sheepskin.

## Conclusion

There are many small scale farmers growing date palms in the area and they did not benefit much from the agricultural policy set up by Government agencies. Then, they try to develop several strategies in terms of cropping systems or in terms of marketing in order to alleviate the main constraints. These strategies have been successful as they could reduce migration flows.

## References

BenZiouche, S., 2000. Analyse de la filière datte en Algérie, étude de cas les daïras de Djamaa et Mghaer. Magister thesis, National Agronomic Institut. Algiers. 350 pp.

Messar, E.M., 1996. Le secteur phoenicicole algérien: situation et perspectives à l'horizon 2010. Options méditerranéennes, série 2, pp. 210-221.

# Developing organic farming in Norway through systemic action research

*H.W. Carlsen[1,2], T.A. Breland[2], G. Lieblein[2] and C. Francis[2]*
*[1]Bioforsk Arable Crop Division, 2849 Kapp, Norway; [2]Norwegian Uni. of Life Sciences, Dept. of Plant & Envi. Sciences, 1432 Ås, Norway*

## A multidimensional approach

Organic farming systems are linked to explicit, multiple goals concerning ecological, economic and social perspectives. It is a challenge for stakeholders concerned with the systems to clarify multiple goals, to elucidate how the goals interact, and to decide how improvements in the organic food system may be brought about without compromising long-term goals of multifunctional and diverse agroecosystems. This particularly applies to grain production in organic farming, as a regionalisation policy practiced for several decades largely has removed the potentials and benefits of mixed crop and livestock farming and increased the distance between producers and consumers.

## Systemic action research

Within the research programme "Organic cropping systems for higher and more stable cereal yields", systemic action research (Checkland and Scholes, 1999; Reason and Bradbury, 2001) is an approach to gain insight into the present situation of the agroecosystem, to envision future wanted scenarios, and to accomplish desirable and workable outcomes. It is important to make this happen in a democratic and participatory way that empowers stakeholders to continue the learning process and system development in a ecologically friendly, socially feasible and economically viable way. Consequently, a farm case study, which involves stakeholders both within and outside the farming system is in progress. The objectives are to *1) promote an on-going self-development and informed action towards shared goals and improved sustainability of currently existing organic farming systems,* and *2) handle multiple stakeholders, perspectives, criteria and methods in concert in order to learn from the process of working with/in complex real-life situations.*

## Methodology, methods and techniques

Within a framework of soft systems methodology (Checkland and Scholes, 1999), interviews (Kvale, 1996), dialogue conferences (Gustavsen and Engelstad, 1986), "Future workshops" (Jungk and Müller, 1989) and multicriteria evaluation (Munda *et al.*, 1994) are being applied. Qualitative and quantitative data from the case farm and pre-knowledge from other studies are used to enrich the database and inform the development of conclusions.

## References

Checkland, P. and J. Scholes, 1999. Soft systems methodology in action. Wiley & Sons, UK.

Gustavsen, B. and P.H. Engelstad, 1986. The design of conferences and the evolving role of democratic dialogue in changing working life. Human Relations 39, 101-116.

Jungk, R. and N.R. Müller, 1984. Håndbog i fremtidsværksteder. Forlaget Politisk Revy, Kbh, DK.

Kvale, S., 1996. Interviews. An Introduction to Qualitative Research Interviewing. Sage Publ UK.

Munda, G., P. Nijkamp and P. Rietveld, 1994. Qualitative multicriteria evaluation for environmental management. Ecological Economics 10, 97-112.

Reason, P. and H. Bradbury (eds.), 2001. Handbook of Action Research. Participative Inquiry and Practice. Sage Publ. UK.

# Farming styles and the future of pig farming in Midi Pyrenees

*Monica A.M. Commandeur and François Casabianca*
*INRA SAD Corte, France*

During the last ten years, two out of three pig farming enterprises in Midi Pyrenees have vanished, and the remaining pig farms face serious challenges. More than 55% of the region's pig production is located in Aveyron and Tarn. Although pig farmers share similar constraints and opportunities, their actions are not uniform.

This study investigates the diversity in pig farming practices and strategies in this part of Midi Pyrenees, adopting the same research methodology as a recent study of pig farming in Bretagne (Commandeur *et al.*, 2006). Its focus is on identifying the most relevant dimensions, which provide the frame of reference, which in turn structures how farmers' express their attitudes and practices. The survey consisted of 25 semi-structured interviews, exploring the scope of diversity in pig farmers' logic.

One of the most relevant dimensions focuses on the farmers' ambitions and expectations. In-depth analysis of this dimension revealed three dominant modes of perceptions. These, in turn are related to the likelihood of succession:

- Mode 1 is characterised by grievances about low income and profitability, an unwillingness to make investments and a general pessimism about the prospects for the sector. These farmers have little prospect of family succession of their farm.
- Mode 2 wishes to continue farming with the means currently available to them and is focussed on the possibility of farm succession. Farmers are pessimistic about the prospects for pig farming, although they do want to continue in their rural area.
- Mode 3 is more varied than the other two, and expresses a degree of optimism. Farmers think that the crises will pass, although structural developments are inevitable. They believe that pig production in the area will continue, and run their farms in anticipation of family succession and their perception of the opportunities.

Analysis of the whole discourse of the interviews led to recognition of five key dimensions that best describe the contrasts in farmers' logic: (i) Animals and Technology, (ii) Labour and Investments, (iii) Ambitions and Expectations, (iv) Socio-Professional Integration, and (v) Images of Production Methods and Products. On the basis of these dimensions, three main *styles of farming* were constructed, and given metaphors that resemble the metaphors used in Bretagne: craftsman (including mode 3), inheritor (including mode 2), and stockman (including mode 1). However, the metaphors represent the diversity among the farmers in the "information space" in Midi Pyrenees. A key focus for structuring local farming strategies is the expected development of a PDO certificate for "Lacaune ham". This perspective is essential for craftsman logic and important for inheritor logic, but rejected by stockman logic. The diversity in farming styles is therefore mainly explained by the contrasts in socio-professional integration and the images of production methods and products.

## Conclusion

Despite the ongoing regression of pig farms and pigs in Midi Pyrenees, the choice for adopting the farm management to new diversified products (or not) – in particular to the expected PDO for "Lacaune ham" – will contribute to further diversification of pig farming styles and production methods. The future of pig farming will also involve further diversification and reorientation of socio-professional networks.

## References

Commandeur M.A.M., R. Le Guen, J.Y. Dourmad and F. Casabianca, 2006. La diversité des styles d'élevage porcins: une approche dans les Côtes d'Armor. Journées de la Recherche Porcine 38, 247-254.

# Transformation of livestock farming in Cameroon: Changes in breeding practices and evolution of farmer profiles

*Patrice Djamen[1], Jean Lossouarn[2] and Michel Havard[3]*
*[1]Institut National Agronomique Paris-Grignon (INA P-G) / SADEL. BP. 293 Garoua, Cameroun; [2]INA P-G, DSA/UMR SADAPT; [3]Institut de Recherche Agricole pour le Développement (IRAD) / CIRAD*

Population growth in Cameroon has given rise to an increased demand for beef (54% of the total meat consumption). It is important to assess the capacity of local farms to meet the challenges of food security within the context of socio-economical changes and shortage of resources.
The objective was to analyse the transformations of cattle faming systems and assess their impact on beef supply. A survey was conducted to characterise socio-economic profile of farmers and transformations in their farming practices with emphasis on exploitation practices and the role of livestock in farms. The study was carried in two contrasted provinces: Adamawa and North. Adamawa is characterised by a strong developing market due to increasing demand of meat from the main cities of the South and neighbouring countries, whereas the North is characterised by inadequacy of resources.
To meet with current challenges of beef supply and inadequacy of resources, cattle rearers diversify their herds and farm activities. In Adamawa, farmers diversify their productions by creating new herds more or less specialized in milk production and/or fattening. In North, the increasing demand of land for crop production has penalised cattle farmers who in the face of competition for resources, have chosen to settle. They reduce and or split up their herds into: home based herds for traction, dairy cows and a few transhumant herds which are the main part of their livestock. At the farm level of both regions, diversification is achieved by activities such as crop production, extra-farm activities (carpentry, trading) or other functions of the cattle business (cattle sales agent or shepherd).
The profile of cattle farmers is evolving. Livestock farming was practised mainly by pastors. Nowadays, three types of cattle owners can be distinguished (Table 1). New farmers (those from other socio-professional categories – OSPC) have marked economic objectives such as milk production, cattle fattening, ranching. The impact of the evolution of the profile of farmers on the functioning of livestock farming systems and beef supply remains very weak. The social function of livestock is still predominating.
The current trend towards more market directed systems is still slow and fragile. The food security of the country and the future of livestock farming depend on effectiveness of the measures (farm inputs, facilities, loan, *etc.*) that should be taken to promote a sustainable development of livestock faming systems.

*Table 1. Functions (%) of livestock in farms according to the profile of cattle owners. Predominant function highlighted in bold.*

| | Role of cattle | | |
|---|---|---|---|
| | Social | Work | Economic |
| Cattle breeders | **75** | 10 | 15 |
| Farmers | 30 | **60** | 10 |
| OSPC | 40 | 5 | **55** |

# Farming future in China: Has time come for financial transfers?

*Michel A.C. Fok[1], Weili Liang[2], Guiyan Wang[2] and Yuhong Wu[2]*
*[1]Centre de Coop. Int. en Rech. Agr. pour le Dév. (CIRAD), France; [2]Hebei Agricultural University, China*

Under exacerbated competition by the globalization process, the survival of farming is at stake. The need to gain productivity is often put forward in developing countries. Farm concentration is taking place in all developed or emerging countries. The involvement in off-farm activities comes out to be critical to farming viability in developed countries (De Hoog and Van Ophem, 2004). In China, these three factors are contributing, but farming is still not guaranteed of a fair future as it is shown by the results of a survey implemented in a cotton production area, in Hebei Province.

## Farming is increasingly little income-attractive

In 2003, the average rural annual income per capita was around US$ 327. It represented only 32% of the average income in urban areas, down from the figure in 1985 (58%). The contribution of farming activities in the global income of rural families also is decreasing (share of around 60%), indicating a commitment in off-farm activities which remains insufficient to catch up the income gap.

## Tiny smallholdings with labor excess

Farming income is restrained by small farm size (less than 0.9 ha) in spite of high level of yield (Table 1). The access to rented land only concerns 25% of the farm holdings whose size remains very small (1.2 ha). With a family size of around 3 to 4 members, labor excess is clear. About 90% of the farm families of our survey are concerned by off-farm activities, but ladies are seldom concerned. It comes out that Chinese farming is a part-time one and conducted by female and old people.

## What would prevent further abandonment of farming?

The increasing income gap should continue attract people out of farming. This phenomenon is starting touching females who less and less accept leaving their husbands migrating alone. It raises the question of who will stay farming in China. It seems that a time has come to make farming further attractive. Financial transfers to the benefit of farming people, as it was implemented elsewhere and as China seems to be able to afford, could be considered.

## References

De Hoog, C. and J.A.C. van Ophem, 2004. Dutch farming families' successful strategies for increasing household income by changing the division of labour between the spouses. In: Les mutations de l'agriculture: conséquences pour les politiques publiques. Paris, France. 8 pp.

*Table 1. Small but high yielding farms.*

| | Farmers having leased land? | | |
|---|---|---|---|
| | No | Yes | Total |
| Percentage of farmers | 75% | 25% | 100% |
| Total cultivated area (ha) | 0.71 | 1.19 | 0.89 |
| Cotton area (ha) | 0.30 | 0.60 | 0.35 |
| Av. yield, kg/ha (std deviation) | 3315 (495) | 3420 (585) | 3378 (524) |

# Rural transport and development: Case of the agro-pastors of the djebel Amour (Algeria)

*Bennoui Foudil*
*Mohamed Khider University, BP 145, 07000 Biskra, Algeria*

Our methodology consisted in quantifying and in analyzing supply and transportation demand for pastors in the steppe and mountainous areas by doing surveys. We recorded very weak vehicle's ownership and agro-stockbreeders equipment (one tractor for 349 ha, housekeeping). The percentage of households having at least one mean of transport was 0.077 in 1997 and 0.115 in 2005. The number of vehicles by household was equal to 0.088 in 1997 and 0.129 in 2005 in the rural district of Brida in central Algeria. Consequently, the pupils give up the school quickly to be employed in agriculture. Because of poverty and because schools are far from their houses, a very limited number of children go to school. Traditional means of transportation for displacements on short distances are used.

Indeed, the owners of means of transport provide a service for transport of people or transportation of goods. Vans are under utilized because they are old, not well maintained, and because of seasonability of the agricultural work. Tractors take part only in one marginal way in transhumants transport and are used especially to carry out work fields and transport the agriculture products, water and building materials. The cost of transport does not constitute a large budgetary burden; the problem lies rather in the absence of transport in convenient time particularly for the transport of people.

Until today in Algeria the literature which treats this question is non-existent. The subject especially becomes extensive at the international forum for rural transport and the development (IFRTD) which collaborates with the experts of the World Bank and other international development agencies. The research results are far from being exhaustive and can be to generalize only if we consider the socio-economic, agronomic and natural conditions that are similar at the national and Maghreb level. The solutions suggested for improvement can be multiple and diverse:

- limitation of the rural displacements while developing their small boroughs and install hangars in order to collect their products;
- encourage rural mobility towards the urban centers while allowing clandestine to transport the rural ones with their products and provisioning on the weekly markets;
- finance the farmer's equipment for purchasing modern means of transport and construction of storage facilities;
- improve the pupils transport and offer them boarding school service.

# The LIFESCAPE approach: Identity and landscape as an incentive for economic development

*Rob Maessen[1] and Peter Nowicki[2]*
*[1]Province of Noord-Brabant, PO Box 90151, 5200 MC 's-Hertogenbosch, The Netherlands; [2]European Centre for Nature Conservation – ECNC, PO Box 90154, 5000 LG Tilburg, The Netherlands*

## Introduction

What hope is there for rural areas in an urbanizing world? Part of the answer lies in finding economic benefits in the maintenance and enhancement of the landscape. The products and services delivered by our natural resources must meet a market demand if they are to be sustained over time. LIFESCAPE – YOUR LANDSCAPE has the objective to test and describe a viable strategy for sustainable rural development, based on these principles, which can be applied throughout North West Europe. The Province of Noord-Brabant in the Netherlands is responsible for leading the project: fourteen partners, five nationalities, and twenty trans-national actions for sharing experiences and working together in one project. The LIFESCAPE project is funded by the European Union as part of a Community initiative to support the EU cohesion policy. The project proposal was approved on 31 March 2005 as part of the INTERREG IIIB Programme for North West Europe.

## The LIFESCAPE approach

The challenge is to strike a new and sustainable balance between our livelihood and our landscape. This will enrich our lives as well as the environment we live in. LIFESCAPE combines the two – "livelihood" and "landscape", the individual benefits and the common good – in one concept. The LIFESCAPE approach is about using our landscape to bring people together to create added value and quality of life.

This project explores innovative approaches for improving regional cooperation and economic development. Acknowledging the landscape as a source of identity and inspiration can bring people together from different sectors to join forces in order to preserve and enhance its quality. The natural beauty of our surroundings communicates an immediate message and an appeal – consciously received, or not yet – that is able to reach across the many social barriers that exist between the networks of modern society. Hence the name: LIFESCAPE – YOUR LANDSCAPE. Pride, enthusiasm and dedication will then become available as a powerful set of tools for sharing experiences and learning by doing.

## A grass roots perspective

The idea for this project brought a group of about twenty people together who believe in the meaning and values of the natural environment for our daily life. Their shared enthusiasm for the concept of LIFESCAPE is driving the success of this enterprise. Each trans-national action provides them with an opportunity to explore the LIFESCAPE challenge under different circumstances. Likewise, it will be the people of the region – entrepreneurs, teachers, farmers, bankers, government officials, and representatives of non-governmental organizations – who will determine the success of creating a stronger awareness of the landscape and using opportunities for economic purposes.

# The influence of climatic and socio-economic conditions on farm performance in Europe

*Pytrik Reidsma and Frank Ewert*
*Department of Plant Sciences, Group Plant Production Systems, Wageningen University, PO Box 430, 6700 AK Wageningen, The Netherlands*

Agriculture is vulnerable to changes in climate and climate variability (Olesen *et al.*, 2002). This vulnerability can be reduced through adaptation. Although it is generally assumed that farmers will adapt to these changes, such adaptive capacity has never been properly quantified. Farm performance and adaptive capacity are not equally distributed. Certain regions and farm types will be more vulnerable than others. To increase understanding of this adaptive capacity, we analyzed the performance of farmers in the European Union. Extensive data on farm characteristics of individual farms from the Farm Accountancy Data Network (FADN) have been combined with climatic and socio-economic data to determine the influence of climate and management on crop yields, income and related variables that indicate adaptive capacity. We applied multilevel regression models that are explicitly designed for statistical analysis of hierarchically structured data and explicitly take the variability at different levels into account.

The results suggest that input intensity, economic size and the percentage of arable land are generally good indicators that enhance the adaptive capacity of farmers. However, the influence of climate, socio-economic factors and farm characteristics on crop yields differs for different crops. High crop yields are not always related to high incomes. This implies that the adaptive capacity of the farms to respond to climate change also depends on the aspect considered. A farmer may be able to adapt to decreasing maize yields, but not to decreasing wheat yields. Even though a farmer isn't able to adapt to changing yields, he may still be able to adapt economically to decrease its vulnerability. Farm net value added per annual work unit is higher in regions with a more favourable climate, but farm net value added per hectare isn't. Farmer's income is not so much determined by crop productivity, but mainly by farm size and subsidies.

Although crop growth is highly influenced by climatic conditions, the actual influence of normal climate variability on farm performance was low. Regional differences in crop yields seem to be related to climatic conditions, but as socio-economic variables are highly correlated with climate variables, this influence can be confounded. Although wheat yields are generally lower in Mediterranean regions, yields of maize are of the highest in the European Union. Analysis of temporal changes indicates that the average farmer adapts its management such that vulnerability to climate variability is relatively low. Actual observed crop yields do not vary as much and often not in the same direction as water-limited and/or potential yields. However, extreme events like the heat wave in the year 2003 can have a large influence on crop productivity (see also: Ciais *et al.*, 2005), and these kind of extreme events are projected to increase to a large extent. Although the average farmer seems to be able to adapt, individual farmers may be vulnerable as there is a large deviation in crop yields and income within regions. In general, large intensive farms will be the least vulnerable to climate change as they have the highest capacity to adapt to their (changing) environment.

## References

Ciais, P., *et al.*, 2005. Europe-wide reduction in primary productivity caused by the heat and drought in 2003. Nature 437, 529-533.

Olesen, J.E. and M. Bindi, 2002. Consequences of climate change for European agricultural productivity, land use and policy. Eur. J. Agron. 16, 239-262.

## Greek agricultural sector: Facing the challenges

*Athanasios Theocharopoulos, Eleni Oxouzi, Venetia Kourkouta and Evangelos Papanagiotou*
*Department of Agricultural Economics, School of Agriculture, Aristotle University of Thessaloniki, Greece*

The agricultural sector in Greece plays an important role in the development of the country with a relatively high contribution to the Gross Domestic Product and the employment rates. Greece has a comparative advantage to the cultivation of many Mediterranean crops, due to its soil characteristics and its climate. The most important products of Greece are olives, wine, cotton, sugar beets, tobacco, many fruits and vegetables, as well as milk and meat of high quality.
Nevertheless, the reforms of the Common Agricultural Policy (CAP) have changed the scenery within the European Union. The agricultural sector has decreased its contribution in the country's GDP from 11% (1990) to 4% (2001) and in most of the cultivated crops a decrease was detected (in terms of land and gross value).
Still, agriculture is considered an important sector of the Greek economy as it contributes to the maintenance of social and economic cohesion of many rural regions around the country as more than 15% of the total work force in Greece is working in the sector. Undoubtedly the new rules imposed by the EU will have a serious effect in the future of Greek agriculture, its work force maintenance and the viability of rural areas.
This paper gives some possible scenarios for the future of the sector in Greece by a quantified analysis of a series of data for the most important crops; over the last 40 years. The data were taken from the National Accounts and the Ministry of Agricultural Development and Food. It was used an econometric model appropriate for time series analysis of data and with the aid of the model was calculated the contribution of each crop and analysed for their future growth. Simultaneously, this paper takes into consideration the decoupled payments that were introduced by the CAP 2003 reform and will be applied in Greece from 2006.
According to the results, the cultivated area (refers on hectares) of the most important crops of Greece is expected to be reduced in next years. It is noticeable that according to the results, the number of farmers and the contribution of agricultural sector in the country's GDP will be reduced. The presentation of strengths, weaknesses, opportunities ant threats (SWOT analysis) of Greek agricultural sector will help advisors and policy-makers to plan the way that Greek agriculture must be follow in order to face the new challenges which are created by the CAP 2003 reform.

# *It Frysk Boerespul*: Historical farmhouses with modern agrarian functions

*Luitzen Walstra and Marit de Vries*
*Van Hall Larenstein BC, PO Box 1528, 8901 BV Leeuwarden, The Netherlands*

The project aims to clarify that the conservation of historical farms may be combined with the preservation of their agricultural functionality. Farms in the scenery of the province of *Fryslân*, characterized by terps (artificial hills) and lakes, have been studied in order to construct a perspective for future development.
For the selection of farms, search regions have been mapped out. A multi-criteria analysis was applied in order to identify areas with scenery of high value. A great number of farmers were addressed and two information meetings were organized for those showing interest. The farms have been visited and validated. The cultural historical value was assessed based on authenticity, age, rarity, location and the combination of value and coherence. Based on this validation, five representative farms located in various search regions, were selected.
The history of a historical farm should be preserved when it is altered. In Frisian farms, building mass is a characteristic phenomenon. The typical Frisian farm types head-rump, head-neck-rump and stelp can all be recognized by their building mass. Frisian farms are characterized by their supporting construction consisting of a wooden frame with coupled tied beams. This construction has a high monumental value and should be preserved. Even after changes in functionality this arrangement should remain visible.
Housing modern agricultural functions in a historical farm implies additional limiting conditions. Farm designs have been used to explore the most favourable realization of functions and the most efficient farm management. Free-stall systems with litter-covered lying places can be fitted in easier than other modern stall systems. The dimensions are more flexible than those of lying boxes, because the animals are housed in groups. However, this system is not often used for dairy cattle. In this system the cows' udders get polluted relatively quickly because excretion takes place in the lying room. This is less relevant to young cattle because they are not milked. For this reason young cattle can well be housed in the old farms. The low height at the place where the cows are housed and the placement of the tied beams may obstruct the dimensioning of cattle housing rooms and may obstruct access by tractors and machinery. The size and location of barn doors may also limit the size of agricultural machinery. The combination of several functions under a single roof puts high demands to the building in order to avoid fire-risks.
Some of the selected farms will function as models. The scenery has a distinguishing open nature. The public function of the model farms offers an excellent opportunity to show the historical development of farmhouses, the yard and the landscape. Old farmhouses have important historical, esthetical and landscape values, contributing considerably to the attractiveness of rural areas. Combining such values with new, so far unexploited, functions contributes considerably to the conservation of cultural heritage as well as the revitalisation of rural areas.

## Acknowledgements

The authors thank the partners of the project (the Province of *Fryslân*, BoerenNatuur, *It Fryske Gea*), which also received financial contributions by the Ministry of VROM, NLTO, Agrotechnology & Food Innovations, and Van Hall Instituut BC.

# Section 4 – Agricultural knowledge and innovation systems in transition

# Introduction to section 4 – Agricultural knowledge and innovation systems in transition

*Kirsten von der Heiden[1], Cees Leeuwis[2] and Uwe Jens Nagel[3]*
*[1]AfoReg, Germany; [2]Communication and Innovation Studies, Wageningen University, The Netherlands; [3]Institute of Extension and Communication, Humboldt University of Berlin, Germany*

While Agricultural Knowledge and Innovation Systems (AKIS) used to have clear contours and institutional set-ups – dominated often by publicly funded agricultural research, extension and education – they have increasingly evolved into much more pluralistic, fuzzy and temporal sets of multi-stakeholder networks. Several factors have been influential in bringing about this shift.

First, the dominant view on agricultural development became more and more contested from the early 1980s onwards. Societal pressure groups and scientists criticized modern agriculture for its detrimental effects on the natural environment and the world ecology at large. Thus, new societal parties, problem definitions and objectives entered the agricultural policy arena, while the old policy objective (increasing agricultural production) had become less relevant in view of substantial overproduction in key sectors. Against this background, the dominant policy view became that AKIS could not remain a purely agricultural affair, and that it needed to open up to new societal players (*e.g.* consumers, nature conservationists, the environmental movement, *etc.*) in order to be able to deal with the new concerns, options and priorities.

Second, on the wave of the generally increased trust in market forces, privatisation of research and extension became an interesting option for leveraging change in the knowledge infrastructure. An additional rationale for embarking on privatisation and decentralisation was that public extension and research organisations were seen as operating in an inefficient, bureaucratic, top-down, paternalistic and inflexible manner, and hence were not responsive to the needs and demands of clients (Umali and Schwarz, 1994; Rivera, 2000). This development went parallel with and was stimulated further by the debate on innovation in which the linear model was contested ever more and the role of users (the demand side) in innovation processes was stressed. Changing theories of innovation, therefore, can be seen as a third major influence that shaped the deliberate reorganisation of AKISs. Nowadays innovations are no longer looked at as consisting of technology only, but rather as successful combinations of 'hardware' (i.e. new technical devices and practices), 'software' (i.e. new knowledge and modes of thinking) and 'orgware' (i.e. new regulations, market arrangements and forms of organisation) (Smits, 2002). In this perspective the social and organisational conditions that may affect the 'survival' of innovations are no longer seen as external and static, but rather as integral parts of the innovation. Thus, innovation depends on effective collaboration in a network of interdependent societal actors. In order to stimulate the emergence of complementary action in a network, new roles for research and extension have been suggested, along with new methodologies and approaches (Gibbons *et al.*, 1994; Röling, 2002).

It is not easy to make well-founded statements about the functioning of the current knowledge infrastructure that is supposed to enhance innovation in agriculture and other forms of land-use. On the one hand this is related to the fact that the situation is far from stable and has not crystallized yet; new developments are taking place on a regular basis. We currently witness, for example, intensified efforts by the EU to shift from supply-side financing (with the government and other users in the role of client) to demand-side funding by means of vouchers, whereby societal stakeholders acquire greater control over the spending of available public money. Another obstacle is that systematic comparative research on the functioning of new arrangements largely is absent. In view of this vacuum, papers in this section set out to describe, analyse and compare current developments in AKISs, with specific attention to forms and processes of institutional and methodological reform.

The contents of this section is organised along three lines:

## Public interests and privatised innovation systems in comparative perspective

An important question that needs to be addressed in the context of current transitions is how public interests and social issues are dealt with – or may be effectively safeguarded – in AKISs that are characterised by increased dominance of market mechanisms. More broadly, some papers present and compare the current set-up and dynamics of AKISs in different European countries, and analyse how their functioning has changed (*e.g.* papers by Labarthe, Leeuwis and Klerkx, Von der Heiden and Wielinga). Special attention will be paid to analysing current linkage mechanisms between research, extension, education and societal stakeholders, and the institutional arrangements that govern interactions in innovation networks. The indications are that coherence and co-ordination between the classical parties in AKISs have declined, which raises the question as to whether (and when) this must be regarded as a positive or a negative development. Discussions will be oriented in part to identifying lessons, strategies and arrangements that are relevant to tackling current problems and challenges within AKISs.

## Changing roles of research, extension and education in stimulating innovation

The new conception that innovation is about fostering new forms of coordinated action among multiple stakeholders, has important implications for the role of change agents and knowledge workers. It implies that conventional repertoires of 'research', 'education' and 'extension' needs to be supplemented with other modes of process support aimed at building networks, developing shared visions and understandings, facilitation of social learning, conflict management, and capacity building (see Leeuwis, 2004). A particular challenge that needs to be addressed is that, notwithstanding the importance of local initiatives and dynamics, it is clear that the potential for innovation at the grassroots level is often constrained by conditions and processes at higher aggregation levels. Thus, new role conceptions for change agents and knowledge workers must anticipate the need to forge new and effective linkages between rural communities and processes and dynamics elsewhere in production, consumption and governance 'chains'. Reflection on the new and old roles that are performed in present day AKISs, and their effectiveness in bringing about innovation is presented by *e.g.* Kenny and Nettle, Spoelstra *et al.* and Brinks and De Kool.

## Methodological lessons

New AKIS configurations and role perceptions have led to considerable experimentation with new approaches and methodologies. Drawing upon these, the papers in this section present recent experiences with *e.g.*, vision building, trans-disciplinary research, demand articulation and steering, networking and community formation (*e.g.* by Francis *et al.*, Davies *et al.*, De Boef and Pinheiro, Proost *et al.* and Vilas Boas and Ferreira). Case specific lessons will be identified, and attempts will be made to arrive at cross-cutting insights (see *e.g.* Neely and Dixon).

## References

Gibbons, M., C. Limoges, H. Nowotny, S. Schwartzman, P. Scott and M. Trow, 1994. The new production of knowledge. The dynamics of science and research in contemporary societies, Sage Publications, London.

Leeuwis, C. (with contributions by A. Van den Ban), 2004. Communication for rural innovation. Rethinking agricultural extension. Blackwell Science, Oxford.

Rivera, W.M., 2000. The changing nature of agricultural information and the conflictive global developments shaping extension. The Journal of Agricultural Education and Extension, 7, 31-41.

Röling, N.G., 2002. Beyond the aggregation of individual preferences. Moving from multiple to distributed cognition in resource dilemmas. In: Leeuwis, C. and R. Pyburn (Eds.) 2002. Wheelbarrows full of frogs. Social learning in rural resource management. pp. 25-47. Royal Van Gorcum, Assen.

Smits, R., 2002. Innovation studies in the 21st century. Questions from a users perspective. Technological Forecasting and Social Change, 69 (9) 861-883.

Umali, D.L. and L. Schwarz, 1994. Public and private agricultural extension: Beyond traditional boundaries. World Bank Technical Paper 247. World Bank, Washington.

# Farming with future: Implementation of sustainable agriculture through a network of stakeholders

*Harm Brinks[1] and Stefanie de Kool[2]*
*[1]DLV Plant BV, PB 7001, 6700 CA Wageningen, The Netherlands; [2]Praktijkonderzoek Plant & Omgeving BV, PB 85, 2160 AB Lisse, The Netherlands*

## Abstract

In the sixties and seventies of the last century the dissemination of new technology, which focussed on high inputs and mechanisation, has resulted in economic benefits for most of the stakeholders in the agricultural network in The Netherlands. Therefore the support for this new technology was strong. In the eighties the innovations moved towards more sustainable farming systems. However, the dissemination and implementation has not been fully successful. Under pressure from environmental organisations and the government in the past decade, the different stakeholders have become aware of the necessity to change towards sustainable farming. As a result of this, all stakeholders signed an agreement in 2004, called the Convenant on Crop Protection. This agreement aims at reducing the environmental risks, caused by pesticides, with 95% in the year 2010 compared to 1998. Part of the agreement is the dissemination and implementation of knowledge on integrated pest management. The project Farming with Future tries to facilitate this process, in cooperation with and with respect for the core business of stakeholders. In the project a dialogue is started with many stakeholders in order to build up a relation and to start up activities that match the special responsibility of the stakeholder within the Convenant. Trough regional networks farmers are actively engaged in testing and improving sustainable measures (Best Practices) and the effective and feasible measures are shared with other farmers. This exchange of information and knowledge takes place in collaboration with stakeholders. The network approach shows to be successful when common interests can be identified. It is important to search for tailor made cooperation between each stakeholder and the project: only those best practices are selected that match the interest and core business of the stakeholder. The new approach demands for enhanced social skills of researchers and extension workers. Therefore it is necessary that the people who will work with the network approach are trained on these skills.

**Keywords:** sustainable agriculture, Best Practices, stakeholder, network

## Introduction

Since the sixties of the last century great changes have taken place in the farming systems in the Netherlands: large scale, strongly mechanised agriculture with a high input of pesticides. These changes were economically driven and therefore beneficial for most of the organisations in the agricultural network: farmers and their unions, the Ministry of Agriculture, the pesticide industry, the food industry and retail. The collaboration between these stakeholders resulted in an effective dissemination and implementation of innovations driven by research. The negative side effect of pesticides on the environment triggered a change of policy in the eighties. The new policy was directed to the development and introduction of sustainable agriculture. Integrated farming systems were developed and showed promising results, both on the environment and farm economics. These integrated farming systems were tested in cooperation with research, extension and farmers in several projects in the last decade of the 20$^{th}$ century. The results on farm level showed comparable results: a large reduction of the input of pesticides without negative economical consequences for the farmer. In spite of the good results, further implementation of integrated farming systems didn't take place. Integrated pest management didn't get sufficient support in the agricultural network. The farmer unions and the pesticide industry have great influence on farm level and did not strongly support the development towards sustainable agriculture. The perceived risks and lack of financial benefits were the main reasons. The past five years, under pressure of the environmental organisations and the Government, the attitude changed and all the stakeholders came to the conclusion that sustainable agriculture is the only way forward towards the future. In 2003, an agreement was signed between the stakeholders: The Convenant on Crop Protection. Parties in this agreement are LTO (farmers union), Government, UvW (water boards), VEWIN (water companies), Nefyto (pesticide industry), Agrodis (suppliers of pesticides) and Plantum (seeds). Within the framework of this agreement all stakeholders took responsibility to contribute to the main objective of the agreement: reduction of the environmental impact of pesticides by 95% in 2010, compared with 1998. Part of the agreement is the dissemination and implementation of knowledge on integrated pest management. The project Farming with Future tries to facilitate this process, in cooperation with the stakeholders and with respect for the core business of these stakeholders.

***The study groups and Best Practices***

Spread over the most important production areas, 34 study groups of farmers are active in testing the Best Practices: in arable farming, flower bulbs, tree nursery, fruits, vegetables, mushrooms and glass house crops. Best Practices are measures that contribute to a sustainable crop protection. The Best Practices that have proven to be effective and feasible are ready for implementation on many farms. Some Best Practices need additional research, whereas others may be rejected for various reasons, such as being too expensive, unfeasible, high risk, *etc.*

*Figure 1. Testing Best Practices. Afgekeurd = rejected; revisie = revision; goedgekeurd = effective and feasible.*

### *Network approach and Best Practices*

Rapid implementation of integrated pest management will only be successful if supported by all (influential) stakeholders. The goal of the Farming with Future project is to stimulate and facilitate the process of implementation of the Best Practices Crop Protection, in cooperation with study groups of farmers and stakeholders. Some examples of Best Practices are: biological control of insects in glasshouses and ware houses, the use of decision support systems, mechanical weed control and improved spraying techniques.

In the agricultural network many different stakeholders, interact with the farming community (see Figure 2). Most important stakeholders for the implementation of Best Practices are those who frequently have contact with and influence on farmers: LTO (the farmers union), including the various LTO study groups of farmers, and members of Agrodis (suppliers of pesticides). These are the preferent partners for Farming with Future. Farming with Future looks for a 'tailor made' cooperation. In the project a dialogue is started with many stakeholders in order to build up a relation and to start up activities that match the special responsibility of the stakeholder within the Convenant. Hereby a "best match' is made between stakeholders and best practices, based on the interest and core business of the stakeholder.

## Experiences and results

Since the start of the project Farming with Future in 2004, many activities have been carried out in the different networks. This has not only led to various successes, but brought some constraints to the surface as well.

### *Successes*

The on-farm testing of Best Practices has shown several positive results with regards effectiveness, feasibility and readiness for dissemination. The next step was the actual dissemination in collaboration with stakeholders and therefore a number of communication activities took place with stakeholders, interested in the specific best practice. Examples of these communication activities are:

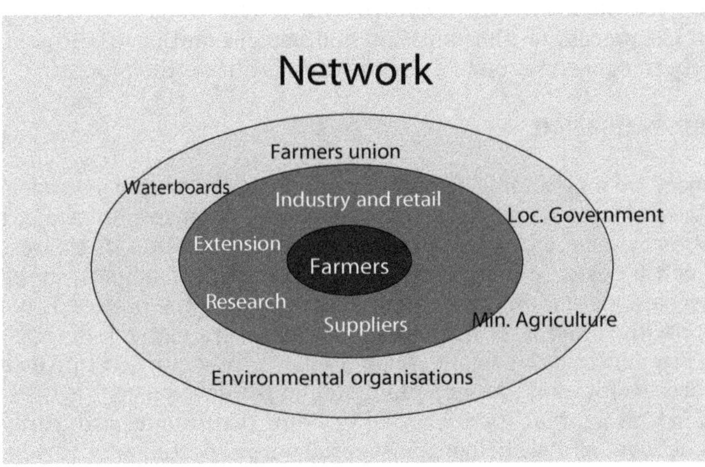


*Figure 2. Network.*

- The dissemination of environmental risk lists of pesticides in collaboration with the farmer union of tree nursery.
- Organisation of "pesticide spraying licence" meetings, topic integrated pest management, in collaboration with farmer unions and pesticide suppliers.
- Development and dissemination of a checklist with measures to reduce the risk of emission of pesticides in the process of disinfection and planting of flower bulbs in collaboration with pesticide suppliers.
- Demonstration and dissemination of biological control of pear psylla with earwigs in collaboration with the bio control industry, the local government and the farmers' union.
- Demonstration and dissemination of spraying techniques to avoid surface water pollution in maize production in collaboration with contract workers.
- Biological control of pests in glasshouses in collaboration with suppliers;
- Seed treatment against aphids in lettuce and mechanical weed control with camera steered machines.

***Constraints***

The successful collaboration with stakeholders in the network depends on the willingness of these stakeholders. This willingness can be partly covered by common interests between the stakeholders and the project. Besides this, the Convenant on Crop protection should contribute to the sense of ownership by the different stakeholders in reaching the convenant's goals. However, it is important to realise that the "Convenant" was signed between trade organisations on a national level and that its objectives are not always shared by the rather independent members on the local level, who are generally the ones that are active in the field. Besides this, the various stakeholders have different ideas on how to reach the objectives.

***Consequences for the organisation and it's employees***

The new approach of dissemination and implementation of innovations has demonstrated a need for new skills of project workers. Whereas in the past it was enough for researchers and extension workers to be an expert on the farming system and innovative farming methods, it has now become equally important to be flexible and innovative in the search for common fields of interests between the project and the different stakeholders. In this process of seeking collaboration it is not an exception that resistance of stakeholders has to be overcome, before any collaboration can be started. As a consequence, social skills of employees have become a key enabler in the process of dissemination and implementation. These social skills should include the ability to cooperate, build a network and facilitate the process.

## Monitoring and evaluation

Goals of the project are a broad implementation of integrated pest management and an active support of all stakeholders that signed the Convenant. An intensive process of monitoring and ongoing evaluation is required to ensure a proper final evaluation of the results. Results in the project are partly deviate from the more technical oriented projects in earlier years (See Table 1). An adequate method for monitoring and evaluating the process is not yet available. Therefore Farming with Future is in the process of developing a method, in cooperation with the department of Communication and Innovation Studies of Wageningen University: the method of "Reflexive Proces Monitoring". An important step in the development is the identification of specific process indicators, that show whether and how the Farming with Future interventions work out in the network of stakeholders. First results are expected later this year.

*Table 1. Example of Stakeholders of a network of Arable crops and qualification of contacts.*

| Stakeholder | | Type of contact | | |
|---|---|---|---|---|
| Company | Type of stakeholder | oct 2004 | mar 2005 | oct 2005 |
| Agerland BV | Pesticide supplier | ☎ | 🕸 | 🕸 |
| Agrarische Unie | Pesticide supplier | ☎ | ☎ | 💬 |
| Avebe | Potato starch company | 🕸 | 🕸 | 🕸 |
| BASF Nederland | Pesticide Industry | – | – | ☎ |
| Kroef B.V. | Potato Industry | 💬 | 💬 | 🕸 |
| Opticrop B.V. | Decision Support System | – | – | 🕸 |
| Aa en Maas | Waterboard | 🕸 | 🕸 | 🕸 |

– no contact so far; 💬 dialogue; ☎ contact by telephone; 🕸 cooperation

## Conclusions

- The described network approach for dissemination and implementation of sustainable farming can be successful, when common interests on the new technologies can be identified with the stakeholders and if the stakeholders take their responsibility.
- It is important to define a tailor made link between each stakeholder and the project: each stakeholder will only support the best practices that fit to its interest and core business.
- The new approach demands for enhanced social skills of researchers and extension workers. Therefore it is necessary that the employees who work with the network approach receive appropriate training to acquire these skills.

# Knowledge development for weed management in organic farm systems in the UK

G. Davies[1], R.J. Turner[1] and D. Gibbon[2]
[1]HDRA, United Kingdom; [2]RULIVSYS, United Kingdom

## Abstract

Farmers advisors, researchers and policy makers, regularly mention weeds as one of the major constraints in organic farming systems. In the UK these represent a range of systems, often adapted to take advantage of specific local environmental and socio-economic situations. Weed management in this context is a very practical activity dependant on a wide range of factors including crop, crop stage, rotation, soil conditions, weather, equipment and/or labour availability. All farmers therefore carry out trials of weeding methods to establish 'what works' and thus develop their knowledge and experience about weeds and weeding. Consequently, a large body of informal knowledge based on farmer experience with weed management exists and farmers place considerable emphasis on communicating and swapping this knowledge with other farmers when learning about weed management techniques and obtaining information on weeds. For this reason they also consistently rank farm walks and open days as the most important way of learning about weeds and weed management. Work in the UK has shown that, surprisingly, farm advisors are often not ranked as important sources of information on weed management and researchers are usually ranked even lower. However, researchers and advisors can potentially contribute a great deal from a vast body of formal research, both conventional and organic, that has been undertaken on weed management. This can be done by trying to construct a participatory and collegiate framework for advisory and research work into weed management, in which all perspectives on weed management are valued. This approach stresses effective communication between the actors as a central part of the process in which researchers, advisors and farmers are encouraged to jointly evaluate both formal and informal knowledge. This field note highlights some of the approaches taken in a recent organic weed management project in the UK.

**Keywords:** organic weed management, participatory knowledge development

## Introduction

Despite the large quantity of information on weed biology and the large number of trials on specific weed control measures undertaken by researchers over many years, organic farmers, when asked, still identify weed management as a priority for further research. A recent Defra funded project on 'weed management in organic farming systems' (OF0315) has began to focus on why such a large research effort has apparently had limited impact in giving farmers the confidence to tackle weed management situations that arise in their organic farm systems and this field note reflects on some of the answers that have arisen to this question.

## Weeds in organic farm systems

The experiences of farmers, advisors and researchers have been used to define the role of weeds in organic farm systems. In summary:

- organic farms are complex systems (many different rotations, enterprises, market systems, values depending on socio-economic and environmental context);
- weeds have a direct impact on desired system outcomes (yield, financial returns, happiness *etc.*);
- weeds are just one aspect of the farm ecology and socio-economics (but an important one! Most farmers have low tolerance of at least some weed types or species);
- weed management technologies and techniques underlie many farm operations (but are not always the only factors taken into consideration).

Weeds are therefore an integral part of all organic farm systems and, from a farmers perspective, need managing to reduce their negative impacts, and, once this is taken care of, to enhance their positive impacts.

## Managing weeds 'organically'

The project has shown that no farmers think that they have all the answers to weed management and it is clear that a range of approaches will work best over a period of time depending on weed type and circumstances. All farmers have adapted their farm systems, within the resources available to them, to manage weeds at some level and most are constantly seeking new knowledge and looking to use and incorporate it into their weed management strategies. A deep understanding of any weed management situation will depend on many factors including field factors (history, crop, weather, *etc.*), farm factors (size, enterprises, rotations, resources, *etc.*), and even national factors (markets, policy, environment, *etc.*).

## Developing knowledge for weed management

Experience during the project has indicated that weed management is about learning for improving complex situations and will necessarily be a mixture of longer tem planning (*e.g.* rotations, crop variety choice) combined with short term reactive or direct measures (*e.g.* harrowing, flame weeding). It is clear from work undertaken as part of the project that farmers and growers spend at least some time finding out information on weeds, informing themselves on how other farmers are managing weeds, and also undertake a great deal of informal experimentation with weed management techniques and methods in their own fields. The outcomes of these experiential learning processes are practical weed management methods and strategies, applicable in specific situations, but often not formally 'evaluated' according to strict statistical rules.

In contrast, most formal research effort has been designed and implemented by 'professional' researchers in order to arrive at generic or overarching recommendations for weed management. These trials, with a strong element of researcher design and control, are statistically evaluated for the conditions under which they were carried out but their general utility for farmers, who require very situation specific knowledge, is often limited. Several authors have reviewed existing information and research efforts for weed management in organic farming systems (*e.g.* Bond and Grundy, 2001).

The central theme for the weed management project as carried out has been to reconcile these two strands of knowledge development and it is towards this end that many of the project resources have been devoted rather than developing technological fixes to particular problems.

## Participatory knowledge development

A crucial question becomes how can researchers (and advisors) best interact with farmers and growers to allow appropriate knowledge development to take place? Our answer is that knowledge development for practical weed management strategies in organic systems requires that farmers are directly involved in processing weed information, combining it with experience and using it to develop new knowledge of relevance to them. This has been done in this project by taking a participatory approach and trying to construct a learning collegiate framework in which all perspectives on weed management are valued.

Although widely accepted for many years in development projects, participatory approaches are relatively new within UK agricultural research and can be regarded as an experiment for exploring constructivist approaches (Woodhill and Röling, 1998) to trying to improve complex systems. This approach has stressed effective communication between all the actors or stakeholders to help them process weed information and knowledge (*i.e.* create a learning atmosphere) by providing access to information, stimulating evaluation of information, and finally to generating and developing new knowledge. Researchers and advisors have increasingly taken on the role of facilitators to information exchange within the project and to helping farmers and growers to develop this information into knowledge, which suits their particular circumstances. It has also been necessary to ask how conventional research processes could be integrated or joined to this more open approach and how the traditional roles of the various stakeholders (farmers, advisors, researchers, funders) might need to altered or realigned

## Project approach: Structure and methods

From the original open remit of the project, which was to define weed problems together with organic farmers and growers, three particular themes have been developed; management of docks and perennial weeds; system approaches to weed management; and knowledge collation and dissemination. These themes themselves arose from an initial round of stakeholder meetings that attempted to define priority actions for weed management in organic farming systems and knowledge on them has been built up through various activities:

*Development of website and database*: an extensive literature review on weeds and weed management largely carried out by scientists has been made freely available to all stakeholders through the site (www.organicweeds.org.uk) as a source of knowledge on weeds and weed management. The database is flexible enough to allow for the input of farmer and advisor knowledge and experience, which can be set alongside the scientific knowledge in order to make the information more relevant. The site is therefore a database of current information and knowledge on organic weed management practice in the UK.

*Development of case studies*: specific weed management stories from working farms have been elaborated and are available for critical analysis. They are a unique source of farmer knowledge and experiential learning which allow farmers to share knowledge and provide a learning opportunity for converters and young farmers. All are available on the website alongside the database on weed information.

*Farm walks and open days*: on specific topics have been organized and used to discuss and develop knowledge (*e.g.* farm walks focusing on dock management, annual weed management in arable systems *etc.*). They have proved an invaluable opportunity for mixed groups of farmers, advisors and researchers to exchange ideas and experience. Special attempts were made to capture 'farmer stories' at these shared field experiences and these are available through the reports on the website. Farmers have especially valued these practical experiences and discussions.

*Talks, discussions, meetings, focus groups and workshops*: the project has helped facilitate of large number of organised talks, farmer discussions (often with field walks), meetings and workshops themed by weed, farm type or crop depending on requests. Many of these were for existing farmer groups but the project also helped facilitate focus groups of researchers, advisors and farmers, which evaluated and reflected on the main project themes, ensuring that the project remained relevant to the concerns of farmers. An annual stakeholder meeting was held to which farmers and advisors were invited an asked to explore weed issues and reflect on the project progress.

*On-farm trials and surveys*: the project has facilitated a number of on-farm trial themes and surveys for farmers and/or farmer groups (themed by weed, topic or crop). Some of the work has provided valuable information where researchers were able to help monitor the effect of weed management practices over a period of time.

*Other outputs*: have included bulletins, leaflets, press articles and conference papers. They have mainly been aimed at improving communications, diffusing information and reflecting on project achievements.

## Project approach: Reflections on effectiveness

The various methods employed by the project have generally been effective in allowing individual participating farmers to reflect and act to resolve specific weed management situations. However it is clear that not all methods suit all actors and that it is important to have a range of different approaches to sharing information that allow all stakeholders to participate.

Whilst the development of a freely accessible website offers a cheap method of storing a large amount of information in a flexible way it is also clear that many farmers do not have time or the inclination to regularly access this information. The project has subsequently spent a lot of time developing alternate routes for information sharing, especially summary leaflets and factsheets. Farm walks and other meetings are also clearly a good way of stimulating information exchange on practical weed management but have been criticised for only being applicable to the individuals who actually participate. The project has spent some time trying to record and share knowledge developed at these types of meetings and it might be necessary to spend more time cultivating links with the agricultural journalists in order to allow them to channel messages to a wider audience.

Although many farmers have been keen to participate in on farm trials to validate weed management techniques results have been limited. Often the value of the trials and surveys has been as a starting point for gaining experience, which has rapidly been superseded by events Researchers have been disappointed because of the lack of 'formal statistical data and results'. Farmers also learn a lot about weed management through observation of fortuitous events or modifying existing machinery and the project has not had a lot of success in documenting these processes and more emphasis should be placed on this in the future.

Whilst it has been possible to fund the current project it has also become apparent that the nature of the project, an open-ended enquiry into weed management, does not fit easily within existing funding structures which tend to demand fixed length projects delivering scientifically 'robust' outputs. Future work will require flexibility on behalf of funders, researchers and farmers.

## Conclusions

Actual weed management practice in organic farming systems will depend on a large number of contingent factors and thus knowledge development to help farmers improve weed management situations best places farmers at the centre of the research process. This demands a reorientation of the conventional roles of advisors and researchers to facilitate information exchange in a flexible format and to help farmers develop information in a manner that provides them with appropriate knowledge or solutions. Other stakeholders, such as funders, may also need to adjust their expectations of what researchers deliver as part of scientific projects.

## Acknowledgements

The project was an inclusive project and could not have been as successful without the willing participation and contributions of organic farmers and growers (in focus groups, at stakeholder meetings, as field day hosts and participants, as case studies), advisers, researchers from other institutions, and last but not least the support of our research managers and funders (Defra) for taking a novel approach.

## References

Bond, W. And A.C. Grundy, 2001. Non-chemical weed management in organic farming systems. Weed Research 41, 382-406.

Woodhill, J. and N.G. Röling, 1998. The second wing of the eagle: the human dimension in learning our way to more sustainable futures. In: N.G.Röling and M.A.E. Wagemakers (eds.). Facilitating Sustainable Agriculture. CUP, pp. 46-71.

# Social construction of knowledge: A leading perspective in a participatory research training programme in Santa Catarina, Brazil

*Walter S. de Boef*[1] *and Sergio L.G. Pinheiro*[2]
[1]*Department of Crop Science, Centre for Agricultural Sciences (CCA), Federal University of Santa Catarina (UFSC), Florianópolis, Santa Catarina, Brazil;* [2]*Empresa de Pesquisa Agropecuária e Extensão Rural de Santa Catarina S.A (EPAGRI), Florianópolis, Santa Catarina, Brazil*

## Abstract

This paper presents the preliminary experiences of a training programme establishing participatory research teams in the State of Santa Catarina in Brazil. A foundation to the participatory research training programme is that through research "with" instead of "for" people, researchers and extensionists share perceptions about reality, values and interests, and thus through joint action socially construct knowledge. This approach changes the focus of research from control to relations, value systems diversity, endogenous development, autonomy and social learning. The training programme has 56 participants from a diversity of organizations; they form nine regional teams. They participate in a process of about two years with 5 sessions. These session are (a) concepts and theory; (b) diagnosis, understanding local realities and capacities; (c) designing in partnership an action plan; (d) project implementation and monitoring progress and learning, and (e) institutionalization and mainstreaming. The action learning programme is realized within the World Bank financed Micro-Watershed Project. The paper concludes with major barriers in mainstreaming and institutionalization within the leading organizations and particular the difficulty to stimulate public-civil society partnerships within participatory processes in Santa Catarina State.

**Keywords:** participatory research, Santa Catarina, design of training programme, agricultural research and extension, public organization, civil society partnerships

## Introduction

The professionalism of agricultural scientists and extension workers is often described as barrier in moving to more participatory approaches in research and extension. Public–civil partnerships are promoted as a mechanism moving professsional paradigms in agricultural research. This paper presents a participatory research programme that used a 'social construction of knowledge' perspective. Primary goals are through action learning foster public-civil society partnerships and to provoke agricultural professionals shifting from a 'blueprint' to 'process' paradigm in agricultural research and development. The training programme is organised by the Santa Catarina State Institution for Agricultural Research and Rural Extension (EPAGRI) through the Micro-Watershed Project 2 (MB2) financed by the World Bank and the State Government of Santa Catarina. The authors are members of the coordination team and responsible for the design, facilitation and execution of the training programme.

## Santa Catarina State and micro-watershed 2 programme

Santa Catarina State is located in south Brazil; it is characterized by small scale family farmers. In the early mid 20th century (the colonization of the state), "traditional" type of agriculture was practiced, characterized by a diversification of crops and livestock production mainly for subsistence. Since the sixties, a first transformation process resulted in a "partial modernization" of production targeting commercial crops (tobacco) and animal (pig and poultry) production. This development was stimulated by the globally oriented agro-industrial conglomerates. This process resulted in undesirable social (increasing exclusion of small scale farmers) and environmental impacts. Since the nineties, many rural families have migrated to urban cities. Rural exodus, poverty, urban violence, environmental degradation, health problems and other social-environmental effects have reached levels increasing social-economic inequity and ecological unsustainability in both rural and urban settings.

Agricultural and environmental research and rural extension in Santa Catarina can be characterised by a major orientation to industrial agriculture targeted at export markets. Formal education in agricultural colleges and universities prepares agricultural professionals using primarily blueprint models to research and development. Research and education institutions have their foundation in the agricultural revolution and expansion since the seventies. Within public research and extension agencies (as EPAGRI) the Transfer of Technology paradigm even though debated since the eighties and nineties is predominant and uncontested. The demands of small scale farmers are hardly catered for. Civil society organisations through NGOs and social movements have been criticizing the model and public sector orientation. They have developed parallel structures for research and extension with hardly any linkages with the public sector.

A change in professional and institutional orientation to research and extension as has been advocated by authors such as Chambers (1993) and Pretty (1995) has encouraged the MB2 project is stimulating a joint learning approach in its participatory research programme. These authors emphasize joint action and public, private and civil society partnerships. Partners in society join in learning and action referred to as the social construction of knowledge within local development processes (Maturana and Varela, 1995). A key objective of the MB2 Project is to promote collective action and solidarity among diverse institutions and stakeholders in Santa Catarina as means to escape from individualism and competition, environmental degradation, inequity and economic exploration of rural families. The MB2 project aims to promote sustainable development and to reverse rural exodus of the territory. In its conception, it is to be implemented through partnerships of the State Government with Municipal Governments, universities, cooperatives, farmer associations, NGOs and other institutions. The principal operational strategy is to plan and implement its activities at the local community level and link those to the project economic, social and environmental activities (see also www.microbacias.sc.gov.br).

The Project has various components addressing agricultural and environmenttal research and rural extension. The participatory research component aims to support research and extension capacities and stimulates a change in orientation from more conventional passive and functional to more interactive modes of participatory interactions between researchers, extension officers and rural communities. The current paper reports on design and partial results to a specific training component associated to this programme component.

## Participatory research training programme: Principles and approach

The training programmes promote the use participatory and qualitative research approaches in which knowledge is socially constructed and through facilitated learning processes. Participatory research has the following foundations: (i) understanding of the subjectivity of knowledge, (ii) recognition of unpredictability of human and complex systems, (iii) the premise of complexity and focus on relations (in life, systems, nature, society and human beings), and (iv) recognition of the human nature of learning about the world and the resulting human organisations. Relations are emphasized in participatory research processes, *i.e. participatory research with* in contrast to *participatory research for* rural communities and local partners. The research pathway is changed and values life and the human being, and embraces human's critical capacity to dialogue and learn.

Teams of researchers and extensionists from public and civil society organisations join in teams. Thereby the linear model and separation of research and extension (as dominant within EPAGRI) is broken. Partners from local government and civil society organisations (farmer associations, NGOs) and other scientific organisations (universities) are joining. Teams learn through various steps in a reality situation to conduct a participatory research process.

In the design of the training we recognised that a multitude of methodologies exist with clearly defined thematic foundations, regional backgrounds and school of thinking. Instead of following them in a rigid manner, the teams are trained to be critical, to compare them and learn from their background and experiences. Methods are approached as orientations. The training programme has analyzed and uses the a broad range of existing participatory research and development methodologies as orientation. They are applied, shaped and adapted to the regional context in space and time. Themes are addressed include social and cultural issues, income generation, dairy production, natural resource management, soil fertility, forest management and chain development. The approach equips teams with the capacity to deal with a diversity of situations; they will be effective in the process of facilitating sustainable agricultural and rural development.

In order link teams as much as possible with local reality, Santa Catarina State was divided in 8 major regions with 10 regional teams for participatory research. An interdisciplinary and multi-stakeholder approach is used that emphasises the involvement within each team of professionals with a diversity of backgrounds and working for various stakeholders. The team consists of 5-6 members; they take a shared leadership role and become within their sub-region responsible for mainstreaming participatory research approaches.

The training is linked with diagnosis and design processes in one particular micro watershed and associated rural community. Training in isolation would remain conceptual and distant from reality and project modalities. Some training components are conducted in the reality of one of the participating team's local partners (community) setting. The teams are trained through a series of five one week workshops. In that manner, they experience various phases of participatory research processes. Based on the training an internal monitoring and evaluation system is being established that facilitates learning, allows for adapting processes to diverse realities and local demands, and allows for management to follow project activities according agreed modalities (human resources, financial resources, project logical framework).

## Participatory research training programme: Methodological steps

Two training workshops were organised on *concepts and theory* that are discussed based on experiences and cases. The teams were exposed to a diversity of methods. They shared experiences and elaborate lessons learnt from cases and experiences. Participants obtained a shared understanding of concepts and theories that support participatory research in rural and agricultural development and natural resource management. Another critical of this phase was to form teams and throughout through coaching improve their facilitation skills (learning by doing).

Two training workshops were organised on *diagnosis in a field work setting*. The workshops were conducted within an identified location. A participatory rural appraisal (PRA) was conducted within a community associated to the MB2 project. This proved to be the first a first critical step for each team to start working with, instead of working for a rural community. The outcomes of the appraisal conducted during the training contributed to the work of the team hosting the event. After the workshop, each team conducted a similar PRA in their associated micro-watershed.

Two training workshops were organised to support *the design of participatory research projects*. Each team presented the outcomes of the regional PRAs. A community consultation was conducted in the actual context of the team's hosting the event. During this community meeting results of the PRA were verified and the partnership between the team and the community was strengthened through joint decision making. The teams entered a process of design a research project outlining in particular its consecutive steps. Community representatives participated during some steps of the design process.

The teams submitted the project proposals to course coordination and each team started to *implement their project*. MB2 provides the financial resources for implementation. The projects are considered pilots within the larger framework of the MB2 participatory research component. The strategy is taken that they turn into reference points in Santa Catarina. Table 1 demonstrates the diversity of themes addressed. The regional teams become gradually involved in stimulating process approaches and institutionalising the new "research with" approach within EPAGRI and MB2.

Two state level *synthesis workshops* are going to be organised including all teams. They share the experiences, *monitor and evaluate* the effectiveness in using a process approach to participatory research, and based on learning modify and adapt the approach. Issues such intellectual property rights, institutional and professsional barriers encountered while working in participatory processes are discussed. These discussions are aimed to contribute to EPAGRI and MB2 developing strategies and policies for *mainstreaming and institutionalising* participatory research and process approaches; *i.e.* to transform the organisation and partners from working for to working with rural communities.

Mainstreaming and institutionalisation is the subject of *two state wide seminars* with EPAGRI and MB2, and partner organisations' managers. Partial results of the training and pilot projects are presented as strategic instrument facilitating the required institutional change.

## Initial lessons learned

The training programme has progressed to the stage where the teams have started with the implementation of their projects. It is therefore early to draw general conclusions on the effectiveness of the training programme in contributing to EPAGRI and partners in MB2 going through a paradigm shift. However some initial lessons can be learnt. The fact that research projects have been developed following a process approach can already be considered an important output, particularly understanding the diversity of nine projects within Santa Catarina.

The training programme through its intensity and action learning process turned out to be a process of "natural selection" of only those researchers, extensionists and partners committed to continue participating in the training programme and finally forming the regional cadre. The orientation on local innovations as starting points has advocated the researchers and extensionists in avoiding returning into the dominant ToT model and passive or functional modes of interaction. The action learning process has increased the capacity of the participants through socially construct knowledge within a local setting.

Most members of the regional teams and training programme coordination team (among which the authors of this paper) experience a reluctance and resistance attitudes to the change process among leadership within a larger institutional setting of EPAGRI and MB2. They encounter professional attitudes that respond in an uncooperative manner to the training programme. These attitudes are founded in existing bureaucratic and political structures and interests that are part of conventional research in top-down and control management. We responded to this reluctance by emphasising and negotiating for securing safe learning environments within the sessions and pilot projects. We heavily in group dynamics within the teams, to stimulate joint learning, and focus on local experiences. We emphasised local change first before addressing institutional and professional change.

Another hindrance encountered is directly related to the public–civil society partnership that is considered critical in participatory research. The partnership between the regional teams and local community associations has been established primarily through MB2 and EPAGRI affiliated organisations. Partnership with other institutional public entities and non-governmental organisations turned out to be very limited. This partially embedded in the political setting of Santa Catarina (and Brazil) where public, private and civil society work in parallel structures oriented either with a commercial industrial or small scale farming agro-ecological and development orientation with few informal but hardly any formal linkages.

## Acknowledgements

The authors wish to acknowledge the contributions of the participants, our local partners and colleagues in the Participatory Research Training Programme organised by EPAGRI within the MB2 project. The training has been financially supported by the State Government of Santa Catarina and World Bank.

## References

Chambers, R., 1993. Challenging the professions. Frontiers for rural development. London, Intermediate Technology Publications.

Maturana, H.R. and F.J. Varela, 1992. Tree of Knowledge. The biological roots of human understanding. Boston, Shambhala.

Pretty, J.N., 1995. Regenerating agriculture. Policies and practice for sustainability and self-reliance. London, Earthscan Publications.

# Participative approaches and extension practices towards sustainable agriculture: A methodological approach based on sociology of translation

*Philippe Fleury[1], Christoph Kirchengast[2], Valérie Mieville-Ott[3], Nathalia Magnani[4] and Sandrine Petit[1]*
*[1]Service d'Utilité Agricole à Compétence Interdépartementale, Groupement d'Intérêt Scientifique Alpes du Nord, SUACI, GIS Alpes du Nord, 11 rue Métropole, 73000 Chambéry, France; [2]Institute of Sociology, University of Innsbruck, Austria; [3]Institute of Agricultural Economics, Swiss Federal Institute of Technology, Zurich, Switzerland; [4]Human and Social Sciences Department, Trento University, Italy*

## Abstract

This methodological paper analyses the elaboration and the implementation of participative local projects aiming at sustainable agriculture and rural development. The methodology is based on the sociology of translation (Callon, 1986). The processes initiated from the elaboration to the implementation of an action plan towards sustainable agriculture are tackled as social dynamics building new networks, requiring different translations and crossing successive phases from the formulation of the problem to the implementation of actions. We examine the progressive enrolment of different stakeholders, the role of non-human actors like leaflets or documents and the mobilisation of different systems of knowledge. The general value of the method has been tested on a set of diversified actions: creation of a collective structure for exchanges of services between farmers, recovering of abandoned grasslands and landscape maintenance, local product valorisation and meat chain, production and commercialisation of wood energy. These actions have been implemented in four different areas across the Alps, in Austria, France, Italy and Switzerland. Firstly, we present the methodology and its validation on this set of situations. Then we discuss the key outputs in terms of implementation of projects aiming at sustainable development.

**Keywords:** participative approach, sustainable agriculture, local development, collective action

## Introduction

Our paper discusses what action is on the basis on an in-depth analysis of implementation of small-scale projects in the frame of participative groups targeting sustainable development. Resorting to the concepts of sociology of translation and actor network theory, we question the standard way of conceiving action. Conventionally, the design and implementation of projects is seen as a linear process, with successive phases of management such as definition of objectives and risks, planning of actions, implementation, adjustments and evaluations of results referring to the level of achievement of objectives. This conception of successive stages is based on the standard sequence: problem finding, problem shaping, problem solving.

The sustainable development concept has introduced the notion of local governance in rural development. Our hypothesis is that the implementation of the local governance principle (multi-stakeholders and participative approaches) in small-scale projects provides an alternative view on the phases of project management. Focusing on the phase of actions implementation towards sustainable development, we examine how complex the process is. We notice in particular how problem formulation is permanently overlapping with problem solving. This alternative proposal implies to elaborate different tools for extension in the field of rural development.

This paper is related to a research and demonstration European project so-called IMALP: implementation of sustainable agriculture and rural development in alpine mountains.

## The methodology

### *The IMALP project*

The IMALP project involves four pilot areas across the Alps (Moyenne-Tarentaise, France; Val d'Hérens, Switzerland; Murau, Austria and Val di Sole, Italy). IMALP initiated an experimental situation where a group of local stakeholders (farmers, elected officials and NGO members) design and implement a consistent set of actions towards sustainable agriculture. In each area, four to eight actions managed by an action group are implemented.

To survey the collective dynamics arisen, we attended the meetings of the local and action groups over a period of three years. The main objective was to evaluate the capacities of their members to negotiate in a collective way an agreement about the goals and the means of change towards sustainability. We reported: who participates in the different meetings, what are the matters discussed (consensual and conflictual topics), interactions between members, modes of negotiation, use of different tools and methods of facilitation, way of decision making, major outputs (concrete actions, leaflets, documents, *etc.*).

To analyse the processes occurring in the implementation of actions, we tested a method designed from the sociology of translation.

### *The concept of translation*

The actor network theory (Law, 1992) is constructivist and refuses to accept essentialist explanations of changes (it was a good idea and relevant means for its implementation were used). Central is the concept of translation in which stakeholders attempt to create a network. During the process of translation, the identity of actors, both human and non-human, the possibilities of interaction and the margins of manoeuvre are negotiated and delimited (Callon, 1986). Four "moments" of translation, which can overlap, are discerned:

1. *Problematisation:* what is the problem? Who are the relevant actors?
2. *Interessement:* how to seal partnerships? Getting the actors interested and negotiating the terms of their involvement. The primary actor works to convince the other actors.
3. *Enrolment*: stabilisation and coordination of actors' role, actors accept the roles defined during interessement.
4. *Mobilisation of allies:* designation and representativeness of spokesmen, a constraining network is working and many human and non-human actors act in common.

## The case studies

To discuss the ability of the method to grasp the complexity of collective actions towards sustainable development we test it on a set of four actions.

### *Working conditions in Moyenne-Tarentaise, Savoie, France*

This action addresses one major problem of sustainability of mountain farming: work overload of farmers. The objective is the construction by farmers of their own solutions. They refined the problem and decided to design a collective set-up to develop exchanges of services between farmers.

***Landscape management in Val di Sole, Trentino, Italy***
Landscape maintenance is a key problem of sustainability in mountain areas to which agriculture can give a relevant contribution. The objective of the action is the establishment of networks between farmers, municipalities and landowners with the aim of recovering abandoned grasslands in the valley.

***Local product valorisation and meat chain in Val d'Herens, Valais, Switzerland***
This action is related to economic sustainability of agriculture. The objective is to increase the value-added of quality meat production by the way of a local chain of marketing and commercialisation. This project required a complex social construction involving farmers, butchers, restaurant keepers, research institutions and extension services.

***Production and commercialisation of wood energy in Murau, Styria, Austria***
In 1992, 14 farmers and a monastery founded a cooperative for biomass heating. Their objectives were to increase the value-added of forest management and the use of local renewable energy (wood chips). Currently the cooperative has enlarged its collaborations and is becoming a significant regional energy supplier.

## Results

The different actions highlight the dynamics of collective actions towards sustainable development. The shifts from consensus to controversial issues, the gradual interessement and enrolment of stakeholders, the role of leaflets and documents are a series of processes of major importance.

***Working conditions in Moyenne-Tarentaise***
This action highlights the role of controversy. The work loads were a consensual matter at the beginning as everyone thought to be overloaded of work. Then, in speeches, the workloads were related to the work organisation: some farmers could be overloaded as others do the same job quickly and have free time. This controversy allowed to refine the problem and leaded to wonder if mutual help and exchanges of services would be feasible. A study evaluating the offer-demand of services underlined the differences between farms for seasonal tasks related to different altitudinal levels. Finally, the group envisaged a cooperative to exchange services between farmers. A reproblematisation went along the search of solutions. Firstly, the collective set-up to exchange services had an objective of mutual help between farmers. In a last stage, they refined its objective and put forward economic arguments based on machinery equipment (Petit *et al.*, 2006).

***Landscape management and recovering of abandoned grasslands in Val di Sole***
A consensus between farmers and local administrations has been swiftly reached. Local administration saw in this action a way to improve the tourist attractiveness. For farmers it was a way to expand their pasture area. The action was however problematised by the landowners. They were external to the action group but their alliance was necessary. Landowners constructed the proposed solution, the farmer intervention to clear the grasslands, as a problem for the integrity of their property. In most cases, the mediation of the local administrations solved the difficulties: their publicly recognised authority managed to gain the alliance of the landowners and to help the reformulation of an individual risk in a collective gain. Non-human actors played a major role in the construction of an agreement between human actors: the expected impacts

of the action were discussed with the help of simulations of landscape evolutions, a geographic information system is used as a common tool to assess the results.

***Local product valorisation and meat chain in Val d'Herens***

Several phases of social construction and negotiation about the objectives, the means and the formulation of collective rules occurred. The first large problematisation "to sell better local products" has evolved in different reproblematisations along the process: local products became meat local products, then cattle meat products and finally Hérens cattle meat products. After that, a clear identification of the product, the design of a strategy of marketing and the need of scientific references on the fattening capacity of the Hérens cattle implied an evolution of the action group and of its network. Other actors emerged like butchers – looking at the good image of the product–, restaurant keepers – looking for a diversification of their menus –, cantonal administration – supporting legal construction of quality standards and marketing tools –, and national agronomic research institutes – interested by research on extensive practices of fattening with rustic breeds. Finally, the relationships between the different stakeholders have been stabilised and officialised by the setting up of an inter-professional association.

***Production and commercialisation of wood energy in Murau***

In recent years, this cooperative has evolved to a significant regional energy supplier. Intense collaborations with different economic stakeholders have been a key factor. A local plumbing company promised support by focusing the promotion of woodchip heating stations, but challenged the quality and quantity of woodchips. The improvement of the quality of chips was a condition for the enrolment of the plumbing company. Consequently, the cooperative improved their technical assets for storage and delivery. In 2005, the cooperative signed a contract with a fish zoo to build and operate a heating station. This is a major change in the business strategy. Instead of only selling the chips, the cooperative now sells the heat produced in their energy stations. Consequently, the members of the cooperative take a higher risk and responsibility, but it enables them to gain a higher value-added. Currently the cooperative is in negotiations with a regional dynamite factory for providing and operating a biomass energy station. This enlarged operation would allow producing also electric energy out of wood biomass, which would mark a milestone towards regional energy autarky. As the starting-up of actions, the scaling-up implies also phases of reproblematisations, enrolment and mobilisation.

## Discussion

***Problematisation and re-problematisation***

A striking fact emerging is that re-problematisations occurred in a short path of time: at the beginning when the group defines objectives, then during the implementation of action. Three types of explanations can be proposed. In the phase of action implementation, concrete decision making refines the problem, requires new knowledge and leads to permanent adjustments of actions. Re-problematisation is also due to changes in the overall context: reform of policies, market fluctuations, consumers' demands. Finally, re-problematisation is also related to the evolution of the action group and of its network. Stakeholders from outside often initiate new re-problematisations as they bring in new perspectives, new kinds of knowledge and networks.

***From interessement to enrolment***

Taking action is a social process, which needs to establish a network of different stakeholders (farmers, consumers, elective representatives, food chain stakeholders, NGO's, institutions, *etc.*). If interessement is a key stage for action progress, enrolment turns out to be crucial. Participants have to become actors and individuals strategies have to be articulated with the implementation of a collective strategy. The implementation succeeds when enrolment is extended in a consistent network of stakeholders and consolidated by institutionalised rules. The establishment of new collaborations can trigger phases of innovation and possibly take actions to the next level.

Interactions between human and non-human actors are key elements of enrolment. Leaflets, price-scales, memorandums, maps, *etc.* play a role as non-human actors. For their elaboration, actors have to reach a common view to be circulated towards an external audience. In other cases, non-human actors, as maps, photographs help to design a common diagnosis. This is a phase of reconstruction of knowledge, values and knowledge) and relationships between human and non-human actors.

***From enrolment to mobilisation of allies***

When enrolment is extended to a consistent network of stakeholders inside and outside the local territory and when they take action in a common way, the action can be considered in a successful stage. Afterwards, the success lies in the capacity to evolve with adaptations enable to involve new actors and to maintain the initial actors. The consolidation of action is related to the management of both internal cohesion and external enlargement. Collective action requires several translations, it is a permanent process, it may fail and is never completed.

## Conclusion

Conception and action are not successive phases. Collective action towards sustainability is a social process where goals, values, knowledge, relationships are re-constructed continuously (Hatchuel, 2000). Our results pinpoint two major points. Firstly, action produces knowledge and action starts always without full knowledge of a problem. Secondly, social concerns as interessement of stakeholders, involvement of institutions, definitions and breakdown of duties and responsibilities are of major importance. The consequences are a refining and in some cases a redefinition of the problem and of the objectives.

The concept of translation is of high scientific significance and seems relevant to analyse this social process. The concrete resulting challenge is to produce tools to accompany extension and advisory services within this framework. These tools could be an alternative to the standard methods of project management conceiving generally changes as a straight ahead process from a current situation to an objective and underestimating the interessement and enrolment phases. These alternative tools focused on the management of the social construction of a project will have to take into account: 1) the process itself evolving with iterative phases and involving human and non-human actors; 2) its outputs both in terms of concrete actions, capacity building and networking abilities of stakeholders.

## Acknowledgements

The project IMALP benefits from financial support of the European Commission (contract QLK5-CT-2002-0109, 5$^{th}$ framework programme for Quality of Life and Management of Living Resources) and of the Swiss confederation (project 02.0364, Secrétariat d'Etat à l'Education et à la Recherche). The institutions involved are: Groupement d'Intérêt Scientifique Alpes du Nord, coordination, France; Istituto Agrario di San Michele all'Adige, Italy; Institute of Sociology, University of Innsbruck, Austria; Institute of Agricultural Economics, Swiss Federal Institute of Technology, Switzerland.

## References

Callon, M., 1986. Éléments pour une sociologie de la traduction. La domestication des coquilles Saint-Jacques et des marins-pêcheurs dans la baie de Saint-Brieuc. L'Année Sociologique 36, 169-208.

Hatchuel, A., 2000. Intervention Research and the production of knowledge. In: Cow up a tree. Knowing and learning for change in Agriculture. Case studies from industrialised countries. Learn group. M. Cerf et. al. (eds.). INRA editions, Paris, pp. 55-68.

Law, J., 1992. Notes on the Theory of the Actor-Network: Ordering, Strategy and Heterogeneity. Systems Practice 5, 379-393.

Petit, S., L. Dobremez and P.H. Fleury, 2006. Working conditions: a challenge for sustainability of mountain family farming systems. Diagnosis at farm level and responses designed by farmers in the Alps. In: Langeveld, H. and N. Röling (eds.) Changing European farmingsystems for a better future, Wageningen Academic Publishers, Wageningen, The Netherlands, pp. 272-276.

# Creating a Nordic regional research network in agroecology: Links to MSc education

*Charles Francis[1], Tor Arvid Breland[1], Geir Lieblein[1], Lennart Salomonsson[2], Ulrika Geber[2], Nadarajah Sriskandarajah[3], Vibeke Langer[3], John Porter[3] and Juha Helenius[4]*
*[1]Norwegian University of Life Sciences, Norway; [2]Swedish Agricultural University, Sweden; [3]Royal Veterinary and Agricultural University, Denmark; [4]University of Helsinki, Finland*

## Abstract

Challenges facing agriculture and food supply in Nordic countries are unique to an ecoregion with short growing season, limited arable land, and long distance from import suppliers. Among the long-term challenges that will influence food availability in the region are scarcity of fossil fuels, changes in production conditions due to global warming, preserving ecosystem services in multifunctional rural landscapes, fragility of national and regional subsidies, and recognition of food as a human right. Research on broad and critical topics is not simple, and often moves beyond the comfort zones of disciplinary scientists. Thesis projects are most valuable if they provide a growth and learning experience for students and faculty as well as contribute in some way to resolving major issues that will impact food systems and quality of life for people of the region. Most frequently, MSc and PhD student research projects address narrow questions for which methods are known and answers can be easily predicted. With limited research funding this is a luxury we cannot afford. Research and education programs should address high-priority issues in the agriculture and food sector, both broad and narrow, as well as on-farm production questions and constraints in local food systems. Agroecology provides an organizing framework to set priorities and address critical issues regarding environmental impacts, alternative farming strategies, economics of production, and social viability of future farming and food systems. Limited budgets and human resources make it imperative that regional networks design coordinated efforts to identify and examine priority concerns for long-term sustainability of farming and food systems. This is one goal of the network AGROASIS, a group of researchers and students who are linked through the regional agroecology MSc education program centered in Norway and run by systems educators from five countries. The process of setting research priorities and seeking support is described.

**Keywords:** research priorities, ecological agriculture, food systems, action research

## Introduction

Most agricultural research projects and thesis topics are directed at specific questions, often informed by the special interests of individual scientists and students, with greatest emphasis on increasing food productivity. Research is rarely approached in the explicit context of the broader food and environmental system. We have a rich history of reductionist, specialized research that is well supported by accepted experimental designs and easily available statistical analysis packages. Yet the funding environment is changing. In the EU the major emphasis is moving toward multifunctional landscapes and rural development. A recent publication in the U.S. envisions agriculture with "further gains in food and fiber production ... [that] also provides other benefits, such as enhanced public health, clean water, more diverse wildlife, rural amenities, and social wellbeing" (National Research Council, 2003). Our research and education programs need to reflect and adapt to this new reality.

Agroecology, as the study of the ecology of food systems (Francis *et al.*, 2003), provides a useful platform on which to build programs with network collaborators in the Nordic region. Faculty members in the network have both research and education responsibilities, and there is a strong desire to link future research to ongoing education programs in the emerging learning landscape. We have at least ten new MSc programme students each year and PhD students interested in systems research, The need for more coordinated efforts has been identified, and a process put in place to organize projects, seek funding, and implement a region-wide focus on the most important issues that will face the region in the long term future. The network has been described by Breland and Francis (2005). Here we provide more details on the research record to date with focus on completed degree projects, a process for setting future research priorities, broad questions to be addressed, and an implementation strategy that is under way.

## Research projects in MSc student programs

As of the end of 2005 there were 18 completed MSc thesis projects, with focus on details of field production systems, overall system analysis, and marketing. Other topics included education at different levels and policy issues. The students came from 11 different countries. Taken as a group, we could say that topics shown in Table 1 represent our collective research priorities to date. We all recognize that these topics are the result of at least 18 separate decisions of students in consultation with their advisors, and not a coordinated research strategy.

*Table 1. Completed MSc thesis projects in agroecology, 2002-2005.*

| Topic | Student | Year |
|---|---|---|
| Role of school garden in organic food system | Hafnlaug Gudlaugsdottir | 2002 |
| Role of restaurants in the development of organic farming in Norway | Kenneth Lenga | 2002 |
| Phosphorus uptake in six wheat and barley cultivars | Jon Magne Holten | 2002 |
| Policy interventions to motivate conversion to organic farming | Leo Kivitar | 2002 |
| Farmer as teacher – the farm as a learning ground | Frøydis Lindén | 2002 |
| Influence of different compost materials on plant growth | Hanne Weichel Carlsen | 2003 |
| Research design and communication in production system experiments | Stefan Adler | 2003 |
| Cover crops in cauliflower production: IPM and yields | Wendy Hall Waalen | 2003 |
| Sustainable farming methods: use of roughage by laying hens | José González Plaza | 2003 |
| Facilitating organizational change in an organic dairy cooperative | Dorte Christensen | 2004 |
| Developing organic farming in Colombia: case study in Bogotá | Laura Cantillo Rojas | 2004 |
| Potentials for food production & wastewater treatment in Havana | Martina Hammer | 2004 |
| Food miles & attitudes of farmers and consumers in farmers markets | Kjartan Åsebø | 2004 |
| Planning communication about farm nature plans | Pernille Vils Axelsen | 2005 |
| Emergy evaluation of grazing meat cattle in the Argentina Pampas | Gloria Rótolo | 2005 |
| Developing Estonian organic sector: importance of information flow | Karme Petrutis | 2005 |
| State of knowledge of small-scale organic production in Cameroon | Diobe Alobwede | 2005 |
| Agriculture intensification and soil quality/nutrient flow in Nepal | Puspa Raj Tiwari | 2005 |

The research provided valuable information on farming system practices and design. For example phosphorus fertiliser use efficiency could be improved by using cultivars of barley and wheat that had longer root hairs, especially when grown in low soil P levels (Holten, 2002). Use of some available compost materials in cucumber production could reduce the impacts of plant pathogens and reduce disease (Carlsen, 2003). Control of insect damage in cabbage fields can be enhanced by promoting beneficial insects in the right cover crop treatments in Norway (Hall, 2003). Both practices could reduce pest management costs. For economic and environmental reasons, livestock should be fed fibrous feeds that are not consumed by humans. One study showed that commercial poultry diets could use grass meal to replace some cereals and thus make production more sustainable (Gonzalez, 2003). Blackwater and composting treatment for later use on vegetable production was seen as a viable option in Havana's urban gardens (Hammer, 2004). Surveys of farmers in Nepal (Tiwari, 2005) and in Cameroon (Alobwede, 2005) revealed ways that farming practices influenced soil fertility and impacted the environment with different levels of intensity of input use. Although these studies concentrated on components of the system, they were conducted in a whole systems context that reflected an agroecology orientation.

Because agroecology includes focus on whole food systems, there is a high level of interest in marketing. The potentials for restaurants to provide price incentives and demand for increased production of organic food were studied in Norway (Lenga, 2002). Farmers markets were a new innovation in nine cities in 2003 in Norway, and both producer and consumer interest in where and how food is produced was studied through surveys at the markets (Åsebø, 2004). Although there was demand for organic foods in the Bogotá region, the major constraint was difficulty in conversion from conventional methods to certified organic production (Cantillo, 2004). Likewise, the complexities of getting actual production and consumption demand for organic milk in better balance in Denmark was found to require joint planning by producers and consumer groups (Christensen, 2004). Studies of consumer demand and marketing are vital to understanding this vital step in the food system, and contribute to the ambitious goals of several Nordic countries to reach target levels of 10 to 15% organic food within the next decade.

Research methodology available for component studies is often not appropriate for analysis of larger farming systems. Production systems studies were examined to compare their statistical designs in 48 experiments across the region (Adler, 2003). One of the newer tools available to evaluate whole farms is emergy analysis, a method used to study cow-calf grazing production systems in the Pampas of Argentina (Rótolo, 2005). The policy environment within which farming operates, especially in the Nordic region where there is a high level of national and EU support, is an important factor in providing an economically supportive situation for farmers to convert to organic production (Kivitar, 2002). Better appreciation of the constraints under which farmers operate, and having appropriate tools for research, can contribute to analysis of systems and design of changes that will make farming more profitable and environmentally sound.

Education and efficient information flow are seen as key factors in the promotion of system changes. Farmers have long been recognized as experts in their specific practices and systems, and with additional training can be effective teachers on the farm (Lindén, 2002). School gardens were found to be an effective method of raising awareness of importance of organic foods in Iceland (Gudlaugsdóttir, 2002). Promoting better information exchange between producers, advisors, and inspectors is important to joint planning in the farming sector in Estonia (Petrutis, 2005) and in Denmark (Axelse, 2005). We consider these education activities an essential continuation and practical application of the learning methods and materials from the agroecology intensive semester courses.

## Developing a Regional Research Agenda

There is no question about the value of research as an educational exercise for students and for its contributions to science. Several articles from the MSc research have been submitted, and some published by recognized international journals. Yet we are concerned that the research does not fit together, and may not address the highest priority issues. We could call many of our short-term research projects examples of "agroecological trivial pursuit" that produces a thesis but little long-term impact. For this reason, we are setting up regional priorities for an integrated research program.

Priorities are being established by the network by first setting criteria on which to base decisions about the broad focus of research. Among the primary constraints for the long term include scarcity of fossil fuels, conversion of available farmland to other uses by society, impacts of agriculture on the environment, and reduced support payments for agricultural production. Future opportunities include greater support for multiple ecosystem services from rural landscapes, growing appreciation of the need for a secure and safe food supply, increased interest in local foods, and more concern about preservation of the cultural landscape. Rural development as it involves agriculture and the food system is a prime concern. Within this context, we are seeking relevant student research projects and setting up guidelines to identify topics. The areas of most interest to EU and to national governments are most likely to receive funding support.

Several broad criteria have been identified by the regional working group and published in the NOVA 10$^{th}$ Anniversary book (Breland and Francis, 2005) and include these:

- Projects must be science based and recognize the challenges of complexity.
- Systems studies are most likely to provide solutions to complex problems.
- Long-term questions are difficult but answers may provide important solutions.
- Meeting society's needs must include close involvement of all stakeholders.
- Action research will likely lead to desired change in society.
- Ecological principles provide guidance to design of future systems.
- Changes in university structure will be needed to facilitate future programs.

Our intent in setting up these criteria and working toward a list of priority projects is not to restrict creativity nor infringe on academic freedom. To the contrary, we hope to achieve some of the same efficiencies that are realized by well focused private research organizations that put their resources and best people to work on high priority questions that will produce a payoff for the investors. In our case we are working toward the collective human good, and hope to set up a research program that will provide appropriate experience as well as lead to collective results that can make a difference in this complex world. We must be sensitive to the perceived needs of society, and thus enhance probability of research funding. Here are some examples of topics that can be approached using a systems perspective, using both biological and social science methods, and that meet the criteria above:

- Integrated crop/animal systems to reduce chemical fertilizer applications.
- Emergy analyses and comparisons of alternative farming systems and products.
- Resilience and production stability in diverse crop/animal systems.
- Compatibility of diverse farming systems with multifunctional rural landscapes.
- Optimum levels of farming system diversity for providing ecosystem services.
- Contingent valuation of environmental costs and benefits in farming systems.

- Potentials of agricultural subsidy programs to promote eco-agriculture.
- Education of society on the economic benefits of ecosystem services.
- Production resilience and food security in global versus local food systems.
- Scale-specificity of farming and food systems research topics and results.
- National priorities for organic food and distribution of benefits of food systems.
- Education of communities on importance of food through local food councils.
- Food security and food equity in a globalised economy.
- Structure and function of the agricultural university and food security.
- Innovative education in the learning landscape for future food systems.

It is our obligation to integrate, synthesize, and apply research to solve contemporary food systems challenges. We would be failing in our mission if we did not use ecological principles to design future systems, establish scenarios for the future and assess their consequences, and help our students learn to ask new questions that have yet to be asked. To peer over the horizon is to anticipate challenges, and thereby to design a favourable future that improves the human condition, rather than being reactive to change and only adapting to what is dictated by strong multinational funding interests. Our students deserve the best efforts we have in helping them prepare for a complex and unpredictable future.

## Conclusions

Research priorities today are most often established by scientists whose world view is informed by the details of a specific discipline, with methods constrained by the modal activities taking place within that narrow framework. Agroecology broadens this vision to include biological and social science perspectives and so-called hard and soft systems methodologies. We can establish priorities and direct students toward thesis projects that will address key questions, and help identify support for their research. It is often essential to work at higher levels of spatial scale than field or farm to better understand the multifunctional agricultural landscape, the scope of rural development, and the challenges facing the food system and food security. Students and scientists who work within the agroecology framework, using a wide range of methods including action research, will provide solutions to current food systems challenges and provide insight on more secure and equitable systems that will serve the region well in the future.

## References

Breland, T.A. and C.A. Francis, 2005. Future collaborative agroecology research network in the Nordic Region. In: NOVA University Network: 10 Years, P. Jensén (editor). NOVA University Network, Alnarp, Sweden, pp. 58-59.

Francis, C., G. Lieblein, S. Gliessman, T.A. Breland, N. Creamer, R. Harwood, L. Salomonsson, J. Helenius, D. Rickerl, R. Salvador, M. Wiedenhoeft, S. Simmons, P. Allen, M. Altieri, C. Flora and R. Poincelot, 2003. Agroecology: the ecology of food systems. J. Sustainable Agric. 22, 99-118.

National Research Council, 2003. Frontiers in agricultural research: food, health, environmnent, and communities. Committee on Opportunities in Agriculture, National Academies Press, Washington, DC, USA.

# Extension as a knowledge partner in farming systems research: Early lessons from *"FutureDairy"* Australia

*Sean Kenny[1] and Ruth Nettle[2]*
*[1]The University of Sydney, Private Bag 3, Camden, 2570 NSW, Australia; [2]The University of Melbourne, Parkville, 3333 Vic, Australia*

## Abstract

FutureDairy is a new initiative developed to help Australia's dairy farmers manage the challenges they are expected to face over the next 20 years. It is investigating the issues associated with large increases in forage and milk production/ha along with new technologies to improve labour efficiency and lifestyle. The project is designed around partnerships between researchers, both technical and social, extension agents and farmers. Such partnerships aim to co-develop the knowledge and practices required for managing technologies to meet industry challenges. These "knowledge partnerships" are viewed as a potential way to "fast-track" learning about the adaptation of technologies on-farm and are being considered for their suitability for other farming systems RD&E in Australia.

The extension role in FutureDairy includes: establishing and supporting effective partnerships between commercial farmers and technical research; brokering the different knowledge types within the partnership; and, building networks with local extension and farming groups associated with commercial farms. Issues emerging in this role include managing diverse farmer expectations and needs; working through the form of partnership required between farmers, extension and researchers; and redefining the role of extension in real time. Such issues are discussed with a view to the development of a framework for effective extension practice as a broker of knowledge partnerships in farming systems research.

**Keywords:** dairy, extension, knowledge partnerships, technology, future farming systems

## Introduction

The terms of trade for the Australian dairy industry have fallen by 1.8% per year over the past decade (ABARE, 2004). Despite a four fold increase in production per farm since the early 80's (Dairy Australia, 2005), productivity growth has been modest at 1.2% over the same period (ABARE, 2004). Such figures suggest that increased farm production is almost being matched by the inputs required to achieve this production and that new technology is providing little in the way of improved farm efficiency. With pressure on the terms of trade for Australian dairy farmers likely to continue, many are questioning what the most efficient future production systems will look like (Garcia and Fulkerson, 2005). The availability and cost of labour along with the challenges associated with efficiently feeding dairy cows are seen as the key issues of concern for dairy systems of the future, and form the focus for the FutureDairy project.

## The FutureDairy Project: A response to industry challenges

The FutureDairy project centres on three modules of research around forages, feeding and innovations (Garcia *et al.*, 2006). The forages module aims to produce more home-grown feed from a given parcel of land. The target is to utilise over 40t DM/ha/yr sustainably using a Complementary Forage Rotation (CFR) (Garcia and Fulkerson, 2005). The feeding module is researching alternative pathways to increasing production per hectare by 30-50%. The key question relates to intensification of the farm business – if dairy farms have to increase output to overcome declining terms of trade, and given the competition for land resources faced by many dairy producers, what is the most efficient way to use bought in feed? Finally, the innovations module is investigating a number of innovations that have the potential to improve farm efficiency, labour management and/or lifestyle. The key innovation under investigation is automatic milking systems (AMS). Investors come from a diverse group including industry bodies (Dairy Australia, Regional Dairy industry development corporations), State government departments, academia (Universities of Sydney and Melbourne) and private companies (DeLaval, C-DAX, Kahn).

As issues surrounding future production systems for increased industry productivity involve technical, economic and social domains of knowledge, a multidisciplinary research design was considered necessary. The central strategy of FutureDairy is to create a platform for different disciplines and knowledge types to co-develop knowledge for improving farm and industry productivity. These 'knowledge partnerships' (Nettle and Kenny, 2006) are facilitated through a research-extension interface with links to five commercial dairy farms (partner farms). This interface is enhanced by links to farmer discussion groups (support groups) and is coordinated by the project's extension leader. Critical to the partnership is social research, whose role it is to capture the decision making process of the farmers around the issues of concern, document learning's of the knowledge partnership, and critically analyse the effectiveness of this RD&E model.

A key challenge for such large, multidisciplinary projects is the role the extension discipline plays. Recent literature advocates learning *and* negotiation processes amongst inter-disciplinary teams in technology development (Leeuwis, 2000). The need for such processes highlights the importance of a designated facilitator of the innovation process, competent in both technical and socio-organisational domains. Campbell (2001) describes such a 'social agriculturalist' as the new extensionist, a professional intermediary between science and practice who recognises the implications of technological change on handling information and social learning. This mediator also informs debates, illuminates trade-offs, and works as a go-between, brokering information. However, it is one thing to describe the role of the 'new extensionist', and another to make it work in an RD and E environment dominated by the agribusiness paradigm (Nettle, 2003). What does the every-day practice for the "new extensionist" in multi-disciplinary teams entail and how does the extension role contribute to innovation? The remainder of this paper aims to address these questions through the documented experiences to date of the extension and social research leader's of FutureDairy.

## Doing FutureDairy: The challenge for extension

The extension role in FutureDairy has primarily revolved around: a) establishing and supporting effective partnerships between commercial (partner) farmers (located in vastly different production and social landscapes of Australia) and technical research (located at one site); b) brokering the different knowledge types of farmers, extension and researchers (both technical and social); and c) building networks with local extension and farming groups associated with partner farms One key reflection for each element of extension's role is discussed below.

### *Establishing and supporting partnerships*

In general the transfer of technology paradigm (TOT) (Rogers, 1983) has dominated agricultural RD and E in Australia. During the 1990's, the espoused focus of extension shifted more toward working and learning together with farmers through developing an understanding of their unique context and building effective responses from this position (Hamilton, 1995; Paine, 1997; Kenny, 2002). The FutureDairy project utilises this latter approach, with farmer knowledge supporting the understanding of issues associated with adaptation of new technology. One persistent challenge relates to shifting the focus within the knowledge partnership away from technical science as the primary discourse, to seeing each discourse within the partnership as equally legitimate. The expectation 'science will provide the answers' still persists and a shift toward the notion of 'co-development' has caused tension within a number of the partnerships. For example, one partner farm and their support group became frustrated when their expectation of new technology was not met early in the project. This was despite significant efforts during the engagement process and initial activities with the farmers to reinforce how each practice is contributing to the knowledge partnership. This issue was resolved through revisiting the nature of the project questions and explaining the farmers' unique contribution. This suggests that a new model of knowledge construction (co-development) needs to be learned, particularly in an environment were alternate paradigms (ie: TOT) dominate. Thus the extension role involves facilitating learning around a shared concept of what co-development of knowledge "looks-like", whilst operating in a tension filled environment.

### *Brokering knowledge types*

As stated previously, the 'new extensionist' needs to be competent in both technical and socio-organisational domains. This is no exception within FutureDairy where extension, in order to fulfil its role as 'knowledge broker', must have the capacity to mediate between multiple knowledge types. An example of this is the process of planning for a complementary forage rotation (CFR) on the project's partner farms. Within the FutureDairy knowledge partnership, technical research has provided knowledge concerning agronomic requirements of growing a CFR whilst the farmers have provided insights into the effects of scale (large herd issues), regional constraints (timing, climate), perceptions of risk and the preferred division of labour for sowing and harvesting the maize components of the rotation. This has meant a far richer picture is emerging of the technical, economic and social requirements of growing a CFR than would have been gained through demonstrating agronomic best practice alone. Extension in this situation provided the farmers with the initial insight into the technology, established an appropriate environment for the technical scientists to share their knowledge around the CFR and developed a detailed understanding of the farm business which enabled 'customisation' of the CFR via modelling potential crop area scenarios. It also provided a means of feeding back the farmer insights around managing the CFR to the researchers, thus ensuring the farmers experience was considered in ongoing research activities. This process effectively formed a case study for social research on how the knowledge partnership functioned around planning for the CFR, insights of which have been critical in further developing our understanding of farmer decision making and the effectiveness of the partnership.

### *Building networks*

The network which forms the basis of the knowledge partnerships within FutureDairy consists of over 50 people, in four vastly different physical and social landscapes, spread across 2000km within Australia's eastern states. These "networks within a network", are predominantly based on good will, verbal agreements and in-kind support. Maintaining, developing and supporting the network becomes the mechanism for brokering knowledge and also the main focus for extension practice in the project. This has presented two main challenges. First is logistical, in that keeping informed of farm progress and ensuring an approach tailored to local cultural and advisory arrangements requires significant travel and communication with local support people. Second is institutional, in that the project relies on the commitment of participating organisations, irrespective of contractual obligations. For example, six months into the knowledge partnerships, one participating organisation withdrew funding for a critical FutureDairy network member. Although an issue for project management, extension as the over-seer of the network has to work to manage the impact this will have on the functioning of the network and knowledge partnerships at the local level. These challenges suggest that facilitating effective networks across varied farming and advisory "landscapes" requires an equal focus on networks at the "local" and the "project central" levels. Such a focus requires a significantly different set of skills to be employed by extension.

## Towards a model for extension in farming systems RD&E

Extension therefore has a unique role as a mediating practice in industry level knowledge management. As such its competent performance is based on a capacity to share in a number of competent performances with a range of other practices like farming, researching and policy making. In so doing extension requires its own body of theory and methodological resources to continuously improve its practice and that of other practices in which it shares a competent performance (Paine *et al.*, 2004). This presents challenges around managing expectations of extension as they currently stand whilst at the same time operating within an alternate paradigm. Although a model for the 'new extension' in farming systems RD and E may be in its early days, early insights suggest that such a model will need to resolve the question of how to effectively resource and manage the relationships required for knowledge partnerships to function effectively. Such relationships will have to be built upon:

- An acknowledgement of the tension filled process associated with ushering in a new paradigm for generating new knowledge around technology adaptation and innovation. Such an acknowledgement will involve a process of negotiating and agreeing upon new a power dynamic within RD&E projects.
- A new technical competence for extension based on a better understanding of the broker role and how to track and modify its performance.
- More definition (through documentation) of the new role of extension in farming systems RD&E so that capacity for managing farming systems projects can be built.

The experience of extension in FutureDairy and the way the role of extension is effectively being redefined in real time is seen as a platform for building theoretical and methodological resources to improve the performance of extension (mediating practices) in farming systems RD&E.

## References

ABARE, 2004. Production systems, productivity and profit. Australian Bureau of Agricultural Resource Economics.

ADIC, 2002. Strategic Review of RD&E. Australian Dairy Industry Council, Melbourne.

Campbell, A., 2001. Managing Australia as if we are here to stay, for good. Proceedings, Landcare Conference, Goondawindi.

Dairy Australia, 2005. Dairy Situation and Outlook - 2005. Dairy Moving Forward, Melbourne. (www.dairymovingforward.org.au).

Engel, P.G.H., 1995. Facilitating innovation. An action-oriented and participatory methodology to improve innovative social practice in agriculture. Dissertation. Department of Communication and Innovation Studies. Wageningen University, Netherlands.

Garcia S.C. and W.J. Fulkerson, 2005. Opportunities for future Australian dairy systems: a review. Aust J. Exp. Agric. 45, 1041-1055.

Garcia S.C., W.J. Fulkerson, R.A. Nettle and S.N. Kenny, 2006. FutureDairy: a national, multidisciplinary project to assist dairy farmers manage future challenges - Early learnings. Proceedings of the National Dairy Science symposium, Mt Gambier, Sth Australia.

Hamilton N.A., 1995. Learning to learn with farmers: An adult learning extension project. Phd Thesis, Wageningen, The Netherlands.

Kenny S.N., 2002. A process for practice: learning for change in the dairy industry. MAgSc Thesis, Melbourne University, Melbourne.

Leeuwis, C., 2000. Learning to be sustainable. Does the Dutch agrarian knowledge market fail? European Journal of Agricultural Education and Extension: international journal on changes in agricultural knowledge and action systems 7, 79-92.

Nettle, R.A., 2003. The Development of a National Dairy Extension Strategy: A Literature Review for Dairy Australia. Parkville, Institute of Land and Food Resources, University of Melbourne.

Paine, M.S., R.A. Nettle and S. Coats, 2004. Countering Knowledge Markets in Agri-systems. Organisational Knowledge and Learning Conference, Austria. March, 2004, pp. 15.

Paine, M.S., 1997. Doing It Together – Technology as Practice in the Dairy Sector. Phd Thesis, Wageningen, The Netherlands.

Paine, M.S., 1999. Improving the Management of Technology Development Through Mediation (Part A). Journal of Education and Extension 6, 93-110.

Rogers, E.M., 1983. Diffusion of Innovations, 3rd Edition, Free Press, New York.

# Technological innovation in extension services and institutional change: Learning from a historical comparison between France and the Netherlands

*Pierre Labarthe*
*Institut national de la Recherche Agronomique (INRA), UMR SAD-APT, Paris, France*

## Abstract

This paper refers to a part of my ongoing PhD research which deals about the dynamics of agricultural extension systems in Europe. Agricultural extension is facing nowadays some major changes. The two most important changes are: a trend for privatisation and/or commercialisation of services, and the impact of new technologies for information and communication. In this paper, I propose an historical analysis of extension systems in France and the Netherlands, for the period of the post second world war decades. The main idea is to show that changes within agricultural extension systems can be analysed by focusing not only on innovation trajectories in extension services, but also on the institutional frameworks of extension systems. In this perspective, I propose to combine evolutionist and regulationist economic theories applied to an institutional analysis of agricultural extension systems. The results show that, during post second world war decades, there was a virtuous relation between innovation in services aimed at producing standardised knowledge, and a distribution of extension for a selective basis of farms. This distribution was the result of the institutionalisation of compromises between the State and groups of farmers. Thus, the beneficiaries of extension systems were farmers who both composed the productive basis of agricultural sector in each of the two countries, and were represented by the main farmers' unions. Since the 1990s, the existence of such a virtuous relation can be discussed. The new configuration of innovations within extension services may have a potential for an enlargement of the beneficiaries of agricultural extension. Nevertheless, the commercialisation of services to farmers changes the conditions for access to services, and the institutional framework which could allow agricultural extension to meet goals linked to rural development - through changes in beneficiaries or themes -is questionable (Labarthe, 2006a).

**Keywords:** agricultural extension services, institutional arrangements, technological innovation and R&D

## Introduction

Technical extension stands usually in agricultural sector for *"the exchanges between farmers and technicians about the management of their production, the management of information necessary to this production, the organisation of activities and the distribution of resources (labour and capital) which are necessary for this activity"* (Laurent *et al.*, 2006). Since the 1990s, there have been deep changes in the structure of extension systems in European countries. These changes are: a trend for privatisation and/or commercialisation of services (Rivera, 2000), and the impact of new technologies for information and communication (Leeuwis, 1993). The extent and the diversity of these changes are so important that they raise the question of the path of agricultural extension systems. Can these changes be analysed as a new configuration within trajectories of service's innovation? Are they rather the expression of changes in the distribution of means for production of knowledge, or the emergence of new institutional arrangements that run this distribution? In order to produce knowledge related to such a question, I propose to build an institutional analysis of technical extension systems in two countries: France and the Netherlands during post second world war decades (1945-1990).

## For an historical analysis based on a comparison between two countries: France, and the Netherlands

In order to analyse the dynamics of agricultural extension, an historical institutional analysis is proposed, by combining régulationist and evolutionist theories applied to a representation of agricultural extension as a service activity. Technical agricultural extension can be described as a service activity, following the definition of services of Gadrey (1994): suppliers of agricultural extension *possess or control technical and human capabilities and sell (or propose on a free basis) to farmers the right to use these capabilities during a certain period.*

A first motive power of the dynamics of agricultural extension systems could be that innovation in services design trajectories for agricultural extension. Gallouj (1999) has proposed a taxonomy of innovation in services based on a fonctionnal approach of services. Such a taxonomy suits to the description of the different innovations within technical extension: innovation based on new competences and capabilities of advisers, innovations in logistic and material operations (increasing mechanisation of activities or economies of scale), innovations in informational operations (reduction in cost for communication, and for use and transfer of information), innovations in relational operations (innovation co-produced directly in the interaction with beneficiaries) and innovations in knowledge operations (standardisation of methods for the production or technological knowledge). These different kinds of innovations can exist simultaneously within technical agricultural extension services. Nevertheless, two hypotheses can be assessed. First of all, within firms or even national sector of technical agricultural extension, some dominant types of innovations can exist for a period of time. I call *innovation configuration* a combination of dominant types of services' innovations for a certain period of time. Second of all, it can be thought that a configuration of services' innovations in a certain period of time can determine the configuration of innovations in the following period of time. Thus, path-dependency can exist and design *services' innovations trajectories*, which can be defined as a succession of *innovations configurations* across different periods of time.

Nevertheless, can the path-dependency of technical extension systems for agriculture be explained exclusively by the innovations' trajectories of services? As agricultural extension is a resource for farmers, can't its evolution also be explained by institutionalised arrangements about the distribution of this resource? Agricultural sector is characterised by its big number of small production units which present a strong diversity in terms of economic size, production systems, and even production activity. This diversity structures the farms into different groups or social groups of farms (Laurent and Rémy, 2000). These different groups can have needs for different knowledge and different solutions in front of a same problem. Furthermore, there can be conditions for access to technical agricultural extension, and particularly to front-office services' relations which enable contextualising knowledge. First of all, access to services' relation can imply investments (financial investments or investments of time) that induce threshold effects. Second of all, access to services' relations can be dependant on the belonging of farmers to social groups. According to Du Tertre (2002), *"services' relations are at the heart of intangible services [...] and imply a recognition of the other which is part of the social belonging of each".* If human resources available for technical extension are limited, situations of rivalry between groups of farmers can exist for access to services. Thus, systems of agricultural extension are the expression of arrangements, by creating rules and institutions for the distribution of extension's services and human resources. It can be considered that these rules are the result of institutionalised arrangements (following Delorme, 1984), which express balances of power between groups of farmers.

The question is then the one of the forces for change within agricultural extension systems. Do these systems follow trajectories of innovation of services, or are they built on institutional arrangements? In order to propose elements of answer to such a question, I chose to compare two national situations: the situations of France and the Netherlands. On a one hand, these two situations offer a great diversity in terms of systems of agricultural extension, but also in terms of recent evolution of these systems. In France, one of the main suppliers of agricultural extension are since the 1960s the chambers of agriculture. Services, goals, management, and financial support of chambers of agriculture have deeply changed in the last decades. Netherlands is an emblematic country for commercialisation of technical agricultural extension in Europe. On another hand, the two countries are also contrasted in terms of distribution and diversity of farms according to their structures, economic sizes, and production systems. Thus, the proportion of farms of great economic size is much more important in the Netherlands than in France. The research is based on a historical institutional analysis. The methodology is detailed in Labarthe (2006b).

## 1945-1990: Innovation trajectories of services and distribution of services between farmers: the virtuous dynamics of agricultural extension

Even though the contexts between the two countries were different, there were two major similarities between extension systems which were both aimed at contributing to increase the productivity of agricultural sector.

1. There has been an intensification of services for farmers, which can be measured thanks to the diminution of the number of farmers per adviser which was more important than the diminution of farmers, especially in France.

2. There has been a standardisation of the production of technical knowledge and references incorporated in agricultural extension. Indeed, at the core of the functioning of extension services, there was a very strong cooperation between applied research and extension organisations. This cooperation was embodied in joint experimentations (within experimental station or directly with networks of farms), joint publication, and meeting on a regular basis. This had enabled a co-management of the work of acquisition and capitalisation of technological knowledge for feeding agricultural extension services. It could be hypothesised that it is precisely these links between technical extension and applied research which were the main source of innovation for agricultural extension during this period of time. The configuration of innovation is characterised by the superposition to innovations based on the competences and capabilities of advisers and of innovations in the methods used to produce and accumulate knowledge within extension's organisation.

Building such organisations and linkage between organisations has been possible thanks to public investments both for agricultural extension back-office and for applied research[1]. Data about the distribution of extension services built on these investments are very scarce in the

1 In the Netherlands, it was also public investments, except for experimentations which were partly financed by volunteer taxes from farmers. In France, agricultural extension (chambers of agriculture) was financed by land taxes and by a national fund for agricultural development fed by taxes on agricultural products which were co-managed by the State and farmers' unions.

two countries. About the distribution of services between different social groups of farmers, only one statistical analysis of data set is available at a national level[2] (SCEES, 1983). This study shows that in 1982, only a minority of farms (only 17% of the sample) benefited from extension services from chambers of agriculture. The farms which had access to extension services from chambers of agriculture were characterised as *"full-time farmers – part-time farmers are quasi excluded – young, more often educated, and with responsibilities in farmers'union"* (SCEES, 1983: 11). Thus, there were in France strong inequalities between groups of farmers for access to agricultural extension. These inequalities were the expression of the institutionalisation of arrangements between the State and groups of farmers, which had settled an appropriation of the resources of agricultural extension by a limited but productive group of farmers. Indeed, in the 1960s, a law had delegated the responsibility of agricultural extension from state department to the chambers of agriculture. This transfer can be analysed as an institutional arrangement. Agricultural extension was then under the control of a farmers' union representing mainly one social group of farmers: professional farmers who ran family farms. Thus, these farmers were the target group and the main beneficiaries of agricultural extension in France: they were a productive basis for the development of the sector, and also the most important contributors to investment in agricultural extension between farmers through the product system.

In the Netherlands, some compensatory mechanism used to exist, which have enabled some small farms or less-modernized farms to have access to agricultural extension. For instance, it existed a specific service[3] dedicated to small farms which proposed technical and managerial extension, either to specialise and intensify their production system, or rather to quit agriculture through a safe transition (Somers, 1991). This specificity of the Dutch situation can be explained on the one hand by modalities of implications of farmers' unions in the construction of extension systems, and on the other hand by the representation of the actual diversity of farms by farmers' union. In the Netherlands, the organisation of farmers into farmers' unions and the extension services were clearly segregated. It was the State which financed and realised the activities of technical extension through its provincial offices. It has enabled the State to develop specific services for farms (small ones for instance) or even regions less modernised and that possessed a strong potential for increase of productivity (Devienne, 1989). If farmers' unions did not have a direct responsibility on agricultural extension, they have contributed, through a consultative role, to the fact that a diversity of potential beneficiaries for agricultural extension had been taken into account. This has been possible thanks to a plurality of farmers' unions in the Netherlands. Indeed, there used to exist three farmers' unions in the Netherlands, which represented a geographical and structural diversity of farms (Frouws, 1991).

## Lessons for an analysis of agricultural extension nowadays

There are two major changes nowadays about agricultural extension in the two countries: 1) a trend for greater privatisation and/or commercialisation of services, which change the conditions of access to services for farmers; 2) new innovations in services, with the impact of NTIC combined to the progress of agronomic modelling, which could also change the productivity of services. Thus, both of these changes could have some strong consequences on the distribution of services for farmers. It is all the more important to study these consequences as the new functions acknowledged to agriculture by rural development policies settle the

2 It was realised by the central service for economic and statistical studies (SCEES, 1983) of the French Ministry of agriculture in 1982 with a representative sample of about 2500 farms.

3 *Dienst voor kleine Boerderijen (DKB)*: service for small farms.

question of the distribution of the beneficiaries of agricultural extension (Laurent *et al.*, 2006). In that respect, the lessons of the historical analyse proposed in this paper are twofold: 1) inertia is to be taken into account: extension systems are partly determined by trajectories of services; 2) these trajectories are not only technological: there could also be the expression of institutional arrangements. As suggested by Boyer, *"it might be possible that, in certain settings, some institutional arrangements could determine the orientation and speed of technical change, as well as the opposite"* (Boyer, 1990).

## References


Boyer, R., 1990. Les problématiques de la régulation face aux spécificités sectorielles. Cahiers d'Economie et de Sociologie Rurales 17, 39-76.

Delorme, R., 1984, Compromis, Etat inséré et crise de l'Etat inséré, Critiques de l'économie politique 26-27, 149-160.

Devienne, S., 1989. Systèmes de production du lait et politique agricole aux Pays-Bas. Thèse de doctorat. INAP-G., Paris.

Du Tertre, C., 2002. Activités immatérielles et relationnelles: quels nouveaux enjeux de régulation pour les secteurs et les territoires. Géographie, Economie, Sociétés 4, 131-153.

Frouws, J., 1991. Les syndicats agricoles aux Pays-Bas. In: B. Hervieu and B.-M. Lagrave (editors). Les syndicats agricoles en Europe. L'Harmattan, Paris, 186-207.

Gadrey, J., 1994. Les relations de services. In: J.-D. de Bandt and J. Gadrey, Relations de service, marchés de service. CNRS Editions, Paris, 23-42.

Gallouj, F., 1999. Les trajectoires de l'innovation dans les services: vers un enrichissement des taxonomies évolutionnistes. Economies et Sociétés, Série EGS 5, 143-169.

Labarthe, P., 2006a. Performance of Services and Unequal Access to Agricultural Extension, Study case in Ain (France) and Zeeland (Netherlands). Journal of Extension Systems (in press).

Labarthe, P., 2006b. Dynamique du conseil technique agricole: entre trajectoire d'innovation des services et inertie institutionnelle. Géographie, Economie, Société (in press).

Laurent, C. and J. Rémy, 2000. L'exploitation agricole en perspectives. Le courrier de l'environnement de l'INRA Juin 2002, 5-22.

Laurent, C., M. Cerf and P. Labarthe, 2006. Agricultural extension services and market regulation: learning from a comparison of six EU countries. European Journal of Agricultural Education and Extension 12 (in press).

Leeuwis, C., 1993. Of computers, myths and modelling. PHD Thesis.Wageningen., Agricultural University of Wageningen.

Rivera, W.R., 2000. Confronting global market: public sector agricultural extension reconsidered. Journal of Extension Systems 16: 33-54.

SCEES, 1983. Le conseil en agriculture. Structures et Environnement des Exploitations Agricoles 128.

Somers, B.M., 1991. Small farmers and agricultural extension. Communication and innovation studies. PHD thesis. Wageningen, Université de Wageningen. 235 pp.

# (Re)-innovating tools for decision-support in the light of farmers' various strategies

*Anne-Mette S. Langvad and Egon Noe*
*Danish Institute of Agricultural Science, Department of Agriecology, Blichers Allé, Postbox 50, DK-8830 Tjele, Denmark*

Present Danish agriculture is characterised by an ever-increasing specialisation and an ever-strengthened orientation towards efficiency, acting as a challenge to the advisory service offered. Simultaneously public subsidy to develop this service is decreasing and public objectives are changing towards creating a viable and environmentally friendly agricultural production. Together these developmental trends point towards an increased demand for well-suited, differentiated tools for decision support.

Until now agricultural knowledge systems and decision-support tools have largely been developed with no special involvement of the farmers as end-users and as if farmers were uniformly suited for support. However, the changing conditions have led to an increased acknowledgement of how Danish farmers cannot be considered a homogenous group with similar needs for advisory service. The internet-based computer programme, Plant Protection Online (PPO), can serve as an example that captures the problem with decision-support particularly well:

## Failing usage of decision-support systems: Towards an explanation

Despite the fact that PPO is constructed to be managed by farmers individually only 3% of Danish farmers subscribe to it, leaving much potential for further distribution. For various reasons failed subscription, cannot be explained by subscription fees, lack of disposure or PPO professional content.

Lack of clear-cut explanations for farmers' failed subscription to tools for decision support is in fact a worldwide problem. McCown describes that 20 years of experience with this fact has only led to few attempts at explanation and no firmly researched based recommendations for changed principles of construction (McCown, 2001; McCown, 2002). Among the limited amount of explanations is a Dutch research documenting that different farming styles call for different tools to support the decision-making process (Leeuwis, 1993) and a Danish research result that farmers have different expectations to decision-making tools depending on styles of farming (Noe and Halberg, 2002). On the basis of this research we take the point of departure that the modus of Danish farmers' decision-making is not only disjointed with the conditions for systemic decision-making, but also different for various groups of farmers.

In order to explore the validity of this hypothesis and its potential consequences we have developed a methodological framework based on two core principles: Continual multi-disciplinary and stakeholder involvement in the innovation and re-innovation of decision support – and usage of an ideal-type concept as the analytical tool to both identify decision-making strategies and to qualify the practical re-construction of decision-support.

## Multidisciplinary involvement as a precondition for developing decision-support

The principle of continual multidisciplinary and stakeholder involvement builds on an earlier developed *theory* of poly-ocularity (Noe *et al.*, 2003, 2005). As a *concept* poly-ocularity conveys that every single perspective contributes with nuances not conceivable through other perspectives. Like binocularity is an optic precondition for perception of depths, poly-ocularity adds dimensions to an object, which are invisible to the single, normal 'eye' or perspective (Bateson, 1979; Maruyama, 2004). It follow from the concept, that the more disciplinary or stakeholder perspectives that are enrolled into the discussion of decision-support the deeper and more complex a comprehension of decision situations ensues.

Even if PPO is only concerned with appealing to reduction of pesticides by way of farmers' cost savings, farmers make their decisions with reference to an indefinite number of criteria. Poly-ocular, non-redundant complexity is important for PPO software, because the system aims at representing real life decision situations. To avoid a discrepancy between recommendations to system development and farmers needs and aims, the system development process must take upon itself the great challenge it is to communicate across various perspectives, both with respect to analysis and practical reconstruction.

## Ideal-types as analytical tools to uncover the logic of decision making

The ideal-type is the central *analytical* concept considering the methodological framework for (re)-innovation of decision support. In spite of its' name, the ideal-type is in no way ideal in the normative sense of the word. Instead it is a descriptive analytical concept by which to grasp, what are the unique and collective characteristics within a group. More specifically, what the ideal-type describes is the logic of reasoning within the operative procedures underlying decision-making at a meso-theoretical, group level. As a tool it helps understand and describe how farmers motivate their decisions on grounds of, what seems for non-farmers to be, reasonable courses of action.

Max Weber, the originator of the concept, has developed a methodology by which to determine the existence of an ideal-type: First it is necessary to establish a statistical probability that a common logic can actually be induced. This so called 'causally adequate' induction of an ideal-type is followed by a qualitatively based interpretation of the logic used. First then, Weber considers the ideal-type explained in a meaningfully adequate way (Weber, 1995, 1998).

In order to determine ideal-typical ways of decision-making, the latter has to be made operative: Farmers' decisions are made in a context of objective structures of the farm, *e.g.* its animal production, plant production, the number of hectares of the farm, its capacity of machinery, *etc.* On these grounds the actual decision is shaped by the farmer's individual priority of values. Subjectively prioritised values, with regard to plant breeding, can *e.g.* be the degree to which, farmers value to keep fields clean, the importance of minimizing working hours or weight put on harvest gross proceeds. A third element of the decision-making process concerns obtained knowledge and sources of knowledge used by the actual decision-maker. Differences in prioritising sorts of advisory and types of newsletters may, in this respect, be a factor of import. Finally, decisions make the foundation for action. Thus, action taken is an indirect indicator of the logic of reasoning connecting objective structures with values and knowledge and forming the decision in place.

## Empirical analysis

The present methodological framework makes use of Weber's ideal-type methodology. Thus, the empirical groundwork for a causally adequate indication of decision strategies has materialized in a survey, which in turn has differentiated between farmers on account of farm characteristics, farmers value prioritising, sources of knowledge and action in relation to planning and actual use of pesticides on crops. In line with the outlined methodology, agronomists, economists and sociologists of the PPO-project have jointly contributed to the formulation of questions included in the survey. The statistical technique underpinning the differentiation is the so-called *categorical principal component analysis*, which, based on an inductive ideal, exposes internal structures of data. Selected variables are used to construct a system of co-ordinates due to their relationships of correlation. Subsequently additional variables are presented within this system. In this case the procedure graphically indicates three different types of decision-making strategies for plant breeders.

However, the key to an understanding of how actions are motivated spring not from clusters of descriptive elements, but from their logic of interrelation. To explain farmers' strategies, they must be qualitatively analysed as regularities of rationales pertaining to groups of farmers.

Therefore four groups of farmers have been selected among survey respondents to participate in focus-group interviews. Three of the groups refer directly to the three strategies indicated graphically by statistical procedures. The latter group has been formed in order to clearly illuminate variations between strategies. In line with the principle of poly-ocularity agronomic and economic researchers have played a co-interviewing role through focus group moderation.

## Decision maker strategies in relation to the PPO system of decision support

The exact typification of each decision-making strategy is made on grounds of characteristic ways in which the participants in each of the three decision-maker focus groups draws connections and make coherence between individual elements in their decision-making. Rationales peculiar to each group are meticulously filtered from general statements by way of cross comparison among all groups. What remain from this analytical process are three logics of reasoning in their essences.

Table 1 illustrates some of the ideal-type characteristics pertaining to Danish plant breeders put in relation to PPO's central preconditions for decision-support. Juxtaposition of system characteristic with essential ideal-typical characteristics of groups of farmers, make it easier to detect contrasts between system and farmer logics. Thereby, reasons why PPO does not make an appeal stand out.

As shown in Table 1, to make exact recommendations for action, PPO demands exact calculations of pest in each single field as input data to the system. This procedure for action, however, does not correspond to any ideal-typical farmer's strategy. System-oriented strategists plan for crop-rotation on a long term basis, and at just calculations are put aside to the benefit of diminishing future amount of weeds. Experienced-based strategists base their cultivation on a close personal knowledge of each individual field, thus substituting counts of weeds with close, continual observations of the fields. Strategists, who contract for their decision-making with the advisor, *i.e.* the advisory-contracting decision-makers, prioritise other operations than plant breeding. Typically, they evaluate weed in study groups or together with the advisor. Because out-sourcing is inharmonious with upholding competences for personal calculations they wouldn't by themselves manage PPO decision-support.

PPO's knowledge base consists of a large amount of agronomic and risk economic data. This kind of knowledge suits the system-oriented strategist perfectly, but still, long-term crop-rotation is of utmost importance to this ideal-typical decision-maker. This is why the system must

*Table 1. Ideal-typical decision-maker strategies and their relation to the PPO system of decision-support.*

| | PPO | System-oriented decision-maker | Experience-based decision-maker | Advisory-contracting decision-maker |
|---|---|---|---|---|
| Actions | Pest has to be calculated in the field and delivered as input, in order for PPO to return solutions. | Weed intensity is controlled by long term strategy and planning. Registration of weeds is made extensively. | Knowledge of weed intensity is based on knowing the fields in general. Registration of weed by forthgoing visual evaluation. | The advisor calculates weed prevalence, elaborate and adjust to plans. |
| Knowledge | Generalised scientific knowledge as a precondition for instructions. Hidden criteria for generalisations | Generalised scientific knowledge is obtained from various sources and juxtaposed to personal experiences when planning. | Generalised scientific knowledge is obtained from many sources put to use if they combine with something already experiences or will be experimented with carefully and selectively. | Requisite knowledge of pest and pesticides is outsourced. Farm manager focus on other branches of production. |
| Value-rationality | Pest is treated with a minimum of pesticides. This procedure saves costs and make for an economical value-rationale. | Future multiplication of pest is avoided and safety margins kept high. This makes for an economical value-rationale. | The challenge of punctuality is the centre of experience-based weed-control. This makes for an economical value-rationale. | Crop protection is adjusted to more valuable tasks of the farm, which makes for an economical value-rationale. |

present complete transparency for every generalising step, that it makes. The experience-based decision-maker acknowledges PPO's recommendations, with no specific attention to the criteria for exact generalisation behind. Nevertheless this type accepts no knowledge without question but will put recommendations to use through 'learning by doing'. Among the three types, the advisory-contracting oriented decision-maker takes the PPO knowledge base the least critically. Unfortunately, this type's decisions are outsourced, whatsoever.

Actions and knowledge are combined within PPO according to a specific economic value-rationale of cost savings by way of reduction of pesticides. It is with without question that any farmer would opt at running his farm in an economically optimal way, but the way in which to reach optimality is in close connection with primary goals and challenges, as well as ways of knowing and obtaining knowledge at the farm. Differences within economical value-rationality stand out clearly in Table 1.

Accordingly, the reason why PPO is not able to really support farmers' decision-making lies in the fact that PPO forms a strategy of its very own.

## Recommendations for (re)-innovation of decision support

By using ideal-types as analytical tools for identification of farmers' different strategies for decision-making we have reached an empirically grounded explanation for failing usage of decision-support. Ideal-types, however, can be further be used to guide reconstruction. Thus,

1. for decision-support systems to respond to farmers' decision-making practise they should integrate into ideal-typical logics of reasoning;
2. decision-support systems should never be adjusted to a single ideal-typical rationality, but be constructed with plural logical reference.

Essences of logic are captured within types, but people are placed with varying distance to them. Besides practise has no determinism, but its logic always marginally changing. Because people change with reference to strategic elements already existing in their surroundings, the best long-term assurance to tools is their ideal-typical references.

As a third consequence of our methodological framework we would like to point to the necessity of poly-ocular system construction. (Re)-innovating tools for decision support in the light of ideal-typical farmers' strategies is a complex translation process, that has to be performed in co-presence of specialised analytical and practical perspectives. The highly differentiated knowledge of scientific disciplines make nuanced analysis and modelling possible but disciplines form rationales that do not accord with practise.

Decision-making systems, no matter how well informed, are technical artefacts, that cannot acquire their own rationale. They have no meaning in and of themselves, but are frameworks established for decision-makers to generate meaning *through* them.

## References

Bateson, G., 1979. Mind and Nature: A Necessary Unity. Bantam Books.

Leeuwis, C., 1993. Of computers, myths and modelling. Netherlands, Grafisch bedrijf Ponsen & Looijen BV.

Maruyama, M., 2004. Polyocular vision or subunderstanding? Org. Stud. 25, 467-480.

McCown R.L., 2002. Probing the enigma of the decisionsupport system for farmers: Learning from experience and from theory. Agr. Syst. 74, 1-10.

Noe, E. and N. Halberg, 2002. Research Experience with Tools to Involve Farmers and Local Institutions. In: Developing More Environmentally Friendly Practices. K. Hagedorn (ed.). Environmental Co-operation and Institutional Change. Cheltenham: Edward Elgar Publishing.

Noe, E. and H.F. Alrøe, 2003. Farm enterprises as self-organizing systems: A new transdisciplinary framework for studying farm Enterprises? Int. J. of Soc. of Agr. & Food 11, 3-14.

Noe, E. and H.F. Alrøe, 2005. Combining Luhmann and Actor-Network Theory to see Farm Enterprises as Self-organizing Systems. Cybern. and Hum. Know. (in press).

Weber, M.,1995. "Makt og byråkrati". Oslo: Gyldendal.

Weber, M., 1998. The Protestant ethic and the spirit of capitalism. Los Angeles, Roxbury.

# New knowledge arrangements in the Dutch agro-ecological innovation system: Tensions between privatisation and innovation discourses

*Cees Leeuwis and Laurens Klerkx*
*Communication and Innovation Studies, Wageningen University, The Netherlands*

## Abstract

This paper discusses insights about the success and failure of 'new knowledge arrangements' in agro-ecological innovation systems. It outlines two major conceptual bodies that have been used to underpin 'new knowledge arrangements': the economically oriented discourse about privatisation, and an innovation theoretical body of thought. This leads to the identification of new roles and services for both scientists and communication specialists, and to a typology of different kinds of 'new knowledge arrangements'. After this more conceptual part, the paper summarises empirical observations about the functioning of the current innovation system. It is concluded essentially that the 'new knowledge arrangements' deriving from the privatisation discourse tend to be counterproductive in fostering new arrangements needed from the viewpoint of innovation theory.

**Keywords:** privatisation, innovation systems, communication, knowledge

## Introduction: the demise of the publicly financed EER system

Currently, 'new knowledge arrangements' for stimulating innovation in the agricultural sector emerge in the vacuum that was left after the rapid breakdown of the publicly financed 'Education, Extension and Research (EER) tryptich' that was highly successful in increasing agricultural production in the Netherlands. After the negative environmental and spatial consequences of this 'success' became apparent, the EER tryptich collapsed due to a.o. a breakdown in societal consensus about desirable directions for agricultural development. Two bodies of thought have been influential in shaping 'new knowledge arrangements' that are experimented with in the Dutch innovation support system: the discourse of privatisation and that of (system) innovation.

## Privatization and the provision of Knowledge Intensive Innovation Support Services

On the wave of the generally increased trust in market forces, privatisation became an interesting option for leveraging change in the knowledge infrastructure. Thus, economic concerns and terminology (such as efficiency, economic goods, supply and demand, knowledge markets) entered the policy arena. Much effort went into thinking about the pro's and con's of markets and the special characteristics of knowledge as an economic good (see *e.g.* Umali and Schwarz, 1994; Katz, 2002). In connection with the latter, it has become clear that in innovation contexts knowledge and information products are only rarely well described and 'sold' in a straightforward manner. In many instances we see that it is unclear in advance which knowledge will be needed and prove relevant in a certain context, while at the same time we know that in any situation much relevant knowledge is implicit and difficult to explicate and formalise (Nonaka and Takeuchi, 1995). Thus, 'knowledge products' often are loosely described services that are embodied in a person (or organisation) that is known to have (access to) certain experience, knowledge ability, skills, credibility and trustworthiness. Such services can take the form of an advisory visit or consult, supervision of a change trajectory, collaborative inquiry and research, *etc.* Therefore, we will speak from hereon about 'knowledge intensive innovation support (KIIS) services' rather than of 'research and extension' (see also Smits, 2002).

### *The changing role of government*

Dominant present thought is that KIIS service provision is a pluralistic affair, with intermingling funding and provision by both public and (newly emerging) private parties. Within such a system, government intervention can be aimed at promoting the public interest and assuring social welfare through ensuring the delivery of specific services to specific audiences (Katz, 2002). In the Netherlands we see that the state still funds particular KIIS activities on a 'public funding, private delivery' basis. Public funds for KIIS service provision can be channelled to end-users either directly or indirectly (Katz, 2002). The direct way, demand-side financing, consists of giving funds directly to the end-users of KIIS services, either in vouchers that represent a certain monetary value, or through refunding investments after proof of an actual transaction between the KIIS service provider and the farmer. In this way the state remains -relatively speaking- at a distance. The indirect way, supply-side financing consists of outsourcing or contracting-out services to private companies, who then execute a service mandate for the state. This essentially means that the state becomes a client (Leeuwis, 2004).

In addition the state can mobilise several 'safeguarding instruments' in order to exert influence on the nature and quality of any services delivered by private extension organisations. Three such safeguarding instruments are available: (1) promoting competition between different providers, (2) setting of legal rules and contracts, and (3) promoting institutional responsibility and product quality (WRR, 2000). Thus, the state becomes essentially a *client* and/or a *market supervisor*.

## Changing views of innovation and innovation support

Over the years, ideas about innovation and change have evolved considerably. The earlier mentioned linear model of innovation has been refuted and criticised by many (Kline and Rosenberg, 1986; Röling, 1988). Nowadays innovations are no longer looked at as consisting of technology only, but rather as successful combinations of 'hardware' (*i.e.* new technical devices and practices), 'software' (*i.e.* new knowledge and modes of thinking) and 'orgware' (*i.e.* new regulations, market arrangements and forms of organisation) (Smits, 2002). In this perspective the social and organisational conditions that affect may affect the 'survival' of innovations are no longer seen as external and static, but rather as integral parts of the innovation. Thus, innovation depends on effective collaboration in a network of interdependent societal actors, which implies that it is no longer useful to look at 'adoption' as something that happens at an individual level. In order to arrive at complementary action, three interrelated processes are important: the building of networks, social learning in the network (Röling, 2002) and negotiation in order to deal with conflicts that occur during an innovation trajectory (Leeuwis, 2004).

### *The changing role of communication specialists*

In the linear model of innovation, communication mainly was regarded as a mechanism for enhancing the adoption and diffusion of innovations by individuals. Communication specialists were seen as 'extensionists' who operationalised their work in the form of persuasive technology transfer campaigns, advisory communication and the support of horizontal communication among farmers. The new conception that innovation is about fostering new forms of *coordinated action* among multiple stakeholders, has important implications for the role of communication specialists. It implies that the conventional repertoire of 'extension' forms needs to be supplemented with other modes of communicative process support aimed at building networks, developing shared visions and understandings, facilitation of conflict management, capacity building, *etc.* From theories about network building, social learning and negotiation several important facilitation tasks can be derived (see Leeuwis, 2004).

***The changing role of science***
The new societal context and modes of thinking about innovation have important implications for the role of scientists as well. Since the breakdown in the dominant consensus regarding agriculture, scientists are operating increasingly in complex situations where uncertainty is high and where different values and interests are at stake. In such situations, Funtowicz and Ravetz's (1993) argue that there is a need for 'post-normal' approach to science, instead of a strategy in which science is only applied for the 'solving of puzzles' or the giving of situation-specific advice. With post-normal science, scientists are intensely involved in societal processes, debates and innovation. In a similar vein, Gibbons *et al.* (1994) speak of the need for scientists to shift from 'Mode 1' to 'Mode 2' science.

Operating in a 'post normal' (or Mode 2) manner has important implications for scientists. When looking more in detail at the contributions that scientists can make (apart from playing a role in facilitation) we can discern four basic roles:
1. help explicate implicit assumptions, knowledge claims and questions;
2. engage in joint fact-finding and uncertainty reduction;
3. providing independent feedback on problems and opportunities;
4. reflexive monitoring of the innovation process.

Playing a role as outlined above requires different modes of operation by scientists than are currently dominant. Current organisational set-ups, epistemic cultures, competencies, assessment procedures, funding arrangements and other reward systems within universities and research institutes are often not conducive to such new ways of working (see Grin *et al.*, 2004; Leeuwis, 2004).

## Types of new arrangements

As outlined above, political, financial and organisational conditions in the agricultural knowledge infrastructure have altered considerably. The above discourses suggest that new knowledge arrangements are needed in four (closely intertwined) spheres:
a. the *knowledge intensive innovation support services* that are being provided and/or deemed necessary, such as new facilitation services;
b. in the *financial* sphere, where new procedures for supply-side or demand-side funding of services are being developed, involving a.o. voucher systems, tendering and bidding procedures for contracting out, and also new public/private funding and share-holding constructions;
c. in the sphere of *supervising the market* for KIIS service delivery, involving a.o. new legislation and quality control systems;
d. in the sphere of creating internal *institutional conditions* that will allow universities, research institutes and other organisations involved in service provision to work effectively in the new environment.

## Initial observations regarding the functioning of the current innovation system

It is not easy to make well-founded statements about the functioning of the current knowledge infrastructure. On the one hand this is related to the fact that the situation is far from stable and has not crystallised yet; new developments and experiments are taking place on a regular basis.

Another obstacle is that systematic comparative research on the functioning of new arrangements largely is absent. We need to rely on occasional research reports, participant observation and informal discussions with actors involved. On the basis of this, we will list some tendencies that may be characterised as 'positive', and a rather larger number of problematic issues.

***'Positive tendencies'***

- there is indeed considerable dynamism and experimentation with new arrangements in today's knowledge infrastructure;
- after a period of increased centralisation we now witness the re-emergence of regionally bound knowledge institutions;
- new knowledge arrangements are accompanied with a much more interactive and participatory language and philosophy than the 'old' EER tryptich;
- new categories of stakeholders (*e.g.* environmental movements, animal welfare organisations and nature conservationists) have indeed become involved;
- the use of contracts has resulted in increased clarity about tasks to be performed and products to be delivered;
- the government is now in a better position to steer (and account for) themes and topics that are to be tackled with public funds, and also to enforce co-operation.

***'Problematic aspects'***

- there no longer exists a recognisable and accessible knowledge infrastructure with a clear point of entry and transparent procedures for getting something done;
- with 'projects' as the dominant organisational form, new knowledge arrangements and intermediary organisations tend to short-lived and fail to co-ordinate activities;
- the transaction costs for getting initiatives funded can be very high, leading to increased overheads and tariffs and reduced pro-activeness;
- contracts and output oriented financing schemes tend to reduce flexibility and by this the space for learning and experimenting in ongoing projects and programmes;
- increased competition among KIIS service providers leads to shielding-off of knowledge and obstacles for co-operation;
- KIIS service providers tend to be in a much better position to shape the contents of publicly funded projects than societal stakeholders, causing ownership problems;
- the quality of problem diagnosis and demand articulation tends to be low due to lack of time, competencies and funds allocated to this;
- there are fundamental problems in formulating 'demand driven' projects, since 'demands' tend to be conflicting and are discovered mainly as processes evolve;
- conventional roles of 'research and extension' still dominate, while new services are insufficiently articulated as 'products' that can be supplied and demanded;
- policy maker still frame problems in terms of 'lack of knowledge' (instead of lack of co-ordinated action) and develop 'knowledge policy' rather than 'innovation policy';
- although the quality of facilitation in complex problem situations is often limited, KIIS service providers often are not in a legitimate position to facilitate;
- many new arrangements focus on innovation development, while well-established and routine-like linkages for 'outscaling' are lacking.

## Conclusion: Contradictory discourses

In essence, the above analysis suggests that the two discourses which have shaped the bringing about of new arrangements (*i.e.* the discourse of privatisation and that of innovation) do not seem to match well. The idea that applied agricultural knowledge can be treated as a good, that can be exchanged effectively and efficiently through a market for KIIS services, may only be valid in the case that 'proven' and easily adaptable innovations are already available. In case this does not apply, however, the overall impression is that such markets may well complicate the very innovation processes that are necessary to arrive at such 'proven' innovations. This because commercialisation of service delivery can easily lead to interaction patterns that hinder the flexible co-operation, learning ability, pro-activeness and creativity that is necessary in order to enhance innovation. In conceptual terms, we can say that the notions of 'supply' and 'demand' are in many ways not applicable to innovation processes. In an innovation process it tends to be rather unclear who 'supplies' and who 'demands' knowledge and information, as successful innovation requires the integration of relevant (but often still implicit) insights from several parties. In addition, we see that the kinds of services being demanded and supplied often still resemble conventional research and extension, while more fuzzy facilitation services that would be more in line with new theoretical perspectives on innovation are not articulated and/or difficult to get funded under the new regime.

## References

Funtowicz, S.O. and J.R. Ravetz, 1993. Science for the post-normal age. Futures, 25, 739-755.

Gibbons, M., C. Limoges, H. Nowotny, S. Schwartzman, P. Scott and M. Trow, 1994. The new production of knowledge. The dynamics of science and research in contemporary societies, Sage Publications, London.

Grin, J., F. Felix, B. Bos and S. Spoelstra, 2004. Practices for reflexive design: lessons from a Dutch programme on sustainable agriculture. Int. J. Foresight and Innovation Policy 1, 126-149.

Katz, E. (with contributions by A. Barandun), 2002. Innovative approaches to financing extension for agricultural and natural resource management – Conceptual considerations and analysis of experience. LBL, Swiss Center for Agricultural Extension, Lindau.

Kline, S.J. and N. Rosenberg, 1986. An overview of innovation. In: R. Landau and N. Rosenberg (Eds), 1986. The positive sum strategy: Harnessing technology for economic growth. National Academic Press, Washington, pp. 275-305.

Leeuwis, C. (with contributions by A. Van den Ban), 2004. Communication for rural innovation. Rethinking agricultural extension. Blackwell Science, Oxford.

Nonaka I. and H. Takeuchi, 1995. The knowledge creating company: How Japanese companies create the dynamics of innovation. Oxford University Press, Oxford.

Röling, N.G., 1988. Extension science: Information systems in agricultural development. Cambridge University Press, Cambridge.

Smits, R., 2002. Innovation studies in the 21st century. Questions from a users perspective. Technological Forecasting and Social Change 69, 861-883.

Umali, D.L. and L. Schwarz, 1994. Public and private agricultural extension: Beyond traditional boundaries. World Bank Technical Paper 247. World Bank, Washington.

WRR (Dutch Scientific Governmental Advisory Council), 2000. Het Borgen van Publiek Belang. SDU Uitgevers, The Hague.

# Immaterial resources: A new concept to formalize information mobilized by beef cattle farmers in their husbandry activity

*M.-A. Magne[1], S. Ingrand[2] and M. Cerf[3]*
*[1]Institut de l'Elevage, UMR Métafort, INRA-SAD, Equipe TSE, 63122 Saint-Genès Champanelle, France; [2]UMR Métafort, INRA-SAD, Equipe TSE, 63122 Saint-Genès Champanelle, France; [3]INRA, UMR SAD-APT, 78850 Thiverval Grignon, France*

## Abstract

Since 1990, agronomists have been studying the technical aspects of farm management and proposed models which include the formalization of farmers' reasoning. In this paper we propose a new framework for the analysis of livestock systems to formalize the information mobilized by the farmers to design and monitor their farming activity by taking into account the meaning they assign to it. To achieve this, we draw on approaches borrowed from the social sciences: 1) the framework of the activity analysis, which focuses on the actors and the meaning they assign to the information they use for their productive action; 2) the notion of immaterial resources borrowed from the management sciences, which enables us to include the contribution of information to livestock system performance whereas agronomists usually focus on the contribution of material resources (herd, pasture and workforce).

**Keywords:** farming system, information, immaterial resource, activity

## Introduction

We propose a new analytical framework for livestock systems to formalize the information mobilized by livestock farmers in designing and monitoring their livestock husbandry activity, as constructs resulting from a socio-cultural process of attribution of meaning by these farmers. In this way we hope to contribute to a better understanding of the technical management processes of livestock systems with a view to developing new management aids. Our proposal is based on a review of studies undertaken by livestock systems scientists on information and decisions in livestock systems as well as on work by social scientists to analyze activities in the workplace and understand the relationship between knowledge and action. In practical terms, we propose to formalize the information system of beef cattle farmers based on the concept of immaterial resources whose heuristic value we shall specify.

## The analysis frameworks of farming systems: A crude formalization of information

In zootechnics of livestock systems, the formalization of the farmers' information system has not been researched as such and is, therefore, crude. The farming system is defined as "a set of elements in dynamic interaction organized by a farmer for the purpose of valorizing resources through domestic animals" (Landais, 1987). The "object" towards which the system is directed is consequently the herd. Only material resources (principally forage) are taken into account: information is not considered as a resource for the livestock system. Nevertheless, in their representation of the dynamic functioning of livestock systems, Osty and Landais (1991, cited in Landais, 1992) considered the information aspect: they distinguished information coming from the subsystem of production elaboration from that which comes from outside. By proposing to distinguish different types of elements that contribute to the making of technical choices (decision-maker objectives, indicators to check the way the production process is proceeding, references to assess the situation, decision rules and social standards of behaviour), Landais (1992) identified different objects of the information system but did not specify their definition, nor their decision-making function, nor how they relate to each other. Now, beyond identifying them and their function, the issue is to define how and why livestock farmers construct and update their information system.

Perrot (1991) based his work on formalization of information systems in cattle farming but proceeded from experts' statements, *e.g.* the statements of farm advisers. Consequently, his study does not make it possible to grasp the meaning farmers assign to information and why it is relevant according to their technical standpoint and their social, economic and historical background. Lasseur (2001) proposed the concept of local farming system to understand farming practices and their transformations as products of social interactions. He showed how interactions between farmers contribute to the social construction of a shared standard of "correct functioning of their system". But this approach does not make it possible to understand the role of "prescriptors" in this construction. In addition, this work provides few elements on information enabling farmers to identify, express and resolve problems. It also remains unclear on the factors motivating farmer integration with networks of actors, whether peers or prescriptors. Nevertheless, it does contribute to the study of the relation between livestock systems and their environment recognized as a necessity in recent work on systems flexibility (Ingrand *et al.*, 2006).

Finally, few studies have dealt with information systems in the zootechnics of livestock systems in relation with the development of traceability (Havet *et al.*, 2005) or the new information and communication technologies (Slavík, 2004). Slavík (ibid) proposed a preliminary grid to differentiate information according to its origin, content and supports. Havet *et al.* (ibid), who analyzed farmer management writings in farms are more interested in the various functions they fulfil for the farmer. However, none proposes an overview of the farmer's information system and the way in which this system is constructed, mobilized and stabilized.

All these studies contribute partially to the formalization of information systems. A large part relates to a procedural approach to the rationality of the decision-maker as developed by Simon (1956) and overlooks the fact that information is a socio-cultural construct. Others subscribe to this latter viewpoint but leave aside the analysis of the individual appropriation of socially constructed standards. On the whole, they pay little attention to interactions with agricultural prescriptors and the way in which these contribute to shaping livestock practices. Moreover, they do not deal with what we call the decisional activity, *i.e.* the activity focused on preparing and controlling the farming activity. This decisional activity integrates both material dimensions (taking samples for analyses, using a computer to store and interpret data) and immaterial dimensions linked to the socio-cognitive processes at work (understanding of the situation, creation of solutions, choice of a satisfactory solution and evaluation).

For this, we propose to focus on the "human" pole of the "human-herd-resources" triptych and thus recenter the study away from the productive activity (execution of tasks) on the decisional and socio-cognitive activity. To understand this, we propose the concept of immaterial resources we will explain and define now.

## Information seen as a social, historical and cultural construct meaningful for the actor

We turned to the social sciences to understand information and its processing as the product of social, cultural and technical interactions. We paid greater attention to studies which consider information both as a vector of the meaning constructed through these interactions, and a producer of meaning for the actors who mobilize it. Although these studies may originate from different disciplinary backgrounds and theoretical frameworks, they are all based on the analysis of activity or the activity system (Engeström, 1996), its construction and its control by the actor, who is also culturally and historically constructed as well as being situated in an environment with which he interacts. They consider that: i) cognition cannot be disembodied from action and goals which are assigned to it by the actor (individual or collective); ii) the meaning that actors assign to the information they mobilize is the result of complex processes linked to reflexivity and the relationships they have with other actors as well as with the other objects relating to their activity and their environment (Linard, 2001); iii) the object of the study is not so much the coding, transmission and decoding of information as the use of information, *i.e.* the information interpreted, embodied in action, situated in a context and integrated into a project (Charlet *et al.*, 2001). In short, these studies are based on the recognition of a continuity between cognition and action. We focus on those that emphasize the constructive activity of actors, *i.e.* their capacity to create the resources for their action by attributing properties to the available resources in relation to their intention to act on the situation. The resources are constructed by and for the carrying out of the action, its control, and the processes of learning and appropriation of the environment that are associated with it.

Some studies mainly focused on the construction of meaning in actions and interactions Theureau (2004) defined "the autonomous actor" *i.e.* who "has an asymmetrical relationship with his environment, in the sense that he interacts only with what, in this environment, interests him or is a source of disruptions for him". Linard (2001) also showed that the meaning the actors attribute to actions is linked to the meaning they assign to their environment and to their own internal state. Studying the internalizing processes in actors of information stemming from the environment and which have as yet no meaning for them, Linard (ibid) emphasized the motivations that animate the actors, whether their motivation relates to the performances of the activity or to their personal development. For Greimas (1966) such motivations are taken into consideration in the actantial model he developed to describe the pursuit of objects by the actor. He defined the attitude of the actor towards the sought object by five verbs: wanting, having to, knowing how to, being able to, thinking – and showed that actors assess their action relatively to the recognition received for it from a moral or social authority.

Other studies focused on the construction of meaning in the processes of appropriation and reconstruction of objects mobilized by the actors to carry out their activity. Rabardel (1995) thus emphasized the instrumental dimension of the object (the attribution of a use scheme to the object), but also underlines the mediating role of the object between the actors and the goals (results and motives for the action) which guide their activity. Far from being neutral, these objects contribute to the performances of the activity. That is what Seifert and Messing (2004) showed about human mediations (the relational feature) in service activities. Leplat (2004) stressed the importance of dissociating the mediating support of the object and its primary function to understand whether lack of mobilization of this object results from support and/or function inadequacy.

Hence, should we not consider information as a resource that livestock farmers construct in the course of action and interactions to create and control their farming activity? Should we not question the meaning conveyed by such information, as well as the meaning attributed to it by the farmers? For this we will talk about immaterial resources. However, this notion is defined and used by us in a different way to that of researchers who focus their study on learning organizations and knowledge management. Indeed, in the latters' approach, immaterial resources concern the organizational knowledge, competences and knowledge of actors whether internal or external to the firm (Bounfour, 1998), but little is said about the ways of mobilizing and constructing them. We define immaterial resources as the set of symbolic mediations (observations, human interactions, measurement devices, written supports...) which are relevant for farmers to manage their livestock, develop their farming practices and assess their expertise.

## Immaterial resources of livestock systems: Understanding the relations between their supports, origin, content and function

We propose to formalize the information system of farmers using the concept of immaterial resources. Thus, we attribute to information a meaning and a relevance that depends on the goal assigned by the farmers to their farming activity. We set information in a utilization perspective: the term resource refers to what is mobilized by farmers to support their decisional activity. The meaning given to the resource can be motivated by and for the development of the biotechnical system or the own development of the farmers. We point out that the information is chosen, used, renewed and developed just as are the material resources more classically studied in livestock farming systems science. We insist on the dimension of the livestock activity we are interested in, *i.e.* the decisional activity. Immaterial resources consist of all that goes to support this activity. We qualify them by four components: their supports, their origin, their content and their function. Two hypotheses underlie this choice: i/ the underlying supports, origin and content of the immaterial resource are not neutral in the interpretation made by the farmers; ii/ the function of the immaterial resource accounts for the meaningful relationship the farmers attribute to each of the three previous components and to the development of their farming activity.

The content of the immaterial resources are the fields relating to the technical management of the livestock system (nutrition, genetics, health...) and refer to a set of farming practices. We favour a global view of the livestock farming activity to understand the way in which farmers identify the problems and resolve them. The origin of the immaterial resources may be: i/ the biotechnical sub-system, farmer experience (stabilized immaterial resources); ii/ the different actors (farmers, experts) and organizations that exert an influence on the farmers' technical management. We need to go beyond a simple internal versus external characterization in order to understand how the information's origin contributes to their meaning for the farmer. The type of support of immaterial resources enables us to take into account the fact that: i) the relational aspect (human support) contributes to constructing the contents of the actors' interventions with the farmers; ii) there are different kinds of mediation that stabilize the immaterial resources mobilized by farmers (written and human supports, computers). Identifying the functions of the immaterial resources aims at understanding how farmers mobilize them to control their livestock farming activity ex-ante and ex-post. The ex-ante control enables to identify the immaterial resources which define for the farmers the correct functioning of their system. The ex-post control enables to identify the immaterial resources, to question those stabilized until then or to stabilize those newly mobilized.

## Contributions to the zootechnics of livestock systems

The framework we propose introduces a new approach to decisional processes. It is inspired from features borrowed from the activity theory. This is why we bring in the decisional activity of farmers to understand the socially, historically and culturally situated work of preparing and assessing production action. This analytical framework can help researchers in the zootechnics of livestock systems to: i) formalize farmer information systems; ii) gain reflexivity about the knowledge produced so far and help define new research topics. Indeed it helps highlight the scientific knowledge internalized by the farmers, the way it occurred, to what ends and with which results; iii) identify relevant partnerships to be developed for the design of support tools for the technical management of livestock farms through better understanding of the origins and supporting media of immaterial resources.

## References

Bounfour, A., 1998. Le management des ressources immatérielles. Maitriser les nouveaux leviers de l'avantage compétitif. Paris, Dunod, 266 pp.

Charlet, J., Reynaud C. and R. Teulier, 2001. Ingénierie des connaissances pour les systèmes d'information. In: C. Cauvet and C. Rosenthal Sabroux, Ingénierie des systèmes d'information, Hermès, pp. 277-316.

Engeström, Y., 1996. Work as a testbench of activity theory. In: S. Chaiklin and J. Lave (Eds.), Understanding practice: perspectives on activity and context. Cambridge (MA), Cambridge University Press.

Greimas, A.-J., 1966. Sémantique structurale. Paris, Larousse.

Havet, A., N. Joly and A. Cochet, 2005. Les écrits comme supports de raisonnement dans les exploitations d'élevage. Renc. Rech. Rum. 12, 327-330.

Ingrand, S., E. Chia, C.H. Moulin and J. Cessieux, 2006. Commercial flexibility of beef cattle farms. In: Langeveld, H. and N. Röling (eds.) Changing European farmingsystems for a better future, Wageningen Academic Publishers, Wageningen, The Netherlands, pp. 252-256.

Landais, E., 1987. Recherches sur les systèmes d'élevage - Questions et perspectives, INRA-SAD, 75 pp.

Landais, E., 1992. Principes de modélisation des systèmes d'élevage - Approches graphiques. Les Cah. Rec. Dév. 32, 82-95.

Lasseur, J., 2001. Pratique d'élevage et gestion de l'espace: la caractérisation des systèmes d'élevages locaux. Mémoire d'ingénieur de l'ENSA Montpellier, 75 pp. + annexes.

Leplat, J., 2004. Procédures d'objectivation dans un entretien de recherche. @ctivités 1, 195-216.

Linard, M., 2001. Concevoir des environnements pour apprendre: l'activité humaine, cadre organisateur de l'activité technique. In: E. Delozanne and P. Jacoboni, Interaction homme-machine pour la formation et l'apprentissage humain. 8, pp. 211-238.

Perrot, C., 1991. Un système d'information construit à dire d'experts pour le conseil technico-économique aux éleveurs de bovins. Thèse de doctorat, INA PG, 211 pp + annexes.

Rabardel, P., 1995. Les hommes et les technologies: Approche cognitive des instruments contemporains. Paris, Armand Colin.

Seifert, A.-M. and K. Messing, 2004. Écouter et voir dans un monde technique: Effets de la discontinuité des horaires sur l'activité de travail des infirmières. PISTES 6, 17 pp.

Slavík, M., 2004. Changes in Information Systems in Czech Agriculture. JAEE 10, 189-199.

Simon, H., 1956. A behavioural model of rational choice. Quarterly. Journal of Economics 69, 99-118

Theureau, J., 2004. L'hypothèse de la cognition (ou action) située et la tradition d'analyse du travail de l'ergonomie de langue française. @ctivités 1, 11-25.

# Farming systems and its cousins: Including and transcending to make a difference

*C.L. Neely[1] and J.M. Dixon[2]*
*[1]Holistic Management International, 1010 Tijeras Ave., NW Albuquerque, NM 87102, USA; [2]CIMMYT, Apdo. Postal 6-641, 06600 Mexico D.F., Mexico*

## Abstract

The reality is that there are no shortages of good approaches or intentions that have emerged over the last 35 years to advance sustainable development in rural areas. With the greater acceptance of participatory and multi-stakeholder processes, farm to landscape linkages, and practice to policy strategies, it is time to take stock of what is working and why. How do we take our best learning forward to truly make a difference in the lives of poor people in rural areas? In this paper, we look at approaches and innovations that have emerged from farming systems in the evolution of good practice for people-centred research and development. Approaches have evolved largely through ad hoc practitioner experimentation and learning, often in diverse field settings. Whilst some methods have been imported from other approaches or sectors, most method improvement occurs endogenously. Nevertheless, farming systems is marked by continuous improvement, albeit mostly fine-tuning, of methods. As we include and transcend to the next generation of approaches, we must clarify what we wish to accomplish within the current context of the MDG targets and what good processes and practices will look like to get us there.

**Keywords:** learning, institutions, multi-stakeholder, transdisciplinarity, systems

## Introduction: What do we want to know?

With some 35 years of experience, farming systems approaches can be found at the root of many of the innovative and emerging methodologies that researchers and development agents are using today (Collinson, 2000; Dixon, 1990; Norman, 2002). New methodologies are often catalyzed as gaps in our ways of working are recognized. As we learn, we tend to include what is working from past experiences, letting go of those that do not, and then transcend to operationalize our next level of knowledge (Wilbur, 1996). This evolutionary path from the early days of farming systems approaches to today has been a fascinating one from which we have derived immense gains in research and development effectiveness. However, for those of us whose work focuses on the sustainable improvement of lives and livelihoods of the marginalized, it is difficult to swallow the fact that progress toward the global goals of reducing poverty, hunger and maintaining the environment has been far less than hoped. A question to explore concerns what we can learn from our learning, and the methodological evolution and adaptation such that the evolutionary leaps are greater leading to more pronounced impact.

The paper draws on the experience of some approaches currently in use and emphasized aspects, noting that the development community faces a huge challenge of accelerating development in order to contribute to the Millennium Development Goals. We also look at mechanisms for capturing and expanding the learning in an effort to better inform the next generation of approaches.

## How far has transcendence taken us?

As farming systems approaches have evolved, we find ourselves with a myriad of "named" or "labelled" approaches (Dixon, 2003) indicative of the aspect of sustainable development that each were grappling with when they came into being. These approaches have, in turn, brought to light insights related to institutions and policy; participation (Lilja and Bellon, 2006; Morris and Bellon, 2004), multi-stakeholder partnerships and peoples' rights; environment and agro-ecosystems (Röling and Wagemakers, 1998); and multidisciplinary and multi-sectoral mechanisms and their interdependence, making them increasingly difficult to tease apart. Indicative examples of a wider set of farming systems related approaches are Sustainable Livelihoods, Agroecology, Territorial Approaches, Holistic Management and Gestion de Terroir. These particular approaches were chosen from among the more comprehensive, holistic approaches currently in vogue. Ecoagriculture, Conservation Agriculture, Community Based Natural Resources Management, Landcare, and Good Agricultural Practices each under the sustainable agriculture and rural development umbrella could also be considered.

In preparing this paper, the authors rapidly reviewed 12 approaches including those mentioned above with a view to parameters such as origin, sectors and disciplines, attention given to food security and safety, poverty reduction, income generation, natural resources conservation, social justice, cultural appropriateness, stakeholder engagement, empowerment, gender, systems approaches, development pathways, local institutional involvement, decision making and practice to policy linkages, scale, innovation and project phase. While different parameters are stressed for different approaches, within most approaches, over time there has been an expansion of parameters addressed with emerging issues or knowledge availability. We must continue to blur the lines among available "named" approaches while seeking new knowledge to provide the best possible options for use within divergent contexts and scenarios.

Four conditions have been described for addressing complex global problems to provide systemically transformative solutions (Generon, 2006) including new alliances, problem solving processes with full participation of a wide range of stakeholders, ownership of solutions by stakeholder institutions and personal commitment to make a change. In the next paragraphs we briefly look at the forward motion toward addressing complexity within available approaches.

### *Participation, multi-stakeholder partnerships, and peoples' rights*

While an emphasis on participatory approaches has become, in one sense, household (and grantsmanship) rhetoric, there is still broad divergence in the degree and nature of its integration into the stages of research and development project lifecycles. Nonetheless the pendulum has swung towards the empowerment edge of participation which has continued to be enriched as the long term benefits of more democratic, people-centred multi-stakeholder processes and partnerships come to be clearly recognized (UN, 2002; Neely, 2003; IFPRI, 2005; Dixon and Singh, 1995) and while transparency and power sharing are not guaranteed, they bring opportunities for awareness raising, shared communication, community envisioning, creativity and social entrepreneurship, creative solutions to areas of common interest, and more interactive learning in addressing complex problems. These approaches have been deepened recently with rights based approaches that consider civil, cultural, economic, political and social rights and access to development processes, institutions and information.

### *Institutions and policy*

The role of institutions in rural development, and particularly that of extension has been transformed with enhanced participation, linkages among various stakeholders and a trend toward facilitation of farmer leadership, innovation and information management (*e.g.*, Swiss Center for Agriculture Extension, Farmer Fields Schools, and Linked Local Learning) (Getz and Neely, 2005; Linked Local Learners, 2006; LEARNing Project, 2006). Additionally, we have moved from voicing the need for 'enabling policy environments' to exploring ways to make improved policy part of the process. Participatory policy making is receiving greater attention represented by the enhanced role of civil society in formerly government-only deliberations (UN, 2003; FAO, 2006) and more recent examples of local good practice and policy change based in empowered negotiations between communities, government and private sector (Vermeulen *et al.*, 2005; IFPRI, 2005).

### *Environment and agroecosystems*

Since UNCED, numerous approaches have partially incorporated natural resource management and conservation into the project cycle. Agroecology, Ecoagriculture, and Holistic Management have stressed empowerment associated with good farming strategies and farmers' organization, landscape linkages across agriculture and environment sectors and decision making based on that which can be supported sustainably (Altieri, 1983; Scherr and McNeely, 2003; Savory,1999). Environmental concerns continue to be considered a luxury. For example, a review of successful sustainable livelihoods cases found attention to the natural resource asset lacking (Neely *et al.*, 2005). In its 2005 report, the World Resources Institute reinforces the importance of ecosystems as wealth-creating assets (WRI, 2005).

### *Multidisciplinary and multi-sectoral dimensions*

Farming systems approaches were at the forefront of moving from multidisciplinary to interdisciplinary approaches through, *interalia*, farmer-researcher linkages. Recent debates on transdisciplinary and holistic approaches are leading toward further break down of sectoral and disciplinary barriers to enhance our knowledge and effectiveness (Gibbons, 2001; Savory, 1999; Getz and Neely, 2005). A holistic approach allows the people who manage resources to draw on available expertise, knowledge, and approaches toward sustainable outcomes (Savory, 1999).

## What processes of learning and sharing benefit our capacity to transcend?

Learning, generally characterized by action, reflection and dissemination, is a daily feature of many development activities. Action Learning, leadership and co-creation approaches are applied in business sector and food systems (Sustainable Food Laboratory, 2006). Generon is applying the U process to complex problem solving based on three stages – those of perceiving and action that include sensing (observing), presencing (allowing inner knowledge to emerge) and realizing (action with a natural flow) (Generon, 2006, Senge *et al.*, 2004). Pasquale and Sternin (2005) identify the learning and opportunities that emerge through unintended consequences and the use of a positive deviance approach to change by identifying and leveraging innovators and contrasts the benefit of participants acting into a new way of thinking rather than thinking into a new way of acting.

Chambers (1997) identifies various "streams" of rural development approaches which embody elements of local learning: participatory rural appraisal; rapid rural appraisal; farming systems; agro-ecosystems analysis; action-reflection research; and applied anthropology. These approaches continue to evolve whilst sharing insights and, to a great degree, field methods. Learning also occurs through reviews and workshops, programme monitoring and evaluation and mentoring. Development practitioners and researchers continue to experiment with a variety of local learning models, which depend on the local dynamics of individual human and social expectations and capacities (Barlow *et al.*, 2002; Röling and Wagemakers, 1998; LEARNing Project, 2006), with occasional attempts at linking these local efforts (Lightfoot and Scheuemeier, 2003). However, at national level these have been limited and even less so at the transnational or global levels – one of the recent exceptions being the IFSA Global Learning Opportunity convened in Rome in 2005.

## What next?

The evolution of farming systems towards more holistic approaches has been a clear and needed response to an ever growing complexity of povery reduction. There is a need for drawing on effective existing approaches in a more holistic framework and investing less in building new approaches or, worse, re-labelling approaches that can create barriers to learning. Successful learning for sustainable agricultural and rural development requires effective incentives, mechanisms and organisational capacity for facilitating learning and adaptation for change. Medium to long term regional learning, development and mentoring programmes are attractive investments for effective learning on best development practice and accelerated and sustainable agricultural and rural development.

## References

Altieri, M.A.,1987. Agroecology: The scientific basis of alternative agriculture. Westview Press, Boulder Co. 227 pp.

Barlow, R., D. Clark, A. Crawford, M. Paine, G. Sheath and J. Weatherley, 2002. Guidelines for farming systems research & learning. Dairy research and Development Corporation. Melbourne, Australia.

Chambers, R., 1997. Whose reality counts? Putting the first last. Intermediate Technology Publications, London, UK.

Collinson, M., 2000. A history of farming systems research. CABI Press, Wallingford, UK. 432 pp.

Dixon, J.M., 2003. The evolution of farming systems approach: From bed sheets to bloggers. Australia Farming Systems Association, Toowoomba, Australia. 7 pp.

Dixon, J., 1990. Ways forward for the farming systems approach in Asia. Paper presented at the Asian Farming Systems Research and Extension Symposium, Asian Institute of Technology, Bangkok, Thailand. 15 pp.

Dixon, J. and B. Singh, 1995. People's participation in sustainable agricultural and rural development, FAO NGO-SARD Workshop (September), FAO-RAP, Bangkok, Thailand. 9 pp.

FAO, 2006. FAO SARD Initiative: People shaping their sustainable futures. http://www.fao.org/sard/en/init/index.html, January 2006.

Getz Escudero, A. and C. Neely, 2005. Exploring holistic approaches for communication in agricultural research, education and extension systems: Examining the concept of transdisciplinary integration. Working Paper. Food and Agriculture Organization, Rome, Italy. 25 pp.

Gibbon, M., 2001, The potential of transdisciplinarity. In: Thompson, J. Klein, W. Grossenbacher-Mansuy, R. Haberli, A. Bill, R. Scholz and W. Welti (editors), Transdisciplinarity: Joint problem solving among science, technology and society. An effective way for managing complexity. Birkhauser Verlag, Berlin, Germany.

Generon, 2006. The U-Process: A social technology for addressing highly complex challenges. http://www.generonconsulting.com/ttp://www.generonconsulting.com/, January 2006.

IFPRI, 2005. Proceedings of an international dialogue on pro-poor public private partnerships for food and agriculture. Washington, DC. September 28-29, 2005. 20 pp.

LEARNing Project, 2006. http://www.inra.fr/learning/learning-aims.htm, January, 2006.

Linked Local Learning, 2006. www.linkedlearning.net, January, 2006.

Lightfoot, C. and L.B. Scheuermeier, 2003. From good research to right research. BeraterInnen News 2/2003, 70-73.

Lilja, N. and M. Bellon, 2006. Analysis of participatory research projects at the International Maize and Wheat Improvement Center. Research Report, CIMMYT, Mexico. 43 pp. (in press).

McNeely, J.A. and S.J. Scherr, 2003. EcoAgriculture. Strategies to feed the world and save wild biodiversity. Island Press, Washington D.C. 324 pp.

Morris, M.L. and M.R. Bellon, 2004. Participatory plant breeding research: Opportunities and challenges for the international crop improvement system. Euphytica 136, 21–35.

Neely, C.L., K. Sutherland and J. Johnson, 2004. Do sustainable livelihoods approaches have a positive impact on the rural poor: A look at 12 case studies. FAO Livelihoods Support Program Publication No. 16. 32 pp.

Neely, C.L., 2003. Priorities of stakeholder decision makers. Working Paper, SANREM CRSP. University of Georgia, Athens, GA. 29 pp.

Norman, D., 2002. The farming systems approach: a historical perspective. In Proceedings of the Seventeenth International Farming Systems Association Symposium, Orlando, USA. 12 pp.

Pasquale, R.T. and J. Sternin, 2005. Your company's secret change agents. Harvard Business Review. May 2005, pp. 1-10.

Röling, N and A. Wagemakers, 1998. Facilitating sustainable agriculture: Participatory learning and adaptive management in times of environmental uncertainty. Cambridge University Press, New York, USA.

Savory, A. and J. Butterfield, 1999. Holistic Management: A new framework for decision making. Island Press, Washington D.C. 617 pp.

Senge, P., C.O. Sharmer, J. Jaworski and B.S. Flowers, 2004. Presence: Human Purpose and the Field of the Future. Society of Organizational Learning. Cambridge, MA. 289 pp.

Sustainable Food Laboratory, 2006. www.glifood.org, January, 2006.

UN, 2002. Report on the World Summit on Sustainable Development. 26 August-4 September, 2002. 167 pp.

Vermeulen, S., J. Williams and L.T. Phi, 2005. Tools for influencing policy. LEISA Magazine 21 Issue 4, 6-8.

Wilbur, K., 1996. A brief history of everything. Shambhala Publications, Inc., Boston, MA. 330 pp.

WRI, 2005. The wealth of the poor: Managing ecosystems to fight poverty. World Resources Institute 2005 Report. 200 pp.

# How to lay the golden eggs? Experiences from collaboration between industry and research in socially accepted poultry meat production in The Netherlands

*Jet Proost[1], Ferry Leenstra[2], Kees Lokhorst[2], Hilde van Dijkhorst[1] and Cees Leeuwis[1]*
*[1]Communication and Innovation Studies Group, Wageningen University and Research Centre, The Netherlands; [2]Animal Sciences Group, Wageningen University and Research Centre, The Netherlands*

## Abstract

Like other agricultural sectors the poultry meat production in the Netherlands faces far reaching changes for the near future. In this paper experiences are presented from 'Prolific poultry meat production', a project aiming at stimulation of socially accepted poultry meat production and processing systems in the Netherlands. Starting with a strategic problem orientation from a method called Sustainable Technology Development, project management together with farmer representatives decided halfway the project on a different strategy. An important argument to make this change was the insight that the problematic situation in the sector was not so much a matter of lack of innovative ideas, but rather one of brittle relations.

Management and farmers representatives choose this project to be placed in the ongoing developments in the sector and that 'research activities' were to be performed together with relevant stakeholders. In this paper the authors describe the dilemmas of a sector, invited to take up an active role in needs assessment and articulation of questions, and a consortium of researchers who had to take up different roles, apart from technical research. Ownership and shared commitment appeared to be essential in the implementation of ideas for the future, which were generated in a 24-hours conference with a wide range of stakeholders participating.

**Keywords:** poultry meat production, sustainable technology development, appreciative inquiry, social learning

## Introduction

In The Netherlands primary poultry production comprises approximately 4,000 farms, with about 100 million birds. The annual egg-production is more then ten billion, while poultry meat production is about 900,000 tons. Over 70% of the production is exported, mainly to EU-countries (Commodity board PVE, 2003). The sector is internationally oriented and trade is important. Within 500 km (triangle London, Paris, Berlin) 150 million prosperous consumers are present. In the Netherlands production for the world market is not cost-effective because of relative high costs of labour and other production factors. In several areas in The Netherlands a high population density coincides with a high density of production animals. This creates problems of spatial planning and environmental pollution, and it is a risk factor in introducing a disease in places with an extreme high density of poultry farms.

On the other hand, consumers in North Western parts of Europe have more and more specific demands with regard to food quality and production methods (*e.g.* an animal friendly husbandry) (Leenstra, 2005). In general Dutch poultry meat production has its major future in fresh products with a high added value. New structures are required, and chain reversal, individual entrepreneurship, chain concepts and cooperation are key issues (Leenstra, 2004).

Like other agricultural sectors the poultry meat production in the Netherlands faces far reaching changes for the near future. Economic results have been problematic for some time, and consumers are more critical towards production methods and have a strong demand for transparency. International competition has become stronger. In 2004 the Ministry of Agriculture, Nature and Food Quality funded a two-year project to induce innovation by cooperation between research and poultry sector. 'Prolific poultry meat production' is a project aiming at stimulation of socially accepted poultry meat production and processing systems in the Netherlands.

The purpose of the project was to develop a realistic vision for the future (2020) together with stakeholders in a dynamic, consumer oriented chain of poultry meat production (broilers, ducks, and turkeys), paying equal attention to people, planet and profit. Balance between and good performance on the three 'P's' (Profit, Planet and People) are the challenges for poultry production worldwide and in The Netherlands in particular (Leenstra, 2004). Issues are product quality and differentiation, reduction of ammonium emission, transparency in the production chain to prevent food scandals and disease outbreaks and the impact of intensive animal farms on the perception of quality of the rural area. Finally consumer awareness on animal welfare, food safety and animal disease problems should be reflected in their buying pattern.

The project was part of a multiyear Research Programme on new systems for animal husbandry, implemented by Wageningen University and Research centre. In the classical way researchers would have taken up the central question for innovations in this sector and formulate their own research projects. In 'Prolific poultry meat production' project management together with farmer representatives decided on a different strategy. They chose this project to be placed in the ongoing developments in the sector and that 'research activities' were to be performed together with relevant stakeholders. In an environment where research is seen as a commodity, this was an exceptional stand point. In the following the authors describe the dilemmas of a sector, invited to take up an active role in needs assessment and articulation of questions, and a consortium of researchers who had to take up different roles, apart from technical research.

## Conceptual background and methodology

In 2004 the project started in the same way as the other projects in the Research Program on designing new systems for animal husbandry. In an interactive discussion with sector representatives, the so called initiators, and during a discussion with interested people, a strategic problem orientation was performed for the poultry meat sector. Based on this result 11 cases were elaborated in a quick scan of the various problem areas.

This approach is part of the 'STD method', which stands for sustainable technology development (Bos and Wolleswinkel, 2002; Grootveld *et al.*, 2000). A main characteristic of this method is interaction which a large group of stakeholders, like relevant experts, government, and social organisations to denominate common problem definitions. An important element is the exploration of a vision for the future 25 years from now. This visioning is then brought to the present through a process of backcasting, which is followed by actual project planning. Content matter expertise is dominant in this approach and innovations are design oriented aiming at technological change, although also focus is expected on structure and cultural aspects. The researchers who were invited to write the position papers on the 11 cases were asked to involve stakeholders, but due to time constraints and also not knowing how to fit in the experiential knowledge from stakeholders resulted in rather theoretical essays.

In this way the project stayed in the domain of the researchers and the scope of this problem based approach was perceived as too limited. Another concern was that the "STD method" takes a time horizon of 25 to 40 years later, while stakeholders mostly have a time horizon of 1 to 3 years. Keeping in mind the intention that this project had to be one with the stakeholders, project management changed the strategy and left the "STD method" for the positive action approach. The focus was set on mutual learning and stimulation of discussion, which took place in a 24-hours conference where about 15 invited professionals from poultry industry (from farmers up to retail) and 15 invited professionals from research and policymaking came together to discuss the future possibilities for the sector. The papers about the 11 cases were used as background material. Strengths and opportunities for future poultry meat production systems, as seen from the various perspectives of the conference participants were starting point.

During the 24-hours conference the methodology of Appreciative Inquiry (AI) was used. AI is an affirmative approach, aiming to make people open up to find their strengths and qualities, which helps building solid common ground. It involves systemic discovery and reflection, and a collective search for what can be. Interventions may give way to people's imagination and changes never thought possible are surprisingly and democratically mobilized (Cooperrider and Whitney, 1999). Most of these approaches are grounded in a constructionist perspective, meaning that groups and organisations are perceived as living, human constructions. People's experiences are point of departure, rather than theories or models, which stimulates the learning process by which people find their way (Proost, 2006 forthcoming).

The key principles of affirmative approaches are: (1) Every system works to some degree, and key is to find out the positive, the living forces of this system, (2) knowledge generated by inquiry should be applicable, meaningful, feasible and relevant to the stakeholders, (3) systems are capable of becoming more than they are, and (4) the process and outcome of inquiry are interrelated and inseparable, and make the process a collaborative one. An AI process usually comprises of four stages: Discovery (what gives life, appreciating the best of what is), Dreaming (what might be, envisioning impact the world is calling for), Design (what should be, co-constructing the ideal) and Destiny (sustaining for empowerment, learning and adjustment).

During the 24-hours conference the participants expressed their appreciation of the strengths of the poultry sector. For more than 50 years the sector survived and provided an income for numerous farming families and agri-businesses. The dreaming phase resulted in lots of ideas, which were discussed and amalgamated. Finally the dreams led to nine major ideas of which, after a selection process, seven had an owner from industry.

In this project multiple stakeholders were involved who share interdependencies in order to find common ground, the basis for collective action. Quite often these stakeholders operate from their respective domains and bring in existing networks, with sometimes powerful vested interests. In addition multiple stakeholders act at different scales. Therefore in this methodology meaningful conversations are very important to share information, but also to build trust. The dialogues during the conference opened contacts and new relationships were built. In strategies for social learning and innovation processes, networking among multiple stakeholders around collective action is a key element. It is often the shared action that defines the network. Building networks has been recognised as one of the most important elements of social learning in innovation processes (Leeuwis, 2004).

There are two major differences between the two approaches used: the "STD method" and Appreciative Inquiry:

1. Time dimension: it appeared problematic for stakeholders to make a jump in time of 25 years or more, while pressing problems ask their attention in the present. In addition they have ideas on how to solve these pressing problems and they want to go into action straight away.

2. Relation between stakeholders and experts. In the "STD method" the process is expert driven, while in the positive action approach ownership of the process is with the stakeholders. It is a continuous process with built in moments for reflection, to see where the stakeholders are at and how people experience the process so far. This is the basis for formulation of new actions.

After the 24-hours conference project management was occupied with two questions in the process: how can we keep the positive energy on the sub-project initiatives and how can we keep an overview on what is happening in these sub-projects.

## Ownership and shared commitment

In the previous paragraph is explained that over time the strategy in the 'Prolific poultry meat production' project was changed. There were several arguments for this change of strategy, but one very important was the insight that the problematic situation in the sector was not so much a matter of lack of innovative ideas, but rather one of brittle relations. The focus of the project gradually changed from technological change towards social change. Ownership and shared commitment became the key elements to work on.

To protect ownership of ideas and prevent researchers or policymakers 'walking away' with ideas generated from dialogue in the conference, the rules in the project were very strict. The contribution of researchers in developing the ideas towards realisation was financed under the strict condition that researchers could only work on questions formulated by stakeholders. In the 12-month period after the 24-hours conference the project ideas were elaborated. Some resulted in a 'project', others merged into other initiatives, and finally a few proved not to be feasible. Two factors appeared essential in further stature: ownership and shared commitment.

People who take ownership show responsibility for the task taken upon them. They are committed to the task, because they believe in a project or an action. They are motivated from a personal belief that bringing an effort is relevant for them and their status in the group. According to Senge commitment builds when people are an active part of the experience of creating something they value together. Human beings like to see tangible results of their efforts (Senge, 1996). In 'Prolific poultry meat production' stakeholders lost track of what others were doing and what all endeavours created for the future of the sector. Furthermore some stakeholders got the feeling that they had to do it on their own, they felt a big gap between demand and supply of stakeholders and research. Some showing commitment, joined other initiatives in the sector where they could find a better linkage between their input and the results related to their concerns. Others reduced the scope of their project to something they could handle together with day to day activities on their own farm or in their company. Nevertheless new connections between partners were formed, and positive results or new insights were 'harvested'.

The issue of shared commitment also emerged during in depth interviews with people from the sector about the added value for poultry meat products. These interviews showed different ways of thinking and strategies which stakeholders have. Some have the opinion that individuals or groups can make a difference and that people should be proactive. Others see the market as decisive factor in order to solve the problems farmers are facing. These belief systems have substantial implications for the feeling of shared commitment and on the ownership of ideas/ activities or projects. In this project these differences were not further elaborated.

Although theory offers insights for the follow up of an AI conference, the project logic of 'Prolific poultry meat production' didn't specifically pay attention to it. A more laissez-faire approach was taken up for the follow up, to refrain researchers from directing sub-projects, but this resulted in connections between people and sub-projects that were too loose to bond, enforce and stimulate.

## Conclusions and discussion

In the project 'Prolific poultry meat production' industry and research pooled their forces to look for innovations with promising prospects for poultry production of broiler, turkey and duck. In the two year period of this project some interesting lessons emerged, which can be of use in new projects. The experiences of both stakeholders and researchers with this process were evaluated in personal interviews and in workshops. Some of the lessons learned are presented here:

- The decision to change the project strategy halfway proved to be successful. The positive action approach fitted better to the experiences and ideas of the stakeholders. The issue was more about social change than on technological change. Connections between people for collaborative action proved to be more important than the "golden egg".
- One clear conclusion from the evaluation workshop with researchers was, that industry does not have sufficient insight in what research can provide them with, or that they have problems with the articulation of their questions. Consequently besides articulation of the questions from industry, also display of services by research institutes is essential. Services can be facilitation, support and organisation in projects and use of researcher's networks, also abroad. Agricultural research institutions in The Netherlands have currently a set up that hampers flexible response to questions from industry.
- Then there is the question of representation. People who usually accept invitations for workshops and conferences may meet the same faces over and again. People who show up act in personal capacity, while commitment of their organisation of institution cannot be taken for granted. To acquire this commitment specific interventions are needed.
- Some stakeholders afterwards had the feeling they were forced into leadership of a project (as a results of a very energetic 24 hours conference), calling it stakeholder participation. Cooperation between stakeholders and researchers would have been a better focus than the reversal of roles.
- The 24-hours conference was well appreciated by the participants. They even used it as an example in other activities. Straight after the conference people were anxious to start project activities, but in every day reality their other chores took up all attention. Because of the lack of proper follow up they lacked the inspiration from other projects and felt that they alone had to accomplish a task, where many others, not participating, would benefit from.

The example of 'Prolific poultry meat production' provides us with information where spaces for innovative action emerge and how they can be used effectively for interaction and collective action. Representatives of the sector and of NGO's cannot be forced into experiential learning. Positive experiences will make them proceed. However, researchers and knowledge institutes (at least in the current situation in The Netherlands) are in different positions: research funding will be increasingly dependent on the ability of researchers and their institutions to interact with stakeholders and indeed work together in formulating and solving problems in an interactive, non-linear way, instead of formulating research projects and carrying those out as a kind of 'table-tennis' were questions fly in one direction and solutions in the other.

## References

Bos, B. and A. Wolleswinkel, 2002. Onderzoek en Transitie, verslag en conclusies naar aanleiding van een workshop op 15 februari 2002. Wageningen UR, Animal Sciences Group.

Commodity Board of Livestock, Meat and Eggs, 2003. Livestock, meat and eggs in The Netherlands. www.pve.nl.

Cooperrider, D.L. and D. Whitney, 1999. Appreciative Inquiry: A positive revolution in change. In: The Change Handbook. Group methods for shaping the future. P. Holman and T. Devance (eds.) Berret-Koehler Publishers, California, USA, pp. 245-261.

Leenstra, F., L. den Hartog, I. Enting, *et al.*, 2004. Poultry production in the 21st century: Future of Dutch poultry industry, results of a national debate. Paper for the XXII World's Poultry Congress, Istanbul, June 8-13, 2004.

Leenstra, F., 2005. Future of Dutch poultry industry, results of a national debate. In: Feedinfo News Service, 11/02/2005. www.feedinfo.com.

Leeuwis, C., 2004. Fields of conflict and castles in the air. Some thoughts and observations on the role of communication in public sphere innovation processes. Journal of Agricultural Education and Extension 10, 63-76.

Proost, M.D.C., 2006. Connectivity and inspired action for change and transformation. In: Scaling up, Social Learning, Social and Institutional Networks in Rural and Farming Systems. Concepts and methods for setting Agenda in Participatory Research and Extension. B. Hubert, *et al.* (eds.). INRA, France (in press).

Senge, P., A. Kleiner, C. Roberts, *et al.*, 1996. The fifth discipline fieldbook. Strategies and tools for building a learning organization. Nicholas Brealey Publishers, London.

Van Grootveld, G., L. Jansen, E. van Spiegel, *et al.*, 2000. Sustainable Technology Development.

# "Ownership" in trans-disciplinary projects

*S.F. Spoelstra, P.W.G. Groot Koerkamp and A.P. Bos*
*Animal Sciences Group, Wageningen UR, The Netherlands*

## Abstract

In projects aiming at contributing to sustainable development of livestock production, ownership was identified to play a crucial role. This applies to formal ownership in financial terms, its perception as well as ownership in terms of motivation and commitment of stakeholders involved. Examples based on projects are given and discussed.

## Introduction

The Dutch Ministry of Agriculture, Nature and Food Quality commissioned two programmes to Wageningen University and Research centre aiming at contributing to sustainable development of the livestock production sector. The first programme was carried out in the years 1999-2002 (Spoelstra *et al.*, 2002; Grin *et al.*, 2004) and the second runs from 2003 to 2006. These programmes find their origin in the societal debate on issues associated with livestock production including environmental impact, animal welfare, animal health, epidemic animal diseases and food safety and the subsequent policy formulated by the Ministry. The projects carried out in the framework of these programmes aimed at influencing target groups. These included the primary sector and other relevant actors. In the Dutch context the goals of the programmes are referred to as contributing to "system innovation" or "transition" in the livestock sector. These notions refer to the idea that radical new ideas and concepts are needed for sustainable development to overcome the problems that have arisen from "simple modernisation" (Beck, 1997).

In the projects carried out, a pattern in the research approach gradually developed. Projects were carried out in a trans-disciplinary interactive setting involving researchers from various disciplinary backgrounds as well as partners from industry, NGOs and the government. Secondly, "top-down" as well as "bottom-up" approaches were followed. Top-down approaches typically comprised an interactive design of visions of a desirable future (Spoelstra *et al.*, 2002) and future artefacts ("concepts"), including animal housing systems and monitoring systems for animal health and welfare. Approaches chosen were inspired by various recently developed theories on the role of technology in sustainable development such as Sustainable Technological Development (Weaver *et al.*, 2000), constructive technology assessment (Grin *et al.*, 1997), interactive societal design (Leeuwis, 1999) and Strategic Niche Management (Roep *et al.*, 2003). The function of the visions and artefacts developed were meant to be at least twofold. Firstly, by involving stakeholders in the interactive design process; their demands and values should become included in the concept. Secondly, the concepts function as sources of inspiration for bottom-up initiatives ("examples"). The linking of concepts with bottom-up initiatives occurred by eliciting, by back-casting or by identifying existing fitting initiatives. In all cases the examples were meant as a source of inspiration to create a multi-level learning environment. Some projects were based solely on identified existing sustainable ways of farming developed by farmers themselves ("novelties"), which could serve as bottom-up initiatives for further development.

It was realised that, in our market-oriented society, it is predominately the private sector that implements and develops sustainable ways of production. From this point of view it was felt essential that notably the bottom-up projects were aligned with the motivation of private partners in the projects. In the projects this motivation was often referred to as"ownership" and also comprises elements as commitment and entrepreneurship along with economic ownership in the sense of availability of budget. In the projects several mechanisms were identified to play an important role in advancing or hampering the progress of projects that were associated with ownership. We will shortly introduce some examples to illustrate this.

## Programme level

The programmes were commissioned by the Ministry to the research organisation as part of a government policy to stimulate sustainable development of the livestock sector. Advice on the research programme was obtained by a steering board installed by the Ministry. In addition to representatives from the Ministry, members of the steering board represented organisations in the livestock sector (commodity boards, farmers' organisations), NGOs, some private enterprises and research organisations. In an enquiry about goals and an assessment of the quality of the programme it was revealed that the members of the steering board did not consider this their task. Representatives of the livestock sector (target groups) considered this the prerogative of the Ministry who had commissioned the programme (financial ownership). In contrast, civil servants of the Ministry in the steering board took the position that this should be done by the representatives of the livestock sector, the target group of the programme. Consequently, both the livestock sector and the Ministry became only distantly involved in the programmes. Thus a vacuum in ownership was created, which led to *de facto* ownership of the research team having the budget as well as the motivation to perform the trans-disciplinary research. Though the programme aimed for interactivity this was not achieved on the level of the steering board. Here the familiar traditional situation was maintained in which research is considered an autonomous societal activity often referred to as Mertonian science (Merton, 1973).

## Top-down projects

The general non-commitment of the steering board demanded that separate steering or guidance groups were formed in the projects that aim at the development of concepts by interactive design. These in general showed more involvement, but in nearly all projects the research team was dominant. Thus, despite the focus being on stakeholder participation and the intention to reach some closure for designed concepts, the eventual result was felt by all participants to be very much a product of the research team. Again situations arose in which the research teams and the results were pushed back into their "autonomous domain".

It was felt that this situation limited the chance for follow-up activities and further experimentation and learning in a practical context. In the research teams this was referred to as the problem of ownership and more precisely the problem of how to shift ownership from research to private parties.

With hindsight, three routes can be distinguished which were attempted to shift ownership.

***Tendering***
Consortia of private and non-governmental organisations were invited to submit their plans inspired on the concept developed for research support to further development. This approach has led to some success.

In one project an animal welfare organisation, supported by animal breeding companies, went for an interactive design of a piggery system primarily based on animal welfare. The animal welfare organisation involved adopted the final concept for a Comfort Class piggery (De Greef *et al.*, 2003) they found support for further development and implementation by the pig farming section of the national farmers' organisation. Together they advocated the ideas and raised budgets to build a proof of principle Comfort Class piggery. However, the budgets being from public sources precluded experimenting in the private sector. Thus the proof of principle experiment was carried out on the premises of a research institute. In fact as well as in the perception of the stakeholders the proof of principle tended to return thereby to the classical research domain. To maintain a connection with practical farming, a learning network of pig farmers was established around the proof of principle experiment. One of the farmers involved is now planning to make investments along the lines of the Comfort Class piggery.

In the example of the Comfort Class piggery a proof of principle was seen by stakeholders as a necessary step between concept and practical adaptation. When technical risks of a concept are high, verification and further development in a research environment may be indicated. It was found, however, that a proof of principle had several drawbacks. Firstly, it shifts ownership back to the research domain. Secondly, it tends to halt the development, because the target group tends to sit and wait for research results and thus momentum is lost. Thirdly, only a limited set of conditions can be investigated in a proof of principle and bears the danger that the system tested becomes normative for adaptation instead that the original concept is adapted for learning.

### *Eliciting*

In the project "Laying Hens Husbandry (Projectgroep Wageningen UR, 2004) two concepts for the housing of layers were interactively designed based on wishes and demands of birds, citizens/consumers and farmers. Though 3000 poultry farmers frequently received information on the project through a farmers' magazine, and 30 of them indicated a wish to be actively involved in the project, none of them intended spontaneously to invest in such a sustainable development immediately after the project. Despite the fact that the project explicitly posed the question 'To whom can we pass on the baton?', leaving the action for the next step to the poultry sector, the project team carried out several activities to stimulate private parties. For this an intensive road show was carried out, presenting results and ideas to several groups of farmers, contributions and presentations to several occasions (fairs, congresses, commodities, student groups, childrens' television programme). An inventory into opinions on and effects of the project was performed among 50 people from the sector, politics, societal organisations and science. Further publications in national newspapers and farmers' magazines were issued and there were many personal contacts and discussions, especially by the project leader. All this certainly contributed to an incubation of ideas and perspectives.

Eventually four farmers actively showed their interest in a follow-up of the project: they planned to invest in new facilities and adapt ideas from the concept and each of them applied for development funds. Their plans were published in several public media, which resulted in the attraction of other business parties form the poultry sector, such as farmers, egg-trading companies, developers of housing systems, veterinarians, breeding companies, *etc.* Probably two facts were essential in their movement: the technical feasibility of one or more of the ideas (*e.g.* the solution how to build a round poultry house), as well as the awareness that a new market-driven chain could be arranged (with limited cost increases and additional revenues).

The concepts and ideas deviated from standard laying hen housing building routines, and standard material could not be used and had to be redesigned and in part be custom-made. Also conflicts with existing regulations on a municipal level had to be overcome. The farmers involved searched financial support from funds supporting innovations in agriculture to cover the extra costs and risks. Initially, also severe contradictions between such an innovative approach and the strict rules for funding based on a linear stepwise innovation model (*e.g.* innovations are developed and tested in a research environment) were encountered. Thus alignment was obtained between farmers' interests and a learning situation in practice. The role of research has now become threefold: designer to develop specific solutions for unique situations, discussion partner and supplier of knowledge to farmers involved, identifying and addressing structural constraints, *e.g.* in laws and regulations and supporting a network of learning among farmers.

### *Identifying*

In the projects mentioned above, the interactive design of a concept preceded the bottom-up phase. In a number of projects no concept was developed. The starting point in these projects was to identify farmers that one way or another had developed a more sustainable way of farming. By a kind of "hearsay" and "snowball" approach innovative farmers were identified, contacted and interviewed on their novelties (Wolleswinkel *et al.*, 2004). Around selected novelties networks of interested farmers were formed to improve the understanding and recognition of their novelties and to identify and possibly lift constraints for further development. Bottom-up concepts were developed by explicating the novelties, involving and addressing other stakeholders.

The research team had the ambition to support the development of the farmers and not to take over the lead. This resulted in discussions between farmers and research team on payments to be made to farmers for their involvement, on leaving the initiative (ownership by farmers) to farmers for defining actions and stakeholders to address. This resulted very limited involvement of other stakeholders than farmers in these networks, thereby limiting the learning capacity of the network.

## Lessons learned

The general goals in projects can be characterised as developing new concepts and examples in practice that fit in sustainable development. It was considered essential to involve stakeholders to such an extent that further actions were taken by the target group after the project had been finished. Alignment of the motivation of specific target groups and trans-disciplinary research activities were considered essential.

However, on all levels (programme level, top-down projects and bottom-up projects) this appeared often difficult to achieve.

On all levels the general assumption by stakeholders was that research - even if explicitly and repeatedly communicated otherwise - is a separate domain. Notably the combination of government commissioning to a research organisation and familiarity with the Mertonian research model led to a strong attitude of distant involvement.

In general, a similar perception on the role of research was present in bottom-up projects. Stakeholders tended to interpret concepts as expert knowledge rather than sources of inspiration and by the unconventionality perceived of concepts; these concepts were often rejected as unpractical. Government and organisations of the sector tended to interpret bottom-up experiments as forerunners with others to follow. This again being a familiar research situation from the past when agricultural development was mainly based on disciplinary research and driven by technology.

In some instances the fact of government commissioning in itself raised suspicion of hidden policy targets and the possibility of future government enforcement of the concepts involved. The main fear being that participation could implicate cooperation to a further decreasing competitiveness of the sector due to even more regulations on societal issues, increasing the costs of production. This situation is very recognisable, because most of today's legislation is based on earlier research.

Tendering appeared to be the most effective way to shift ownership on to the level of steering groups. Most often it concerned representatives of organisations that have a clear mission (as a person and/or an organisation) and breaks with the more bureaucratised routines of traditional institutions of the livestock sector. In case of NGOs difficulties have been seen in shifting ownership further to entrepreneurs.

The ownership in bottom-up projects did not seem dependant on how these projects were formed or identified. The decisive element is probably the direct linkage with persons that express their motivations to be realised and their problems to be solved.

In these projects it was obvious that learning of the concepts occurred in the projects. New ideas were exchanged and picked up in the networks. However, at the moment actors involved considered any investment they "became" entrepreneurs, and developed strong preference for exclusivity as shown by a more distant position to the research team and by looking for subsidies to meet the additional costs of investment.

## References

Beck, U., 1997. The Reinvention of Politics. Polity Press, Cambridge.

De Greef, K.H., W.G.P. Schouten, C.M. Groenestein, R.G. ten Hoope and M. de Jong, 2003. Husbandry systems from the animal's point of view: Pigs. In: Book of Abstracts of the 54th annual meeting of the European Association for Animal Production.

Grin, J., H. van de Graaf and R. Hoppe, 1997. Interactive technology assesment. Rathenau Institute, The Hague.

Grin, J., F. Felix, B. Bos and S. Spoelstra, 2004. Practices for reflexive design: lessons from a Dutch programme on sustainable agriculture. Int. J. Foresight and Innovation Policy 1, 126-149.

Leeuwis, C., 1999. (Ed.) Integral design: innovation in agriculture and resource management. Mansholt Studies 15. Mansholt Institute / Backhuys Publishers: Wageningen/ Leiden. 277 pp.

Merton, R.K., 1973. The sociology of science: theoretical and empirical investigations. University of Chicago Press.

Projectgroup Wageningen University and Research Centre, 2004. Laying Hens Husbandry – towards a happy hen life, proud farmers and a satisfied society. Wageningen UR, Lelystad / Wageningen, 30 pp.

Roep, D., J.D. van der Ploeg and J.S.C. Wiskerke, 2003. Managing technical-institutional design processes: some strategic lessons from environmental co-operatives in the Netherlands. Neth. J. Agric. Science 51, 195-217.

Spoelstra, S.F., J.W.A. Langeveld, P.W.G. Groot Koerkamp, J. Luttik, G.B.C. Backus and J. Grin, 2002. Sustainable Technology Development applied to livestock industry in the Netherlands. In: Pre-proceedings fifth IFSA European Symposium. Farming and rural systems research and extension. Florence (I), April 8-11. P90-99. Facolta di Agraria, University of Firenze, Florence, Italy.

Wolleswinkel, A.P., D. Roep, K.J. van Calker, S.J.G. de Rooij and F.P.M. Verhoeven. 2004. Atlas van innoverende melkveehouders. Report Wageningen UR, Wageningen. 163 pp.

Weaver, P., L. Jansen, G. van Grootveld, E. van Spiegel and P. Vergragt, 2000. Sustainable Technology Development. Greenleaf Publishing, Sheffield.

# Coupling goal-based modelling to farm innovation

*Barbara Sterk[1] and Martin K. van Ittersum[2]*
*[1]Plant Production Systems and Communication and Innovation Studies, Wageningen University, The Netherlands; [2]Plant Production Systems, Wageningen University, The Netherlands*

## Abstract

Theoretically, cross-pollination between prototyping research and goal-based farm modelling seems beneficial for farm innovation. However, there are no practical examples of the integration of the two methodologies. We analyzed two case-studies, one in the Netherlands and one in Uruguay, to reveal the hindrances and to identify viable functions for goal-based farm modelers in farm innovation processes.

## Introduction

A capacity for mobilizing and using science and technology, *i.e.* linking knowledge to action, is increasingly recognized as an essential component of strategies for promoting sustainable development (Kates *et al.*, 2001). Both at a philosophical and experiential level, the consequences of this view for the nature of research have been explored (Cash *et al.*, 2003; Reece and Sumberg, 2003; Edward-Jones, 2001; Keating and McCown, 2001; Funtowicz and Ravetz, 1993). One of the crucial factors that has allegedly been missing from much agricultural research has been communication between researchers and the users of the research. One potential way to alleviate this problem is to encourage a greater degree of participation in the research process. In this frame, we focus here on one specific group, researchers using computer modelling as their main methodological tool.

In large parts of the world, farmers are stimulated to innovate their farm strategies in order to retain a 'license to produce'. Many modelers of agricultural systems have claimed potential usefulness of their work in innovation processes (for an elaborate discussion of these claims see Keating and McCown, 2001; Cox, 1996). Since the 1990s, cropping system models have been successfully used in the farming context. Particularly, the ASPRU group in Australia has been involved in studies to examine which biophysical and social factors have to be considered in making generic simulation models applicable to location-specific problems and appealing to farmers with farm-specific interests and issues (Carberry *et al.*, 2002; McCown, 2002). However, less progress was made in the use of *goal-based* modelling at *farm level* in innovation processes. Cropping system models are particularly powerful in addressing plot scale issues, or for analysis of relatively simple cropping systems, comprising only a few crops. They are, however, less suitable for redesigning entire farming systems and complex crop rotations, in which yield-defining, yield-limiting and yield-reducing factors (Van Ittersum and Rabbinge, 1997) strongly interact and determine ultimate production options. Goal-based farm models may be defined as objective-oriented tools for exploration of alternative, promising farming systems. Directed by the formulation of conflicting objectives (*e.g.*, economic profit, minimal nitrogen leaching) and the potentials of a system, the consequences for farm management of a preference for one of the objectives are explored. Thus, farm design is a consequence of objective specification and the potentials of the system. Goal-based modelling studies were carried out in the Netherlands and elsewhere (*e.g.*, Dogliotti *et al.*, 2005; Veysset *et al.*, 2005; Ten Berge *et al.*, 2000).

Parallel, the prototyping methodology was developed to support the generation of innovative farming alternatives on-farm in Dutch agricultural research (Vereijken, 1997). The main features of prototyping are: (i) quantification of multiple societal goals; (ii) designing as organizing principle; (iii) design ánd on-farm testing. The methodology was implemented in a number of Dutch and two EU-funded projects on both experimental and commercial farms. Theoretically, prototyping research and farm system modelling could benefit from cross-pollination. Incorporation in prototyping projects would offer farm system modelers opportunity to connect model results to action. On the other hand, the prototyping approach could benefit from modelling as well. Farm system models provide means to expand and formalize expert knowledge and to integrate this and scientific agro-ecological knowledge at farm level. These model qualities could reveal design (im)possibilities otherwise perhaps overlooked. Yet, we observe that little cross-pollination between the modelling and prototyping efforts took place in practice even though the methodologies were applied in parallel and in one country. This observation fits earlier research (Sterk *et al.*, 2006) in which we found that it is not trivial to find niches for the application of goal-based farm models. The type of outcomes (design options) of goal-based models appeared not to match questions of farm managers optimizing their farm performance in a frequently observed manner, *i.e.* monitoring and learning from their own and other farmers' practices. In this earlier research, we focused on the potential link between emerging questions in innovation processes to goal-based modelling. However, the conclusions of this earlier research do not exclude the possibility that other applications, such as the evaluation after a project, or extrapolation to other conditions of a design, could be interesting for innovators. It is these alternative purposes of goal-based modelling that we would propose to explore in this paper.

## Case studies in the coupling of goal-based modelling to farm innovation

To identify viable, alternative applications for goal-based models, two supplementary case studies were investigated: (i) prototyping practice in the Netherlands and (ii) a recent effort to use a goal-based farm model in a Uruguayan on-farm innovation project. As observed in the introduction, cross-pollination of prototyping and goal-based modelling failed to occur in the Netherlands. However, though the methodology was extensively documented, the practice of prototyping has never been described. Consequently, it is impossible to make any statements about the reasons for the lack of cross-pollination, nor is it feasible to develop strategies to stimulate cross-pollination in the future on basis of the existing documentation; hence the selection of the first case study. The second case study is similar to the first in the sense that both concern collaboration of scientists and farmers to test ideas on-farm. But in the Uruguayan case study it was the same person coordinating the on-farm innovation project and modelling. In the Netherlands, modelers and project workers were different persons and usually institutionally separated. For this difference, the combination of the two case-studies enriches the analysis.

To collect the data, qualitative research methods were applied: semi-structured interviews, observation and study of internal and external project documentation. The researchers attended gatherings, such as project team meetings and bilateral encounters between project team members and farmers (so called kitchen table meetings). Informal conversations (*i.e.*, unstructured interviews) were held with the participants during and after these activities to uncover their interpretations of what was going on.

### *Prototyping practice in the Netherlands*

The first Dutch experiences with the prototyping approach originate from the early eighties when three arable farming systems –integrated, organic and conventional- were laid out on an experimental farm in Nagele in the Flevopolders. The main reason for the initiative had been a call for an "entire farming system approach" (El Titi, 1992) in reaction to limited adoption of Integrated Pest Management by farmers. After a decade of research at the experimental farm in Nagele, two projects with commercial farms were initiated, *i.e.*, an Integrated Arable Farming and an Organic Arable and Open-field Vegetable Farming project. Prototyping on other experimental farms and more projects with commercial farms followed (Langeveld *et al.*, 2005). In the prototyping projects with commercial farmers, three stages can be distinguished, labeled: (i) acquaintance stage; (ii) experimenting stage; (iii) external communication phase. The acquaintance stage encompasses the large investments in credibility building between the project team and participating farmers and amongst the project team and farmers' groups themselves. The experimenting stage is characterized by intensive and regular communication between the project team and farmers about possible improvements, actual activities and effects and hence has a strongly localized character. In these first two stages no computer models are used. Stage 3 consists of all efforts to communicate and extrapolate project output to those interested apart from the project team and participating farmers. In this stage, issue-specific simulation models, addressing questions such as the relation between nitrate leaching and nitrate in groundwater, are increasingly mobilized.

***On-farm experimentation in Uruguay***

In Uruguay, both experts and farmers perceive physical and biological soil fertility as an important yield-limiting factor. As part of his thesis work, Dogliotti developed a goal-based farm model (Dogliotti *et al.*, 2005) to explore options for improvement of vegetable farmers' income in the Canelón Grande micro-watershed in South Uruguay, while reducing soil erosion and improving physical and biological soil fertility simultaneously. In 2004, Dogliotti and several colleagues initiated an on-farm innovation project to address these same issues in an experiential manner in collaboration with commercial farmers. The methodological approach bears resemblance with Dutch prototyping practice. Although the first field year just finished, elements of the acquaintance and experimenting stage (see previous paragraph) can be recognized already. Dogliotti intends to somehow integrate his earlier work with goal-based modelling in the project. Till date, the complete goal-based farm model was not mobilized. However, though not planned, Dogliotti used one component of the model generating crop successions to develop farm-specific rotation plans. These plans were then evaluated with the entire project team and subsequently proposed to the participating farmers. The project team discussed the plans without questioning the model behind it. The farmers were unaware of the fact that a model component was used in the planning.

## How goal-based modelling can contribute to farm innovation

In an earlier paper, we hypothesized that the chances of capitalizing on modelling expertise are likely to be higher when researchers with such expertise are a logical and more or less permanent component of ongoing trajectories than when these researchers come from outside to purposefully search for a niche. Comparison of the Dutch prototyping and Uruguayan on-farm experimentation cases provides some initial support for this hypothesis. The analysis of Dutch prototyping practice reveiled that in a collaboration with farmers, credibility building requires a large investment. The introduction of a model(er) requires credibility development as well, both with the project team and with the farmers. In Uruguay, Dogliotti started to use a model component, unplanned and without meeting major objections of the rest of the project team. These observation suggest that the investment is considerably lower when the modeler is part of the project team. Besides, he could only take this unplanned initiative because he had responsibilities in the project other than modelling. As yet, this analysis can merely serve as focus for further research because the Uruguayan project is still in its infancy. Moreover, also in the Uruguayan case the complete goal-based model was not mobilized yet. However, there seem good possibilities that this will happen in the future.

## Acknowledgements

This work was supported by the C.T. de Wit Graduate School for Production Ecology and Resource Conservation and the Mansholt Graduate School. A travel scholarship was granted by the Netherlands Organization for Scientific Research.

## References

Carberry, P.S., Z.Hochman, R.L. McCown, N.P. Dalgliesh, M.J. Foale, *et al.*, 2002. The FARMSCAPE approach to decision support: farmers', advisers', researchers' monitoring, simulation, communication and performance evaluation. Agricultural Systems 74, 141-177.

Cash, D.W., W.C. Clark, R. Alcock, N.M. Dickson, N. Eckley, *et al.*, 2003. Knowledge systems for sustainable development. Proceedings of the National Academy of Sciences 100, 8086-8091.

Cox, P.G., 1996. Some issues in the Design of Agricultural Decision Support Systems. Agricultural Systems 52, 355-381

Dogliotti, S., M.K. Van Ittersum and W.A.H. Rossing, 2005. A method for exploring sustainable development options at farm scale: a case study for vegetable farms in south Uruguay. Agricultural Systems 86, 29-51.

Edward-Jones, G., 2001. Should we engage in farmer-participatory research in the UK? Outlook on Agriculture 30, 129-136.

El Titi, A., 1992. Integrated farming: an ecological farming approach in European agriculture. Outlook on Agriculture 21, 33-39.

Funtowicz, S. and J.R. Ravetz, 1993. On Post-Normal Science and its Ethical and Epistemological Implications. Futures 25, 735-755.

Kates, R.W., W.C. Clark, R. Corell, J.M. Hall, C.C. Jaeger, I. Lowe, J.J. McCarthy, H.J. Schellnhuber, B. Bolin, N.M. Dickson, *et al.*, 2001. Sustainability science. Science 292, 641-642.

Keating, B.A. and R.L. McCown, 2001. Advances in farming systems analysis and intervention. Agricultural Systems 70, 555-579.

Langeveld, J.W.A., H. Van Keulen, J.J. De Haan B.M.A. Kroonen-Backbier and J. Oenema, 2005. The nucleus and pilot farm research approach: experiences from The Netherlands. Agricultural Systems 84, 227-252.

McCown, R.L., 2002. Locating agricultural decision support systems in the troubled past and socio-technical complexity of 'models for management'. Agricultural Systems 74, 11-25.

Reece, J.D. and J. Sumberg, 2003. More clients, less resources: toward a new conceptual framework for agricultural research in marginal areas. Technovation 23, 409-421.

Sterk, B., M.K. Van Ittersum, C. Leeuwis, W.A.H. Rossing, H. Van Keulen and G.W.J. Van de Ven, 2006. Finding niches for whole-farm desing models – contradictio in terminis?. Agricultural Systems 87, 211-228.

Ten Berge, H.F.M., M.K. Van Ittersum, W.A.H. Rossing, G.W.J. Van de Ven, J. Schans and P.A.C.M. Van de Sanden, 2000. Farming options for The Netherlands explored by multi-objective modelling. European Journal of Agronomy 13, 263-277.

Van Ittersum, M.K. and R. Rabbinge, 1997. Concepts in production ecology for analysis and quantification of agricultural input-output combinations. Field Crops Research 52, 197-208.

Vereijken, P., 1997. A methodical way of prototyping integrated and ecological arable farming systems (I/EAFS) in interaction with pilot farms. European Journal of Agronomy 7, 235-250.

Veysset, P., D. Bebin and M. Lherm, 2005. Adaptation to Agenda 2000 (CAP reform) and optimization of the farming system of French suckler cattle arms in the Charolais area: a model-based study. Agricultural Systems 83, 179-202.

# Chances of success in investing all AKIS/RD's efforts in informal linkages under privatised extension conditions

*Kirsten von der Heiden*
*AfoReg, Germany*

## Abstract

When investigating solutions and model approaches to sustainable knowledge exchange on agro-environmental issues and experiences in the Agricultural Knowledge System - AKS in Brandenburg, Germany, the specific privatised extension conditions have to be taken into consideration. Aiming to learn from the strengths and weaknesses of privatised extension systems within AKS in other countries, literature studies and expert interviews have been drawn on. This paper focuses on learning from the Dutch for theoretical further development of the concept for the Brandenburg Agricultural Knowledge and Information System for Rural Development - AKIS/RD, within the author's current doctoral thesis in progress. The solution can mainly be found within the AKS as a whole, combined with policy level solutions. Ways of overcoming the significant lack of linkages and omission of functions will be proposed, *e.g.* by establishing informal co-operation of potential partners and networking activities including new role divisions; assessments for future development will be addressed to administrative bodies, researchers and intermediaries. The contemporary issue of EU top-down demands for Cross Compliance extension particularly poses a challenge to privatised systems. Newly systemic approaches are the only sustainable chance of success, including the establishment of ongoing collective learning discourses, the joint multi-stakeholder decision-making towards an "AKIS/RD in life" for Brandenburg.

**Keywords:** AKIS/RD, privatisation, roles/linkage gaps, Cross Compliance extension

## Introduction: the problem

Rivera (2004) proposes AKIS/RD as a further development of AKIS, stressing linkages between the three knowledge institutions: research, extension and education, all related to farmers, and highlighting improvement of co-ordination and networking. AKIS/RD follows the paradigm that agriculture is an aspect of rural development. Compared to other AKS approaches, it puts forward the idea of "systems of co-ordinated rural knowledge and information networks aimed at diffusing development information on non-agricultural subjects" Rivera (2004). Following Sánchez (1996), even to characterise an extension system, diverse parameters would be needed; therefore the most important indicators must be identified. Specific regional and national characteristics have to be taken into account. These criteria, with national examples, might help to identify the key issues for future investigation on an internationally comparable assessment framework which Alex *et al.* (2004) are searching for. The case based problem which is the subject of the paper focuses on suggestions for overcoming inadequacies of linkages and functions in the Brandenburg AKS in achieving sustainable knowledge exchange on agro-environmental issues and experiences.

## Methodological background

The Agricultural Knowledge System - AKS analysis of the Brandenburg system status in 2002 (specified in Nagel *et al.* 2003, Von der Heiden 2004) highlights strengths and weaknesses for future knowledge exchange on agro-environmental innovations. Such analysis has been taken as a starting point for literature studies of international AKS cases, meeting the challenge of privatisation, and expert interviews with experienced national researchers. Part of the solution of the difficulty of handling the manifold differences in framework conditions between international knowledge systems when drawing conclusions from literature study to the Brandenburg case has been to add a feedback system of assessment by respective national experts. Within the author's current doctoral thesis in progress, titled "Public Goods and Privatised Extension (...) - reflections beyond Brandenburg's AKS", an adaptable, semi-structured interview schedule has been developed for flexible national-specific use. Main interview topics have been: specifications, strengths and weaknesses of international extension/ AK systems, new tendencies, advice to other AKSs, integrated versus specialised advisory approach to agro-environmental topics, sustainability and privatisation. Relevant literature was studied before devising the interviews (with experts from the Netherlands in 2001, South Australia/New Zealand in 2003-2004) for knowledge systems in specific national situations. The experts interviewed are familiar with the privatisation processes of the extension systems in their countries so far, as well as being in contact with administrative bodies, advisory services and other organisations involved. Thus, international examples of privatised agricultural extension within AKS, relevant literature and, to reassess written statements, expert estimations are taken as feedback for the author and input to theoretical further development of the Brandenburg case within the overall concept of Agricultural Knowledge and Information System for Rural Development - AKIS/RD. The classification system of Rivera (2001) is taken as the basic principle for analysis of international extension systems, including the characteristic indications for purely privatised extension as "private funding and private delivery".

Table 1 gives an overview of criteria chosen as important for international extension and AKS estimation searching for AKIS/RD options in the thesis.

A summary of most interesting aspects found in the Netherlands, which can be applied to Brandenburg, is given in the paper.

*Table 1. Criteria chosen as important for international extension and AKS estimation.*

| Criteria taken for description and identification of characteristics |
|---|
| Type of system and its classification, motives for choice and methods of decision making.<br>Country specific framework conditions, visions and (inter)national examples.<br>Process of transition, time span and trials.<br>Thematic indicators for expanded stakeholder analysis, potential assessment.<br>Structure or Sub-systems, linkage and functions analysis, role divisions. |
| Estimation and Chances of Success for AKIS/RD |
| Strength and weaknesses of the existing systems' structure and performance.<br>Lacking functions/ roles/ linkages and potential organisations for embeddedness.<br>Interpretive schemes analysis, balance and readiness to take risks, openness for innovation, *etc.*<br>Identification of potential roles and functions needed in AKIS/RD => Indicators for future transformation flexibility and networking activities.<br>Lessons learned and transferability of main aspects.<br>Assessment of trends, visions and follow up. |

## Specific problems to overcome in Brandenburg's AKS

The Brandenburg extension system is the only completely privatised one in Germany. From 2002 on, advisory funding has been cut back. In 1989, nearly all formal linkages broke away, regardless of their importance for future development or acceptance. The privatisation of agricultural extension started in 1991.

Nagel *et al.* (2003) identified some weaknesses: that under given privatised conditions, the present advisory system in Brandenburg will not be able to contribute significantly to national sustainable rural development objectives in Germany or EU agricultural policy goals: it cannot deliver public goods services. To go one step further, the fulfilment of the EU demands addressing Cross Compliance extension will raise new difficulties unless solid structures are installed for presence or virtual group extension. Taking EU regulations, *e.g.* Cross-Compliance-regulation No. 1782/2003, all EU member states as payees are hard-pressed to transform and update their extension systems concepts before January 2007. The lack of financial resources in Brandenburg is a limiting factor and has disastrous consequences for public goods and Cross Compliance extension. Advisors' income is dependent on market mechanisms, thus acceptability issues take on great importance. At the moment, most qualification organisations addressing farmers and advisors must apply for funding of each thematic advanced vocational training a year in advance. Assuming that nearly all linkages in Brandenburg´s AKS are loose (Nagel *et al.* 2003, Von der Heiden 2004), there is a particular lack of linkages between the research and extension sub-systems and of intermediary organisations within an AKS.

## Learning from the Dutch

The Netherlands' reformed extension system acted as an example for Brandenburg from the beginning: the former Dutch government extension service DLV - De Landbouw Voorlichting was privatised in 1989, aiming at separating policy development from executive tasks (for further reading see Leeuwis, Proost, Röling, Wielinga). Following the classification of Rivera (2001), the Netherlands´ extension system privatisation process was partial, because government still funded specific projects. A secure pool of public funding, tending towards specific advisory projects, was still spent on selected organisations and extension products were purchased. In the Netherlands, a revival of the importance of networking has been observable recently. There are several ongoing government-funded programmes, putting researchers into the position of knowledge network facilitator. An example of such networking activity is described in Wielinga (2005). Concerning the new EU demand (Cross-Compliance-regulation No. 1782/2003), Snel (DLV advisor) foresees (Boland *et al.* 2005) the organisation of Cross-Compliance extension in a similar way to project-based public funded advice. Farmers are supposed to select their advisors freely from a pool of authorised private advisory services, accredited by the state.

Results of expert interviews with researchers most familiar with administrative bodies (JP, CW, NR, CR, HW, personal communication 03.-07.09.2001) are:

***Strengths:***
- the step-by-step privatisation: contracting for specific thematic advice; special project funding; providing capital for extension organisations;
- the advisory offers are highly diversified;
- competition results in high extension quality;
- farmers in general are willing to pay for good-quality advice;
- a lot of information at the research level is available.

***Weaknesses:***
- saving money while privatising is not the case: in a federal state, special interests will always lead to spending money;
- the influence of the government on the farmers was lost;
- training functions which were formerly provided by the government have fallen away or become a cost factor;
- the transformation process of the extension system has led to the drastic reduction of staff;
- the most difficult linkages are the ones between both researchers and administration representatives with advisors;
- research information is not reaching the farmers;
- information about farmers' needs, now randomly gathered by advisors, no longer systematically reaches research level,
- farmers who cannot afford advisors´ payment rates try to get their information elsewhere;
- specialist advisors are not linked together.

***Coping with knowledge exchange on public goods issues in particular, constant attention has to be paid to:***
- retaining or renewing the character of the AKIS system;
- all sub-systems´ linkages; - connectedness of stakeholders;
- their awareness of being part of an AKIS;
- their take-over of responsibilities and functions;
- prevention of double payments (as client and as research investor);
- usefulness and benefit of extension and education topics for the client;
- transfer of environmental requirements for farmers;
- consumer information about products;
- farmers' information about markets.

## How to overcome the linkage gaps?

Both systems lack sufficient two-way information lines between major stakeholders such as farmers', extension, education and researchers' sub-systems. Focussing on agro-environmental knowledge exchange, both Knowledge Systems are non-stable. Röling (personal consultation 2001) advises in such cases solving the problem at a higher level: the policy level or the knowledge system level. I hypothesises that a systemic approach is the only sustainable chance of success, including new informal linkages and networking while setting up an expanded stakeholder pool. Thus an external impulse is needed, as already mentioned (Von der Heiden, 2006).

Solution Approach on knowledge system level: For a conceptual solution of the Brandenburg AKIS/RD development, the establishment of new co-operations of potential organisational partners and networking activities, including the task of new role division, are proposed. A survey map and database of potential organisations in Berlin and Brandenburg is given by the author, including following organisation categories:

- qualification organisations: applying for funding according to the Brandenburg´s qualification guideline / providing education on areas of expertise / need identification surveys;
- advisory services: including project management / providing special advice and training;

- intermediaries doing project co-ordination with thematic emphasis on: knowledge exchange / agro-environmental topics / special education and advice / including allocation or monitoring functions;
- scientific and research organisations: including a knowledge transfer approach / NGO research experts / including tutor, educator, advisory tasks;
- associations for landscape management;
- regional offices, public authorities: including allocation or monitoring functions / experts on agro-environmental topics and landscape planning / providing surveys.

On the policy level, facing the European Commission's Cross Compliance Extension call for conceptual solutions in January 2007, the MLUV (Brandenburg's ministry for rural development, environment and consumer protection), is currently developing a concept. The established privatised advisory firms reject the return to dependence on EU, so they propose a solution, which empowers their enterprises to fulfil EU demands. Supported by the MLUV and the agricultural academy, BLAk, their concept would provide a thematic information system addressing the established advisory service. Part of the information input will come from the national Farmers´ Union. To receive EU co-funding for training, BLAk must solve the problem of how to handle the paradox in training regulations which demands that training should be provided jointly for extension workers and farmers in the same group (personal communication with K. Schwartzer, MLUV Brandenburg, 17.01.2006). In the AKIS in Brandenburg, it is worth mentioning a high accountability of educational institutions, applying module-wise funding by the state, the "land"-wide National Farmers´ Union and some administrative bodies which inform farmers about new EU regulations.

## Conclusions

So far, attempts to develop a Cross Compliance extension concept simply reproduce the inadequacies of the past, in terms of lack of linkages, roles and functions. It is clear that a new approach to thematic networking is required in order to satisfy new EU demands. I'd like to address my estimates for future development in Brandenburg to administrative bodies, researchers and intermediaries: a rethinking and multi-stakeholder reformation of the whole AKS situation is needed. I propose an ongoing instalment of collective learning discourses on need based agro-environmental information, potential role definitions, functions and role divisions, networking and further joint development activities for future transformation towards an "AKIS/RD in life" for Brandenburg. The pre-arranged fusion of the Brandenburg ministries of the environment and agriculture provides an important chance for merging network activities. Network to network co-operation is also required for AKIS/RD networking tasks. I agree with Leeuwis and the Wageningen chairgroup that three simultaneous processes deserve particular attention and communicative support: network building, social learning and negotiation. The network approach Wielinga proposes shows lines for construction of coherence, for matching top-down and bottom-up approaches and methodologies for process monitoring.

## References

Alex, G., D. Byerlee, M.H. Collion and W. Rivera, 2004. Converging Views on Institutional Approaches? Ext. and Rural Dev. Disc. Paper 4, Internet publ., p. 14. http://siteresources.worldbank.org/INTARD/825826111129171182/20431824/DiscussionPaper4Final.pdf.

Boland, H., A. Thomas and K. Ehlers, 2005. Expertise zur Beratung landwirtschaftlicher Unternehmen in Deutschland. Eine Analyse unter Berücksichtigung der Anforderungen der Verordnung (EG) Nr. 1782/2003 zu Cross Compliance. In: BMVEL 2005 (editors), pp. 98.

Nagel, U.J. and K. von der Heiden, 2003. Privatising Extension in Post-Socialist Agriculture - The Case of Brandenburg, Germany. In: Alex, G., W.M. Rivera (editors), World Bank online publication Vol 2., 2004, Ext. Reform for Rural Development - Case Studies of Intern. Initiatives. http://lnweb18.worldbank.org.

Rivera, W.M., 2001. The invisible frontier: The current limits of decentralization and privatization in developing countries. In: Brewer (editor), Agricultural Extension: An International Perspective, 2001, Chapter 17, Erudition Press.

Rivera, W.M., 2004. Communication for Rural Development: Challenges to Diffuse Development Information on Non-agricultural Rural Needs. 9th UN Roundtable on Communication for Development, 6-9 Sept. at FAO, Rome.

Sánchez, de Puerta Trujillo, F., 1996. Extensión agraria y desarollo rural. Estudios Ministerio de Agricultura, Pesca y Alimentacion, Secretaria General Tecnica (editor), Serie Estudios no. 123, pp. 481.

Von der Heiden, K., 2004. Missing links towards bi-directional agri-environmental knowledge transfer - the knowledge systems approach of Brandenburg, Germany. In: Proc. on the Intern. Symp. ´Education – way for the prosperity of future generations´, Sept. 2004, Dep. of Edu., Czech University of Agriculture, Prague, pp. 63-70.

Von der Heiden, K., 2006. Lessons learned about instigating change towards an AKIS/RD in Brandenburg, Germany. In: Langeveld, H. and N. Röling (eds.) Changing European farmingsystems for a better future, Wageningen Academic Publishers, Wageningen, The Netherlands, pp. 74-78.

Wielinga, H.E., 2005. Living Innovative Networks of Animal Farmers to Science. In: Oktay E. and M. Boyaci, 2005. Towards a participatory and multi-actors extension system. Proceedings of the 17th ESEE in Izmir, Turkey, pp. 61-67.

# Improving leaders' skills and competencies: The challenge for farming systems in the beginning of the millennium

*Ana Alice Vilas Boas and Milton Bernardes Ferreira*
*Human and Social Sciences Institute, The Federal Rural University of Rio de Janeiro, Brazil*

## Abstract

This paper aims to discuss the skills and competencies of leaders in cooperatives in order to help members and committee' members to overcome their weaknesses and get the better from cooperatives everywhere because some cooperatives have proved to be very effective in providing conditions for their members so they can be more productive, while others are not so effective. A very successful cooperative was identified, looking at: number of members, economic results and marketing. A sample of 36 managers and key figures filled up a structured questionnaire elaborated on the basis of a national survey developed by the Federal Council of Management. The outcomes of the Mixed Cooperative of Farmers from the Southeast Region of Goiás, were compared of this broad survey on the skills of managers and leaders in major Brazilian companies. The most important competencies for cooperative leaders that were identified are: responsibility, decision-making, initiative, honesty, leadership, personal motivation, self-confidence, good personal relationship, capacity for working in group, and negotiation skills. They also mentioned capacity to take risk, communication skills, capacity for planning ahead, innovative spirit and emotional stability as relevant to get reasonable results in this area. Comparing these results with the outcome of the national survey, it was found that six competences are more important in cooperatives than in other places: personal relations, capacity to take risks, personal motivation, negotiation skills, honesty and decision making. The presence of these skills and competencies show the high level of managers' competence in this cooperative, because they are helping the organization to provide positive outcomes for members in the last years. Thus, to prepare leaders to reach this profile in movements like cooperativism and farmers' associations seems to be a significant contribution to improve farming systems and help people to reach better life standards in rural sector.

**Keywords**: cooperativism, farmers' associations, Brazil

## Introduction

The complexity of transformations generated in the beginning of the millennium point out to new innovative decison-making processes in any activity. For this reason, searching for new skills and competencies in every economic sector is relevant, including agriculture. It is also true for cooperativism, since it is seen as an experience to minimize the problems of rural sector, given more security and stability to farmers and growers. Thus, it places specific requirements on known more about improvements for this sector.

According to Silva (1994: 11), cooperatives are characterized by a socio-economic system which search for well-being and social peace for their members and do not have profit as its only aim. This means that cooperatives need profit to grow and help their members to reach a better standard of life. In order to do it, cooperative leaders have to face the challenge of professionalizing themselves mainly because they normally do not have proper skill and competencies to take this activity into account as they are the members themselves.

Some cooperatives have proved to be very effective in providing conditions for their members, so they can be more productive, while others are not so effective. Thus, what is making these successful managers so effective? In this context, to discuss the skills and competencies of leaders in this kind of organisation is relevant to help members and committee members to overcome their weaknesses and get the better from cooperatives everywhere.

## Leaders' and entrepreneurial characteristics

According to De Vries and Miller (Gimenez, 1993), it is relevant to understand the nature of the individuals which integrate a board of directors and managers in any kind of organization because organizational requirements reflect the personality of the most influential managers in any business. In this situation, the organizational results are like a reflex from the organizational powerful actors' cognitive basis (Hambrick and Mason, in Gimenez, 1993).

Caravantes (1979) states managers should have technical, human and conceptual skills. They need also to have an analytical aptitude inside his speciality and facility to deal with tools and techniques in different areas. It is worth to remember, Hersey and Blanchard (1986) conceptualized technical skills as one of the most important competencies in execution of management process, and complemented articulating this skill with human relations and conceptual results. So, technical skills can be understood as the ability to adopt knowledge, methods, techniques, and equipments needed to implement certain specific tasks.

Human skills are related to capacity and discernment to work with people, while conceptual skills have to do with the ability to understand the complexities of organizational global aspects. Connecting technical with humanistic vision, it is possible to have a broad vision about business which can be understood as conceptual skills. Until now, it was not possible to establish a proper scientific psychological profile of leaders and entrepreneurial, for many reasons, among them the differences in samples. However, it was identified some characteristics and aptitudes more frequently present in potential leaders that can be developed to guarantee more chances for success (Filion, in Gauthier, 1999).

It is possible to identify some determining characteristics which lead to certain pattern of behaviours such as needs, skills, knowledge and values. Thus, the launched of a new business or activity is a tool to satisfy their needs, in accordance to their skills and believes (Longen, 1997). In order to have an excellent skill, each person builds it by himself. It cannot be described by means of words only.

According to Longen (1997), skills are related to ability and aptitude to develop physical and mental patterns of behaviour which manifest by the actions individuals have from their previous experiences. Each time this experience repeats um certain situation, the answers incorporate in cognitive system. There are some situations in which the individuals absolve even the method adopted to provide an appropriated answer, developing a new skill that can be used to face new situations.

Gibson *et al.* (1981) stated that values are a group of preferences, aversions, points of view, duties, internal predispositions, rational and irrational judgements, preconceptions and patterns of association that determine the vision of the world that a person has. Personal values are very significant in determining the level of skill and competencies one person may have, even managers.

According to Caird (1988), there is a series of personal tendencies that are normally associated to an entrepreneurial person. These tendencies include the need of success, autonomy, creativity, capacity to take risks, impulses and determination.

## Methodological approach

This paper is based in a fieldwork research carried out by four main steps. First, successful agricultural cooperatives were identified, looking at: number of members and employees, economic results and marketing activities in the whole country. Second, a very successful cooperative was selected, the Mixed Cooperative of Farmers from the Southeast Region of Goiás, known as COMIGO, located in the Central-west region of Brazil.

Third, some leaders were informally interviewed and a sample of 36 managers and key figures filled up a structured questionnaire elaborated on the basis of a national survey developed by the Federal Council of Management. Finally, the outcome was compared to the results of this broad survey on the skills of managers and leaders in major Brazilian companies from different sectors.

## Results and discussion

COMIGO's headquarters is located in Rio Verde, 220 kilometres far from Goiânia, the capital of region and have members from 32 different municipalities nearby. This municipality and its surroundings are leaving under the boom of soybean and cotton production. Facing some difficulties and overcoming them since its launching, this cooperative consolidated itself in the economic state scenery, becoming in less than 30 years in one of the most important source of income for the rural sector and one of the twenty biggest agricultural cooperatives in Brazil.

Nowadays, COMIGO has 4.221 members and 1.510 employees. Its storage capacity is for 788.280 tones. It has also great capacity to accept and store grains to process into other industrialized products. Among members, 25.07% has up to 50 ha; 18.63% from 50 to 100 ha and 56.29% has more than 100 ha of cultivated land. In 1996, for instance, this cooperative commercialized 20 thousand tones of fertilizers and this rate grown to 205 thousand tones, in 2003. It shows the potential to assist farmers in their needs and how they are getting more and more integrated to the cooperative. This cooperative supports research services to improve the quality of grains and other products commercialized by members and offers technical, administrative, educational and social assistance to all members.

### *Cooperativism with professionalism*

The professionalization of cooperative management system is a real need nowadays, which makes them very similar to other businesses. Lack of professionalism is no longer accepted among managers in any kind of economic and even social activity. Moreover, the complexity of cooperative management requires very well prepared managers. Sometimes it is not possible to find them because normally the owners/members of cooperatives are the ones in charge of its management. Apart from that, most of the graduated managers are not ready to work in cooperatives due to their peculiarities.

At COMIGO, the board of directors has worked in very clear pattern with superintendents, managers and employees turning in reality all the interests for the reason it was created, even within a global dynamics. They always have meetings to evaluate the results, apart from the general assemblies and supplementary assemblies when it is needed.

The drawing of the COMIGO managers' profile was possible trough this research carried out among 36 managers. The majority of managers are male (88.8%), and the majority of them are married (77.7%). Almost half of them (52.7%) are graduated, and in this case 41.6% is graduated in Business Management, 22% in Agronomics, 16.6% in Accountancy, 15.5% in Economics, and 11.1% in Law.

Table 1 shows the most important competencies for cooperative leaders that were identified during the research. They are: responsibility, decision-making, initiative, honesty, leadership, personal motivation, self-confidence, good personal relations, capacity for working in team, and negotiation skills. Apart from these, they also mentioned capacity to take risk, communication skills, capacity for planning ahead, innovative spirit and creativity as relevant to get reasonable results in this area. Comparing these results with the outcome of a national survey, it was found that six competences are more important in cooperatives than in industries and other business sector: personal relationship, capacity to take risks, personal motivation, negotiation skills, honesty and decision making, as shown on Table 1 (bold competences).

It seems to happen because this survey was filled up by managers whom are at the same time the owners of the cooperative, since they are part of the members in charge of the cooperative for a certain period of time. On the other hand, the survey carried out by the Federal Council of Management, are based on answers of employees only.

*Table 1. Comparison about Skill of Managers in Brazil versus COMIGO.*

| Skills | Brazil[1] | COMIGO[2] | $X^2$- test |
|---|---|---|---|
| Responsibility | 85.0 | **94.4** | 1.039 |
| Leadership | 81.0 | **86.1** | 0.321 |
| Initiative | 80.0 | **88.8** | 0.968 |
| Capacity for working in team | 76.0 | **80.5** | 0.266 |
| **Decison-making** | 76.0 | **91.6** | **3.202** |
| Creativity | 74.0 | 75.0 | 0.014 |
| Communication skills | 74.0 | **80.5** | 0.571 |
| **Honesty** | 73.0 | **91.6** | **4.739** |
| Self-confidence | 72.0 | **83.3** | 1.773 |
| Capacity for planning ahead | 72.0 | 77.7 | 0.451 |
| **Personal motivation** | 66.0 | **86.0** | **6.060** |
| Innovative spirit | 64.0 | 75.0 | 1.891 |
| **Negotiation skills** | 62.0 | **80.5** | **5.520** |
| **Personal relationship** | 61.0 | **83.3** | **8.152** |
| **Capacity to overtake risks** | 60.0 | **80.3** | **6.868** |
| Work under pressure | 57.0 | 55.5 | 0.039 |
| Emotional stability | 57.0 | 66.6 | 1.617 |
| Good looking | 55.0 | 52.7 | 0.096 |
| Generalist vision | 52.0 | 58.3 | 0.763 |
| Audacity | 52.0 | 58.3 | 0.763 |

[1] Source: Federal Council of Managers (1995).

[2] Source: Fieldwork research – COMIGO.

Managers in charge of cooperatives should have good personal relationship with everybody whom are directly or indirectly involved with the cooperative, should have capacity to take risks, a constant personal motivation to fulfil their duties, and a strong skill to perform negotiations with other members, employees, potential buyers and suppliers (Table 1, bold percentages). Besides, they have to be honest and able to make right decisions, because these skills presented the higher $X^2$-scores (Table 1, bold scores).

The skills and competencies that are at the same level of importance in both surveys are creativity, work under pressure and good looking. These skills, at first glance, do not seem to produce big differences, as confirmed by $X^2$-square.

## Conclusion

Cooperative managers are usually owner of the business that they manage because they are cooperative members at the same time. This situation gives them a different perception from managers of industries, stores and other urban business. As a result, cooperative managers need to develop different skills and competencies to push ahead their business, when they are part of the board of directors, for instance. The presence of these skills and competencies show the high level of managers' competence in this cooperative, because they are helping the organization to provide positive outcomes for members over the last years.

Personal relations, capacity to take risks, personal motivation, negotiation skills, honesty and decision making are the most relevant skills identified in a fieldwork research carried out in a cooperative in Brazil. Thus, improving managers and leaders skills and competencies is relevant to overcome their weaknesses and get the better from their cooperatives which are different from other business.

Thus, to prepare leaders to reach this profile in movements like cooperativism and farmers' associations seems to be a significant contribution to improve farming systems and help people to reach better life standards in rural sector. In summary, members' and leaders' skills and competencies seems to be one important ingredient in this history of success in the central-west region of Brazil.

## References

Caird, S., 1988. A review of measuring enterprise attributes. DUBS, August.

Caravantes, G.R., 1979. Planejamento Organizacional: Dimensões Sistêmico- gerenciais. Makron Books do Brasil, Porto Alegre, FDRH.

Conselho Federal de Administracão. 1995. Perfil do Administrador e Mercado de Trabalho, Brasília.

Gauthier, F.A., 1999. Análise de questionários para a avaliação de perfil empreendedor. Anais 1° ENEMPRE UFSC.

Gibson, J.L., 1981. Organizações: Comportamento, estrutura, processos. São Paulo: Atlas.

Gimenez, F., 1993. Escolhas estratégicas e estilo cognitivo: Um estudo em pequenas empresas. RAC, v. 2, n. 1, jan/abr, pp. 27-45.

Hersey, P and H.K. Blanchard, 1986. Management of organizational behavior: utilizing human resource. 4. ed. New Jersey: Prentice-Hall.

Longen, M.T., 1997. Um modelo comportamental para o estudo do perfil empreendedor. Master Dissertation. EPS. UFSC.

Silva, C.L.M., 1994. Mudança Ambiental e Reorientação Estratégica. Revista de Administração de Empresas. N. 14, December.

# Intermediate roles in privatised extension systems: The Dutch case

*Eelke Wielinga*
*LINK consult, The Netherlands*

## Abstract

The privatisation of formerly state extension services in European countries did not only bring about benefits. At present in the Netherlands several programmes, funded by the Ministry of Agriculture, are ongoing to re-establish linkages that got lost in the post privatisation period. The Dutch experience sheds some light on functions that seem to be crucial in Agricultural Knowledge and Information Systems (AKIS). Are there essential functions that tend to be overlooked in privatised systems?

Different concepts of knowledge, rooted in different rationalities, lead to different answers to this question. Recent Dutch history shows how these rationalities have changed from an expert orientation to dominance of the market and a recent interest in networks. It is hoped that networks will restore connections that got lost in the over-competitive knowledge market.

In this paper I argue that it is impossible to construct a suitable list of essential functions for AKIS. Connection is the key word, but there are many ways to create and maintain it. A better perspective is to make sure that any network has at least one *free actor* capable of doing what is necessary to keep relevant networks healthy, meaning that stakeholders are connected. This calls for special attention for the position of intermediate roles in AKIS.

**Keywords:** extension, linkages, AKIS, networks

## Privatising extension

The trend to privatise State Extension services started in the 80ties. After the examples of New Zealand and England (ADAS), The Netherlands followed in 1989 when DLV (De Landbouw Voorlichting *The Agricultural Extension*) was put on its own feet. In that same year the Iron Curtain fell and formerly socialist countries had to restructure their knowledge systems. Brandenburg (Germany) studied the reformed extension system of The Netherlands for its own reform (Nagel and Von der Heiden, 2003: 33). Other German states followed the move towards privatisation, as well as *e.g.* Hungary, the Czech Republic, Estonia and Poland (partly). In other countries like Switzerland the discussion is already ongoing for years. Not all privatisations were pure in the sense of "private funding and private delivery" (Rivera, 2001): government funding remained important in most cases. The main change was the shift from a government service to a company with commercial goals and employees losing their status as government officer. Did the choice for privatisation bring about the effects policy makers hoped for? At least for the Dutch case it must be said that opinions have changed rather dramatically over time.

## The Dutch case

Reasons for privatising of state extension services usually are reduction of government spending and stimulation of more demand driven knowledge systems. Although this seems obvious, it was not true for the Netherlands (Wielinga, 2001). In the eighties government was criticized for being inflexible and inefficient. Policy development as core task of ministries should be separated from executive tasks such as services to the public, and the number of government workers should be reduced. Yet, my colleagues and me in the policy unit for extension in that time, were highly surprised by the announcement in 1987 that the state extension service would be included in this major reform, because in our opinion extension had become an important policy instrument in the introduction of restrictive measures such as milk quota (1984) and regulations against pollution (from 1984 onwards).

The Dutch state extension service had an international reputation. It maintained very short information lines between farmers, researchers, policy makers and educators, and the influence of farmers organisations and professional groups was strong.

When the negotiations started, the farmers organizations appeared to welcome the privatisation, stating that finally the extension workers should serve their interests again. They would gradually contribute financially in the exploitation of the new extension service, starting with 5% two years after privatisation up to 50% in 2001. The service was split: 700 staff formed the DLV for „first line extension" to farmers, and 550 staff remained under government for the „second line extension" maintaining the linkages in the knowledge system. Soon after the Information and Knowledge Centres (IKC) for animal and plant production were formed, a third one was added for nature and landscape management.

Things did not turn out as expected. DLV started to charge farmers for their services and reclaimed a market share where other private advisory companies had become quite successful as well. DLV bypassed the IKC's by establishing direct links with researchers. For its profile of being at the farmers side DLV kept distance from the IKC's as "policy instrument". At the other side, the policy makers in The Hague claimed the subject matter specialists in the IKC's for policy support only, giving lower priority to the linkage function (Wielinga, 2001). After a range of staff reductions and mergers, in 2005 the remainder of the IKC's was integrated in the Department for Knowledge of the Ministry of Agriculture, Nature and Food.

Upon the initial success of DLV, the ministry reduced its contribution much faster then was agreed. From 1997 onwards the yearly amount had to be applied for in the form of projects with objectives set by government. In 2001 the budget for extension activities for public interests was opened up for all extension service providers. Today DLV is facing heavy competition. It has shrunk to 400 staff, it has almost dissolved its central office and it continues as a network-organization of regional business units.

Nowadays many stakeholders complain about their once so famous AKIS: farmers can't find researchers who solve their problems, politicians blame researchers for keeping knowledge on the shelve, researchers blame bureaucracy because paperwork for getting funds is taking too much of their time and advisors hardly have time to complain, running from one farmer to another for reaching their business targets. Something essential got lost. What was it and how could it happen?

## Changing rationalities

Until the end of the eighties an instrumental rationality was dominant giving high status to science that should lead the way to progress. Extension was to connect science with practice. Knowledge workers (researchers, extension agents, educators) were given much freedom to do what was necessary according to their expertise. The world was a clockwork: the better its mechanics were known, the more obvious it was which way was best. Under the guidance of experts, stakeholders felt jointly responsible for development and overcoming the problems that occurred. This rationality still was leading when DLV and the IKC's were created (Wielinga, 2001).

This changed rather drastically in the nineties when the knowledge workers failed to solve all problems and could not address the diversified needs in agriculture and rural development. The rural playing field had become much more complicated than in the decades before. Where *uncontested knowledge* was lacking, scientists lost their role as reliable referee in conflicts of interests. Policy makers lost their confidence in the farmers leaders who only seemed to aim for preventing painful but necessary measures from being taken. The era of shared responsibility came to an end.

Instead, a strategic rationality became dominant, translating linkages into market relationships. *Knowledge became a product with value* for those who wanted to buy. Researchers became producers and extension workers should act as traders in knowledge products. Farmers were clients and for public good issues like environment government should act as a client funding specified knowledge programmes. Up to date this is still a dominant way of thinking amongst many decision makers. The world was a jungle where only the fittest should survive. Government did not feel responsible anymore for maintaining linkages: too much involvement would only make it difficult to judge the quality of the product. In the market the client should rule: either the farmer or the government in the name of the public. This change of thinking explains the loss of interest in linkages during the nineties. The ministry paid for output of research and extension, and kept itself at a distance.

However, the market does not solve all problems. For many issues of public interest, such as environment, landscape, acceptable production methods, *etc.*, shared understanding is a precondition for collective action. This requires *accepted knowledge* that cannot be provided by experts or imposed by power. It can only emerge from a social learning process with the involvement of all relevant stakeholders. Knowledge is an individual property that might be shared with others after a joint learning process. People should realise that their faith is interconnected in the world as a global village This communicative rationality has been advocated by many scientists (see *e.g.* Röling, 1992) and found its application in numerous participatory methods such as the Rapid Appraisal of Agricultural Knowledge Systems (RAAKS) method (Engel, 1997). These methods have their weakness, however. For example, RAAKS was applied to achieve collective action for reducing health risks in the Dutch animal production sector. All major stakeholders were involved. Confronted with the results, the transport sector blocked publication of the report and further action. Within one year a dirty lorry returning from Eastern Europe caused an outbreak of hog fever, costing the tax payer almost two billion euros. If consensus cannot be reached or actors hold hidden agenda's, it is hard to see what would legitimise further intervention. This is why communicative approaches are often dismissed as being soft and ineffective if power struggle is at stake.

## An ecological rationality

Röling and Jiggins (2000) stressed the need for an ecological rationality that would enable people to balance their behaviour with the carrying capacity of their ecological environment. This rationality has been elaborated in a "Living Networks" approach (Wielinga, 2001), postulating networks of people as living organisms, being part of larger networks and ultimately connected in the global biosphere being a huge living system, referring to the Gaia Theory of Lovelock (1979). Some details of this approach and a methodology for intervention in networks have been explained in the previous IFSA conference (Wielinga, 2004). In this paper I explore what it could contribute in assessing essential functions in the AKIS that got lost after privatisation.

Basic to all forms of life is connection. Living systems will grow autonomously towards more task division and specialisation as long as the mechanisms that structurally couple all elements together grow along with the complexity. In agriculture this growing complexity as an autonomous phenomenon can clearly be observed. In the golden years of Dutch agriculture the relevant stakeholders had developed a fine mazed network of information lines, creating space for an innovative and powerful sector with a shared focus. In the nineties this could not carry on for two raisons. First: the growing complexity could no longer be served by one system managed by the state. It to make way for a knowledge market. Second: the time that the powerful agricultural sector could ignore signals from the outside world was over. Relationships had to change, permitting other legitimate stakeholders to take position. However, in the nineties the dominant market orientation did not favour the search for new connecting structures that would enable collective learning processes between such stakeholders, although at small scale many attempts were made.

Healthy living systems correct themselves. This is how the present revival of the importance of networks in agriculture and rural development in The Netherlands can be understood. Two major efforts should be mentioned here. "The Green Cooperative" regroups all agricultural schools and colleges under the lead of Wageningen University in order to render services to the public. Educators should become networkers establishing links for learning processes that are useful for both students and partners in practice. The other effort puts researchers in the position of facilitators in knowledge networks. In animal production 50 networks of farmers were selected on innovative ideas they had put forward, to be assisted by 35 researchers. In 2006 this program entered its second phase with 33 networks and 28 researchers. Methods, based on the Living Networks Approach, are being applied and further developed, improving the capacity of facilitators to distinguish different stages in the knowledge process as well as interaction patterns between relevant actors, and to enlarge their repertoire of appropriate interventions for keeping the process healthy.

Accepting the world as a living system entails that outcomes of interactive processes cannot be controlled. This is true by the way for all essential things in life, *e.g.* creativity, innovativeness, passion, trust, and love. Striving for it is counterproductive. The alternative is to improve the capacity to respond effectively to changing conditions. Connection is the essential ingredient for doing so. This legitimises intervention. Communication, in the sense of *i.e.* creating awareness or negotiating between different views, belongs to the repertoire of possible interventions. Power play, influencing mutual positions, is part of everyday life. Social learning processes require balanced positions of power of actors who have to take each other seriously. Methods to establish such a power balance belong to the repertoire of interventions as well.

*Box 1. Four rationalities and their consequences for the role of knowledge.*

| rationality | instrumental | strategic | communicative | ecological |
|---|---|---|---|---|
| knowledge | uncontested | valuable | accepted | responsive capacity |
| orientation | expert | power | consensus | connection |
| metaphor | clockwork | jungle | village | living organism |

## Lessons and future perspectives

The challenge for well functioning AKIS systems is to include all relevant stakeholders in social learning processes that lead to relevant knowledge for sustainable development in settings where not only agriculture but also many other interests are at stake (Röling, 1992). It would be beautiful if it were possible to construct a list of functions that are essential for any system, although they might be performed by national/regional actors in different constellations. It would enable researchers to assess actual systems and recommend on lacks that should be filled up. However, such a list could never be exhaustive, given the ongoing diversity and specialisation as well as different historical backgrounds that count. Another option is to reconsider the contents of ideal typical roles, like those of farmers, advisers, researchers, educators, policy makers, NGO's, *etc.* Positions do matter for the range of activities and the legitimacy of certain contributions to social learning processes. A few examples:

- *Researchers* have relative freedom for taking initiatives in social learning processes as long as they bring in their knowledge and use their learning experiences for theory building in a scientific manner;
- *Educators* could use their position for combining training of students as well as adults with contributions to projects with rural stakeholders;
- *Extension workers* (advisors, subject matter specialists, private consultants) work closely with their clients, not only on solving technical problems but also on the articulation of demands, organising synergy with other advisors, assisting farmers (networks) applying for on-farm research, *etc.*
- *Policy makers* could make space for bottom up networking activities and become less hesitant to engage in social learning processes themselves. How to build trust between these stakeholders is part of the learning process. This cannot be achieved if policy makers stay at the side line, only being involved as a client of the results of projects they have funded.

What counts is that stakeholders feel involved in networks, including those that are relevant for public goods issues. Any network requires at least one free actor in the position and capable of doing what is necessary to keep actors connected (Wielinga, 2001, 2004). This could provide a criterion for the assessment of knowledge systems to be worked out further.

## References

Lovelock, J., 1979. Gaia. New York, Oxford University Press.

Nagel, U.J. and K. von der Heiden, 2003. Privatising Extension in Post-Socialist Agriculture – The Case of Brandenburg, Germany. Online publication Vol 2. In: G. Alex and W. Rivera (editors): Extension Reform for Rural Development - Case Studies of International Initiatives, Volumes 1-5, 2004.

Rivera, W.M., 2001. The invisible frontier: The current limits of decentralization and privatisation in developing countries. In: Brewer (editor), Agricultural Extension: An International Perspective, 2001, Chapter 17, Erudition Press.

Röling, N.G., 1992. The emergency of knowledge system thinking: a changing perception of relationships among innovation, knowledge process and configuration. Knowledge and Policy. The International Journal of Knowledge Transfer and Utilisation 5, 42-64.

Röling, N.G. and J. Jiggins, 2000: Agents in Collaborative Management: The Logic of Collective Action. In: L.E. Buck, C.G. Geisler, *et al*, (editors): Biological Diversity: Balancing Interests through Adaptive Collaborative Mgt. Boca Raton: CRC Press.

Wielinga, H.E., 2001. Networks as Living Tissue. A Study on Knowledge, Leadership and the Role of Government in Dutch Agriculture since 1945 *[Dutch language]*, PhD Thesis Wageningen University.'s Hertogenbosch, Uilenreef Publisher.

Wielinga, H.E., 2004: The Response-Ability of Networks: Healthy and Sick Agricultural Knowledge Networks in the Netherlands. In: Christovão, A. (ed.) European Farming and Society in Search of a New Social Contract – Learning to Manage Change. Proc. of the 6th Eur. IFSA Symp., Vila Real, UTAD, Vol. 2, pp. 483–496.

# The development of a monitoring method for innovation projects

*Barbara van Mierlo and Cees Leeuwis*
*Communication and Innovation Studies, Wageningen UR, PO Box 8130, 6700 EW Wageningen, The Netherlands*

Over the last few decades European societies have shifted the services they demand from rural areas. The 1950s to 1970s witnessed the development of agricultural production for cheap and bulky food supplies. From the 1980s onwards citizens increasingly demand beautiful landscapes, clean water and food, and animal-friendly produced meat. To this end, rural systems need radical socio-technological reform which requires not only actors to change their practices, objectives and values, but changes in structures also. A lot of initiatives for innovation towards a sustainable development are taken. However, innovation processes are difficult to manage. Innovation-seeking actors need methods to actively construct, improve and justify their efforts. These considerations led us to conceptualise *reflexive process monitoring*; a method to guide and actively shape on-going innovation processes. This monitoring method focuses on *learning processes* that may contribute to system innovation.

A Dutch rural innovation project "Breeding for the Future" was interested in applying such a method. This hands-on project aims to prevent environmental harm of the use of agrochemicals among others by reducing the use of it. To comply with policy-set standards, the spreading of technical measures in the shape of "best practices" among farmers is stimulated. Some 'external' actors, such as pesticide traders and the agricultural branche organisation, were found to have a large influence on the farmers. Reflexive process monitoring is expected to help to stimulate an innovation process towards integrated pest management while new forms of coordinated action among these heterogeneous actors are being developed.

In this and other projects the method called reflexive process monitoring will be developed and tested. At several moments during a project, assessments of the quality of the learning and innovation processes and the existence of the conditions for these processes, serve to give feedback to the participants in order to encourage fundamental reflection (on values, goals assumptions, structures, bottlenecks perceived and the like). If needed, these assessments can help the project management to improve the interventions to take care of process conditions that may enhance learning and innovation. For instance, to stimulate reflection on values new perspectives may be brought in or to encourage scrutinizing institutional barriers that used to be taken for granted, vision building techniques can be used.

Each monitoring activity itself may very well contribute to the learning process, even those that are mainly meant to analyse of the learning process. Interviewing someone *e.g.* may very well enhance fundamental reflection on values and the dynamics in the environment. At the field, communication scientists will develop a working relationship with the project staff and other actors by taking into account and noting their routines and outlook to the problems. The ultimate goal is to make reflexive process monitoring an integrated part of innovation projects in a way that makes sense to actors in the innovation projects under study, with the aim to enhance the chance that these projects will contribute to a sustainable development.

# Applied research for organic dairy farming

*J.B. Pinxterhuis*
*Animal Sciences Group, Wageningen UR, The Netherlands*

## Organic dairy farming research

In 1997 the Institute for applied research (PV) of the Animal Sciences Group was asked by the Ministry of Agriculture to start research for organic dairy farming. This field was new for the group, and new alliances were needed. A joint research project, Bioveem, was developed with an advisory service for organic dairy farming (DLV) and fellow researchers from the Louis Bolk Institute (LBI), an institute with a long track record of work for biodynamic farming, based on an anthroposophical and holistic paradigm. PV had a strong background of applied experimental research for livestock farming, with predominantly a linear approach of research and knowledge transfer. The challenge in Bioveem was to join the best parts of the two different research approaches of PV and LBI, which resulted in a new structure of research and knowledge transfer.

## Connecting to practice

Bioveem started as a project where researchers, advisors and organic dairy farmers collected on-farm technical and economical data. The performance and experiences of these farmers were used to set up the agenda for a second phase in the project. Key areas of interest for the farmers became topic for on-farm research: factorial experiments, trial-and-error in farm management and explicit descriptions of farmers experiences. This approach is described in Baars *et al.* (2005). Simultaneously, PV continued conducting factorial experiments on the research farm. Topics and experimental treatments came from questions arising in Bioveem and in some cases articulated in ad-hoc workshops with stakeholders.

## From communicating results to implementing in practice

The Bioveem approach appeared to be successful in several ways (De Wolf *et al.*, 2005). Key successes are the great affinity of the farmers with the research and the practical relevance of the topics and results Furthermore, the regional study groups around the Bioveem farmers seemed successful. The results and experiences of Bioveem inspired the other participants to take action on their own farm. For the next years the challenges will be to imbed the Bioveem approach in more research projects, improve its connection with other research approaches of the Animal Sciences Group and to further develop the structure of study groups. For the latter we plan to form study groups around innovative farmers who participate in research projects. Participants are required to implement new (self chosen) strategies or technologies in their management. This is supported and monitored by advisors and researchers. Practical syntheses of results and experiences will inspire many more farmers.

## References

Baars, T, G. Iepema, N. van Eekeren en E. Baars, 2005. Bioveem rapport 11: Werkwijze en methode binnen de Bioveem-aanpak. Animal Sciences Group, Lelystad, The Netherlands.

De Wolf, P., J. Proost, M. van der Voort en F. Hubeek, 2005. Werkende wijs worden. Procesevaluatie van Bioveem in LNV onderzoeksprogramma PO-34 (2001-2004). PPO-AGV, Lelystad, The Netherlands.

# Section 5 – Management of natural resources

# Introduction to section 5 – Management of natural resources

*Chris Stoate[1], Andrea Knierim[2] and Jacques Baudry[3]*
*[1]The Game Conservancy Trust, Allerton Project, Loddington, United Kingdom; [2]Institute of Socio-Economics, Leibniz-Centre for Agricultural Landscape Research, Germany; [3]SAD-ARMORIQUE, France*

Several key themes emerged from the Natural Resource Management (NRM) workshop at the 2004 IFSA conference in Vila Real. The implications for agricultural landscapes are as strong under the current period of CAP reform and declining farm incomes as they were during the period of agricultural intensification. Farming is increasingly under pressure from international trade reform and environmental regulation. What role do farmers have in this new scenario of changing policies? What are the opportunities for farmers, as well as for wider society, in the quest for sound management and use of natural resources?

Individual farmers do not always recognise their roles in the wider landscape, or management of natural resources beyond those associated immediately with cropping, and yet farms differ considerably in their relationship with the environment through differences in farm size, production systems, geographical location and farmer backgrounds and attitudes. This diversity is itself a source of innovation which must not be inhibited by emerging regulations. This farmer knowledge is strengthened through communication and collaboration within farming communities, and farmers have an important role to play in guiding future NRM policy and ensuring that it is practically grounded.

It is particularly important that state interventions such as regulation and agri-environment schemes are compatible with farmers' developing market opportunities. In developing policies, objectives must also be sensitive to scale, from field to farm, to landscape, catchment and region. People from outside the farming community and from outside the farming area have an increasingly important role to play in determining how agricultural landscapes will evolve. Attitudes and behaviour of urban people can often clash with those of farmers and other rural people, but also provide opportunities through recreation and product marketing. What about the role of researchers? Should they lobby within the research realm to promote specific types of research, or in wider society to promote specific values? As researchers, we are not independent of the systems we study.

A common theme is the increasing need for communication, between farmers, researchers and other stakeholders within rural and urban communities. Modelling and GIS provide means of facilitating such communication, as well as improving understanding of spatial and temporal issues within individual communities.

These themes are further developed in the papers of this section. Sustainable integration of agriculture and environmental management is the theme for Jayanthi *et al.* and Jennings *et al.*, reporting respectively on a mixed livestock and cropping system in India and a dairy system in Australia. The diversity of situations in which NRM is paramount is clear, but the need to recognise such diversity at the regional scale and below is also accepted. The extent to which agricultural production and NRM are integrated varies between sites and systems. Van Doorn investigates landscape functions within an agriculturally 'marginal' area of Portugal where NRM is integrated with farming but varies between farm and farmer. Levin *et al.* explore the relationship between uncropped landscape features, in the form of hedges, and the influence of changes in farm and field size in Denmark. Knickel *et al.* provide an example from Germany of an integrated approach to landscape management involving the co-operation of farmers, conservation groups, small food companies, tourism and other interests in developing a multifunctional vision for an agricultural landscape.

Targeted wildlife conservation is a recurring theme amongst post-productivist agricultural policy, but how does this accommodate the structural attributes of farms in different landscapes, and the equally variable attitudes and interests of farmers? Swagenmakers and Knierim and Kächele provide insights into this issue in The Netherlands and Germany respectively. Again, knowledge exchange and dissemination is a key theme, involving the knowledge and skills of farmers themselves, coupled with those of scientists and other stakeholders. Stobbelaar *et al.* investigate the relationship between the type of landscape preferred by local people in the Netherlands, and the farmers who are expected to produce it.

Increasingly, catchment scale management of water is an issue for farmers, requiring improved understanding of the land managed by farmers in the context of the wider environment, and specifically river basins. Co-ordination between farmers and between farmers and other actors is also essential for this process. Souchère *et al.* explore the potential of geographical modelling to improve understanding and communication of NRM concepts in the French Upper Normandy region. Jennings *et al.* use different process tools to engage farmers in a similar issue, while for Baptista and Sousa, the issue is one of water as a resource for irrigation in Portugal. They apply Earth Observation techniques to this increasingly widespread and important issue. Groot *et al.* develop a visual application to engage local farmers and other stakeholders in a different range of interrelated issues associated with landscape and conservation within a farming system.

The section therefore covers a wide range of NRM issues in a range of countries across the world. Throughout though, the aim is to improve the sustainability of NRM and use through integration of objectives and through improved understanding amongst all involved.

# Can earth observation technology improve efficient water use in irrigation systems? A case study of an irrigation scheme in Chaves

*Alberto Baptista and Vicente Sousa*
*University of Trás-os-Montes e Alto Douro. 5000-660 Vila Real, Portugal*

## Abstract

The article is about the problematic of the transferability of DEMETER technology to other irrigation schemes, in particular that of Chaves, in the north of Portugal. The DEMETER project is a research and demonstration project, designed to assess and demonstrate how the integration of Earth Observation techniques in routine Irrigation Advisory Services can improve efficiency in the use of water irrigation. The technology transfer is a complex process that requires a diverse set of conditions and favorable factors to be successful regardless of the type of irrigation system, the political and socio-economic context, the interests and objectives of the actors, the type of irrigators and their needs.

The objectives of this communication are: (i) to analyze the desire and the possibility of transfering the DEMETER technology in the context of an irrigation system of Chaves: (ii) to identify the factors that favor the transference of this technology, for other irrigation system. In this study we use a diverse methodology: bibliographical research; collecting data in the water association; fields visits and observation of irrigation; interviews with the staff of the water association; questionnaire to 107 water users of Chaves irrigation scheme.

The DEMETER technology seems a good solution for some types of irrigation system, but for the Chaves system other changes are necessary.

**Keywords:** DEMETER project, EO Technology, irrigation system, transferability

## Introduction

This paper is about the transferability of DEMETER technology. DEMETER is a research and demonstration project, designed to assess and demonstrate in an operational perspective how the integration of Earth Observation (EO) techniques in routine Irrigation Advisory Services (IAS) can improve efficiency in the use of irrigation water (Calera *et al.*, 2005). The main objectives of the DEMETER project are: to provide IAS with innovative tools to achieve greater efficiency in the use of water; to develop and demonstrate a modular prototype technology for creating new space and IT-assisted IAS. Basically, satellite images of irrigated fields are periodically analysed and crop coefficients derived from the crop's vegetation development. This information is stored in a database and can be accessed by farmers through several different ways such as phone calls, SMS messages, internet, email. Three irrigation schemes have been selected as pilot zones: Barrax, in Spain; Caia, in Portugal; Destra Sele, in Italy. The irrigation schemes of Velestino in Greece and Chaves in Portugal have been selected to evaluate the transferability of this technology.

Technology transfer to the agrarian sector is a complex process that requires a diverse set of conditions to be successful. In this sector, the social and economic issue always plays a very relevant role, sometimes higher than the technical one, and has to be carefully considered. The level of economic income of the activity, the intensity of application of production factors, the level of education and professional skills of the farmers can determine the success of a technology transfer process.

The objectives of this paper are to: a) analyze the technical feasibility of the DEMETER application in the context of an irrigation system in Chaves; b) identify the interest of farmers by such technology; c) list the factors and conditions that, in general, encourage the DEMETER technology transfer.

In this study we use a diverse methodology: bibliographical research; data collection from the water association and the Regional Directorate of the Ministry of Agriculture; fields visits and observation of irrigation; interviews with the staff of the water association; questionnaire completed by 107 water users of a Chaves irrigation scheme.

## Theoretical framework

Transferability describes the process of applying the results of research in one situation to other similar situations. According to Wichmann, 1996, and to many other authors (Judd *et al.*, 1998: 78) it is clear that "agricultural technology is simply not very transferable". Technologies, especially for agriculture production – are usually developed in response to a country's relative factor endowment. Krueger, (1991: 464), cited by Wichmann (1996), states that "the lesson of the past three decades is that when technology is inappropriate, the net benefits may even be negative".

Most of the studies devoted to technical advances do not pay enough attention to the interdependence of the economic, social and technological issues (OECD, 1988: 31). Technological innovation must consider the particular condition of each situation and not simply mimic the technology developed in other countries in different conditions (Hayami and Ruttan, 1971: 4). Some projects failed because they sought simply to transplant an approach from one area to another without paying enough attention to differences in the local context OECD (1999:6). Therefore, it is critical to take into account differences between situations and modify the research process accordingly (Cziko, 1993: 10).

According to Seegers and Kaimowitz (1990) "The issue for researchers is to develop economically viable technology that is readily adaptable to rural society. To be effective, agricultural research must be directed to producers needs and the results made available to producers readily and conveniently". This opinion is in accordance with Mangano (1995), who said "in the sector of irrigation projects many failures can be attributed to direct transfer technology that is successful in the developed countries to a developing country".

According to Tollefson and Wahab (1995), "difficulties with the transfer of agricultural technologies are endemic in most countries and have not been solved by establishing large extension services. There has also been excessive focus on the means of transfer rather than on the transferability of the technology, which comes down essentially to the question of whether the technology can be readily adopted and be of benefit to the end-user and ultimately increase his or her profit margin".

When the technology transfer is related to irrigation, several different irrigation systems and ways of water delivery have to be considered. Horst (1998) identifies five types of irrigation system which determine the operational procedures and staff requirements. The systems can vary from "traditional", with mostly farmer-managed and simplified technology, to automatically controlled system, where water distribution is on the basis of crop water requirements. Each system type has different delivery schedules, and requires its own system in terms of canal capacity and water division structures.

The implementation of an irrigation scheduling system as the aim of a IAS must consider: local needs; motivation of people involved; water pricing; cost of scheduling; management flexibility; the formal education of people involved; the institutional concerns; behaviour adaptation; the changes in technology and; the role of the extension service (Tollefson and Wahab, 1995; Horst, 1998).

The complexity of the process of innovation and technological adoption and the diversity of irrigation systems, are realities to take into account during the phase of conception of the technologies and in its transference to the final users.

## The case study of the irrigation scheme of Chaves

### *Characterization of the irrigation scheme*

The irrigation scheme of Veiga de Chaves is located in the north interior of Portugal. The agro-ecological conditions in this region do not in general support intensive agriculture. The exception is the valley of Chaves where soils have a relatively high production potential.

The irrigation scheme of Chaves is divided into two Blocks. Block I was established in 1948 and recently rehabilitated. From the main channel, the secondary and tertiary channels have been substituted by buried pipes, from where water is delivered at low pressure. area covered is 1007 ha, divided into 4009 parcels (0.25 ha/parcel) belonging to 1852 farmers. The Block II was concluded in the year 2001, covers an area of 459 ha divided into 1448 parcels (0.32 ha/parcel), belonging to 649 farmers. The water source is a small dam in the nearby mountain delivering water under pressure.

The government through public funding built the irrigation scheme infrastructure, but it is the "Associação de Regantes" which is in charge of maintenance and manages water delivery.

It is not possible to charge for the water used by each farmer. Considering the very small size of most of the irrigated parcels, more than one farmer uses each one of the water outlets. Also there is no meter to measure the amount of water delivered by the water outlets. Therefore, each farmer pays a fixed fee per year and ha (100 euros/ha).

According to the questionnaire, more than 90% of the land is irrigated using surface-based methods. There is a limited use of sprinkler systems and trickle irrigation is almost unused.

Even considering the general flatness of the land these irrigation systems require intensive labour. Access to water is relatively easy except for those whose plots lie at some distance from the water outlets.

### *Social and economic aspects*

The level of economic and social development is low, about 62% of the national average. This region is permanently losing population; a decrease of 40% in the last 40 years. The population that remains in the region is concentrated in urban centres. Agriculture still has an important role in the regional economy, but it is not organized and composed of a great variety of crops leading to production without critical dimension.

The regional and institutional situation in Chaves does not favour agricultural activity or investment. Production levels and revenue are low and commercialization of products is difficult. As a result economic activity in this sector is in decline. Regional and local institutions do not encourage entrepreneurship and no attempt is being made at innovation or to keep young people in agriculture. The local Water Users Association has many weak points – principally organizational and financial - and is lacking in personnel and equipment -, conditions which do not encourage change.

The majority of the irrigators questioned is old (93% are over 45), with a low level of education (93% have less than six school years), and live in family units of between one and three members.

The farms studied are small (average 2.8 hectares) and a large proportion of them are divided into four smaller plots. Labour on the farm is mostly from within the family (59%), with outside help only for specific tasks (25%). Paid labour exists only in 10% of the units. The family undertakes the irrigation. There is a big variety of irrigated crops (we identified 25 crops), all of which have a low market value, not least of all due to difficulties in marketing them. The greater part of production (40%) is for domestic consumption. Only 13% of the families sell the major part of the production. Family income derives principally from social security (35% of the units), or social security, complemented by an agricultural wage (28%) and employment in the manufacturing and in service/commercial sector (23%). Only 13% of irrigators lived solely from agriculture.

***Farmers' standpoint about irrigation and problems***

For the farmers, the decision to irrigate is based on a number of factors: the conditions of the plants, soil, climate and length of time since the last irrigation. Farmers tend to over-water in an attempt to avoid the risk of plant dehydration – an entirely rational decision when there is no link between the quantity of water consumed and its cost.

The main problems identified by farmers with regard to the irrigation system are: the lack of tertiary channels, leaks, the price of water, the method of payment for the water.

Users prefer the existing system of distributing water "on demand", reject the implementation of a "timetable" and manifest no interest in the provision of an "advice system". The water users interviewed felt that the current distribution system worked adequately and, due to their considerable practical experience, any outside attempts to advise them on the subject faced technical difficulties, above all because a wide variety of plants are sown. From the farmers point of view the improvement of the irrigation systems requires three fundamental changes: (1) better management by the Association; (2) an increase in the number of irrigation outlets, and; (3) payment on the basis of volume water consumed. For the irrigators, problems such as low incomes and commercialization have greater priority than improvements in irrigation systems.

## Factors and conditions that favor the transfer of DEMETER

This study raises some questions about the advantages or limits on transferring this technology to other contexts. There are sets of factors and conditions that favor the transfer of DEMETER that we have divided into four parts:

1. *Region and Institutional Actors*: Regions with high economic development; Institutional framework conducive to stimulating initiatives; Application of control measures to increase water efficiency; Agricultural extension services to develop irrigated agriculture; Research Centers to solve local problems.
2. *Planners/Designers and Type of Irrigation:* Make the design with the interaction of the future users of the system; Types of irrigation systems that permit changes in flow and subsequently in water level; Irrigation systems that permit single and effective control of water resources; Water distribution based on crop water requirements; Installation of water meters in the water outlets; Automatic delivery and control of water.
3. *Water Users Association:* Strong association, dynamic, recognised/supported by farmers, with a well-functioning accountability mechanism backed by enforced laws and regulations, financial autonomy and their own human resources and equipment; Promotion of modern technology (automation, computerization) with trained staff; Existence of Irrigation Advisory Services (IAS), or good conditions for its implementation; Cost recovery based on volumetric payment and automatic delivery and control of water.
4. *Water Users and Farms:* Respond to local needs and motivation of people involved and behavioural adaptation; Relatively young irrigators with a good level of education; Large scale irrigated areas with big farms and with small number of plots; Relative scarcity and high consumer price of water; Agricultural activities capable of generating substantial income flows; Low diversity of irrigated crops produced, oriented to the market, with high value added; Entrepreneurial oriented irrigators with investment capability; Well equipped farms, especially with regard to irrigation systems; Acceptable cost for the farmer.

## Conclusions

The favourable conditions for the transfer of DEMETER and the conditions encountered in the irrigation scheme of Chaves, show that there exist two quite distinct realities at the socio-economic and institutional level, as well as farm profiles and type of irrigation systems.

The objective of increasing water use efficiency can be achieved in a variety of ways. One solution could be based on DEMETER technology, suitable for certain contexts and types of irrigation systems, in particular in those regions that have (1) a growing agricultural sector, with large scale plots under monoculture, or with only 2 or 3 clearly market-oriented main crops with high potential value-added; (2) where water is scarce and relatively expensive; (3) where consumers pay for the water they use but are organized in a dynamic, well equipped, well-staffed, financially sound and empowering water users associations.

With regard to Chaves, the actors have no interest in substantially changing the water distribution system or improving conditions. Irrigation scheduling to meet the individual needs of thousands of small plots is unrealistic given the existing infrastructure.

Given this fact, it is necessary in this case to look for other solutions to the problems of simplifying water distribution and reducing costs. An improvement in efficient water use must reside in the reduction of the substantial amount of water lost in the primary and secondary channel system. In the tertiary channel system, the use of hoses to take water from the irrigation source to the plots should be encouraged and, in Block II, the use of sprinkler irrigation should be further employed.

## Acknowledgements

The authors acknowledge the contribution of Ana Magalhães and Sónia Abreu, to field work and data analysis. We also thank the Association of Irrigators, water guards and the farmers. Without whose collaboration the study could not have been made.

DEMETER (Demonstration of Earth Observation Technologies in Routine Irrigation Advisory Services) is co-funded by the European commission under its fifth Framework Programme (Energy, Environment and Sustainable Development Programme, contract EVG1 – CT – 2002 – 00078.

## References

Calera, A., A.M. Jochum, A. Cuesta, A. Montoro and P. López Fuster, 2005. Irrigation management from space: towards user-friendly products. Irrig. Drain Systems 19, 337-353.

Hayamy, Y. and V. Ruttan, 1971. Agricultural Development: an International Perspective. Baltimore and London, John Hopkins University Press. 367 pp.

Horst, L., 1998. The Dilemmas of Water Division: Considerations and Criteria for Irrigation System Design. International Water Management Institute, Wageningen Agricultural University.

Judd, A.M, K.B. James and E. Robert, 1986. Investing in Agricultural Supply: The Determinant of Agricultural Research and Extension Investment, Economic Development and Cultural Change 35, 77-113.

Malinen, P., 2001 Problems of Transferability – Good Practice in Local Development Policy. In: Rural development Lessons from the North.

Mangano, G.V., 1995. Applicability of Irrigation Scheduling in Developing Countries. In Irrigation Scheduling; From theory to practice – Proceedings of the ICID/FAO Workshop on Irrigation Scheduling, Rome.

OECD, 1988. News Technologies: Socio-Economics Strategy for the 90´s. Paris, OECD.

OECD, 1999. Best Practice in Local development. The Local Economic and Employment Development Programme (LEED), OECD, 128 pp.

Seegers, S. and D. Kaimowitz, 1990. Relations Between Agricultural Researchers and Extension Workers. Journal for Farming Systems Research - Extension, 29-44.

Tollefson L.C. and M.N.J. Wahab, 1995. Better Research-Extension-Farmer Interaction Can Improve the Impact of Irrigation Scheduling Techniques, in Irrigation Scheduling; From theory to practice – Proceedings of the ICID/FAO Workshop on irrigation scheduling, Rome.

# Supporting discussions on land use: A design instrument that integrates nature conservation, economic performance and landscape quality in agricultural areas

*Jeroen C.J. Groot[1,2], Walter A.H. Rossing[1], Derk Jan Stobbelaar[1,2,3], Henk Renting[3] and Martin K. van Ittersum[2]*
*[1]Biological Farming Systems Group; Wageningen UR, The Netherlands; [2]Plant Production Systems Group; Wageningen UR, The Netherlands; [3]Rural Sociology Group Wageningen UR, Marijkeweg 22, 6709 PG Wageningen, The Netherlands*

## Abstract

Intensively managed agricultural areas in North-Western Europe have been undergoing a shift from solely production oriented use to provision of multiple services and functions. Design and assessment of multifunctional agricultural landscapes can be supported by exploration of the potential to effectively combine economic performance, landscape quality, nature conservation and environmental quality. This paper presents the Landscape IMAGES methodology for spatially explicit exploration of options for multifunctional agriculture in landscapes at a scale of a few km². The framework has been developed to support discussions and informed decision making by local and regional policy makers, land owners and land managers and other relevant stakeholders.

The structure of the Landscape IMAGES framework is elaborated and its functioning is illustrated with a near-real example of a grassland-based landscape with hedge rows bordering the fields. In this landscape, four objectives are being pursued by adjusting land-use intensity and hedge row presence: 1. acceptable agronomic yields for farms; 2. diversification of the botanical composition of fields and hedge rows; 3. variation in plant communities in the fields and half-openness of the landscape; and 4. reduction of nutrient emissions to the environment. For exploration of trade-offs between multiple objectives a heuristic search method (differential evolution) is employed, which yields a large range of alternative, acceptable configurations of the landscape. The framework provides explicit insight in the trade-offs between the objectives and is implemented in a visual application that enables assessment of alternative options. The method can be applied to a range of spatially explicit land-use and nature allocation problems and will be further developed in interaction with stakeholders.

## Introduction

Over the last two decades, attention in policy, land-use planning and research applied to intensively managed agricultural areas in North-Western Europe has been undergoing a shift from solely production oriented land-use to provision of multiple services and functions (Vos and Meekes, 1999), for example maintenance or improvement of landscape structure, sustainable management of renewable natural resources, preservation of biodiversity and contribution to socio-economic viability of rural areas (OECD, 2001).

Due to the changes in land-use objectives for the involved stakeholders, the decision making process concerning land-use at different scales has become to a large degree a spatial planning process, integrating issues of agricultural production and its side-effects (emissions, deterioration of food webs) at field and farm scales, and nature conservation, environmental protection and landscape quality at the regional/catchment scale.

To achieve an integrated view on the required adjustements and innovation in landscapes and land-use systems where complex, uncertain and value-laden issues occur, systems approaches that integrate various issues, stakes of social actors, disciplines and scales are indispensable. They are characterized by a multi-disciplinary approach to problem solving, involving both technical and social sciences, and a high degree of stakeholder participation. The resulting complexity in both land-use planning issues and stakeholder interactions necessitates the use of supporting methodologies to inform stakeholders and policymakers, to design alternatives and to explore scenarios for the future.

The approach of 'discussion support systems' as proposed by Nelson *et al.* (2002) aims to contribute to learning and dialogue between stakeholders about development options. By 'addressing issues of mutual interest, and explicit examination of the consequences of different objectives and preferences' (Struif Bontkes and Van Keulen, 2003), such tools could facilitate social learning, which can be described as an ongoing process of learning and negotiation with particular attention for communication, perspective sharing and collective problem solving.

In this paper we present a spatially explicit, GIS-based land-use optimization methodology named Landscape IMAGES (Interactive Multi-goal Agricultural Landscape Generation and Evaluation System). This approach is developed to assist in the design of multifunctional agricultural landscapes and can contribute to innovation and learning by providing insight in the trade-off between land-use objectives and by visualization of alternative future development options.

## Method

In the model agricultural land-use on the fields and the placement of hedges adjacent to the fields are allocated in a optimal manner, taking into account spatial heterogeneity and spatial interactions. In the current prototype implementation applied to regions dominated by dairy farming systems, the model seeks to maximize (i) gross margin from agricultural production; (ii) nature value of fields and borders; (iii) variation the the landscape in terms of species presence and hedge row allocation (half-openness); and to minimize (iv) emission of nutrients to the environment. Constraints are applied to nutrient input and the proportion of herbage grazed.

Alternative land-use options that can be applied to fields were generated from simplified agroecological relations for grasslands and dairy production systems. To accomodate the implementation of discrete farm management choices and the possible inclusion of nature management packages, a discrete production activity generation approach was adopted.

As indicator for the economic performance of farms, gross margin was adopted. The returns from production per field were calculated directly from the milk production and the milk price. Costs per field were separated into costs related to production (harvesting by grazing or mowing and fertilizer) and transport costs. The financial revenues from nature conservation packages were added to the value of the objective function for economic results. The applicability of conservation packages to individual fields was assessed on the basis of plant species abundance, and harvesting and fertilization regimes.

The species abundance in the grass swards and hedge rows was used as an indicator for nature conservation value. The relationship between nutrient availability and average species presence in grasslands was derived on the basis of data of Oomes (1992).

Landscape quality was related to variation in the landscape, calculated as the weighed sum of (1) the variance of the species number for each field and its adjacent fields and (2) the half-openness of the landscape, represented by the squared deviation from 50% occupation of the proportion of borders occupied by hedges.

The emission of nutrients was directly derived from empirical agroecological relations by calculating the difference between uptake of N by grass and availability of N from natural soil fertility and fertilizer application.

Optimization was performed with the genetic algorithm of Differential Evolution (DE). This algorithm was initially developed for single-objective optimization problems over a continuous domain, and thereafter adjusted for and applied to a range of multiple objective, mixed variable optimization problems (*e.g.*, Xue *et al.*, 2003). Preliminary optimization runs were performed to assess its performance and to find appropriate parameter settings.

## Case study

The Landscape IMAGES modelling approach was applied to a case study area located in the Northern Frisian Woodlands, The Netherlands. In this region, the merging of farmers cooperatives has resulted in the founding of a large regional cooperative that aims to realise a vital regional economy, attractive residence areas, a clean environment and maintenance of landscape and biodiversity. This initiative is supported by local, regional and national governments, and farmers and landscape stewardship organisations. The strengthening of landscape and nature values in combination with agricultural practice is being pursued by the regional cooperative by an array of measures. The regional cooperative and associated local stakeholders have solicited scientific support to evaluate proposed adjustments in the design and assessment of landscape and agricultural land-use practices. A multi-objective landscape and land-use design methodology could contribute to these developments.

The Northern Frisian Woodlands are characterized by a small scale landscape on predominantly sandy soils with dairy farming as the prevailing land-use activity. On some farms a limited proportion of up to 5% of the area is used for forage maize production, while the rest of the area is occupied by permanent grassland, rotationally grazed and mown. The fields with an average size of 2 ha are often surrounded by hedge rows. The average grazing season lasts 6 months from May to October. Grazing systems applied range from day and night grazing to restricted and zero grazing. The selected case study area of 232 ha enclosed by roads is presented in Figure 1. The majority of fields in this area belong to three farms, denoted A, B and C.

## Results

The solution sets obtained after 12,000 generations of improvement by the DE algorithm covers a large range of possible configurations of the landscape in terms of land-use on fields and the placement of hedges on field borders.

The trade-offs between the various objectives at landscape scale are presented in Figure 2. The frequency distribution of species numbers in fields and emission per field for the selected landscapes show the large contrasts (Figure 3). The landscape with low profitablity and high nature value is dominated by fields with high species numbers and low emissions as a consequence of low fertilizer inputs.

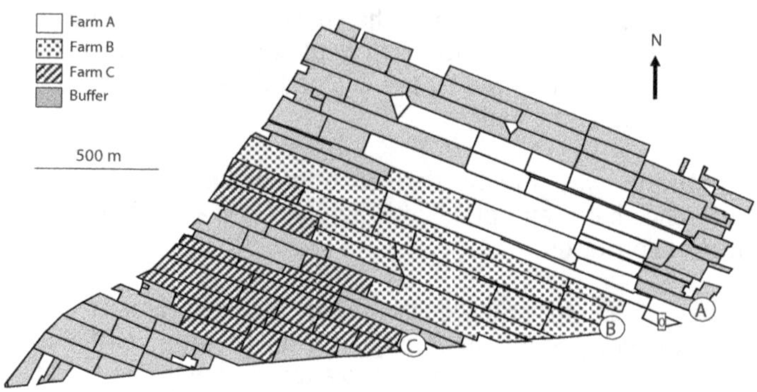


*Figure 1. Landscape of the case study area with fields for farms A, B and C and buffer fields. Circles with letters indicate the position of the farm houses and stables for the farms.*

In addition to the trade-offs between the objectives, considerable arrays of alternative solutions varying in nature and landscape value and emission levels exist for each level of profit from agricultural practice. For example, larger variation in the landscape could be reached (Figure 2b, *e.g.* selected landscapes III and IV) by improved spatial distribution of management intensities on fields and hedges on field borders.

## Discussion

The proposed framework integrates knowledge from diverse disciplines, so that trade-offs between objectives related to different disciplines can be evaluated and discussed. Thereby, assessment and enhancement of multifunctionality of land-use and landscape design could be supported.

The current version of the Landscape IMAGES framework appears to partly meet the requirements for effective model utilization in discussion support by, *e.g.*, parameter, objective and constraint

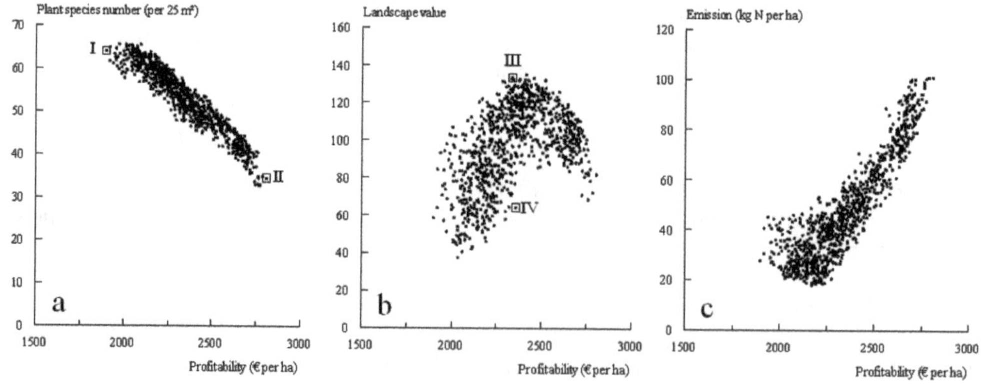


*Figure 2. Landscape scale trade-off curves between profit and average plant species number (a), landscape value (b) and nutrient emissions (c).*

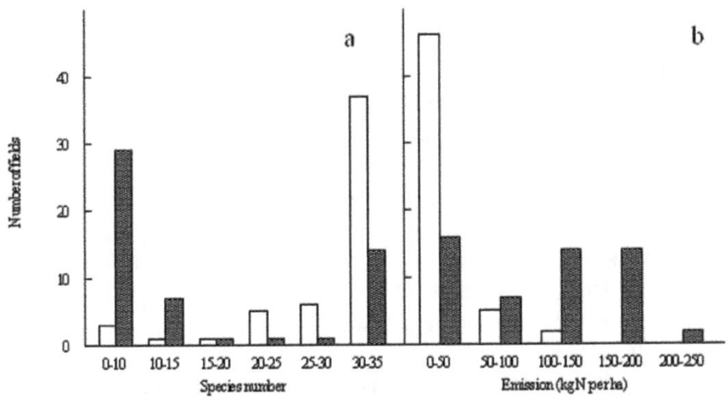


*Figure 3. Frequency distributions of plant species numbers (per 25 m²; a) and emission (kg N per ha; b) per field in optimized landscapes with high nature value (open bars; solution I in Figure 2a), high profitability (shaded bars; solution II in Figure 2a).*

adjustment at the three relevant scales (field, farm, landscape), and selection of dimensions for visualization to enable interrogation of the results. These features enable the assessment of issues of mutual interest and explicit examination of different objectives and preferences (*cf.* Struif Bontkes and Van Keulen, 2003).

The solution sets obtained in the present optimization study offered a large range of possible configurations of the landscape in terms of land-use on fields and the placement of hedges on field borders. These solutions showed trade-offs between various objectives. The generated alternatives offer ample opportunities for discussions with stakeholders on various topics, for instance:

- with farmers about the feasibility of farm management adjustments in the context of the existing farm bio-physical environment, field configuration and endowment, and the farmer's interests;
- with farmers' environmental cooperatives and regional and national governments about landscape scale consequences of changes in management of individual farms and the repercussions of regional rather than farm level upholding of rules and regulations;
- with policy makers and non-governmental organisations on the possibilities to reach their goals.

In future applications, the presented Landscape IMAGES framework could be applied to a range of spatially explicit land-use and nature allocation problems.

## Acknowledgements

The study was carried out as part of the project 'Multifunctional Agriculture: From Farm Practice To Farm Design and Institutional Innovation' in the MFA programme financed by the boards of WUR and INRA.

## References

Nelson, R.A., D.P. Holzworth, G.L. Hammer and P.T. Hayman, 2002. Infusing the use of seasonal climate forecasting into crop management practice in North East Australia using discussion support software. Agric. Syst. 74, 393–414.

OECD, 2001. Multifunctionality – towards an analytical framework. Organisation for Economic Co-operation and Development, Paris. 27 pp.

Oomes, M.J.M., 1992. Yield and species density of grassland during restoration management. J. Veg. Sci. 3, 271–274.

Struif Bontkes, T.E. and H. Van Keulen, 2003. Modelling the dynamics of agricultural development at farm and regional level. Agric. Syst. 76, 379–396.

Vos, W. and H. Meekes, 1999. Trends in European cultural landscape development: perspectives for a sustainable future. Landscape Urban Plan. 46, 3–14.

Xue, F., A.C. Sanderson and R.J. Graves, 2003. Pareto-based multi-objective differential evolution. In: Proceedings of the 2003 Congress on Evolutionary Computation (CEC2003), Vol. 2, IEEE Press, pp. 862–869.

# Nutrient recycling potential of linked components in wetland and rainfed farming systems of western zone of Tamil Nadu, India

*C. Jayanthi, S. Marimuthu, G. Vivek and B. Sivakami*
*Department of Agronomy, Tamil Nadu Agricultural University, Coimbatore, India*

## Abstract

In wetland Integrated farming system (IFS) enterprises like cropping, fishery, poultry, pigeon and goat rearing were tried to recycle the residue and by-products of one component over the other to sustain the productivity, profitability and employment for lowland farms. In one hectare farm, an area of 0.75 ha was assigned for cropping, 0.10 ha for fodder, 0.03 ha to goat shed and 0.12 ha to 3 fish ponds. Three IFS *viz.*, crop + fish + poultry (20 layers), crop + fish + pigeon (40 pairs) and crop + fish + goat (20 + 1) were tried for three years. The highest net return of Rs.1,31,118 was obtained by integrating goat + fish + cropping applied with recycled fishpond silt. The employment was also increased to 576 mandays/ha/year by integrating fish+goat in the cropping. Another study involving grain and fodder crop, fodder trees, perennial grasses and goat rearing in one hectare of rainfed land was carried at Western zone of Tamil Nadu, India. The cropping pattern was modified through integrated approach including both grain (0.25 ha) and fodder sorghum (0.25 ha) along with cowpea. To meet the fodder requirement of goats, subabul and *Cenchrus ciliaris* was raised in 0.28 ha of land, and *Acacia senegal* and *Prosopis cineraria* in another 0.2 ha. The goat unit 20 + 1 goats sheltered in 0.02 ha. The results revealed that short duration field crops and perennial crops can meet the feed requirement of goats for one year. Twenty one goats with kids under deep litter system yielded 11.2 t of valuable manure, which is sufficient to enrich the soil fertility under rainfed situation. Hence residue recycling could be achieved through crop with fish + poultry under wetland system and crop+ agroforestry+goat under dryland ecosystem.

**Keywords:** integrated farming system, wetland, rainfed, recycling

## Introduction

Indian economy is predominantly rural and agriculture oriented where marginal and small farmers constitute 76.2% of the farming community. There is no scope for expanding the individual farm size because of steady explosion of population with shrinkage of cultivated land due to urbanization and industrialization. Therefore, the vertical expansion of agricultural production alone is possible by integrating appropriate farming components requiring lesser space and time and ensuring periodic income to the farmer. For assured regular and periodic income and decent living the farmer has to think of other allied enterprises with better complementarity to cropping activity. This is made possible through optimal crop and livestock mix, consistent with the farm resources immediately available. Hence, the present integrated farming system research was carried out to evolve economically viable and sustainable farming system component integration for marginal wetland and dryland farming situations.

## Methodology

In wetland Integrated farming system enterprises like cropping, fishery, poultry, pigeon and goat rearing were tried with a view to recycle the residue and by-products of one component over other and to sustain the productivity, profitability, employment generation for lowland farms of Tamil Nadu, India. In one hectare farm, an area of 0.75 ha was assigned for crop activity, 0.10 ha for growing fodder grass to feed the goat unit (20+1), 0.03 ha allotted to goat shed and the remaining 0.12 ha allotted to 3 fish ponds. Three integrated farming systems *viz.*, crop + fish + poultry (20 Bapkok layer birds), crop + fish + pigeon (40 pairs) and crop + fish + goat (Tellicherry breed of 20 female and 1 male maintained in 0.03 ha deep litter system) were tried for three years. Polyculture fingerlings of 400 numbers (catla, rohu, mirgal/ common carp and grass carp) in the ratio of 40:20:30:10, respectively, reared in 3 ponds of size 0.04 ha (depth of 1.5 m) each.

Fishes were fed with dropping poultry, pigeon (700 kg poultry/pigeon droppings obtained from 20 Bapkok layers/40 productive pairs of pigeon) sheltered over two fish ponds and goat droppings (3 animals – 800 kg droppings) to assess the feasibility of rearing fish by using different organic manures as feed. Under Integrated Farming System, cropping sequence includes (i) sugarcane (planted) – sugarcane (ratoon) – banana (3 years) (ii) banana - turmeric – rice – banana (3 years) and (iii) maize – rice – sesame – sunnhemp (annual) each in 0.25 ha and bajra - napier grass + desmanthus (perennial) in 0.10ha.

Conventional cropping system comprising (i) rice -rice - black gram (ii) maize – rice - black gram (iii) maize - rice - sunnhemp and (iv) rice - rice - sunnhemp each in 0.25 ha, as practised by the farmers was taken up for comparison. To sustain the productivity of soil through integrated nutrient supply, recycled poultry, pigeon and goat manures and composted crop residue (banana waste and sugarcane trash) as vermicompost each @ 6.25 t/ha were tested along with 100, 80 and 60% of the recommended fertilizer for the sequences of cropping.

Another study involving grain crop, fodder crop, fodder trees, perennial grasses and goat rearing in one hectare of rainfed land was carried out at Western zone of Tamil Nadu, India. Generally, in dryland, farmers raise only fodder sorghum yielding 3 to 10 tonnes of fodder/ha depending upon the rainfall. In the integrated approach, the cropping pattern was modified by including both sorghum grain crop (0.25 ha) and sorghum fodder crop (0.25 ha) along with cowpea. To meet the fodder requirement of goats, 0.28 ha of land was raised with subabul and *Cenchrus ciliaris* and another 0.2 ha with *Acacia senegal* and *Prosopis cineraria*. The goat unit comprised 20 ewes and 1 buck of Tellicherry goats sheltered in 0.02 ha land area. The observations on productivity, economics, employment and resource recycling of the systems were recorded.

## Results

Integrated farming system provides an opportunity to increase economic yield per unit area per unit time by virtue of intensification of crop and allied enterprises. Research results on integrated farming system revealed that integration of crop with fish, poultry, pigeon and goat resulted in higher productivity than cropping alone under lowland. Crop + fish + goat integration recorded higher rice grain equivalent yield of 39610 kg/ha (mean over three years) than other systems. Similarly, as an individual animal component, the goat unit (20 + 1) gave the highest productivity of 8818 kg (mean over three years). This could also provide 11.0 t of valuable manure apart from supplementing the feed requirement of 400 numbers of fish. While assessing the feasibility of rearing fish by using poultry, pigeon and goat droppings as feed, the fish fed with poultry droppings resulted in higher fish yield (825 kg / 0.04 ha ponded water) than the other two sources of feed.

The highest net return of Rs.1,31,118 and per day return of Rs.511 $ha^{-1}$ were obtained by integrating goat + fish + cropping applied with recycled fishpond silt enriched with goat droppings. Higher net return of Rs.3.36 for every rupee invested was obtained by integration of pigeon + fish + cropping applied with recycled fishpond silt enriched with pigeon droppings. The employment opportunity was also increased to 576 mandays/ha/year by integrating fish+goat in the cropping as against cropping alone (369 mandays/ha/year) (Table 1). Combining cropping with other allied enterprises would increase labour requirement and thus provide scope to employ family labour round the year. The poultry, pigeon and goat droppings were utilized as feed initially and at the end of a year after the fish harvest, about 4500 kg of settled silt from each pond were collected. The pond silt was utilized as organic sources to supply sufficient quantity of nutrients to the crops.

Additional nutrients gained by recycling were the highest with poultry manure (65.7, 28.4 and 25.0 kg N, $P_2O_5$ and $K_2O$ respectively) than with goat or pigeon manure. The system as a whole provided opportunity to make use of produce/waste materials of one component as input on another at least cost/no cost at farm level (Jayanthi, 2002).

Dryland farming system results revealed that through short duration field crops and perennial crops the feed requirement of 20 ewes and one buck for all the 365 days can be met. After 5 years, the perennial fodder trees could bear the shock and would yield sufficient quantum of loppings to supplement the feed requirement of 21 adults along with millets, legumes and perennial grass linked. Tellicherry goat is a good breed to build up body weight for every unit of feed secured through different sources of feed under rainfed condition. It is a dual purpose animal, where it gives 80 to 100 ml of milk after satisfying the full requirement of dependent kids on it every day. Twenty productive females could give 45 kids per annum. Each kid at the time of weaning will weigh around 12 kg. Moreover the unit of 21 animals with different stages of kids under deep litter system would give 11.2 t of valuable manure including the bedding material (coir pith) used (2600 kg/year). Thus deep litter manure will be an excellent source of primary, secondary and micro nutrients. This quantum is sufficient to enrich the soil fertility under rainfed situation as farmers hardly apply inorganic fertilizers. This confirms with the findings of Radhamani and Esther Shekinah (2002).

## Conclusion

To enhance the productivity, economic returns and employment generation for family labour, integration of crop with fish + goat or pigeon or poultry could be adopted than cultivating the crop alone under lowland situation. Better residue recycling could be achieved through integration of crop with fish + poultry under wetland system and crop+ agroforestry+goat under dryland ecosystem.

*Table 1. Economics of integrated farming systems.*

| Farming systems | Net return (Rs. $ha^{-1}$) | B:C ratio | Per day return (Rs. $ha^{-1}$) | Employment generation |
|---|---|---|---|---|
| Cropping alone | 37153 | 2.43 | 178 | 369 |
| Crop + Fish + Poultry | 97731 | 3.02 | 400 | 515 |
| Crop+Fish+Pigeon | 98778 | 3.06 | 400 | 515 |
| Crop+Fish+Goat | 131118 | 3.36 | 511 | 576 |

## References

Esther Shekinah, D., 2002. Integrated farming system for sustainable resource management in rainfed vertisols of western zone of Tamil Nadu. Ph.D. (diss) Tamil Nadu Agricultural University, Coimbatore.

Jayanthi, C., 2002. Sustainable farming system for lowland farming of Tamil Nadu. Integrated Farming System Ad-hoc scheme Completion report, 2002.

Radhamani, S., 2001. Sustainable integrated farming system for dryland vertisol areas of Western Zone of Tamil Nadu. Ph.D. (diss) Tamil Nadu Agricultural University, Coimbatore.

# Improving natural resource management by "thinking industry and acting regionally": Introduction and successes of dairy Australia's multi-scaled natural resource management strategy

*J. Jennings[1], P. Day[2], C. Phelps[3] and R. Packham[4]*
*[1]Natural Resource Management Coordinator (NSW), Dairy Australia, University of Western Sydney, Hawkesbury, Australia; [2]National Coordinator Dairying for Tomorrow, 2000-05; [3]National Coordinator Dairying for Tomorrow, 2005-; [4]Former Professor, University of Western Sydney, Hawkesbury, Australia*

## Abstract

Following the Australian Government's National Land & Water Resources Audit (commenced in 1997; NLWRA, 2005), and with support from the National Landcare Program (NLP, 2005), the Australian dairy industry developed its strategy to improve Natural Resource Management (NRM) called *Dairying for Tomorrow* (DfT, 2005). This strategy is inclusive of industry, government and NRM bodies, and functions simultaneously across local, regional, state and national domains of NRM.

To deliver this strategy, DfT identified the need to embrace regional diversity between different - but interdependent - social, industry and natural systems, whilst maintaining a common focus on improving NRM. Two process tools for achieving on-farm change were designed with support and ownership from Australia's eight Regional Dairy Programs (RDP), including local piloting. Flexibility to *regionalise* both tools to reflect the local context within which they are applied in theory and practice has proved critical for effectively addressing NRM issues and maintaining viable on-farm (and hence industry) financial performance.

**Keywords:** natural resource management, dairy, process change

## Context: Regional governance, industry and the environment

Australia is made up of six States (and two non-dairy Territories), each with their own tier of government. NRM programs and projects are increasingly driven by state and federal governments that utilise the recently instituted regional NRM bodies within each state, known as Catchment Management Authorities (CMA). The state of New South Wales, for example, is divided into thirteen CMAs, of which the dairy industry is significantly present in eight.

Given this regionally focused government framework for addressing NRM, the DfT program devised a process structure for desired NRM outcomes to be negotiated between itself (industry) and the CMAs. These industry partnerships with CMAs are able to attract funding and resources from state and federal sources that may be administered and delivered on-farm through the CMA. Additional non-government support is sought where appropriate.

Partnerships created at the catchment scale between the dairy industry and a particular CMA deliberately and explicitly facilitate the inclusion of local farmer input, local, state and/or federal government participation, and a variety of other relevant private sector stakeholders such as milk processors, energy and water companies, agri-suppliers *etc.* These partnerships are intended to build multi-faceted, professional, formalised relationships that work toward achieving mutually agreed NRM targets.

A key benefit of the regionally-oriented focus on NRM is that environmental and social/ community issues are targeted accurately. This is demonstrated by the fact that different regions have different environmental and industry characteristics, and hence the solutions must be differently tailored to be relevant and effective. This point has proved a valuable learning experience in terms of addressing both the condition of the natural environment and the propensity of farmers to achieve better NRM. This regional focus has created a culture of 'acting regionally', and is an approach advocated here.

## DfT design and structure

The DfT program became a reality following the Australian Government's audit of Australia's natural resource base (NLWRA, 2005), which included a national dairy survey of current on-farm practice (to be repeated in 2006 to evaluate progress). One consequence of this audit was a federal funding scheme (Pathways Program; NLP, 2005) that enabled an NRM oriented partnership to emerge between the Australian government and the national dairy industry, Dairy Australia. This funding supported the design and delivery of DfT, including two process tools known as the Dairy Self Assessment Tool (DairySAT) and Setting Targets for Change (STC), as well as the appointment of Regional NRM Coordinators stationed in each major dairy region across Australia (eight in total). These coordinators help broker NRM programs and projects for groups of farmers, while local facilitators and technicians are employed to implement them.

Having designed a two-pronged approach for achieving on-farm change, DfT sought to develop each by piloting them in regional dairy areas and incorporating all critical feedback of all relevant stakeholders, especially dairy farmers. This consultation successfully created a sense of ownership and acceptance of DairySAT and STC as viable NRM tools, as well as fine-tuned their design to best meet the needs of on-farm application and NRM objectives at the catchment scale.

In early 2005, with funding from the federal government's *Sustainable Industries Initiative*, part-time Regional NRM Coordinators were appointed by DfT across Australia. These positions are intended to broker, enhance and facilitate specific industry-NRM relationships at the catchment scale that support locally delivered projects resulting in individual on-farm improvement (management change and on-ground works) that collectively, produce positive NRM sub-catchment and whole catchment-scale outcomes. Typically, an industry-NRM relationship is forged between a single dairy group and their relevant CMA, with support as appropriate from other interested stakeholders. In NSW for example, there are 8 major dairying areas, totaling approximately 900 dairy farms, with the range of farms per region being from as little as 30 to around 300. The initiation, maintenance, and progress of these relationships has largely become the responsibility of the Regional NRM Coordinators, who are well placed to address each region's dairy industry agenda and NRM catchment issues, using the tools of DairySAT and STC.

## Tools for change: DairySAT and STC

DA has developed a considerable emphasis on capacity building and providing supportive tools (such as industry guidelines) for farmers. Many are developed with a view to being compatible with existing tools, such as Environmental Management Systems (EMS), which provides farmers with the choice to utilise whichever method they feel is most appropriate without departing from the broad industry agenda for improving NRM.

DairySAT (2005) is a comprehensive farm level survey that assesses the current NRM performance of an individual farm. It comprises nine environmental categories that are relevant to Australian dairying (effluent, irrigation, nutrients, soils, chemicals, farm wastes, pests and weeds, biodiversity, and, air and energy), with simple tick boxes to identify if a farm is managing below, on par, or above acceptable industry practice. Having identified the biophysical characteristics of a farm, the farmer is supported by a continuous improvement process involving one-to-one facilitation and group workshops that, in addition to addressing technical and biophysical matters, deliberately create a social support network. From these networks farmers source motivation, ideas, technical solutions, peer review and approval, and ultimately an industry-initiated culture of profitable NRM that effectively function as a community of practicing dairy farmers and stakeholders that share a positive NRM ethos.

Once an individual farmer has completed DairySAT (with assistance by a trained DairySAT facilitator, usually taking approximately 2 hours), they can identify where they need to improve their NRM. These priority areas are assessed by each farmer to determine exactly which issue/s will be dealt with first. The participating farmers and their DairySAT facilitator can then collectively identify the highest priority NRM issues across the entire group, and critically, are able to translate their local NRM issues to sub-catchment and whole catchment environmental priorities. This farmer knowledge (documented by DSAT process) provides a highly valuable information base for NRM bodies to support clear industry-defined NRM targets that are realistic in terms of on-farm achievement.

The STC process begins where the DairySAT process ends, by taking the NRM priority areas identified by DairySAT and using them as the basis for engaging formally with NRM bodies to negotiate the setting of targets for on-farm change that will simultaneously improve catchment health. In theory, STC is a step-wise process that engages farmers and dairy stakeholders to: i) identify the impact of industry on the environment; ii) define the preferred catchment and sub-catchment environmental conditions; iii) assess on-farm management practices; iv) identify management change targets; v) develop and implement programs for on-farm change; and finally, vi) evaluate actions and outcomes achieved, and re-assess the cycle of events leading to a fresh round of identifying industry contributions to environmental issues (STC, 2005).

In NSW for example, two farms that participated in a DfT pilot (using DSAT and STC) identified the need to improve river water quality. Their solutions comprised substantial upgrades of their laneways and installing off-stream watering systems. With government and CMA support both farms were able to afford the upgrade and evolve their management styles to be greatly improved both environmentally and economically (herd size capacity was increased while nutrient pollution into water flows was reduced; DEC, 2005).

## Analysis of DfT theory and practice to date

Reflection upon the DfT program's framework indicates that industry motivation to implement NRM improvement critically requires farmer ownership and commitment to NRM issues. Under DfT, the incentives for farmers are threefold: i) financial support for major on-farm infrastructure upgrades and transition to superior management practices; ii) long term structural improvements to farm capacity and profitability resulting from more efficient and more environmentally sensitive farm management; iii) conforming to environmental legislation and, in most cases, gaining a sense of pride in exceeding minimum requirements and being responsible custodians of country for future generations. Once farmers are individually motivated to improve their own NRM, their collective NRM issues are profiled and presented to NRM bodies for support and mutually agreed targeting of priority issues, including design of regional programs, local projects and their resourcing and implementation.

DfT experience has shown the value of first establishing an industry-approved agenda of local NRM issues that are then negotiated and (hopefully) supported by government and NRM bodies to achieve mutually beneficial industry and NRM outcomes. Dialogue between Dairy Australia and the federal Australian government was instrumental in shaping the framework that linked both industry and government agenda regarding NRM (on a national scale), and helped initiate government policy and programs such as the *NLP Sustainable Industries Initiative*. From this national framework other state and regional structures were able to be added, hence facilitating DfT's capacity to work across all scales simultaneously.

Genuine farmer ownership of their NRM agenda is achieved by allowing the regionalisation of both DairySAT and STC, while still maintaining the core national dairy industry agenda of improved environmental management within profitable farming systems. Furthermore, farmer's ownership of NRM issues is extended to NRM and funding bodies through dialogue of catchment targets. In most cases, a CMA is invited to take ownership of the solutions for farmers' NRM needs, but importantly with a high level of confidence that their resource contributions will drive and achieve specific NRM outputs and outcomes – which in general reflects their own internal charters and Catchment Action Plans for investing in environmental outcomes (NRCMA, 2005).

At the farmer level, DfT provides a process framework that can genuinely shift farmers' personal world view on the value of improving their NRM, engage farmers in personal learning experiences, and enhance their individual and collective capacities to better manage their farms for profit and care of the natural resource base. It does this by leveraging funding and resources from NRM bodies and relevant stakeholders to support resource intensive, one-to-one and group-based extension practices. The successful results of this comprehensive extension approach are validated and value-added by enabling NRM bodies to share on-ground industry successes. For the NRM bodies the critical step of appropriately valuing individual farmer learning as a means to achieving NRM outcomes cannot be overlooked, and the tools of DairySAT and STC make this explicit.

## Conclusion

This paper introduces the philosophical, structural and strategic design of the DfT program, which itself is emerging as a valuable model for achieving multi-stakeholder engagement across an array of institutional relationships amongst farmers, industry, the private sector, NRM bodies (regional, state and national), and all levels of government. The DfT approach has been to think conceptually about improving NRM across the entire dairy industry, but to focus actions on regional and local requirements. To date, this has proved a successful formula for achieving on-ground NRM improvements that deliver improved farm profitability and promote catchment scale health for the broader rural and urban community.

## References

DairySAT, 2005. DairySAT – NSW (version 1), available on CD from Dairy Industry Development Company (NSW) or Dairy Australia, Melbourne, Australia.

DEC, 2005. Dairy gains for the environment through increased productivity, Department Environment and Conservation, Sydney, Australia.

DfT, 2005. Dairying for Tomorrow website hosted by Dairy Australia, accessed 15/01/2005 at www.dairyingfortomorrow.com.

NLP, 2005. National Landcare Program website hosted by Australian Government, accessed 15/01/2005 at http://www.daff.gov.au/content/output.cfm?ObjectID=D2C48F86-BA1A-11A1-A2200060B0A04273.

NLWRA, 2005. National Land and Water Resources Audit website hosted by Australian Government, accessed 15/01/2005 at http://www.nlwra.gov.au/about.asp?section=13.

NRCMA, 2005. Northern Rivers Catchment Management Authority – Draft Catchment Action Plan, Grafton, Australia.

STC, 2005. Targets for Change – Setting Resource Management Targets for Australian Farms, Dairy Australia, Melbourne, Australia.

# New ways in the management and valorisation of natural resources? Results of a comparative analysis of a sample of agriculture-environment projects of the *Regional Action* pilot programme in Germany

*Karlheinz Knickel, Melanie Kröger and Sarah Peter*
*Institute for Rural Development Research, J.W. Goethe University, Frankfurt, Germany*

## Abstract

Finding a new balance between societal demands for high environmental quality and the pressures resulting from competition in a world market economy has become a key issue in policy-making. The German *'Regional Action - Rural Areas Shaping the Future'* pilot programme has been implemented in order to gain best-practice models for securing the economic, ecological and social viability of rural areas and for trial-testing a new integrated, bottom-up approach. Managing the countryside is less and less comprehended as being the responsibility of farmers alone. The role and contribution of the various types of actors involved is changing and co-operation and co-ordination is becoming increasingly important, the co-operation between farmers, conservation groups, small processing companies, gastronomy and businesses involved in rural tourism being particularly vital. The integrated approach helps to reconcile diverse interests.

A characteristic feature of the programme is the heightened awareness of the specific regional potentials among regional actors, which helps to discover new possibilities for an amenity-led development of rural areas. Natural resources constitute added value through environment-friendly use while their viability is secured. Conservation practices contribute to the economic competitiveness of agricultural enterprises by enhancing the use of natural resources but also by capitalising environmental assets. A significant number of Regional Action projects aim at valorising the aesthetic and cultural value of agricultural landscapes, for example, by exploiting rural and green tourism potentials.

The projects that are being presented can be interpreted as aiming at a reconstitution of nature-society relations, indicating that agriculture and the potential of rural areas are no longer being evaluated in monofunctional terms.

**Keywords:** rural amenities, valorisation of natural resources, agriculture, sustainable development, multifunctionality, Common Agricultural Policy

## Sustainability, multifunctionality and the modernisation of agriculture: New policies as a reflection of new societal demands

After a long period of classical modernisation policy, meanwhile the Common Agricultural Policy (CAP) is distinguished by a quite new approach and concept of modern agriculture. While the main aim of traditional agricultural policy was the enhancement and maintenance of a sufficient production of food and raw material, at the beginning of the 1990s a comprehensive reorientation of agricultural policy was initiated. More and more attention was paid to ecological and social problems, the side effects of the structural change of agrarian production. The reorientation also reflects the increasing societal reservations about modern agriculture characterised by standardisation, industrialisation, rationalisation, intensification and mechanisation (Kröger, 2006). New political and societal paradigms came into existence: Sustainable development became a generally accepted and promoted cross-sectional aim of regional, national and international policy while the more and more common term multifunctionality points to the systemic nature of the agricultural system, its surroundings and diverse outputs. Cairol *et al.* (2005) emphasize that multifunctionality can be directly related to the three pillars of sustainability, and that it can in this respect be seen as a useful analytical framework. The introduction of these concepts into public debates, decision-making and research also means that the notion of a 'modern' agriculture itself is changing. As this reorientation refers explicitly to the (negative) modernisation effects, the process as a whole can be seen as a process of 'reflexive modernisation' (Kröger, 2006; Beck, 1994).

## Political measures to promote the valorisation of natural resources: The example of the Regional Action pilot programme

Within the new paradigm of a more sustainable and a multifunctional agriculture production remains only one among various functions of agriculture (Knickel, 2001; Pretty, 1998). The guiding idea is that through multifunctionality agricultural enterprises can create a broader basis of income generation and at the same time gain greater appreciation of their outputs from society (Knickel, 2001; Wilson, 2001). Besides food and non-food crops, agriculture produces environmental, social and cultural non-commodity outputs (Hervieu, 2003; Green *et al.*, 2006). The aesthetic and cultural value of agriculturally shaped landscapes in turn can to a considerable extent be valorised, for example, by realising rural tourism potentials (Knickel, 2001; Pretty, 2002). New political measures and programmes reflect these considerations (Knickel and Peter, 2005).

The 'Regional Action - Rural Areas Shaping the Future' pilot programme is an example of a new type of support scheme that specifically addresses the development of economic activities as well as their linkages with the enhancement of environmental quality. It was initiated by the Federal Ministry of Agriculture (BMVEL) in 2001. Regional Action aims at an integrated and sustainable development of rural areas, *i.e.* development initiatives that go beyond the agricultural sector. The active generation of synergy is central to the activities and their combination at farm and regional level. While specialisation in agricultural production and segregation of agriculture from other rural activities had been envisaged in the past, multifunctional and amenity-led development is focused on mutual benefits and 'win-win situations' created by different activities (Brunori and Rossi, 2000; Knickel and Renting, 2000). The expansion of tourism services to provide opportunities for combining environmentally compatible agriculture and forestry in an intact landscape with marketing of regional products, addressing the interconnectedness of positive landscape image, regional identity and the success of new services and products is one of the explicit objectives of Regional Action. The revitalisation and strengthening of urban-rural linkages is a closely related aim promising mutual benefits.

Although different interests still characterise the relationship between farmers, landowners, inhabitants and nature conservation organisations, new forms of co-operation become possible as the integrated approach helps to reconcile diverse interests.

## The role of environmental and natural amenities within the Regional Action programme

The preservation of rural landscapes attractive for living and tourism is an important goal in all of the model regions. The model regions' landscape potentials such as lakes and flower-rich grasslands and infrastructures such as hiking trails offer visitors and the local population possibilities for leisure activities. These interrelations are reflected in the main aims of the projects implemented: For 37 percent of the almost 730 projects that have been implemented in the 18 model regions since the beginning in 2002, the development of eco-tourism plays a major role, followed by agriculture and nature conservation with 34 percent (Table 1).

In 52 percent of all projects, adding value to landscape and other natural resources as well as their enhancement or maintenance is a clearly recognisable, explicit objective. A share of 13 percent of all projects is mainly aimed at adding value to natural resources such as wild plants (berries, herbs, *etc.*) or biomass. 18 percent of all projects are mainly aimed at adding value to regional landscapes, mostly through eco-tourism (16 percent). Maintaining or enhancing the character and diversity of regional landscapes plays a major role for 12 percent of the projects, while a small share of projects *primarily* focuses on the cultural heritage of regional landscapes (5 percent) or maintaining or increasing biodiversity (4 percent).

Due to the integrated, intersectoral alignment, the contents of the projects are addressing development processes that take place at several interfaces. A share of 16 percent of all projects is located at the interface between agriculture and landscape/nature, a share of 5 percent at the interface between nature and the overall (economic) development of the region.

## Projects at the interface of agriculture, nature and landscape

The projects introduced below are intended to give an impression of the character of initiatives located at the agriculture-nature interface.

### *Recultivation of wildflower meadows*

The wildflower project in Ostfriesland (north-western Germany) was implemented between 2003 and 2005. Wildflower meadows used to be an important element of the regional landscape. Due to land amelioration and drainage, and, more generally, the intensification of grassland

*Table 1. Importance of projects addressing relevant interfaces.*

| Project aims at | Number of projects | Share of projects (%) |
|---|---|---|
| maintaining / enhancing biodiversity | 27 | 4% |
| maintaining / enhancing landscape character / diversity | 89 | 12% |
| maintaining / enhancing cultural heritage of region | 33 | 5% |
| adding value to natural resources (other than landscape) | 95 | 13% |
| adding value to landscape (generally) | 12 | 2% |
| adding value to landscape through eco-tourism | 117 | 16% |
| other | 353 | 49% |
| Total | 726 | 100% |

Source: Own compilation based on BMELV, 2006.

use, the diverse fauna and flora vanished. Regional actors stress that flowering landscapes do form the region's image vitally, but have to regret that the variety of flower species actually has dramatically diminished. Forms of quiet recreation of the local population and visitors such as walking, cycling and boat tourism predominate in the region. There are numerous bicycle routes, which the region is famous for, and waterways running through the region, being the two big potentials for regional tourism. The project aims at enhancing the economic, ecological and aesthetic value of wildflower meadows by cultivating new ones using typical regional seed derived from remaining wildflower meadows. It tries to integrate three differing objectives: realising tourism potentials, contributing to biodiversity and improving agricultural incomes through management contracts (BMVEL, 2006).

### *Generation and use of renewable energy in agriculture*

Whereas in the case mentioned above nature conservation is a predominant objective, there are other projects focussing on the exploration and utilisation of regenerative sources of energy as main objective. The initiative 'generation and use of renewable energy in agriculture' (2002 to 2005) in the northeastern Mecklenburgische Seenplatte is an example. It is a project addressing the use of renewable resources and at the same time aimed at maintaining and enhancing regional added value and creating jobs. Investments in an oil mill supporting the use of oilseed rape for energy generation in the agricultural enterprise carrying the project are combined with new services such as information and advice for the population. Information is provided regarding the use of rapeseed oil as environment-friendly fuel and for other farmers who are interested in adopting rapeseed oil production. Additionally, the transparent production process that can be watched by visitors is expected to provide an impulse for rural tourism (BMVEL, 2006).

### *Landscape care with grazing sheep*

Landscape care by using grazing animals, practiced in several model regions throughout the country, is illustrated well by the sheep project (2003 to 2005) in the Wendland region located close to Ostfriesland. The initiative is carried by a cross-sectoral co-operation between agriculture, forestry, small enterprises, municipalities, nature conservation and tourism. Like in the first example are diminishing distinctive landscape features – heathland and typical non-manured meadows (which have been reduced by 90 percent within the last two centuries) – due to afforestation and intensified agriculture the starting point. Objective of the project is the creation of a network of the remaining meadow biotopes, connected by a fixed trail for a newly established migrating flock of sheep. This form of landscape care contributes to the maintenance of endangered plants, insects and other species as opposed to mechanical measures that only help counteract scrub growth. Flower meadows, like in the example introduced above, as well as sheep attract tourists. Besides care contracts, the sheep farming is mostly financed by marketing products such as meat and wool (BMVEL, 2006).

## Conclusions

Projects that simultaneously pursue ecological and social goals, and that aim at a valorisation of environmental and natural resources effectively link the three pillars and goals of sustainable development. Programmes like Regional Action may in this respect generate both positive short and long-term effects for agriculture and rural areas. In countries like Austria and Switzerland has the compensation of environmental goods a longer history and the combination of agri-environmental schemes with other rural development measures ties up to a long-lasting tradition of diversified farms (Kröger, 2006). Agriculture and the potential of rural areas are no longer being evaluated in monofunctional terms. Regional actors perceive a balanced economic development as a precondition for strengthening the role of farmers as producers of services, landscapes and biodiversity. The other way round, high-quality agricultural production with a high added value can be regarded as the key to a balanced overall rural development. Rural areas can in this respect be seen as trendsetters for a sustainable development.

## References

Abler, D., 2001. Multifunctionality: the question of jointness, applying the OECD framework: a review of literature in the United States, background paper prepared for the OECD conference Multifunctionality: applying the OECD analytical framework guiding policy design, 2-3 July, Paris (http://www.oecd.org/agr/mf/doc/USA-Abler-revised.pdf) Accessed: 02/2005.

Beck, U., A. Giddens and S. Lash, 1994. Reflexive Modernization. Politics, traditions and aesthetics in the modern social order. Cambridge: Polity Press.

BMELV, 2006. Projekt-Dokumentation (http://www.modellregionen.de) Accessed 20 January 2006.

Brunori, G. and A. Rossi, 2000. Synergy and coherence through collective action, some insights from Tuscany. Sociologia Ruralis 40, 409-423.

Cairol, D., E. Coudel, D. Barthélémy, P. Caron, E. Cudlinova, P. Zander, H. Renting, J. Sumelius and K. Knickel, 2005. Multifunctionality of agriculture and rural areas: From trade negotiations to contributing to sustainable development. New challenges for research. (http://www.multagri.net/section/deliverable/exec_file.php?doc_id=374) Accessed 20 January 2006.

Hervieu, B., 2003. Multifunctionality, a conceptual framework for a new organization of research and development on grasslands and livestock systems. Paris: Institut National de la Recherche Agronomique (INRA).

Knickel, K., 2001. The marketing of Rhöngold milk, an example of the reconfiguration of natural relations with agricultural production and consumption. Journal of Environmental Policy and Planning 3, 123-136.

Knickel, K. and S. Peter, 2005. Amenity-led development of rural areas, the example of the regional action pilot program in Germany. In: G.P. Green, D.W. Marcouiller and S. Deller (eds.): Amenities and rural development, theory, methods and public policy. New Horizons in Environmental Economics. Northampton: Edward Elgar Publishing, pp. 302-321.

Knickel, K. and H. Renting, 2000. Methodological and conceptual issues in the study of multifunctionality and rural development. Sociologia Ruralis 40, 512-528.

Knickel, K., H. Renting and J.D. van der Ploeg, 2004. Multifunctionality in European agriculture. In: F. Brouwer (ed.) Sustaining agriculture and the rural economy, governance, policy and multifunctionality. Cheltenham (UK) and Northampton (MA/USA): Edward Elgar Publishing, pp. 81-103.

Kröger, M., 2006. Die Modernisierung der Landwirtschaft. Eine vergleichende Untersuchung der Agrarpolitik Deutschlands und Österreichs nach 1945 (in press)

Pretty, J., 1998. The living land, agriculture, food and community regeneration in rural Europe. London: Earthscan.

Pretty, J., 2002. Change in agricultural policy and its consequences, will conservation keep farmers in business? In: J. Frame (ed.) Conservation pays?, BGS Occasional Symposium No 36 British Grassland Society, University of Reading, pp. 15-25.

Wilson, G.A., 2001. From productivism to post-productivism and back again? Exploring the (un)changed natural and mental landscapes of European agriculture. Transactions of the Institute of British Geographers 26, 77-102.

# Activating farmers' interests: Action research for biodiversity conservation and enhancement in Brandenburg

*Andrea Knierim and Harald Kächele*
*Leibniz-Centre for Agricultural Landscape Research, Institute of Socio-Economics, Eberswalder Str. 84, 15374 Müncheberg, Germany*

## Abstract

The paper presents the preparatory phase of an action research project on the encouragement of farmers' provision of public environmental goods in Brandenburg. Before the background of a weak institutional setting, the role of a farmers' association and a concerned research organisation in the process of problem analysis, needs assessment and the operationalization of knowledge are investigated. The initiated one-year dialogue process has resulted in a concerted call for action on knowledge exchange and dissemination and on methodological skills and reflections.

**Keywords:** dialogue process, voluntary change, research–practice partnership, nature conservation, Germany

## 'Rocky Roads' toward biodiversity conservation in Brandenburg's agriculture

Public environmental goods as *e.g.* biodiversity have a strong significance in Brandenburg. This can be seen by the fact that nearly 40% of the state's surface are part of landscape and nature conservation schemes and roughly 22% belong to Natura 2000 sites (MLUV, 2005). The public interest lays in the touristic attractiveness of the state, because tourism is one of the few increasing economic sectors (MW, 2005). Especially related to the 15 large scale protected conservation parks (cover areas between 100 and 1.300 km$^2$ ), there are broad political and public expectations that land users, *e.g.* farmers, should continue to provide beautiful landscape features and wildlife and plant diversity.

However, there are weak institutional structures that support these requirements on farmers' change of land use measures: Within the privatised agricultural extension system, these public interests are not covered by the advisers (Nagel, 2003). A large transdisciplinary research project (1998–2002) on the implementation of more sustainable land use practices in Brandenburg showed that:

- although farmers are aware of environmental problems and possibilities to improve the farm's state, the large majority accorded no (high) priority to this issue;
- private extension agents were not ready to incorporate environmental concerns into their mostly micro-economics oriented counselling;
- slow decision making within the public administration hindered the integration of farmers' ideas and interests within policy making; and
- even existing linkages and knowledge exchange between researchers and farmers decreased during the projects time frame (Nagel *et al.*, 2003).

Summarising, the authors perceive only a fragmented agri-environmental knowledge system, in which the pathways and interfaces for the necessary information and knowledge exchange are not (sufficiently) operational and still cause more problems than present solutions for interested users. Hence, the situation is characterised as a 'rocky road' and doubtfully, the question is posed whether it is realistic to expect, that farmers themselves will become active and demand for information to realise public interests (Nagel, 2002).

## Research's role and potentials for the enhancement of biodiversity

In contrast to the above mentioned findings, today's agricultural politics widely promulgate the concept of a multifunctional agriculture based on a managerial type of farmer who actively searches for and realises new income generation possibilities beside the traditional production (Knierim, 2006; Wiggering *et al.*, 2003). To explore this assumption with regard to the provision of ecological goods and to assess research's possibilities to enhance a more sustainable land use, we - social scientists of a large interdisciplinary research centre in Brandenburg (ZALF) - initiated a dialogue project with the major farmers' association of the state Brandenburg (LBV). With respect to the ZALF's mission to foster a more sustainable use of agricultural landscapes through research and the supply of decision support systems (www.zalf.de), this dialogue process was configured as the preparatory phase of an action research project. Substantial findings of a sequence of natural science and economic research projects, that resulted in applicable practices for a biodiversity friendly farming were used as an example to start the interaction (Berger *et al.*, 2003, 2005).

The action research approach is the appropriate methodological frame because we had proper interests to contribute to problem solving and to increase our understanding of the organisational change processes assumed to be necessary (Albrecht, 1992). Moreover, we considered the relationship ZALF–LBV as one of active learning partners and not as one of analytical advisors towards clients (Liu, 1997).

Our initial assumptions for the cooperation were that:

1. the multifunctionality concept is a promising way for ascertaining future income generation in agriculture;
2. farmers and farmers' association could and should embrace the protection and development of biodiversity as one service towards society; and
3. that social science research should contribute actively within this process of change to enhance knowledge exchange and user orientation in the research subsystem.

## Preparing the stage: Developing a dialogue process

Based on these assumptions the dialogue process was organised as a series of four meetings in the year 2005 (*cf.* Table 1) on which participated both employed members and voluntary representatives of the farmers' association, several researchers from ZALF and at times representatives from the state's ministry of agriculture and other organisations concerned. The four meetings can be considered as complementary facets of a systemic situational analysis, exploring several perspectives from farmers and researchers (Leeuwis and Van den Ban, 2003: 265 cont.). The process started with a general discussion, in which the situational appraisal was given by both partners. Next, a single farm project of a public-private cooperation for nature conservation was presented. Thirdly, crucial research results were displayed and discussed during a field trip. And finally, the state of knowledge, as collected by ZALF, was appreciated and conclusions were elaborated.

*Table 1. Preparing the stage: Characterisation of the four meetings.*

| Date | Type and target of the meeting | Focus of results | Participants |
|---|---|---|---|
| 01.3.05 | Opening Discussion: General situational analysis | Identification of interests and fears; Outline of the components of a problem analysis | LBV employees (2), Members of the farmer association (5), ZALF social scientists, other |
| 01.06.05 | Presentation of farm biodiversity conservation financed by supplement and compensation funds | Single farmer as initiator and driving force of measures and partner for public and private agents | LBV employees (2), Members of the association (4), ZALF social scientists, other |
| 27.09.05 | Field – Trip: Visit of nature conservation oriented set aside on a farm; Additional expert information and discussion | Project related experiences of farmers as partners for publicly funded measures | LBV employees (2), Members of the association (15), ZALF social and natural scientists (3), other |
| 25.11.05 | Presentation: Study on farming related impacts on biodiversity in Branden-burg; Joint assess-ment discussion | Identification of associational deficits and starting points for improvements | LBV employees (2), Members of the association (3), ZALF social scientists |

## Results of the iterative situational analysis

Summarising, the dialogue process has led from a rather general situational analysis to a problem focusing specified description that is now the basis for action. During the first meeting, four fields of barriers to farmers' engagement were identified: (i) political insecurity, (ii) the broad range of self-perceptions among farmers about roles and tasks, (iii) lack of societal feed back or predominantly negative feed back and (iv) incomplete knowledge about the 'real' problem. As a response to this last barrier, it was stipulated that the ZALF partner would prepare a report on the scientific 'state of the art' with regard to the question whether and how farming really affects biodiversity in Brandenburg. The second and the third meeting allowed to enlarge the understanding of the interests and motivations of farmers, that are practically engaged in biodiversity related farming practices. It became obvious that (a) to understand and defend the ecological processes that are aimed at politically, farmers need to acquire considerable knowledge, (b) it is this kind of knowledge which helps farmers to communicate their doing to the local observers and especially to land owners and (c) this knowledge is not acquired automatically when participating in agri-environmental or nature conservation measures. Hence, knowledge exchange has to take place between different actors: researchers and land users, land users and local communities, neighbours and further more.

The state of the art report, elaborated by the ZALF researchers and discussed during the fourth meeting, summarised manifold starting points for improving the biodiversity conservation both on arable and on grasslands. From the LBV perspective, many of these findings are already incorporated within farmers' best practices and actual political requirements. However, it was admitted, that the average farmer might not know about the ecological benefits he is or could be able to provide.

While during 2005, the LBV could increase his participation in the state's related policy making process due to the dialogue process, the organisation did not yet start to work on the societal acceptance and the feed back of social groups and stakeholders like consumer associations or environmental lobbying organisations.

## Discussion and conclusion

With respect to our initial assumptions, the dialogue process showed that:

- multifunctionality of agriculture is a useful entry point into discussion but does not yet reflect farmers' approach to biodiversity conservation;
- the farmers' association might be conscious of the relevance of biodiversity conservation and their members might be ready to realise it as a by-product of farming but only few farmers are already active and self-confident service providers;
- the tasks fulfilled by us through knowledge operationalization, participating with different actors in a dialogue and facilitating complex system appraisal within such a nature conservation related knowledge system constitute a substitute for the non-existing extension system, in which multi-stakeholder learning and network building would clearly be key functions (Leeuwis, 2004).

After this year of dialogue, the farmers' association has started to extend the external interaction (*e.g.* with the ministry) and planned increased internal intervention too (developing own vocational measure for biodiversity conservation). We are invited to accompany this training and dissemination measure through the participatory development of a tool for evaluation and supervision of the organisation's change.

With respect to the action research approach, we consider this dialogue process as an example for a social experiment which reveals the orienting effects of a targeted intervention aiming at voluntary change (Albrecht, 1992). We conclude that the jointly negotiated dialogue process is a solid way to prepare the basis for a substantial contribution of natural and social sciences research to the tackling of practical problems.

## References

Albrecht, H., 1992. Sozialwissenschaftliche Aktionsforschung in Entwicklungsprogrammen: Bedeutung und Bedingungen. In: Zeitschrift für Ausländische Landwirtschaft, Frankfurt, 7 (1963) H 1:4-21 quoted in Hoffmann, V. (ed.): Beratung als Lebenshilfe. Margraf Verlag, pp. 128.

Berger, G.; H. Pfeffer, H. Kächele, S. Andreas and J. Hoffmann, 2003. Nature protection in agricultural landscapes by setting aside unproductive areas and ecotones within arable fields ("Infield Nature Protection Spots"). – J. Nature Conservation 11, 221-233.

Berger, G.; H. Pfeffer, H. Kächele and J. Hoffmann, 2005. Nature conservation in agricultural landscapes by linking EU set-aside with agri-environmental programs. - In: Agricultural landscapes as habitats: agri-environmental measures and set asides. Proceedings of a symposium; 3rd of September 2004, Brussels; Hamburg (Deutsche Wildtier Stiftung), pp. 85-100.

Knierim, A., 2006. Farm Management Systems and Voluntary Action. International Journal for Agricultural Resources, Governance and Ecology (in press).

Leeuwis, C., 2004. Fields of Conflicts and Castles in the Air. In: J. Agr. Ed. Ext. 10, 63-76.

Leeuwis, C and A. van den Ban, 2003. Communication for Rural Innovation. Blackwell Publishing UK, 412 pp.

MLUV (Ministerium für Landwirtschaft), 2005. Naturschutz und Landschaftsschutzgebiete. http://www.mlur.brandenburg.de/cms/detail.php/183316.

MW (Ministerium für Wirtschaft), 2005. Touristische Infrastruktur wird weiter ausgebaut. http://www.brandenburg.de/sixcms/detail.php?id=243075&_siteid=20.

Liu, M., 1997. La validation des connaissances au cours de la recherche-action. In: Etud. Rech. Syst. Agraires Dév. 30, 183-196.

Nagel, U.J., K. von der Heiden and R. Siebert, 2003. Public goods and privatized extension: the rocky road towards agro-environmental extension. - Case study prepared for the workshop: Extension and Rural Development: A Convergence of Views on International Approaches? Sustainable Agricultural Systems and Knowledge Institutions, Agriculture and Rural Development; Washington, DC (The World Bank Group).

Nagel, U.J., 2002. Wissenstransfer in die Praxis. In: K. Müller, A. Dosch, T. Aenis, E. Baranek, T. Boeckmann, R. Siebert and V. Toussaint (eds.): Wissenschaft und Praxis der Landschaftsnutzung: Formen interner und externer Forschungskooperation. Margraf Weikersheim. 375 pp.

Wiggering, H., *et al.*, 2003. The Concept of Multifunctionality in Sustainable Land Development. In: K. Helming and H. Wiggering (eds). Sustainable Development of Multifunctional Landscapes. Springer Verlag, Berlin, Heidelberg, New York. 286 pp.

# Structural development in Danish agriculture and its implications for farmland nature

*Gregor Levin*[1], *Vibeke Langer*[2] *and Pia Frederiksen*[1]
[1]*Department of Policy Analysis, National Environmental Research Institute, PO Box 358, DK- 4000 Roskilde, Denmark;* [2]*Department of Agricultural Science, The Royal Veterinary and Agricultural University, DK-2630 Taastrup, Denmark*

## Abstract

During the last decades, development of Danish agriculture was characterised by concentration of land on fewer and larger farms and consequently increasing farm sizes. On basis of three case studies supplemented by national data we explore relationships between farm size change and farmland nature in terms of field sizes and hedgerow densities. Data point to a significant relationship between farm enlargement and increases in mean field sizes. Furthermore, mean field size is negatively related to densities of hedgerows. For the coming decades, scenarios for Danish agriculture point to a continued increase in farm sizes. We argue for an increased focus on the effects of a continued scale enlargement on farmland nature in terms of changes in field size structure. In order to reduce harmful influences of scale enlargement, we suggest that in relation to changes in field structure, agricultural policies should focus on restrictions on removal of old hedgerows and on subsidies for planting of new hedgerows.

**Keywords:** structural development in agriculture, farm size, farmland nature, field size, field size structure, hedgerow density

## Background

In Denmark, from 1960 until 2000 mean farm size has more than tripled (Hansen, 2001). Scenarios of the agricultural development estimate that the number of farms will decrease by 52% from 79,300 in 1990 to 38,500 in 2010 (Landboforeningerne, 2002). Scenarios also estimate that the number of farms in all size categories up to 100 ha will decrease, while number of farms larger than 100 ha will increase, leading to a dramatic concentration of farmland on fewer farms.

During the last century, farmland nature has suffered as a consequence of the structural development, due to homogenisation of the cultivated area, adjusting it to new technology (Ministry of Environment and Energy, 1995). However, in the mid-1980s the environmental discourse was introduced in agricultural policy and particularly since the early 1990ies, focus on the conservation and restoration of farmland nature increased. However, it remains an important question, whether the continued scale enlargement in agriculture still implies continued homogenisation and impoverishment of the farmland nature.

Over the last 50 years new technology necessitated the enlargement of agricultural fields through the merging of field plots, resulting in increasing field sizes (Benton *et al.*, 2003). In both Denmark (Clausen and Larsen, 1997) and England (Barr *et al.*, 1993; Westmacott and Worthington, 1997), during the last 50 years mean field sizes increased considerably and negative impacts on farmland nature and on hedgerow densities have been documented (Smith *et al.*, 2005; Benton *et al.*, 2003; Robinson and Sutherland, 2002).

Hedgerows have been planted in Denmark with public subsidies since 1880 (Fritzbøger, 2002). The Nature protection Agency however estimates a 40% decrease in dikes and hedgerows during the last 100 years (Prip, 1995). Decreases in hedgerow densities have also been documented for other intensively farmed landscapes in Western Europe (Deckers *et al.*, 2005; Barr and Gillespie, 2000; Haines-Young *et al.*, 2003; Burel and Baudry, 1990). Hedgerows in field divides are sensitive to the merging of fields and it can be hypothesised that merging of fields does influence hedgerow density.

Due to the varying functions of hedgerows, the development has not been continuous in time and space. In addition to agricultural production also other factors like *e.g.* farmers' age (Ackerman, 2003) or aesthetic and environmental functions (Kristensen, 2003; Busck, 2002; Kristensen, 2001) are important for hedgerow dynamics. Moreover during the last century subsidy schemes for hedgerow planting have particularly focused on protection against soil erosion in western Denmark, where sandy soils dominate (Fritzbøger, 2002).

In this paper we focus on scale enlargement in agriculture as driver for field size- and hedgerow dynamics. Combining results from different Danish studies we discuss the effect of a continued scale enlargement on field size- and hedgerow development. Our aim is to draw up a picture of future landscape development for use in policy making on nature conservation in the general farmland outside protected areas.

## Data and methods

The paper draws on partial results from 3 case studies and a national analysis. Focus is on the period from 1995 to 2004, and the main analytical variables used are farm size, field size (the area covered by a single plot of agricultural land use) and density of hedgerows (measured as m/ha).

The first study is based on quantitative interviews with app. 10% (N=340) of Danish organic farmers, exploring land use and field- and farm size in 2001 and landscape activities from 1996 to 2001 (Frederiksen and Langer, 2005). The second study analyses landscape changes on 72 conventional and 40 organic farms in the period 1995 to 2004 using aerial photos. Hedgerow and field size development were followed, while changes in farm size were not investigated. This study has also been used to confirm that hedgerow data for organic farms are representative for Danish farms in general. The third study analyses changes in field and farm sizes from 1997 to 2002 on 234 organic farms, which converted to organic farming in 1997 and were still organic in 2004 (Langer *et al.*, 2005). The 3 studies are supplemented by national data on changes in field- and farm sizes from 1998 to 2004. Furthermore, national data on hedgerow density in 2001 were derived from a national map of hedgerows. Finally, as data on hedgerow change are only available from study 2, we explore field size as an indicator of hedgerow density through this study.

For studies 1, 3 and national data, field sizes are derived from agricultural registers and reflect field units, for which farmers have applied for EU subsidies. Here, fields are administrative units, which however also constitute units of agricultural land use. For study 2, field sizes were registered on basis of aerial photos. Fields registered on aerial photos highly conform to fields recorded in agricultural registers. In this paper we estimate mean field size as the mean size of all fields within one farm unit.

Relationships between hedgerow density, farm size and field size in 2001/2002 are analysed on basis of studies 1, 2 and national data. How these variables change over time, is explored for the period from 1995 to 2004 in three steps. 1) On basis of study 1 we first establish the relationship between farm size, field size and hedgerow density in 2001 for a sample of farms, which are distributed over all major Danish landscape types and have a similar distribution over farm types as at national scale. 2) Based on study 2, we analyse the link between development in field size and in hedgerow densities. 3) Based on study 3 and national data, we explore the link between development in farm size and field size. Finally, based on study 1, we analyse how hedgerow activities are related to farm size.

## Results

From 1998 to 2004, the average farm size in Denmark increased from 44 to 57 ha, while the total number of farms decreased from over 60.000 to about 47.000. In the same period mean field size for the whole country increased from 3.7 to 4.0 ha. As large fields (>8 ha) increased by number, and the total agricultural area has been decreasing, change in mean field size must be linked to the merging of fields.

### *Relationship between hedgerow density, farm size and field size*

Data collected in study 1 show a hedgerow density of 59m/ha in 2001 and a weak but significant negative relationship between density of hedgerows and farm size ($r^2$= 0.12 for log of farm size). There is also a weak, but significant negative relationship between hedgerow density and mean field size ($r^2$= 0.07). Analysing the relationship between mean field size and farm size in 2001 shows a clear and significant negative relationship ($r^2$= 0.39 for log of farm size).

In study 2 the mean density of hedgerows was 39m/ha and a weak but significant negative relationship is found between hedgerow density and farm size ($r^2$= 0.08 for log of farm size) and between hedgerow density and mean field size ($r^2$= 0.04). Also a strong and positive relationship between farm and mean field sizes was found ($r^2$= 0.38 for log of farm size).

National registry data for 2001, including all 53.750 Danish farms, show a significant positive relation between farm size and mean field size ($r^2$= 0.22 for log of farm size)(Figure 1) and significant negative relations between farm size and hedgerow density ($r^2$= 0.03 for log of farm size) and between mean field size and hedgerow density ($r^2$= 0.03).

### *Changes in hedgerow density, field size and farm size*

On the 112 farms in study 2, mean field sizes increased at average from 2.2 ha in 1995 to 3.2 ha in 2002, while hedgerow density increased on average by 12%. 25% of all farms removed hedgerows and 50% planted new hedgerows. While no relationship between change in mean field size and total change in hedgerow density was found in this study, a clear positive relationship between change in mean field size and both density of removed and density of planted hedgerows was found, indicating that farms with larger fields are most active in adjusting their field structure.

Among the 234 farms in study 3 significant increases in mean field size (>20%) were seen on 20% of all farms, and on 35-50% of the farms with considerable or major growth in farm size. There is a significant relationship between change in farm size and change in mean field size from 1997 to 2004, indicating that enlargement in farm area is linked to field enlargement. National data on farm size and field size development from 1998 to 2004 support this: on the 40.385 farms, which could be traced in both the 1998 and 2004 agricultural register there is

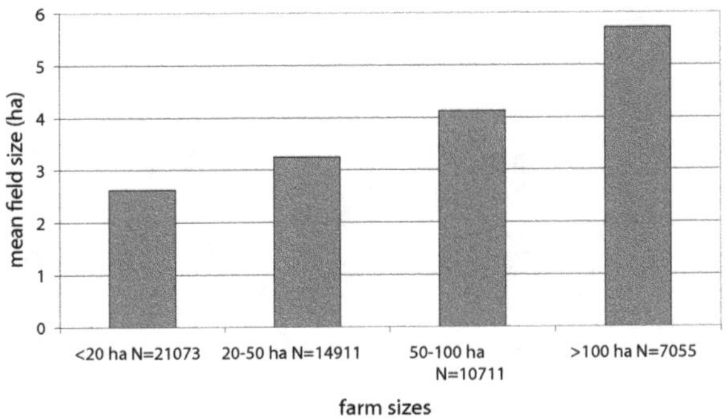


*Figure 1. Mean field size over farm size, Denmark 2001. Source: Agricultural register 2002.*

a strong and significant relationship between change in farm area and change in mean field size ($r^2$= 0.32) (Figure 2).

***Planting and removal of hedgerows on different farm sizes***

From 1996 to 2002 11% of the 340 farmers in study 2 removed hedgerows, but very few of these did not establish hedgerows during the same period. 37% of the farms established hedgerows. On small farms (<20 ha) the net increase in hedgerow length was app. 10 m pr ha of farmed area, while for the other 3 farm size groups net increases were only 2-3 m/ha. Thus, although there is a higher planting activity on large farms (Figure 3), the resulting densities are higher on the area managed by small farms.

## Discussion

Our data show that during the last decade increasing proportions of large farms were accompanied by increasing mean field sizes. Enlargement of farms and fields does not necessarily lead to a decrease in overall hedgerow density but very large farms seem to be most active in both establishment and removal, which could be due to a rationalisation of farm layout. While hedgerow densities increased among all farm sizes, small farms show the relatively

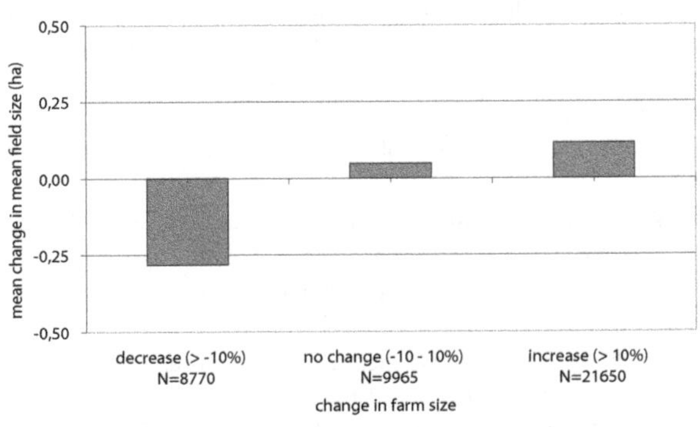


*Figure 2. Change in mean field size over change in farm size, Denmark 1998 – 2004. Source: Agricultural registers 1998 and 2004.*

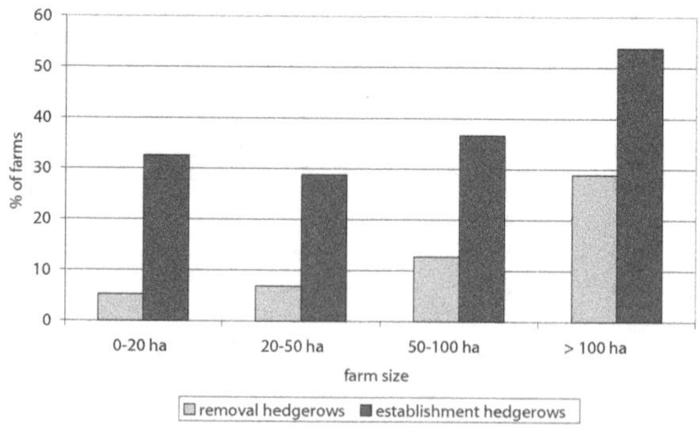


*Figure 3. Farm hedgerow activities 1996 – 2002. Source: Questionnaires with 340 organic farms in Denmark.*

highest increases. As farm enlargement is widely expected to continue in the next decade, this development will probably lead to a continued merging and thus enlargement of fields. This may affect farmland nature by removal and replacement of old hedgerows, as hedgerow removal and establishment is extensive on large farms. Small farms seem to have higher hedgerow densities in general and density of plantings is relatively high. Consequently, worries of farmland nature in relation to hedgerows should be primarily directed towards large farms and the possible removal of old and valuable hedgerows, and policies of advice on farm nature supplemented by continued subsidies for planting new hedgerows and protection of existing hedgerows would appear to be beneficial. However, other landscape elements such as field divides and remnant biotopes cannot be expected to follow the same development. Particularly densities of field divides can be expected to be largely affected by the scale enlargement within agriculture.

## References

Ackermann, H.Ø., 2003. Økologiske landmænds landskabsforvaltning - og faktorerne bag. Institute of Geography, University of Copenhagen, Denmark.

Barr, C.J., R.G.H. Bunce, R.T. Clarke, R.M. Fuller, M.T. Furse, M.K. Gillespie, G.B. Groom, C.J. Hallam, M. Horning, D.C. Howard and M.J. Ness, 1993. Countryside Survey 1990: Main Report. Department of Environment, Eastcote, UK.

Barr, C.J. and M.K. Gillespie, 2000. Estimating hedgerow length and pattern characteristics in Great Britain using Countryside Survey data. Journal of Environmental Management 60, 23-32.

Bentaon, T.G., J.A. Vickery and J.D. Wilson, 2003. Farmland biodiversity: is habitat heterogeneity the key? Trends in Ecology and Evolution 18, 182-188.

Burel, F. and J. Baudry, 1990. Structural dynamics of a hedgerow network landscape in Brittany France. Landscape Ecology 4, 197-210.

Busck, A.G., 2002. Farmers' landscape decisions: Relationships between farmers' values and landscape practices. Sociologia Ruralis 42, 233-249.

Clausen, M.C. and A.B. Larsen, 1997. Småbiotoptæthed i økologisk og konventionelt jordbrug i Østdanmark. Landskabsøkologiske skrifter 6, 61-67.

Deckers, B., E. Kerselaers, H. Gulinck, B. Muys and M. Hermy, 2005. Long-term spatio-temporal dynamics of a hedgerow network landscape in Flanders, Belgium. Environmental Conservation 32, 20-29.

Frederiksen, P. and V. Langer, 2005. Density, structure and management of landscape elements on Danish organic farms. proceeding from a NJF seminar on Organic farming for a new millenium, 15-17 June, Alnarp, Sweden. NJF report 1.

Fritzbøger, B., 2002. Bag hegnet. Historien om levende hegn i det danske landskab. De Danske Plantningersforeninger, Skjern, Denmark.

Haines-Young R., C.J. Barr, L.G. Firbank, M. Furse, D.C. Howard, G. McGowan, S. Petit, S.M. Smart and J.W. Watkins, 2003. Changing landscapes, habitats and vegetation diversity across Great Britain. Journal of Environmental Management 67, 267-281.

Hansen, H.O., 2000. Strukturudvikling i landbruget - forløb, årsager og fremtid. Tidsskrift for Landøkonomi 3, 186-218.

Kristensen, L. S., 2003. Agricultural landscapes - changes and public polices in the 1990s. The Royal Veterinary and Agricultural University, Copenhagen.

Kristensen, S. P., 2001. Hedgerow planting activities by Danish farmers: a case study from central Jutland. Geografisk Tidsskrift, Danish Journal of Geography 101, 101-114.

Landboforeningerne, 2002. Dansk landbrug 2010, Udviklingsscenarier. Copenhagen.

Langer, V., P. Frederiksen and J.D. Jensen, 2005. The development of farm Size on Danish organic farms - a comment on the conventionalisation debate. In: U. Köpke, U. Niggli, D. Neuhoff, P. Cornish, W. Lockeretz and H. Willer (editors). Researching Sustainable Systems. Proceedings of the First Scientific Conference of the International Society of Organic Agriculture Research (ISOFAR), Adelaide, pp. 321-325.

Ministry of Environment and Energy, 1995. Biologisk mangfoldighed i Danmark - status og strategi. Copenhagen.

Prip, C., 1995. Biologisk mangfoldighed i det åbne land. In: Brandt, J. and J. Primdahl (editors), Naturbeskyttelse i det åbne land. Hørsholm, Forskningscentret for Skov og Landskab.

Robinson, R.A. and W.J. Sutherland, 2002. Post-war changes in farming and biodiversity in Great Britain. Journal of Applied Ecology 39, 157-176.

Smith, R.K., N.V. Jennings, N.V. and S. Harris, 2005. Mammal Review 35, 1-24.

Westmacott, R. and T. Worthington, 1997. Agricultural Landscapes: A Third Look. Countryside Commission, Cheltenham, UK.

# The livestock commodity chain: A lever for monitoring pioneer dynamics in the Brazilian Amazon Region?

René Poccard-Chapuis[1], Marcelo Thales[2], Adriano Venturieri[3], Marie-Gabrielle Piketty[4], Benoit Mertens[5], Jonas Bastos da Veiga[3] and Jean-François Tourrand[6]
[1]CIRAD, France; [2]MPEG; [3]EMBRAPA, Brazil; [4]CIRAD/USP, Brazil; [5]CIFOR, Brazil; [6]CIRAD/UNB, Brazil

## Abstract

In the debate about deforestation in the Brazilian Amazon Region, the cattle sector is pointed out as the major cause, and the access to the world market is perceived as a serious threat. However, the reality is far more complex as shown by the analysis of the commodity chains and the emerging land organization in several frontier areas. A multidisciplinary research team has developed an integrated approach in order to study those interactions, which in the Amazon are usually analysed separately. The livestock commodity chain can be divided into local, regional and "export" sub-chains. Each of these three activities owns a specific spatial influence, where it plays a particular role on the emerging land organization. In frontier areas, its action relies upon four successive and interdependent steps: cattle herd increase, emerging organization of the commodity chain, attraction of the frontier area and land organization. A spiral of events can make the frontier area occupation look like a steamroller, leaving behind a smooth space organised for one specific function: cattle and milk production. The understanding of this four-step trajectory might favour the identification of appropriate measures. The starting point is the production of reliable and up-to-date information. Beyond this information gathering, the measures should apply on two key processes in terms of spatial dynamics implication. First, the establishment of commodity chain regulations, to implement tools for municipal land management and to revitalize land use planning in the frontier areas. Secondly, land management particularly in critical municipalities, should rely on a participative implementation of spatially-explicit tools (GIS) with local administration and stakeholders. Finally, without measures for land use planning in frontier areas, the cattle commodity chain will often be the only emerging sector, which will organize the land in the most productive way, with the associated massive forest clearings.

## Introduction

Several controversies surround the issue of cattle ranching and deforestation in the Amazon. Indeed, cattle ranching is one of the main sources of deforestation but also one of the main factor of regional development (Tourrand *et al.*, 2004). Recently, research studies have underlined that livestock production and marketing chain efficiency was the main underlying factor influencing cattle ranchers strategies in the Amazon (Faminow, 1998; Veiga *et al.*, 2004). That is why, it also appears as a major driver of deforestation (Ferreira, 2001). In this context, shown of ecological damages and disrespect of labour legislation, some markets could be closed in the future to the Amazon cattle products, as the demand for environmental and social quality of food products is increasing. The monitoring of the livestock chain impacts in the Amazon agricultural frontiers is thus one of the greatest challenges for the country.

The functioning of the livestock chain is not very well known, nor the spatial process of agricultural frontiers development. Remote sensing and geographic information system (GIS) are useful to identify the correlations between deforestation location and its presumed driving factors (Alves, 2002; Mertens *et al.*, 2002). However, they do not allow to understand the processes behind and the causality links. This may lead to misleading conclusions, as found, for example, in studies stating that cattle ranching is less competitive in region with rain rate lower than 2,000 mm per year (Chomitz *et al.*, 2002), or that road improvement would have, in next 50 years, the same impact than during the last 20 years (Laurance *et al.*, 2001).

To understand agricultural frontiers dynamic in the Amazon, this study focused on the links between the organization of the livestock production and marketing chain and spatial structures, based on some of the results of five multidisciplinary research projects, that have worked on the following issues: (i) cattle ranchers strategies in four pioneering fronts of the Eastern Amazon, (ii) Analysis of satellite Landsat TM and ETM+ image of the last 15 years in the same locations, and (iii) livestock production and marketing chain organization in the State of Pará.

## Three contrasted sub chains, delimiting critical areas

The sub chain concept allowed us to understand the complex organization of the several actors belonging to the chain, from production to consumption, and to define possible levels of intervention (Poccard-Chapuis, 1997). In each sub chain, specific actors' strategies, products and quality attributes are interacting with contrasted impacts on the pioneering fronts. Thus, the final consequences on deforestation may be completely different from one sub chain to another (Poccard-Chapuis, 2004). This systemic concept resulted in the definition of three sub chains: the expedition sub chain, the local and the regional sub chains, keeping in mind that the two last ones show some similarities on territorial issues.

The expedition sub chain is currently the main driver of spatial dynamic of the pioneering fronts. Between 2000 and 2003, the number of slaughtered heads increased from 735,000 to 1,218,000. Such a tremendous growth is linked to new markets access, thank to the development of a new processing plants exporting cooled or frozen products to distant markets[1]. These slaughter-houses are following international quality standards and are located in the backwards of pioneering fronts, leading to their specialization as cattle fattening basins and making cattle ranching an attractive opportunity on a perimeter of some hundreds of kilometres. Cattle fattening takes place in large farms, leading to a strong calves demand. Smallholders benefit from this new market opportunity and invest in calves breeding (Tourrand and Veiga, 2003). In this organization, all types of landholders invest in cattle ranching, generating a strong demand for new lands, in order to grow pastures. Under such strong influence, speculation occur on land markets: pastures feed cattle herds but also guarantee property rights in front of possible land ownership contest, increase land value, allow to rent it for neighbouring farmers, and even facilitate access to credit. Directly or indirectly, deforestation ends up expanding until hundreds of kilometres around the slaughter-houses. However, such organization is limited to some specific counties of the Para State, located in the South-eastern region.

In contrast, the local and regional sub chains only supply small saturated markets and are often unable to generate any production growth, or are generating very small growth close to the largest urban centres. However, they are found in the whole territory and, consequently, they guarantee a minimum market for cattle that does not enter the expedition sub chain, either because of the product low quality, either because of the geographic isolation. These

[1] Because of foot-and mouth disease, the product can not be directly exported on international markets.

sub chains are essential for the numerous less competitive cattle ranchers and represent 42% of slaughtered head in the State in 2000.

## Interactions between the livestock chain and the pioneer territory: Four simple mechanisms

The interactions chain/territory do not only depend on markets access, they follow a four steps cyclical process (Figure 1). If the process is interrupted or stabilized, interactions remains limited, as for local sub chain. But if it is not, it is acting as a road roller, leaving backwards a territory organized around one function only: to produce cattle for the chain.

The first step is the chain organization. In the pioneering fronts of the Eastern Amazon, the initial impulse came from the local demand of the new urban centres. Market access must be large enough to absorb production growth from the first cattle ranchers, migrants, and investors. The second step is the attractiveness of the pioneering front. If the first step succeeds, then the pioneering fronts become attractive for cattle ranching for all the new migrants as for the local actors, who start to systematically invest in the activity on their own properties (*pecuarização* of smallholders, for example). The third step is the structurating of the pioneering front. Its determinants are modified by the two first steps, since they are influencing land market, migrants location and their production and land management strategies. Spatial organization is becoming more efficient for cattle and the pioneering front becomes progressively an efficient productive tool for the chain, as shown by R. Brunet (Brunet *et al.*, 1998). The fourth step is logically the cattle production growth. The combination of the four steps influences the first one: they lead to a better organization of the chain, spreading out, for example, genetic improvement technology or better pastures management, or even influencing public policies as foot-and-mouth eradication program. Investments in secondary processing plants appear (leather products and, in longer term, footwear). This reinforces the region attractiveness, guaranteeing the cycle continuity.

Such dynamic process is encountered in the south of Pará State, where the pioneering front structurating is provoked by and for the livestock production and marketing chain, resulting, amongst others, in high deforestation rate. The forest is nowadays only a small input in this process (space, fertility, wood). There is not anymore strong links between the wood and the livestock chains as it has occurred at the beginning of the Amazon colonization. In this context, the environmental, labour and land legislations are often useless.

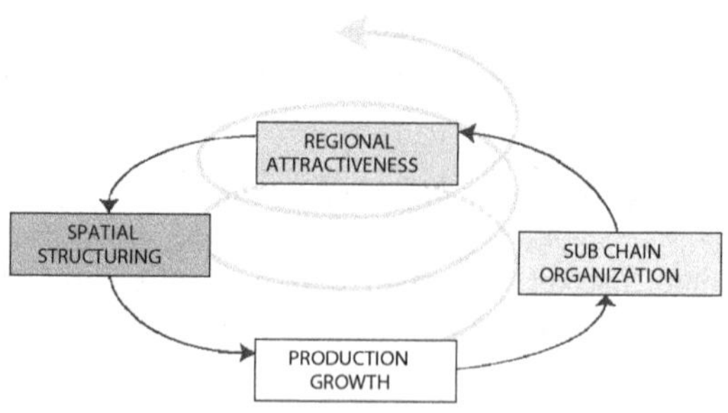


*Figure 1. Interaction between livestock chain and spatial organization of the Amazon pioneer fronts.*

## How to control these interactions?

Each one of the four steps can be documented by a simple information system, requiring cooperation and efficiency of the involved institutions.

For the chain organization and production growth steps, statistical data can be found in several agencies and could be joined together into a sole database. The institutions involved would be the Ministry of Agriculture (MAPA), the State Secretary of Farm (SEFA), and the Agency of Farming Defence of Pará (Adepará), in charge foot and mouth eradication.

For the attractiveness step, it would be necessary to record all the land issue data in the various institutions responsible for land regularization, in notary offices where land transactions are registered, in the Environmental State Secretary (SECTAM) and in Federal institutions responsible for indigenous and preservation areas delimitations, and at INPE (National Space Agency), that registers annual deforestation data in each county. The structurating step can be documented with simple landscape analysis, based on satellite imagery (Mertens *et al.*, 2002; Venturieri, 2003). From such data collection, synthetic indicators could be produced as an input to the monitoring of the regional dynamic.

The sharing of such information could be a common reference for all stakeholders, making possible the identification of consensuses and agreements between them on three issues: (i) the chain regulation; (ii) space management; (iii) and territorial ordering of the pioneering front.

The objectives behind the introduction of chain regulation mechanisms are to preserve some areas of primary forests and to prevent the possibility, at a short or medium term, of closing markets for cattle products of the Amazon because ranchers are not respecting environmental and labour laws, or even human rights (slavery). The vulnerability of cattle ranching in the Amazon has increased during the last years because of environmental and trade lobbying efficiency, the permanence of high deforestation rates, but also because of several cases of slavery found in some large farms of the south of Pará. It is important to notice that, at the national level, the livestock sector depends more on international markets and may suffer heavy losses from a bad picture of the sector in the Amazon. In this context, the region will probably have to reach international standards under the demand of others regions already inserted in international markets.

Livestock production and marketing chain regulation in the Amazon could be supported by farms certification, cattle and meat traceability, a satellite monitoring of deforestation and a systematic update of land and labour registration. Market access could be guaranteed by economic, environmental and social criteria. On these issues, progress have been made by Brazilian institutions, through the development of the SISBOV system, but it remains limited to the traditional cattle production basins (South, Southeast and Centre-West). The Amazon is still considered at a land of pioneers or outlaws, where cattle ranching is not based on modern technology. Consequently, it remains outside of the quality improvement program, despite crucial social and environmental issues there.

## Acknowledgements

Financed by the Program Pilot of the G7 for protection of the tropical forests of Brazil (PPG7), the Inter American Institute (the IAI), the Brazilian Institute of Farming Research (EMBRAPA), the Government of the State of Pará, the program of cooperation between France and Brazil (COFECUB) and the Coordination of Academic Superior Formation (CAPES).

## References

Alves, D., 2002. An analysis of the geographical patterns of deforestation in Brazilian Amazon in the 1991-1996 period. In: C. Wood and R. Porro (eds.). Patterns and processes of land use and forest changes in the Amazon. Gainesville: University of Florida Press, pp. 22-34.

Brunet, R., R. Ferras and H. Thery, 1998. Les mots de la géographie: dictionnaire critique. Paris, Reclus La Documentation française. 518 pp.

Chomitz, K.M. and T.S. Thomas, 2002. Geographic patterns of land use intensity in the Brazilian Amazon. Washington: World Bank. 43 pp.

Faminow, M.D., 1998. Cattle, deforestation and development in the Amazon: an economic, agronomic and environmental perspective. Oxford, CAB International. 253 pp.

Ferreira, L.A., 2001. Le rôle de l´élevage bovin dans la viabilité agro-écologique et socio-économique des systèmes de production agricoles familiaux en Amazonie brésilienne: le cas d´Uruará. Paris: Institut d´Élevage: Inra-SAD. 188 pp.

Laurance, W.F., M.A. Cochrane and S. Bergen, 2001. The future of the Brazilian Amazon. Science 291, 438-439.

Mertens, B., R. Poccard-Chapuis, M.G. Piketty, A.E. Laques and A. Venturieri, 2002. Crossing spatial analyses and livestock economics to understand deforestation processes in the Brazilian Amazon: the case of São Felix do Xingu in South Pará. Agricultural Economics, Special issue "Spatial analysis for agricultural economics: concepts, topics tolls and example", pp. 12-57.

Poccard-Chapuis, R., 1997. Filières bovines et construction de l'espace en Amazonie orientale: l'exemple du municipe d'Uruará sur la transamazonienne. Paris: Université Paris 1. 321 pp.

Poccard-Chapuis, R., 2004. Les réseaux de la conquête: rôle des filières bovines dans la structuration de l'espace sur les fronts pionniers d'Amazonie orientale brésilienne. Paris, Université de Paris X - Nanterre. 435 pp.

Tourrand, J.F. and J.B. Veiga, 2003. Viabilidade de sistemas agropecuários na agricultura familiar da amazônia. Belém, Embrapa Amazônia Oriental, 468 pp.

Tourrand, J.F., M.G. Piketty and J.R.D Oliveira, 2004. Élevage bovin, déforestation et développement régional: le cas du Sud du Para, Amazonie Brésilienne. Bois et Forêts des Tropiques, Seine, n. 280, pp. 5-16.

Veiga, J.B., J.F. Tourrand, M.G. Piketty, R. Poccard-Chapuis, M.A. Alves and M.C. Thales, 2004. Expansão e trajetórias da pecuária na Amazônia: Pará, Brasil. Brasília, Editora da Universidade de Brasília. 162 pp.

Venturieri, A., 2003. A dinâmica da ocupação pioneira na rodovia transmazônica: uma abordagem de modelos de paisagem. Rio Claro: Universidade estadual de São Paulo. 167 pp.

# Re-inventing sustainable cropping systems in Costa Chica, Mexico

*Diego Flores Sanchez[1], Jobert Kleine Koerkamp[2], Walter A.H. Rossing[2], Egbert A. Lantinga[2] and Ariena H.C. van Bruggen[2]*
*[1]Colegio de Postgraduados, Carretera México – Texcoco km. 36.5, Montecillo, Estando de México, CP 56230, México; [2]Biological Farming Systems Group, Wageningen University, Marijkeweg 22, 6709 PG Wageningen, The Netherlands*

## Abstract

Economic benefits by increased market integration stimulated changes in traditional agricultural land use by smallholders in Costa Chica, a remote part of Guerrero State, Mexico, resulting in more input-dependent, simplified production systems, but also soil degradation and declining yields. This situation has resulted in a spiral of unsustainability. Nevertheless, there is a group of farmers that are conscious about these problems, and they are interested in agro ecological innovations addressed to improve the current cropping systems by means of compost/animal manure, crop rotations including legumes and cover crops, and a transition to organic production. This contribution reports early results of participatory agronomy aimed at searching for ways to revert the degradation of the resource base.

**Keywords:** nutrient cycling, farm diagnosis, cropping systems, sustainable systems

## Introduction

In the short term, sustainable development from an economic perspective often conflicts with sustainable development from an agro-ecological perspective. Economic benefits originating from increased market integration stimulate changes in traditional means of agricultural land use, usually resulting in more input-dependent and simplified production systems. Where simplification results in unacceptable environmental degradation rebalancing of production systems is called for. How to generate relevant information and organize research – farmer community interactions in such situations represents a formidable scientific challenge.

Sustainability concerns of farmers in a remote part of Guerrero State, Mexico are starting point for a research project based on interactive re-inventing of sustainable cropping systems. Farming systems in the municipality of Tecoanapa in the Costa Chica area of Mexico are based on a few crops only, traditionally in mixed cultivation. Animal production is limited and mainly concerns self-sufficiency. Roselle, of which the flower bases are sold for *e.g.* tea and juices, constitutes a major cash crop but prices decreased in recent years. During the last decade mixed cropping systems have been replaced by monocultures usually without cover crops, with high value crops being grown on the best soils with minimal rotation. A group of farmers has acknowledged these problems, and depending on farmers' preferences they are interested in innovations to improve the quality of the soils (in particular physical fertility), quality of the products (in particular roselle), increase yields (in particular maize) and develop alternative market channels. This situation has prompted a research and development project aimed at development of scenarios of sustainable farming systems by means of an interactive process between farmers and researchers. Sustainable farming systems require development of agricultural farming systems based not only on economic performance, but also based on contribution to objectives in areas such as environment, health and well-being, rural scenery and nature (OECD, 2000). An agro-ecological systems approach can help to design sustainable farming systems, and agro-ecological engineering approaches aimed at design and exploration of alternative land use systems at various scales may support the identification of appropriate land use options (Hengsdijk and Van Ittersum, 2003). Aim of the project is to investigate how current systems can be redesigned to meet goals of resource use efficiency, resource conservation and acceptable economic return. Phases in the project comprise diagnosis, redesign and on-farm testing, at field and farm scales. This contribution describes first results.

## Approach

The municipality of Tecoanapa (16°48′ N, 7°11′ W) comprises 52 communities. The diagnosis phase involved surveys at different spatial scales, involving a selection process that in the first steps was based on qualitative information mainly, and used increasingly more quantitative information. Aim of the diagnosis was not so much to obtain a representative socio-economic and bio-technical picture of the municipality, but rather to acquire insight in the diversity of natural resource conditions and the associated management by farm households.

The first step in the diagnosis involved the selection of two communities contrasting in terms of current quality of the resource base. Based on talks to local experts, such as the director of the local agricultural college, the mayor, administrators, locally resident scientists, two contrasting communities were selected, Las Animas and Xalpatlahuac. Las Animas was generally seen as experiencing more resource degradation than Xalpatlahuac. Also, social structures in the two villages differed, with more social control being in place in Xalpatlahuac than in Las Animas. Contacts with community leadership and farmers were established through various surveys and workshops on *e.g.* compost production, and the first author lived in the villages during large parts of the year during two years before starting the study. Regular visits by students from the *Colegio de postgraduados*, one of the main universities in Mexico, further enhanced social linkages between research project and the communities.

The second diagnosis step was to establish diversity at the livelihood and field scale in the two communities. During this 'general survey', farm walks were conducted with 14 farmers from Las Animas and 16 from Xalpatlahuac, and the variety of crops, and cropping and nutrient management systems was recorded. A structured interview was held with each farmer on perceptions and behavior regarding production activities, perceptions of land quality and its management, and what they think about opportunities for shaping formal knowledge and local innovation.

The third step in the diagnosis involved selection of 4 farmers from each community for a 'verification survey' involving a structured formal survey in order to characterize the farming systems in more detail. These farmers were chosen based on the results of the general surveys. They differ in terms of availability of land, cropping systems, cultural practices and socio-economic strategies in their farming systems.

Farm and field level data were collected such as yields of the main crops (maize and roselle), soil samples for textural classification and chemical properties. These data were used to map resource flows (biomass and nutrients) between spatial components of the farm and nutrient balances. In addition, data on the socio-economic aspects of the households were collected, such as wealth, labour and sources of income, and data on institutional setting in each community. Data collection has just been completed and first results are shown in this paper.

## Results

### *Soils, crops and cropping systems*

Soils in Xalpatlahuac (700m asl) and Las Animas (600m asl) are of volcanic origin and classified in general terms as regosol. Their organic horizons are scarcely developed; as a consequence their fertility is limited. The farming systems in both communities are organized in small production units, land holdings ranging from 1.5 to 9 hectares and the number of fields per farm varies from 1 to 5. Common crops are maize (*Zea mays,* L.), roselle (*Hibiscus sabdarifa,* L.), beans (*Phaseolus vulgaris,* L.), squash (*Cucurbita pepo,* L.), and cassava (*Manihot sculenta,* L.). These crops are grown during the rainy season (June to November). Maize is generally grown for home consumption and cobs or grains are stored to satisfy the families' need for the year. Roselle, squash (seeds), and cassava are cash crops. Different cropping systems were encountered. Most commonly, maize is intercropped with roselle, squash and beans (squash and beans are planted at low densities), though there are fields in which maize, roselle and cassava are grown as monocrops. For instance, 15 farmers had both monocrops and intercrops, 13 farmers had only intercrops, and 2 farmers had only monocrops. The use of intercropping systems was a response to scarcity of land and as a means to reduce risks due to biotic (pests and diseases) and abiotic (typhoons) factors.

Crop rotation is not commonly practiced. The major reason for the absence of crop rotation is the first and most important objective of every farmer to ensure that his family has sufficient food. However, at the moment 17 farmers include fallow periods of 1 to 5 years on their fields to improve soil fertility.

### *Crop management*

Related to the intensive use of the lands is the increase in the population of weeds. In the past, hand-weeding was practiced. Nowadays, more than 80% of the farmers use herbicides. Figure 1 shows the variation in the use of the herbicide Paraquat, around the technical recommendation of 2 l ha$^{-1}$. The lack of other options to control herbicides has generated a high dependence on this input.

Fertilizers are widely used and applied in the planting hole. However, on steep hills there is a risk of run-off. Figure 2 presents the variation in application rates of N and P in maize-roselle systems. The variability has several reasons: a) identified differences in soil types (fertility,

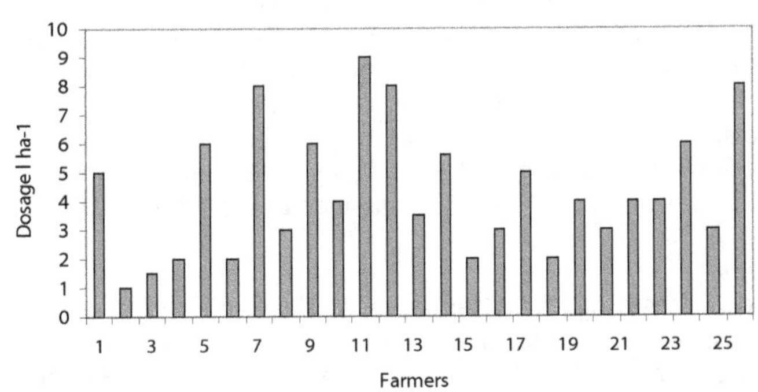


*Figure 1. Dosage of herbicide (Paraquat) (l ha$^{-1}$) used by farmers in Las Animas and Xalpatlahuac (2005).*

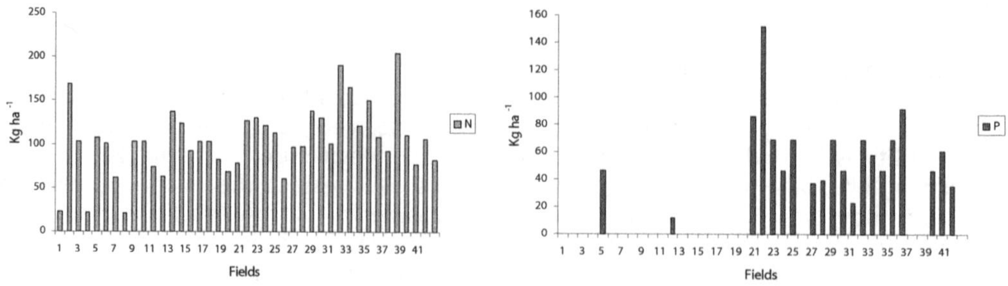


*Figure 2. Fertilizer use (kg ha$^{-1}$) on 41 fields in Las Animas and Xalpatlahuac (2005).*

color and texture) within one field by the farmers, and b) the availability of cash. The farmers receive 109-69-00 (kg N-P-K) per ha as a subsidy from the municipal government, but this holds only for 2 ha. However, most farmers have a greater area, and they are therefore forced to "ration" the fertilizers, to apply the fertilizers on the less fertile field only or to buy more fertilizers. The farmers are pretty much dependent on fertilizers to allow good yields by saying that the land is "tired" and it is impossible to get yields without fertilizers. The use of organic manure as a means to augment soil fertility is not a common practice so far. This practice is only done by those who own animals. In cases where farmers do apply manure, it is done in an irregular manner and in quantities that vary extremely (from 150 up to 500 kg ha$^{-1}$).

Due to land scarcity, some farmers are forced to practice agriculture on extremely steep slopes. Without cover crops and terraces, erosion poses major concerns since it causes not only loss of nutrients but, and perhaps more important, loss of growing medium for the crops. The majority of crop residues is left on the field after harvest in order to improve soil fertility and quality and because of practical reasons. During maize harvest, only the cobs are taken away from the field; the plant itself is left behind. At the harvest of roselle, the branches with the flowers are cut and carried to a shaded part of the field where the calyxes are removed. Afterwards, the branches are thrown back on the field.

### *Animal component*

The farmers usually keep one or two draught animals like donkeys for transport of materials (fertilizers, grains, forages) and some chickens, pigs, goats, and/or cows for home consumption and additional income. In Las Animas, 7 out of 14 interviewed farmers did not own any animals, in Xalpatlahuac 3 out of 16. Several types of animal husbandry systems were found, all based on grazing, or cut-and-carry. It is common practice that animals graze on communal fields or other farmers' fields. In the dry season animals are kept on fields where crop residues have been left after harvest. Some farmers put up fences to exclude strange animals. Animal production, although practiced on a small scale, is an economically important activity.

### *Farmer innovations*

Some farmers are experimenting with alternatives to the traditional cash crop roselle. Trials have been set up with cassava and gladiolus, and hybrid maize varieties have been adopted by some farmers which can be marketed better than the traditional criollo landraces. In Xalpatlahuac, one farmer is experimenting with "nescafé" (local name for velvet bean) as a green manure crop already for more than 10 years. As a result, soil structure and moisture content are now much better than in neighbouring fields, and this is attracting broad attention from the community.

As a follow-up of workshops, a number of farmers are active in making compost from animal manure and diverse carbon sources. The municipality has created a composting facility but this is, however, currently underutilized due to ownership issues.

## Discussion

The aim of this diagnosis phase is to allow for the analysis of the production process/systems, soil fertility and resource flows. In this first stage, variability in the different components of the farm management such as cropping system, fertilizer and herbicide usage has been observed. An integral analysis by means of nutrient flow diagrams, nutrient balances, yields, biomass, analysis by a nitrogen model will be carried out in order to have more elements and to explain the variability among fields. Understanding and interpreting this variability generates knowledge and tools to design various scenarios or prototypes for each agro-ecological situation with relation to the socio-economic situation of the farmers. These data will be reported in the poster. An important fact is that the farmers are willing to participate in the re-design of their current cropping practices and systems in order to improve the fertility of their soils and yields. The farmers already have some ideas and innovations on how to improve their soils, and they are willing to experiment with alternatives such as using compost, cover crops and green manure crops. Based on these ideas and the quantification of the resource flows the alternatives will be designed in close interaction with farmers so that the alternative practices can be carry out under supervision of the farmers and using analyses and modelling tools from research on the effects of green manures, compost and manure on soil fertility. Undoubtedly, these alternatives will have a local adaptation, and this will enable the farmers to produce a vision of a future based on more ecologically sound farming. This experience will provide a framework for application to other eco-regions.

## Acknowledgements

Special thanks are due to the International Foundation for Science for their infrastructural support. The project is financed by the Mexican Council for Science and Technology (CONACYT).

## References

Hengsdijk, H. and M.K. van Ittersum, 2003. Formalizing agro-ecological engineering for future oriented land use studies. European Journal of Agronomy 19, 549-562.

OECD, 2000. Multifunctionality – towards an analytical framework. Paris.

# Differentiation of territories according to the location of agricultural activities: Methods for regional scale approaches

*V. Souchère[1], C. Mignolet[2], C. Schott[2], E. Bienaimé[3] and M. Benoît[2]*
*[1]UMR SADAPT, BP 1, 78850 Thiverval Grignon, France; [2]INRA SAD, 662 avenue Louis Buffet, 88500 Mirecourt, France; [3]INRA, Equipe SIG, Centre de Recherche de Nancy, 54280 Champenoux, France*

## Abstract

Some ecological processes as runoff or contamination of groundwater and surface water by diffuse pollution are largely the results of changes in agricultural practices (land use and technical management) but also of the way they are organised within watersheds. Their regulation imposes a spatial coordination of the cropping systems within farms and more often among neighbouring farms.

This paper proposes a research procedure to study, with the help of different statistical data, agricultural spatial dynamics at different spatial scales which correspond to different levels of intervention. Two methods are presented. The first is used to model the crop sequences diversity and location whereas the second ensures to analyse geographical proximity between farmers and farms in order to assess possibilities of coordination between farmers. Results shows an important spatial differentiation of the cropping patterns and crop sequences at the regional scale and reveals that long distances, the large surface areas farmed and the high number of external farmers may constitute major obstacles to the creation of ad hoc cooperative processes.

**Keywords:** diversity of cropping systems, geographical proximity, spatial coordination, regional scale, catchment area, statistical data

## Introduction

There is a wealth of information on the role of spatial interaction between adjacent agricultural fields as drivers of ecological processes. However, environmental problems are largely the result of changes in agricultural practices but also of their spatial arrangement within catchment area. Therefore, their regulation imposes to analyse the processes at stake in order to determine the relevant spatial entity at which to coordinate cropping systems within farms and more often among neighbouring farms. Erosive runoff in Upper-Normandy corresponds to this situation because it is characterized by a disconnection between upstream areas producing runoff and downstream areas where runoff creates erosion; its control requires concerted management over the entire catchment area. Geographical proximity between farmsteads and the catchment areas in which the farmers operate is a significant factor in this awareness, and hence in the collective actions that need to be set up locally.

Since 1999, twenty two watershed management committees have been set up by the public authorities in order to promote coordination and create opportunities of dialogue between the different actors concerned. Their role is essential as farmers have a limited knowledge of the water pathways through the different sections of the catchment. But to manage actions, the watershed management committees need tools adapted to catchment scale that can be used to study the impacts of agricultural spatial dynamics and to estimate geographical proximity between farmsteads and the catchment area. Many scientists have studied ecological processes controlled by agricultural practices at the scale of a few ten or a few hundred hectares watersheds, in which methods based on direct investigation in farms, can be implemented. In this paper, we propose a research procedure developed at larger geographical areas (a few thousand of $km^2$). At this scale one must use information mainly gathered by the French agricultural statistical system to characterize cropping systems. The first part of the paper is devoted to a method to model the crop sequences diversity and location based on data mining and statistical mapping methods applied on a French statistical database called *Teruti*. This is relevant as different crops and crop sequences are associated with differing risks of soil erosion. In the second part, we present a method to analyse geographical proximity between farmers and farms on the basis of data drawn from CAP (Common Agricultural Policy) documents. This forms a basis for an approach to farmers to mitigate soil erosion within catchments.

## How to model the diversity of the crop successions at the scale of the catchment area

### *Material and methods*

The *Teruti* data are collected by the French agriculture administration on the whole metropolitan territory. They represent the land use of the country on a one year basis. Two levels of resolution are achieved. A first sample consists in selecting aerial photographs. The French territory is divided into a grid of 3820 squares in which, four aerial photographs are selected. Secondly, on each photograph, a 6 by 6 grid determines a cluster of 36 points that are field checked every year in June. The land-cover category of these points is logged in a matrix in which the rows are the sites of the country and the columns the time slots (from 1992 to nowadays). Finally, one *Teruti* point is roughly 100 hectares.

Crop successions are identified by a data searching method using Hidden Markov Models (HMM), developed in collaboration with IT specialists (Mari *et al.*, 2000, 2002; Le Ber *et al.*, 2006). This method consists in extracting patterns from the *Teruti* surveyed points, on which we assume that land cover change according to a Markov process. The HMM allows us, to calculate the probabilities of occurrence of ordered sequences of soil occupancy (Mignolet *et al.*, 2004). To provide quantified and homogeneous information on the crop successions of each watershed of the Seine-Maritime department, we calculate the probability of occurrence of triplets of three successive crops from the *Teruti* data from 1992 to 2001. The dominant triplets of successive annual crops are used to represent the diversity of crop successions in the Seine-Maritime department. This diversity is then modelled by a watershed typology built by a principal component analysis and an ascendant hierarchical classification using the Ward's algorithm, applied on the table of percentages of arable land per triplet of successive annual crops and per watershed.

### *Results*

Statistical analyses define a classification of watersheds into five types ordered according to a decreased gradient of the number of crops and crop successions (Figure 1). The first two types concern all the coastal watersheds localised in the western half of the Seine-Maritime. Starter crops are mainly spring crops with high added value (sugar beet, flax, potato) and also pea and maize. The cereal proportion is quite low within the cropping pattern, especially in the first type. It is the reason why a lot of crop successions include starter crops that follow each other during two or three successive years. The most typical crop succession is the "flax – sugar beet – wheat" one, either in a three-year form or in a five-year form when it is preceded by "pea – wheat" or by "potato – wheat". The second type has a higher proportion of four-year successions (as sugar beet – wheat – maize – wheat), and also some three-year ones of the type "starter crop – wheat – barley" (*e.g.*: sugar beet – wheat – barley). The third and fourth types are composed

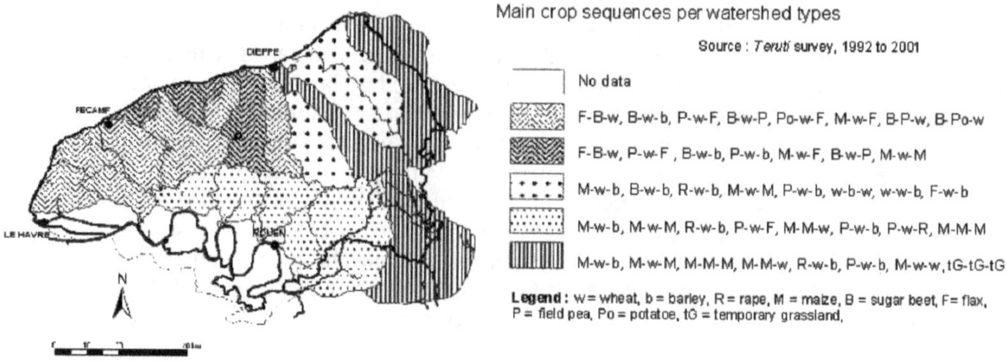


*Figure 1. Main crop sequences per watershed types.*

by eight watersheds localised in the south or east of the department. They all present, as shared characteristic, a higher proportion of cereals and a predominance of maize as starter crop, even if pea, sugar beet or rape are present in this type. As a consequence, the main crop successions are maize based successions (as maize – wheat – barley, maize – wheat – maize – wheat...) and three-year successions of the type "starter crop – wheat – barley" (maize – wheat – barley, rape – wheat – barley). The third type is also characterised by sugar beet – wheat – barley and flax – wheat – barley successions, as the fourth one presents four-year successions coming from the combination of the different upper mentioned starter crops (pea – wheat – flax – wheat, pea – wheat – rape – wheat, maize – wheat – rape – wheat). Lastly, the fifth type is represented by three watersheds localised in the east of Seine-Maritime. Their crop successions are simpler, mainly based on maize; rape, pea and temporary pasture are present in lower proportion.

## How to analyse geographical proximity between farmers

### *Material and method*

Despite the significant statistical apparatus that France has to describe its territory and agriculture, it isn't easy to obtain information on geographical dispersion of farmers because we need a set of data enabling us to link useable agricultural area (UAA) to a specific territory. Such data can be found in applications for compensation submitted in the framework of the Common Agricultural Policy (CAP). Since 1992, farmers are obliged to set aside a part of their land to compensate for the drop in the price of crops qualified for indemnities. The farmers annually declared the area of their different crops. These areas are allocated to the municipality where the fields are really located. So, the CAP data, give the surface areas, for every municipality, farmed by farmers who have their farmstead in any other municipality in the region.

Joannon *et al.* (2004) have identified three main criteria to discuss geographical proximity between farmed fields and farmsteads. The first criterion is the surface of each municipality under external control *i.e.* the percentage of UAA farmed by farmers who have their homestead outside the considered municipality. These farmers are called "external farmers" and the municipalities where their homesteads are located are called "external municipalities" in the rest of this paper. The second criterion is the number of external farmers who farm the surface under external control in a given municipality. However, for the sake of confidentiality, CAP data are aggregated to the communal scale and provide only the number of external municipalities. But, it is nevertheless reasonable to assume that a high total number of actors will correspond to a high number of municipalities. So, the second criterion taken into account is the number of external municipalities. The last criterion is the mean distances between the considered municipality and the municipalities in which external farmers live. Since this distance is the Euclidian distance from the centre of a municipality to the centre of another municipality, it is systematically under-estimated.

### *Results*

The analysis of the three criteria presented above had been performed on 1409 municipalities of the Upper-Normandy with the 2001 CAP data coupled to the geographic data base of the Institut Géographique National. Graph A in Figure 2 shows that the UUAA of 70% of the municipalities in Upper-Normandy is farmed by farmers from between 6 and 15 different external municipalities. Moreover, the fields of approximately 16% of the municipalities are farmed by farmers from over 16 different municipalities. This adds up to a high number of external farmers. Graph B shows that only 0.2% of the total number of municipalities have their UAA under no external control. It appears that close to 75% of the municipalities in Upper-Normandy have between 30 and 99% of their UAA under external control and 3.4% of the municipalities have their surface areas under external control equal to the total useable agricultural area of the municipality. Graph C shows that fewer than 14% of the municipalities in Upper-Normandy have fields farmed by external farmers who live close by, *i.e.* at a maximum of 5km. The most common range of mean distances in the region is 5 to 10 km. This concerns over 60% of the municipalities. Note that in over 28% of cases the distances travelled to farm lands in another municipality are in excess of 11 km.

Combining the three criteria presented above, we built an index in order to analyse geographical proximity at the scale of watershed management organisations. This index I is calculated by Equation 1, where $d_i$ the distance between the studied municipality and the external municipality i, UAA the total useable agricultural area of the studied municipality, $UAA_i$ the useable agricultural area of the studied municipality controlled by the external municipality i and N is the number of external municipalities. The values of this index calculated for each municipality of the Seine Maritime department varies between 0 and 94; the higher the value is, the lower the geographical proximity is. Figure 3 shows that this index can be used to identify situations in which coordinating the farmers would potentially be more difficult than in others to get to act towards erosion mitigation at the scale of watershed management organisations. Insofar as we consider that long distances, a large surface area under external control and a high number of external municipalities constitute major obstacles to the creation of ad hoc cooperation processes.

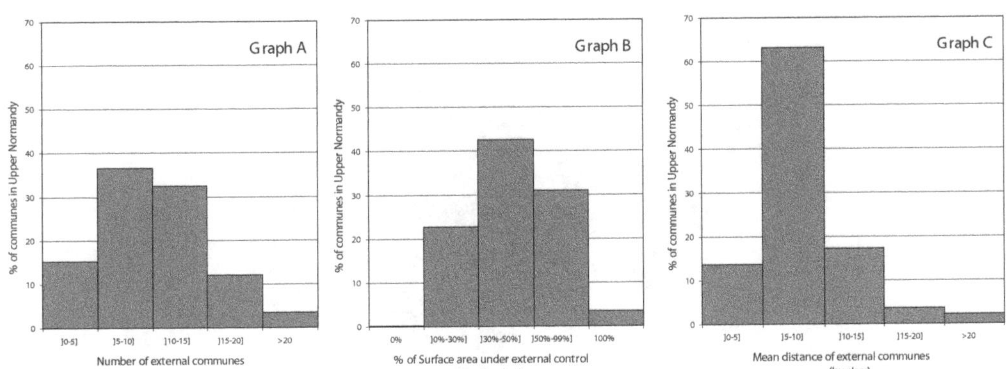


*Figure 2. Analysis of geographical proximity criteria at the scale of Upper-Normandy.*

$$I = N \frac{\sum d_i \times UAA_i}{UAA} \quad (1)$$

## Conclusion

In the first part, we have presented a procedure allowing to define homogeneous catchment areas according to similar combinations of crop sequences during the last ten years. As a perspective, we need more information to describe the sensitivity of dominant crop sequences to runoff within each catchment area before we are able to map regional erosion risk for cultivated soil in France. In the second part, we have presented an index for analysing relations of geographical proximity between the farmers and the land they farm, in order to assess possibilities of coordination between farmers within catchment area. This index provides indications of where the coordinator of the watershed management structure can set up a local coordination process to fight efficiently against erosive runoff. As a perspective, we intend to calculate this index replacing the total UAA with the surface areas of specific crops which are known to be more sensitive to erosive phenomena. This is possible CAP data allow knowing surface area under external control for crops which are qualified for indemnities.

## References


Joannon, A., A. Torre, V. Souchère and P. Martin, 2004. The determinants of local collective action regarding erosive runoff: An analysis of farmers' geographical proximities in Upper-Normandy, France. Int. J. Sustainable Development 7, 302-320.

Le Ber, F, M. Benoît, C. Schott, J.F. Mari and C. Mignolet, 2006. Studying crop sequencies with CarrotAge, a HMM-based data mining software. Ecological Modelling 191, 170-185.

Mari J.F., F. Le Ber and M. Benoît, 2000. Fouille de données par modèles de Markov cachés. Journées francophones d'ingénierie des connaissances; 197-205.

Mari, J.F., F.I. Le Ber and M. Benoît, 2002. Segmentation temporelle et spatiale de données agricoles. Revue Internationale de Géomatique 12, 439-460.

Mignolet, C., C. Schott, M. Benoît, 2004. Spatial dynamics of agricultural practices on a basin territory: a retrospective study to implement models simulating nitrate flow. The case of the Seine basin. Agronomie 24, 219-236.

SCEES, 2002. L'utilisation du territoire en 2001. Nouvelle série de 1992 à 2001. Agreste, Chiffres et Données Agriculture 141. 83 pp.

*Figure 3. Index of geographical proximity.*

# Opportunities and constraints of farming styles to meet the landscape preferences of rural inhabitants

*Derk Jan Stobbelaar[1,2,3], Florien Kuijper[3], Henk Renting[3], Jeroen C.J. Groot[1,3], André Jellema[1] and Walter A.H. Rossing[1]*
*[1]Biological Farming Systems Group, Wageningen University and Research Centre, Marijkeweg 22, 6709 PG Wageningen, The Netherlands; [2]Plant Production Systems Group, Wageningen University and Research Centre PO Box 430, 6700 AK Wageningen, The Netherlands; [3]Rural Sociology Group, Wageningen University and Research Centre PO Box 8130, 6700 EW Wageningen, The Netherlands*

## Abstract

More and more people are living in the rural areas because they want to enjoy the countryside. Farmers play an important role in providing the qualities attached to the countryside. Rural residents of the Northern Friesian Woodlands (The Netherlands) were interviewed about their landscape preferences. They all like the landscape they live in with its small plots and hedgerows and see an overall progress in landscape quality. They dislike the large modern barns, maize cultivation and boring green grasslands. All types of farmers can provide some of the preferred qualities, but farmers belonging to the farming styles Calm farmer and Stayers perform the best in this respect. Farmers' cooperatives can play an important role in connecting the various landscape qualities of individual farmers on region level.

## Introduction

Landscape is becoming increasingly important for inhabitants and visitors of the rural area. Therefore, besides food production, the production of valuable rural landscapes can be regarded as a new goal for the agricultural sector (RLG, 2005). As a result of this emerging demand, farmers are - individually or organised in environmental co-operatives - more and more presenting themselves as providers of this 'green service' for society.

In order to match the demand of the landscape 'consumers', mainly inhabitants of the area, with the 'supply' of farmers (co-operatives), the question can be asked to what extent landscape maintenance that is being implemented by farmers is in accordance with this demand. De Bruin and Van der Ploeg (1991) showed that farming styles are linked to the landscape preferences of the farmers involved, and that these farmers strive to change the landscape on their farm in accordance with their preferences. Farming styles are thereby a possible analytical concept to analyse and influence the interrelations between landscape preferences of inhabitants, farm strategies and landscape images of farms. Therefore, our research question is: *what are the opportunities and constraints of different farming styles to meet the landscape demands of rural inhabitants?*

This research was conducted in the Northern Friesian Woodlands, an area in the north of the Netherlands which is characterized by small, linear and long stretched plots, reflecting a history of peat reclamation. Plots are divided by hedgerows on wooded banks, which traditionally had a function as field boundary and for providing wood. Nowadays these hedgerows are valued because of their ecological function and as a cultural landmark in the landscape of the region, which furthermore is characterized by wide, open spaces (Renting, 2004). The landscape is well-preserved by farmers and predominantly intact.

## Methodology and outline

To answer the research question, the following steps were taken and reported in this contribution:

- First, literature research into the psychology of landscape preferences was done, to be able to formulate hypothetical preferences of inhabitants of the research area.
- Second, these hypothetical preferences were tested by conducting a pilot survey among ten inhabitants. Besides testing the hypothetical preferences another goal of the pilot survey was to test the questionnaire for a larger survey and to offer a first insight in landscape preferences. Respondents were selected by key informants from NLTO (Northern Agricultural Organisation) and a nature education organisation. The questions involved three themes: 1. attractiveness of landscape, compared to other landscapes in the province of Friesland, taking into account overall beauty, special landscape features and possibilities for outdoor activities, 2. policy regarding landscape maintenance and 3. opinion on farmers' contribution to landscape maintenance. All respondents are inhabitants of the Friesian woodlands with a known interest in landscape and nature. All were aged over 45 and had been living in the area for a long time.
- Third, a brief description of farming styles in the Northern Friesian Woodlands and their implications for the landscape quality is given. Fourth, these landscape qualities are matched with the answers from the pilot survey.

## Theoretical landscape preferences

Landscape psychological research distinguishes 'shallow' and 'profound' preferences of landscapes. Shallow preferences relate to physical features of landscape as such. Profound preferences take into account associations, memories, symbolic interpretations and geographic coherence of the landscape (Hendriks and Stobbelaar, 2003). Positively valued landscape features with respect to shallow preferences are (Arriaza *et al.*, 2004): 1. water, 2. wilderness, 3. mountains, 4. vegetation, 5. colour contrast. Schupbach *et al.* (2004) mention – in order of importance - 1. character 2. naturalness and 3. variety. Naturalness is interpreted as the presence of natural elements as swamps and shrubs, but also extensively managed grassland. This is in line with Coeterier (1996) who argues that experiencing naturalness is not only related to natural or wild landscape elements, but on the contrary, can be a feature of all landscape elements.

Kaplan and Kaplan (1989) found that complexity in landscapes is valued (together with mystery, coherence and legibility). Coeterier (1996) adds to this point of view that should contribute to the unity of landscapes. He also states that the amount of elements that contribute to the perceived unity of the landscape (called satisfiers) can not easily be too much, whereas a very small amount of elements that do not contribute to the unity (dissatisfiers), can ruin the landscape quality.

A basic theory in landscape research, the savannah theory, stipulates that humans prefer half open landscapes with good views, dispersed patches of trees and shrubs, grassy non swampy soil, but with surface water and the absence of danger and threat (Van den Berg, 1999). Absence of danger and threat also refers to the possibility to understand the landscape and to get an overview of it.

When this theoretical knowledge is applied to the Friesian Woodlands, the hypothesis can be formulated that this landscape will be highly valued in terms of shallow preferences. We see a half open landscape, with grassland, water, shrubs, a variety of colours that refer to a specific time of the year and possibilities to view in the distance. The landscape has a natural character with sufficient variety or complexity. Very few dissatisfiers are present.

## Actual landscape preferences

All respondents praise the characteristic landscape of small plots and hedgerows and see an overall progress in landscape quality. Half of the respondents even consider the landscape the most important factor determining the quality of their living environment. All respondents consider hedgerows and small pools as most characteristic features of the landscape, followed by farmyards, shrubs and grassland. Arable land is considered a-typical for the area.

When asked about 'dissatisfiers'; elements that are considered annoying and not fitting in the landscape, large modern farm sheds, silo's and electricity lines were most frequently mentioned. Maize cultivation was mentioned as well.

Most respondents considered the hedgerow structure just good as it is, and are satisfied with the current maintenance activities. With respect to the grassland, all respondents answered that the grass is 'boring' and 'too green', due to high fertilizer use, some referring to traditional grasslands with lots of herbs and flowers. Nevertheless, most respondents are moderately positive about the role of farmers in landscape and nature conservation, a statement that is confirmed by earlier research in the region (SNM, 2005). They predominantly responded neutrally to the statement that farmers should do more in this respect, half of them adding that 'farmers already do a lot'. Also remarkable is that most respondents were positive on increasing subsidies to farmers for contributing to landscape value. The promotion of region-specific farm buildings and farmyard planting was also considered important, while improving the accessibility of the region by creating more walking and biking routes was less valued.

When asked their opinion on creating passages in the hedgerows or occasionally removing a hedgerow completely to improve agricultural efficiency, respondents differed, being either completely positive or (moderately) negative.

In short: the respondents generally asked for a landscape with hedgerows, traditional farmyards with region-specific buildings and trees, and diverse grassland with flowers and herbs. They also considered farmers as important maintainers of the landscape, who should receive subsidies for those activities.

## Farming styles

Agricultural activity in the Woodlands is dominated by dairy farming, but shows a variety of farm household strategies or 'farming styles'. Farming styles are the total of interrelated, share ideas of farmers about the organization of production and the developments of the farm (Van der Ploeg, 2003) and represent different modes of ordering natural and social aspects within farming activities (Renting, 2004). Recent research in the Friesian Woodlands indicates that farm strategies are differentiated according to the level on which nature and landscape may be integrated with farming activities (Swagemakers and Wiskerke, 2006). Some farms cannot integrate nature and landscape on owned farm land, but have possibilities to take up management activities on additional rented land in nature reserves. Other farms have more possibilities for management activities, as long as these are limited to field boundaries and do not affect the management of agricultural fields. Again other farms see extensive possibilities to integrate nature and landscape, also when situated on parcels, and effectively build their farming strategy on the integrated and sustainable use of natural resources.

These variable possibilities to integrate nature and landscape are related to different underlying farming styles. According to De Bruin and Van der Ploeg (1991) four farming styles can be distinguished in the Woodlands region: (1) Business farmers, (2) Breeders, (3), Calm farmers, and (4) Stayers. *Business farmers* aim at large farms, both in terms of scale and intensity, in order to produce milk at the lowest costs per unit. Possibilities to integrate nature and landscape are limited, and are only taken up when these can be developed in a cost-effective way that compensates productivity losses. *Breeders* base their strategy on high milk yields per cow and premium prices for high fat and protein levels. This style has fewer problems to work within the small-scale landscape, active management of hedgerows may easily be integrated, but there are little possibilities to adjust the management of fields (due to loss of fodder quality). *Calm farmers* develop their farm gradually in order to avoid indebtedness and keep overall cost levels low, while the available family labour is an important starting point for farm development. This style knows low levels of external inputs and relatively extensive land-use, resulting in a wider range of possibilities to integrate nature and landscape management. Lastly, *Stayers* tend to have small farms and a low milk yield per cow. They aim to optimally valorise their own labour and actively search for complementary income sources. The implications of this particular style for nature and landscape are differentiated. Some representatives actively take up management activities as alternative income source, while in other cases limited labour availability (allocated to other activities) hinders the integration of nature and landscape management.

Farming styles therefore have different implications for nature and landscape. This is illustrated by Figure 1, which represents farming styles according to their possibilities to work in small-scale landscapes (macro nature of the landscape) and the preferred micro-production conditions in terms of type of farm buildings and type of pasture. Breeders and Calm farmers see the most possibilities to integrate their farm management within the small scale landscape as opposed to business farmers whose style requires larger parcel sizes. On the other hand, Calm farmers and Stayers express the highest preference for rich and varied grasslands and region-specific farm buildings, while other styles tend to have cubicle sheds and homogenous grasslands.

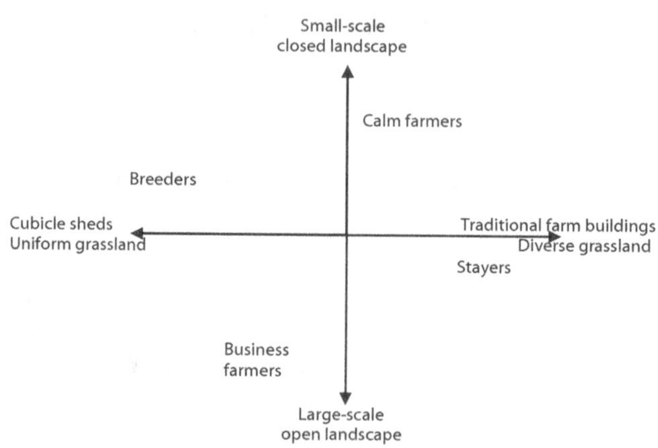


*Figure 1. Styles of farming according to the dimensions 'micro production conditions' and 'macro nature of the landscape'.*

## Conclusion and discussion

The preferences of respondents in the pilot survey are in line with the theoretically expected outcomes. The inhabitants value natural grasslands and the dense, complex hedgerow structure. They also like the region-specific buildings and farm yard planting. Large barns and sheds are disliked. Being a dissatisfier, these elements can lower the perceived landscape quality dramatically.

Farming styles can potentially contribute to landscape qualities, although specific opportunities are differentiated. Some styles, like Calm farmers and Stayers, have good opportunities to add to preferred landscape features like diverse grasslands, region-specific buildings and ecologically managed hedgerows. Others, like Breeders and to some extent Business farmers, see fewer possibilities but can still contribute to basic landscape qualities by the management of linear elements.

Especially promising is the role of the farmers' cooperative for nature and landscape management that is active in the region. It could play an important role in the collective development of landscape networks at regional scale that interconnect the qualities offered by different farmers from different styles. In such networks fields and boundaries with high nature quality on some farms can be linked by linear elements of basic landscape quality managed by others. The positive attitude of inhabitants to the creation of passages in the hedgerows suggests that plans could also include landscape adjustments to improve farm efficiency.

New research will be conducted among a larger and more aselect population of inhabitants and tourist visitors. The preliminary results show that a closer look should be taken at the role of dissatisfiers in the landscape. Also the role of Stayers, being a very diverse group including hobby farmers needs closer examination.

## References

Arriaza, M., J.F. Canas-Ortega, J.A. Canas-Madueno and P. Ruiz-Aviles, 2004. Assessing the visual quality of rural landscapes. Landsc. & Urban Plan. 69, 115-125.

Coeterier, J.F., 1996. Dominant attributes in the perception and evaluation of the Dutch landscape. Landsc. and Urban Plan. 34, 27-44.

De Bruin, R. and J.D. van der Ploeg, 1991. Maat houden. Bedrijfsstijlen en het beheer van landschap in de Noordelijke Friese Wouden en het Zuidelijk Westerkwartier. Landbouwuniversiteit Wageningen, Vakgroep Agrarische Ontwikkelingssociologie.

Hendriks, K. and D.J. Stobbelaar, 2003. Agriculture in a legible landscape. How conventional and organic farms contribute to landscape quality. (In Dutch, with English summary). Chair Landuse Planning, Wageningen University. 268 pp.

Kaplan, R. and S. Kaplan, 1989. The experience of nature: a psychological perspective. Cambridge University Press, New York. 340 pp.

Raad Landelijk Gebied, 2005. Nationale Landschappen: Vaste koers en lange adem. Raad voor het Landelijk gebied, Amersfoort. Publicatie RLG 05/1. 80 pp.

Renting, H., 2004. Mapping the nature conservation potential of agricultural land-users. 3rd WUR/INRA seminar on Multifunctional Agriculture, Paris, 17-18 Sept. 2003.

SNM, 2005. The Netherlands can be so beautiful; the perception of 52 areas by inhabitants (In Dutch). Stichting Natuur en Milieu, Utrecht, 45 pp.

Swagemakers, P. and J.S.C. Wiskerke, 2006. Integrating nature conservation and landscape management in farming systems in the Frisian Woodlands. In: Landscape research to landscape planning: aspects of integration, education and application. Springer, Dordrecht, The Netherlands, pp. 321-334.

Van den Berg, A.E., 1999. Individual differences in the aesthetic evaluation of natural landscapes. Rijksuniversiteit Groningen, The Netherlands. 149 pp.

Van der Ploeg, J.D., 2003. The Virtual Farmer. Past, Present and Future of the Dutch Peasantry.Van Gorcum. Assen.

# Farmer-driven protection of meadow birds in the Friesian Woodlands (northern Netherlands)

Paul Swagemakers
Rural Sociology Group, Wageningen University, The Netherlands

## Abstract

Though more land managers and stakeholders are involved in Natural Resource Management (NRM), farmers play a decisive role in the protection of meadow birds. In the Northern Friesian Woodlands (northern Netherlands) farmers together with scientists, politicians and other stakeholders have developed a 'system innovation' that results in increasing numbers of meadow birds in farmers' fields. For that, the good use of manure is a key factor. In this paper I will focus on how protection of meadow birds, here the Black-tailed Godwit in particular, is successfully integrated in farming systems. I will describe social dynamics, explain the relevance of heterogeneity of farming systems, and show how ecological objectives at farm and regional level can be made transparent.

## Introduction

In this paper, the focus is on farmer-driven protection of meadow birds. Different from scientists who often present and implement outcomes of research carried out under controlled or at least verifiable experiments, farmers explore and discover in practice, learn under different, specific conditions. Within the context of farm site specific conditions farmers continuously observe, interpret, change and evaluate their farming systems. Step by step they adapt their farming systems. If and how they adapt depends strongly on the objectives they have, resulting in heterogeneity in farming systems. Consequently, I hypothesize farming systems to perform differently with regards to the protection of meadow birds. Careful empirical analysis can bring out potentials; considering those farming systems in which farmers successfully integrate the protection of meadow birds might raise insights complementary to current scientific insights.

First of all, I will briefly introduce historical, societal and political context of agricultural development and the protection of meadow birds. This is followed by a description of the case study area. After that I will briefly outline the research methodology, present current insights and tentative research results on the protection of meadow birds in the case study area, and finally give some concluding remarks.

## Agricultural development and meadow birds in farmers' fields

In Europe, the modernization of food production oriented on increasing scale and intensification of land use cause environmental pollution and the destruction of natural habitats and biodiversity (Baudry *et al.*, 2003). Destruction and fragmentation of foraging and nesting habitats for meadow bird species in farmers' fields result in a general decline in number and range of these species (Beintema *et al.*, 1997; Duncan *et al.*, 1999). In the Netherlands special attention is paid to the protection of the Black-tailed Godwit, due to the fact that during the past decade its population dramatically declined (Kleijn *et al.*, 2001). Protection of this bird specie is especially important as 80-90% of the northern European population breeds in the Netherlands (Beintema and Müskens, 1987).

## Case study

While the population of the Black-tailed Godwit generally declines, in Gerkesklooster, a village situated in the outskirts of the Northern Friesian Woodlands (northern Netherlands), farmers succeed to increase meadow bird populations in their fields. Over the past 12 years in particular the population of the Black-tailed Godwit increased.

In the region, the protection of meadow birds is incorporated in agricultural practices throughout history. According to Friesian traditions, farmers and other stakeholders work together in order to monitor and protect meadow bird populations through local groups (Beintema *et al.*, 1995). In these groups, voluntary bird conservationists play an important role. Voluntary bird conservationists for example help farmers looking for the nests and indicate what measures farmers can apply for the chicken to survive; what currently is called mosaic management protects meadow birds in the region by tradition.

Since the start of a three year national program to protect the Black-tailed Godwit, a so-called 'Black-tailed Godwit Circle' in the area north of Gerkesklooster was implemented. This implied a formalization of the local bird protection groups, which are nowadays composed of farmers (associated in environmental co-operatives), bird conservationists (the so-called 'Vogelwacht') and nature managers (representatives of regional associations for nature and landscape protection). Farmers, the different stakeholders and land managers reflect on and share insights and so increasingly gain experience with integrating the protection of meadow birds in farming systems.

In this paper I will focus on farmers' experiences and insights and explain the concepts they use. Because concepts in use in practice seem yet far from congruent and consistent, I aim at clarifying the concepts rather than proofing and claiming their truth.

## Methods

For the explorative research, case study research (Yin, 1984) forms the core method. The material presented is based on qualitative field research done in the summer of 2004 and 2005 that builds upon research methods as presented in Swagemakers and Wiskerke (2005) developed within the Wageningen School for Rural Sociology (Wiskerke, 2004). Further ethnographic research (Handwerker, 2001) is to be continued and can be valorized by statistical analysis (Nooij, 1995) later on.

The field research consisted mainly of conducting in-depth interviews with farmers and other stakeholders in the region, recorded on tape. Afterwards, full transcripts were made. Ordered thematically the transcripts became building blocks for an overall story on the use and reproduction of natural resources by farmers in the region involved in Natural Resource Management (NRM).

## Current insights on Natural Resource Management

Farmers in the Friesian Woodlands increasingly base their farming systems to the natural resources in the region again. From a systems approach developed by farmers together with scientists, politicians and other stakeholders (Stuiver *et al.*, 2003) they adapted their farming systems initially in order to reduce losses of nitrogen by re-balancing soil-plant-animal interactions (Verhoeven *et al.*, 2003). Farmers adapted their farming systems to such a large extent that one can speak of a 'system innovation', *i.e.* a range of adaptations in the farming system that are applied through time (Wiskerke and Ploeg, 2004). Further empirical analysis can bring out presence and potential of the process of re-balancing.

Based on the interview material I describe concepts farmers consider relevant and that form building blocks for an overall story on the use and reproduction of natural resources, and to the protection of meadow birds in particular. About that a spokesman says:

'Here in the fields you find nature. Birds, but also all that is in the soil. One has to get to the bottom of the soil. There you find life. About the influence of manure on soil life I always have been telling we should know much more. Why meadow birds nest in particular fields is because a meadow bird has knowledge of what lives in the soil.'

Going deeper into the subject, farmers refer to the importance of having a healthy soil that contains a high level of organic matter or humus and has an open structure. 'Management of the soil,' as someone explains, 'results in resilient soil,' resilient soil being an indicator for the organic matter in the soil.

In order to keep the organic matter content up, farmers manure their fields with high-quality manure, *i.e.* slurry with a high C/N ratio and a lower concentration of inorganic nitrogen compounds. Also, the high-quality manure stimulates soil life; worms take care of the organic matter being transported from the surface into the soil where it is converted into nitrogen. By stimulating soil life, farmers maximize the delivery capacity of the soil, and especially that of nitrogen. The more soil life is created, the faster the conversion process of organic matter into nitrogen, and the higher the delivery capacity of nitrogen of the soil.

In this way, farmers were able to reduce the input of artificial fertilizers without significantly loss of grassland production. As a result, a different color of the grass was achieved, the grass containing less protein and having more structure. The different nutritional value of the grass entailed an improved fodder quality, for which the traditional Friesian cattle was in use or being reintroduced. Currently, the newly balanced fodder composition to feed the cows together with the additive Euromestmix forms the base for the production of good manure (Goede *et al.*, 2003), that is the core of the 'system innovation' that despite decreasing the input of artificial fertilizers keeps the grassland production up.

In the past, applying large amount of artificial fertilizers combined with good grass growing weather caused the grass sometimes grew very fast and being too long, the grass fell flat. Firstly such grass is not attractive as fodder to feed the cows and secondly cause the Black-tailed Godwits having difficulties to find the nest with chicken, or, by heavy rainfall, the nest becoming to wet (Kruk *et al.*, 1997). These days, farmers use the high-quality manure. For a better use and development of soil life, opposite of what legislation in the Netherlands prescribes, the manure is spread on the surface of the land. Also, it is spread more often and in little bits at a time causing the grass to grow constantly. Initially developed for improvement of grassland production, this practice simultaneously results in improved nesting habitats for the Black-tailed Godwit. Further on, it results in improved foraging habitats for the Black-tailed Godwit; injecting toxic manure (*i.e.* manure with a high concentration of inorganic nitrogen compounds) into the soil would cause one-third of the worms in the soil to die and the rest to disappear deeper into the soil for at least one week. In that period, the Black-tailed Godwit would not be able to feed himself. Hence, spreading the high-quality manure on the surface keeps worms available for the Black-tailed Godwit. In addition, it is a healthy soil that is created, which has a high water containing capacity that, as an adult Black-tailed Godwit uses its bill to search for, among others, worms (Beintema *et al.*, 1995), makes the Black-tailed Godwit indeed find the worms. After all, applying manure as explained above makes farmers less dependent on hiring a contractor and more flexible to decide when it is the best moment to manure the fields. That, in turn, is relevant for further grassland management; the good use of manure in early spring determines grass growth later on in the season and influences the possibilities for mosaic management that helps chicken to survive when the mowing season starts.

So far, by giving some examples I reflected on how the protection of meadow birds is embedded in farming practices. These examples illustrate that once one starts recognizing the importance of stimulating soil life one also begins to understand the importance of the good use of manure for improving of the foraging and nesting habitats of the Black-tailed Godwit. Without doubt those adaptations in the farming system are to be accompanied with supplementary measurements for the protection of meadow birds.

## Preliminary research results on Black-tailed Godwits' nests and use of manure

In the region, the local voluntary bird conservationists collected data on the occurrence and nesting success of meadow birds since 1993. Analyzing those data, an increase in the number of Black-tailed Godwits was found. Unintended, the work of the volunteers put things in a different light: farmers no longer cause a decline of meadow birds, but actively construct habitats and help to protect meadow birds resulting in an increase of the population of Black-tailed Godwits in their fields.

Next, the relation between farming systems and number of meadow birds in farmers' fields was further examined in the 500 hectares large area for which the 'Black-tailed Godwit Circle' was implemented. Preliminary evaluation of farming system performance indicates that the improvement of manure regarding its C/N ratio and efficiency in nitrogen management applied on the farm correlates positively with the number of meadow birds (unpublished internal report).

Furthermore, researchers of CLM (Centre for Agriculture and Environment) did a quick scan on the number of worms in farmers' fields and found in fields where meadow birds nest many worms were present (unpublished data). The positive correlation between number of worms and number of meadow birds indicates the importance of the good use of manure; by that soil life is stimulated, providing a potential habitat for meadow birds.

## Concluding remarks

Although often agricultural production causes environmental pollution and destruction of natural habitats and biodiversity, there are farming practices that form good habitat conditions for meadow birds. For foraging and nesting habitats of the Black-tailed Godwit, insights based on farmers' experiences indicate that the good use of manure is a crucial factor. Further on, the good use of manure contributes to nitrogen use efficiency (NUE) resulting in nitrogen surplus reduction (Verhoeven *et al.*, 2003). According to Groot *et al.* (in press) those farmers in the Friesian Woodlands reaching the largest improvements in NUE also obtained the best economic results. In brief, economic sustainable agriculture can go well together with the improvement of the environment and the construction of natural habitats and biodiversity.

## References

Baudry, J., F. Burel, S. Aviron, M. Martin, A. Ouin, G. Pain and C. Thenail, 2003. Temporal variability of connectivity in agricultural landscapes: do farming activities help? Landscape Ecology 18, 303-314.

Beintema, A.J. and G.J.D. Müskens, 1987. Nesting success of birds breeding in Dutch agricultural grasslands. Journal of Applied Ecology 24, 743-758.

Beintema, A.J., O. Moedt and D. Ellinger, 1995. Ecologische atlas van de Nederlandse weidevogels. Schuyt & Co, Haarlem. 352 pp.

Beintema, A.J., E. Dunn and D.A. Stroud, 1997. Birds and wet grasslands. In: D.J. Pain and M.W. Pienkowski (editors), Farming and birds in Europe: the common agricultural policy and its implications for bird conservation. Academic Press, London, pp. 269-296.

De Goede, R.G.M., L. Brussaard and A.D.L. Akkermans, 2003. On-farm impact of cattle slurry manure management on biological soil quality. NJAS – Wageningen Journal of Life Sciences 51, 103-133.

Duncan P., A.J.M. Hewison, S. Houte, R. Rosoux, T. Tournebize, F. Dubs, F. Burel and V. Bretagnolle, 1999. Long-term changes in agricultural practices and Wildfowling in an internationally important wetland, and their effects on the guild of wintering ducks. Journal of Applied Ecology 36, 11-23

Groot, C.J., W.A.H. Rossing, A.L. Lantinga, 2006. Evolution of farm management, nitrogen efficiency and economic performance on Dutch dairy farms reducing external inputs. Livestock Production Science (in press).

Handwerker, W.P., 2001. Quick ethnography. AltaMira Press, Walnut Creek, CA, 299 pp.

Kleijn, D., F. Berendse, R. Smit and N. Gilissen, 2001. Agri-environment schemes do not effectively protect biodiversity in Dutch agricultural landscapes. Nature 413, 723-725.

Kruk, M., M.A.W. Noordervliet and W.J. ter Keurs, 1997. Survival of black-tailed godwit chicks Limosa limosa in intensively exploited grassland areas in The Netherlands. Biological Conservation 80, 127-133.

Nooij, A.T.J., 1995. Variabelen en modellen: multivariate analyse in het sociaal - wetenschappelijk onderzoek. Boom, Amsterdam. 301 pp.

Stuiver, M., J.D. van der Ploeg and C. Leeuwis, 2003. The VEL and VANLA environmental co-operatives as field laboratories. NJAS – Wageningen Journal of Life Sciences 51, 27-39.

Swagemakers, P. and J.S.C. Wiskerke, 2005. Integrating nature conservation and landscape management in farming systems in the Friesian Woodlands (N-Netherlands). In: B. Tress, G. Tress, G. Fry and P. Opdam (editors). From landscape research to landscape planning: aspects of integration, education and application. Springer, Dordrecht, pp. 321-334.

Verhoeven, F.P.M., J.W. Reijs, and J.D. van der Ploeg, 2003. Re-balancing soil-plant-animal interactions: Towards reduction of nitrogen losses. NJAS – Wageningen Journal of Life Sciences 51, 147-164.

Wiskerke, J.S.C., 2004. Rural sociology in the Netherlands: past, present and future. Agrarwirtschaft und Agrarsoziologie 02/04, pp. 39-63.

Wiskerke J.S.C. and J.D. van der Ploeg (editors), 2004. Seeds of transition. Essays on novelty production, niches and regimes in agriculture. Van Gorcum, Assen. 356 pp.

Yin, R.K., 1984. Case study research: design and methods. Sage, London. 160 pp.

# Linking landowner type and landscape functions: A proposal for an integrated approach from southeast Portugal

*Anne van Doorn*
*Department of Biophysical and Landscape Planning, University of Évora, Portugal*

## Abstract

In this paper some findings of the European research project VISTA, (*Vulnerability of Ecosystem Services to Land use change in Traditional Agricultural landscapes)* are presented for a Portuguese study area. One of the objectives of the work presented is the identification of landscape functions in relation to observed land cover and in relation to the type of landowner.

The study area, currently considered as marginal in terms of agricultural production, is located in the southeast of the Alentejo region and characterized by an agro-silvo pastoral system. For the area, land cover is identified through aerial photo interpretation with the use of an extensive land cover classification. Subsequently ecosystem services were allocated to the land cover classes in a qualitative way, on the basis of expert knowledge and information from interviews with landowners.

The interviews also served to obtain information on land use and management and to identify different types of landowners. Because of the availability of spatial information of the property structure the information on land use and landowner typology could be related with observed land cover.

Combining the information on services, allocated to the different land cover classes and the landowner typology by using the land cover as the common denominator, results in an evaluation of which ecosystem services are supported by which types of landowners. The presented study is a proposal for a methodology to link the concepts of landscape functions with landowner types in a practical way.

**Keywords:** land cover, landscape functions and services, landowner typology, interdisciplinary assessment, Alentejo

## Introduction

The management of today's European rural landscapes and natural resources requires an approach on the landscape level which takes into account landscape multifunctionality (Brandt and Vejre, 2003). The multifunctionality of a landscape is considered to be the co-existence of different landscape functions in a dynamic relation within a single landscape. These functions provide in turn goods and services, a coherent list of these is presented by De Groot *et al.* (2002). In spite of the number of research papers written about landscape functions and services, *practical* examples of how to identify existing linkages between landscape functions and land use systems are still missing and remain a challenging subject of research.

This paper presents a proposal for a methodology of how this issue could be tackled, by using land cover as an interface to link both aspects. On the one hand land cover is an empirical entry to the description of a function (Brandt and Vejre, 2003); on the other hand land cover is the visible expression of past and present land use systems. In this paper land use systems are expressed in terms of landowner types. It is assumed that different landowner types have different management regimes and thus different types of land use, which is reflected in the composition of land cover of their property. This territorial based approach that links landscape functions with landowner practices through the spatially explicit information on land cover as common denominator, is considered to be especially applicable on studies on landscape scale.

## Material and methods

### *Study area*

At present the region is characterised by its peripheral location and the less favoured conditions for agriculture, because of poor and shallow soils and the summer droughts. The population density is very low and has been progressively decreasing last decades. Agriculture, forestry and hunting are main sources of income for the rural population. Besides elderly peasants, which normally own less than 50 ha, urban landowners possess large properties, of more than 200 ha, which are mainly used for hunting reserves and forestation projects. Forestation of former cultivated land, supported by regulation n° 2080 of the Common Agricultural Policy, is one of the most visible and significant land cover changes that can be observed over the last decades.

### *Data collection and analysis*

For the identification of the land cover, ortho-photographs of the year 2000 were used. Interpretation of the images was done on screen, with a minimum map unit of 0.5 ha. The land cover classification contained 8 land cover categories (Table 1), taking into account the different types of Montado in terms of tree and shrub cover. The result was a thematic map of land cover on a 1: 25,000 scale. Through oral information and data of the regional agricultural administration the property structure of the area could be added to the spatial database. In this way data on land cover and landowner data could be directly connected.

The list of landscape functions and services provided by De Groot *et al.* (2002) was adopted. From each of the 4 landscape functions, production, habitat, information and regulation, a set of related services was selected. This set of 13 services is not supposed to be complete, but represents as much as possible the significant services for the study area.

*Table 1. Land cover classification.*

| Land cover classification | |
|---|---|
| 1 | Arable / grassland |
| 2 | Natural pasture |
| 3 | Montado < 30% t.c. < 20% s.c. |
| 4 | Montado < 30% t.c. > 20% s.c. |
| 5 | Montado > 30% t.c. < 20% s.c. |
| 6 | Montado > 30% t.c. > 20% s.c. |
| 7 | Forest plantation |
| 8 | Scrub |

t.c. tree cover, s.c. shrub cover

Subsequently the services were allocated to the land cover classes in a qualitative way, on the basis of expert knowledge and information from interviews with landowners. Only positive relationships were considered: when a land cover category hosts or supports a service, indicated with a '1'. All other relationships, negative or neutral, are indicated with a '-'. Of course, more advanced scaling methods can be applied if needed.

Since the study area is a peripheral rural area, the landowners and farmers are the most important agents when dealing with landscape functions. In the case study area 28 landowners were interviewed about their present land management. The interview data served two goals: 1) to obtain data about the farmers in order to constitute a farmers typology and 2) to understand their land management and to relate observed land cover to typical functions and services of the landscape.

Data on age, education, main source of income, property size, gross cattle unit, and distance from property to living place *etc.*, were used to carry out a cluster analysis of the farmers. A hierarchical clustering with average group linking was applied.

A matrix approach (De Chazal *et al.*, submitted) is used to quantify, link and integrate information on ecosystem services and landowner types, the data on land cover serves as the common denominator.

## Results

### *Land cover and landscape functions*

Montado occupies the largest area (3-6), accounting for 31%, most of this area is the type with the scattered trees and the understorey consisting of arable land (22%). The more dense type is less present (3%). The forest plantations (7) also occupy almost a third of the area (30%). Arable land/ grassland covers 13% of the area, scrub about 17% and natural pasture 9%.

The associations between land cover classes and ecosystem services are reflected in Matrix 1. A '1' reflects a positive association between land cover class and selected service which means that the land cover category at present hosts and/or supports the service. There is no ranking between services; they are all to be considered equally important. For example, the service 'cereal growing' is associated with land cover class 1, arable /grassland and 3 & 5, the Montado classes without a shrubby understorey. The service 'honey production' is mostly associated with those land cover categories with shrubs containing aromatic woody species, like categories 2, 4, 6 and 8. Hunting is done basically all over the place, with the exception of the forest plantations.

Every land cover category is associated with various services. Outstanding are the categories of the Montado, which are associated with up to 9 services.

### *Landowner typology*

The clustering of the interview data resulted in 4 groups of landowners, which are, with associated parameters, reflected in Table 2. Type A is the type with the highest age and the lowest education. The landowners belonging to this type live close to their properties and most of them have left agriculture or will do so in the near future. They do not have income (anymore) from their property. Type B are the 'real' farmers, considerably younger than the type A landowners and with a somewhat higher level of education. Their properties are much larger than the ones of type A and have the highest number of animals. Their main

*Matrix 1. Land cover categories linked with services, numbers of land cover classes correspond to the ones in Table 1.*

| | Landscape functions | | | | | | | | | | | | |
|---|---|---|---|---|---|---|---|---|---|---|---|---|---|
| | Production | | | | | | | Information | | | Habitat | Regulation | |
| Land cover | Cereal growing | Livestock grazing | Hay making | Timber production | Cork production | Hunting | Honey production | Scenic value | Nature education | Local sense of place | Conserv. of flora & fauna | Prevention from erosion | Accum. of org.mat |
| 1 | 1 | 1 | 1 | - | - | - | - | - | - | 1 | - | - | - |
| 2 | - | 1 | - | - | - | 1 | 1 | - | 1 | 1 | 1 | 1 | - |
| 3 | 1 | 1 | 1 | - | 1 | 1 | - | 1 | 1 | 1 | 1 | - | - |
| 4 | - | 1 | - | - | 1 | 1 | 1 | 1 | 1 | 1 | 1 | 1 | 1 |
| 5 | - | 1 | - | - | 1 | 1 | - | 1 | 1 | 1 | 1 | - | - |
| 6 | - | 1 | - | - | 1 | 1 | 1 | - | 1 | 1 | 1 | 1 | 1 |
| 7 | - | - | - | 1 | - | - | - | - | - | - | - | - | - |
| 8 | - | - | - | - | - | 1 | 1 | - | 1 | - | 1 | 1 | 1 |

agricultural activity is livestock breeding, mainly sheep (for meat), and cereal growing. Type C landowners have the largest properties and are designated as innovative landowners. Their management is characterized by a variety of activities, livestock breeding, hunting, forestry, nature protection, honey production *etc.* They have the highest level of education and are the youngest landowners.

*Table 2. Landowners' typology and associated parameters used in cluster analysis.*

| | Type | Age (year) | Edu. (year) | Distance home–prp. (km) | Prop. size (ha) | # Animals | Main agr. activity |
|---|---|---|---|---|---|---|---|
| A | Retired farmer | >65 | <4 | <10 | <100 | <30 | none |
| B | Active farmer | 40-65 | 2-9 | <30 | 100-1000 | 30-500 | livestock/ cereals |
| C | Innovative landowner | 30-40 | >5 | >30 | 100-1000 | 50-100 | livestock/ hunting/ forestry |
| D | Landowner on distance | >35 | >4 | >50 | 100-500 | 0 | forestry |

The last type is a group that lives far away from their land and therefore they often have forest plantations but rarely livestock. They are usually middle-aged landowners with medium sized properties. They want to keep the land, which they normally inherited, because of the family connection and/or the investment of capital.

### *Average land cover distribution per landowner type*

The categorization of the interviewed landowners into these four types, allows us to estimate the average property size and its land cover distribution for each of the landowner types. For this purpose we crossed the land cover data, the data on property structure and the categorization into the 4 landowners' types. The result is reflected in Matrix 2, the distribution of land cover categories, in percentages of the property, is related to each landowner type. For example the average distribution of land cover classes of a property of a type B land owner is as follows (second line Matrix 2): 21% arable/grassland, 5% natural pasture, 28% Montado with dispersed trees, without shrub, *etc.*

### *Multiplying matrices*

Subsequently, the Matrices 1 and 2 can be multiplied: the entries on the first column of Matrix 1 are multiplied by the entries on the first row of Matrix 2 and outcomes are summed, the entries of the second column of Matrix 1 are multiplied with the entries of the second row of Matrix 2 and the outcomes are summed, *etc.* This calculation results in Matrix 3, and reflects how many percent of the property of a type of landowner supports or hosts a service. These percentages are rough estimates and should not be read as exact data. More interesting is the ranking among the landowner types, *i.e.* which type is most associated with which services. Numbers in bold correspond to the highest percentages among the 4 landowner types.

We can read from the matrix that production functions are mostly supported by type B landowners, especially because of the large proportion of arable/grassland on these properties. Honey and timber production are most supported by respectively type C and D landowners, because of high percentages of shrub area and forest plantation. The services related to the information function, like scenic value and local sense of place, are mostly supported by type A and B landowners, mainly due to the relatively large proportion of Montado on their properties.

The innovative landowners and the landowners on a distance support more services related to the regulation function like the prevention from erosion and the accumulation of organic matter. This is probably related with their extensive management that results in large areas of shrub vegetation.

*Matrix 2. Types of landowners and their average distributions of land cover categories in percentages of the average property size. Numbers of land cover classes correspond to the ones in Table 1.*

| land owner type | Land cover | | | | | | | |
|---|---|---|---|---|---|---|---|---|
| | 1 | 2 | 3 | 4 | 5 | 6 | 7 | 8 |
| A | 0.15 | 0.04 | 0.31 | 0.08 | 0.05 | 0.01 | 0.27 | 0.10 |
| B | 0.21 | 0.05 | 0.28 | 0.09 | 0.04 | 0.04 | 0.19 | 0.10 |
| C | 0.04 | 0.10 | 0.12 | 0.06 | 0.01 | 0.00 | 0.53 | 0.14 |
| D | 0.09 | 0.00 | 0.07 | 0.05 | 0.00 | 0.02 | 0.60 | 0.16 |

*Matrix 3. Linking landowners types with ecosystem services.*

| | Landscape functions | | | | | | | | | | | | | |
|---|---|---|---|---|---|---|---|---|---|---|---|---|---|---|
| | Production | | | | | | | Information | | | Habitat | Regulation | |
| Land owner type | Cereal growing | Livestock grazing | Hay making | Timber production | Cork production | Hunting | Honey production | Scenic value | Nature education | Local sense of place | Conserv. of flora & fauna | Prevention from erosion | Accum. of org.mat. |
| A | 0.46 | 0.64 | 0.46 | 0.27 | **0.45** | 0.58 | 0.22 | **0.44** | 0.58 | 0.64 | 0.58 | 0.22 | 0.18 |
| B | **0.49** | **0.71** | **0.49** | 0.19 | **0.45** | **0.60** | 0.28 | 0.41 | **0.60** | **0.71** | **0.60** | 0.28 | 0.23 |
| C | 0.16 | 0.33 | 0.16 | 0.53 | 0.19 | 0.43 | **0.30** | 0.19 | 0.43 | 0.33 | 0.43 | **0.30** | 0.20 |
| D | 0.16 | 0.24 | 0.16 | **0.60** | 0.15 | 0.31 | 0.24 | 0.13 | 0.31 | 0.24 | 0.31 | 0.24 | **0.24** |

In spite of the limited number of landowners and the local scale of the study, the proposed methodology is expected to be a useful approach in linking landscape functions and services to land use systems. To further enhance its applicability, more advanced and systematic methods to relate land cover to landscape functions should be developed. Also a ranking among services could be introduced in order to distinguish the more significant services.

Typifying land use systems is done is this study through a landowner classification, but could also be done at another scale with the use of different parameters.

## References


Brandt, J. and H. Vejre, 2003. Multifunctional landscapes - Volume I: Theory, Values and History. Centre for Landscape Research, Roskilde, Denmark.

De Groot, R.S., M.A. Wilson and R.M.J. Boumans, 2002. A typology for the classification, description and valuation of ecosystem functions, goods and services. Ecol. Econ. 41, 393-408.

Pinto-Correia, T., 1993. Threatened landscape in Alentejo, Portugal: the 'Montado' and other 'agro-silvo-pastoral' systems. Landsc. & Urb. Plan. 24, 43-48.

# Why and how design monitoring processes of farmer practices in European landscape?

*Marc Benoît[1], Michel Passouant[2], Claudine Thenail[3], Laurence Hubert-Moy[4], Pierre Dupraz[5], Christian Brassac[6], Florence Le Ber[7], Jean-François Marie[8], Christian Bockstaller[9], Olivier Barriere[10], Dominique King[11], Christine Le Bas[12] and Pierre Ruelle[13]*
*[1]INRA station, SAD, Mirecourt, France; [2]Cemagref-CIRAD-engref, UMR tetis, France; [3]INRA, SAD, Armorique, France; [4]Université Rennes 2, UMR CNRS 6554, LETG, France; [5]INRA, SAE, ESR Rennes, France; [6]Université Nancy 2, Lab Psy. Lor., équipe CODISANT, France; [7]ENGEES, CEVH, France; [8]INRIA, LORIA, UMR-CNRS 7503, France; [9]INRA, Environnement et Agronomie, UMR INPL-ENSAIA-INRA, France; [10]IRD, Milieux et Environnement, US 166-coordination Roselt, France; [11]INRA, Environnement et Agronomie, UR Science du sol, France; [12]INRA, Environnement et Agronomie, Infosol, France; [13]Cemagref, UMR G-Eau, France*

Sixty years after its launching through the "Marshall Plan", the European agriculture revolution is up again, but with some strong contradictions: water pollution, landscape uniformization, ethical crisis (Fresco, 2000). These harmful side-effects of agriculture could be aggravated if the evolution of agricultural practices continues following the current trends towards greater concentration, intensification and technicality.

We focus our poster on agricultural practices, from their choice by farmers decisions to their effects, as they continuously remodel our agricultural landscapes. The approach of farming systems as landscape "builders" is a new one, but its background is the vision of land as resource for agriculture (De Wit, 1992; Lardon *et al.*, 1990). But we have a large lack of datas and in information system helping stakeholders in their negociations.

These farmer practices changes appear to be strongly linked to all the severe environmental questions in Europe, as we developp in last IFSA Congress (Benoît *et al.*, 2005). They conceal large geographical disparities which may explain the spatial differences observed in levels of pollution. The object of the present paper is to propose a research procedure of farmer parctices monitoring at regional scale and to illustrate it for some european situations: Münich watershed protection, the Seine basin, four specific régional zones in France.

The designing of monitoring of farmer practices is a central challenge for stakeholders and researchers on farming system changes if we want to protect natural resources. For us, the main questions are:

- How to build a *generic framework* for those datas? This question refers to collective building of facts and argues studied as situed and distributed knowledges.
- How can we organize *informatical datas bases* linking cropping system, farming system, soil, communes, environnemental zones? We propose to use the UML informatical framwork to organize the levels and their relationships.
- What should be the *rules for datas use*? This central question will be used to improve the status of information uses during farming system changes
- How can we take into account the *"situated and distributed" dimension of those datas?* This question helps us to identified all the stakehloders involved in farmer changes for natural resources management.
- For those monitoring, what kind of *complementarity* have we to build between remote sensing, statistical datas, local surveys?

We propose to expose this question in two parts:

1. An european synthesis on farmer practices monitoring at regional scale and some examples of recent research works.

2. The french new research program developped on this topic: COPT (designing monitoring of farmer practices) in "Agriculture et développement Durable" project.

## References

Benoît, M., E. Kernéïs, Y. Pons, A. Gerbaud, S. Herrmann and F. Le Ber, 2005. Agricultural land design modelling : a methodological proposal. IFSA congress (workshop 3). Villa Real. 12 p.

De Wit, C.T., 1992. Resource use efficiency in agriculture. Agricultural Systems, 40, 125-151.

Fresco, L., 2000. Crop science: scientific and ethical challenge to meet human need. Paper presentated at the 3rd International crop science congress, 17-22 august 2000. Hamburg (Germany). 11p.

Lardon S., J.P. Deffontaines J. Baudry and M. Benoît, 1990. L'espace est aussi ailleurs. In : Brossier J., B. Vissac and J.L. Le Moigne (éd.) Modélisation systémique et système agraire. Décision et organisation. Paris : INRA, 321-37.

# Interactive Technology Assessment: The ideal and the unruly practice

*S. Goddijn[1], M. van Bavel[1], B. Bos[2] and G. van der Peet[2]*
*[1]LEI, Wageningen University and Research Centre, PO Box 29703, 2502 LS The Hague, The Netherlands; [2]Animal Sciences Group, Wageningen University and Research Centre, PO Box 65, 8200 AB Lelystad, The Netherlands*

## The ideal: Charactaristics of ITA

Many current problems in the agro sector can be described as 'unstructured problems', featuring both uncertainty about important facts and dissent about target values. For such problems, Grin *et al.* (1997) developed Interactive Technology Assessment (ITA). Characteristic of ITA is an interactive design process around a proposed solution route seeking synthesis instead of exchange of interests. This takes place among others by means of the systematic identification of the competence theories of the different parties involved. ITA is unique and ambitious in its approach, but is seldom applied in practice. This article hopes to provide deeper insight into the practical applications of ITA. We will primarily focus on how vulnerable the method is for suboptimal limiting conditions. ITA thus assumes the willingness of stakeholders to exclude power politics from the process and the ability of analysts to operate strategically (Grin *et al.*, 1997). In practice, this will not always be possible to achieve. What consequences does this have for the effectiveness of the method? This question is important for the method itself, as well as for the nature of the resulting solutions. We discuss this in relation to a practical case, in which we search for a common monitoring system for animal health in the Netherlands.

## The unruly practice: Monitoring of animal health

The results show that lack of sense of urgency among the participants made it difficult to achieve the necessary amount of interaction, and made it necessary to focus on aspects of the solution perceived as most urgent. This led to much attention for the division of responsibilities, which created energy but also brought politics into the development process. This made it impossible to exclude power politics from the discussions. In this setting, researchers are required who are sensitive, factually competent and most important strategically very strong. In the interactive design process, decisiveness and progress are important, but are not compatible with precision. The iterative character gives the process great corrective capacity. The results show that ITA is able to give people insight into each other's approach and to bring people together, bringing solutions closer. However, it is clear that the safe environment on which the method is based is almost impossible to create, thus making it more difficult to achieve real synthesis.

## Reflection: ITA in SNM perspective

Whereas ITA searches for the optimal solution supported by all important stakeholders, another (theoretical) approach, Strategic Niche Management (SNM) is based on the idea of trying out a range of associated innovations in an environment (niche) which is to a certain extent protected from the dominant regime. In the situation of SNM, researchers are able to focus the needs and preferences of one or more clear 'problem owners' and exclude participants which lack sense of urgency. Concluding, from the SNM perspective it seems relative easier to fulfil the pre conditions of a 'clear problem owner' and 'exclusion of power politics in the process'.

## References

Grin, J., H. van de Graaf and R. Hoppe, 1997. Interactieve Technology Assessment. Een eerste gids voor wie het wagen wil. Rathenau Instituut. 43 pp.

# Farmer's choice and land productivity between direct seeding and transplanting in rainfed lowlands in Northeast Thailand and Northwest Cambodia

*A. Kamoshita[1], S. Hayashi[1], H. Ikeda[1], J. Yamagishi[1], A. Kotchasatid[2], B. Jongdee[2], S. Chea[3], M. Ouk[3], S. Leang[4], S. In[5], C.C. Sophorn[5], N. Kojima[6], M. Hamano[6] and K. Tokida[6]*
*[1]University of Tokyo, Japan; [2]Ubonratchathani Rice Research Center; [3]Cambodian Agricultural Research and Development Institute, Cambodia; [4]Prek Leap National School of Agriculture; [5]Battambang Provincial Department of Agriculture; [6]Japan International Cooperation Agency*

## Objective

Rainfed lowland rice farmers have to adjust their rice planting methods to unpredictable rainfall patterns during early rainy season (IRRI, 2002), but few studies are available as for how farmers choose to plant rice under different field and weather conditions either by transplanting or direct seeding. Farmer's choice and grain yield among cultivation methods of direct seeding and transplanting were studied by experiment, field survey, and farmer interview, at 2 rainfed lowland sites, Ubonratchathani, Northeast Thailand and Battambang, Northwest Cambodia. University of Tokyo collaborated with the national and provincial agricultural research institutes or department, with the emphasis on understanding of and learning from farmers' practices and delivering alternative rice production technologies.

## On-farm direct seeding

Direct seeding is generally increasing in Northeast Thailand during 1990's due to labor shortage, while yearly trends in Northwest Cambodia is not yet clear. Through the farmers' interview and field survey, large variation in percentage of direct seeded fields (13-100%) out of total rice cultivated areas were found among farmers and villages within the study site in Northeast Thailand, as well as wide ranges of sowing time over 2 months from early April to mid June. Remote fields from farmers' houses were often direct seeded, and their grain yield was often lower due to less intensive management and more weed infestation, compared with transplanted fields. Some of the direct seeded fields at lower toposequence which get moist earlier during rainy season turned out to be high yielding in case of Northeast Thailand. At farm level, drought during early rainy season caused delayed transplanting of aged seedlings or more severe weed infestation on direct seeded fields, and perceived by farmers as most serious production constraint.

## Potential productivity of direct seeding

However, in the weed controlled field experiments in the two studied sites, maximum yield was generally higher in direct seeding than transplanting. The most striking cultivation methods that have impressed the employed farmers or local counterpart technicians/scientists were (1) mixed sowing of rice and mungbean to have suppressed weeds in Northeast Thailand, (2) effective use of herbicides to control weeds in Northwest Cambodia, and (3) superior grain yield of direct seeding than transplanting in both sites.

## References

IRRI, 2002. Direct seeding: research issues and opportunities. Edited by S. Pandy *et al.*, IRRI, Philippines. 383 pp.

# Improving nitrogen management on arable farms in a participative project in The Netherlands

*J.W.A. Langeveld[1], F.J. de Ruijter[1], P. Dekker[2], O.A. Clevering[1] and B.M.A. Kroonen-Backbier[2]*
*[1]Plant Research International, PO Box 16, 6700 AA Wageningen, The Netherlands; [2]Applied Plant Research, Wageningen University and Research Centre, PO Box 430, 8200 AK Lelystad, The Netherlands*

## Introduction

Groundwater nutrient concentration in The Netherlands exceeds standards set by EU and national legislation. Agriculture is the main source of nitrogen (N) emissions. 'Farming with a future' helped farmers to reduce nutrient emissions, combining research on experimental farms with systems research on commercial (pilot) farms. This paper discusses farmer-researcher interaction and changes in N management on arable pilot farm groups in the northeast (NE, sand and reclaimed peat soils), southeast (SE, - mostly deep - sandy soils), and southwest (SW, old marine clay soils). A survey showed that N use efficiency is highest in NE, and reduced by high fertilisation levels in SE and autumn manure application in SW (Table 1). The project provided guidance to pilot farmers, communication between researchers and extension officers on the one side and pilot farmers on the other side being based on group discussions and principles of open dialogue and two-way communication.

*Table 1. Major characteristics of arable farming groups.*

| Group | Number and average size | Main attitude | Main problems | Progress made |
|---|---|---|---|---|
| NE | 5 farms (90.7 ha) | Self assured, economic and technical focus | Declining economic results | Not much progress made |
| SE | 4 farms (34.6 ha) | Uncertain, technical environmental focus | High N application, high nitrate levels | Considerable progress crop level |
| SW | 5 farms (59.2 ha) | Reluctant, strong economic focus | High N application, surface water quality | Considerable, mainly last years |

## Results and discussion

Farmer participation included annual evaluation of farm performance, N monitoring, on-farm trials, lectures and provision of feedback to extension officers and researchers. During the project large differences were found in attitude and progress made (Table 1). Farmers in NE focused on technical innovations and economic results, farmers in SE showing more interest in technical and environmental aspects, while those in SW had a strong economic focus, which is in line with the findings of the preliminary survey. Farmers in NE exceed recommended N application levels for potato but remained below such levels for other crops. Farmers in SE applied less and farmers in the SW more N than was advised. Farms in NE complied with national MINAS N objectives at the start of the project. Farmers who did not comply at that moment showed considerable progress during the project, although in the SE not sufficient to comply with nitrate objectives. These findings are in line with researchers' expectations. Although progress has been made with respect to N management, especially in SW and (to a lesser extent) SE, more progress is needed to comply with environmental legislation (*e.g.* Water Framework Directive). This will require a combined effort from farmers as well as researchers and extension officers.

# Analysing the link between technical practices and social organisations: The example of a French rare breed management set up, the "Landais" sheep breed

*A. Lauvie[1], A. Audiot[1], F. Casabianca[1] and E. Verrier[2]*
*[1]INRA SAD, Sicomor, BP52627, 31326 Castanet Tolosan Cedex, France; [2]INA P-G, France*

## The French context of rare breeds management

After World War II, dynamics of agriculture towards high productivity have led to the selection of a few breeds and to the decrease of the use of a large number of local breeds. The diversification of the stakes surrounding those breeds (from set up mostly focused on limiting the genetic variability loss to wider interests expressing various values of the populations) have led to a diversification of the stakeholders. We consider a management set up as a network including stakeholders, technical devices, knowledge and know-how surrounding the animal population, with the aim of managing it. As a consequence, we must combine analyses of individual's perceptions, collective action, as well as genetic issues at the population's level.

## Analysing a management set up in its whole

We investigated the management set up of the "Landaise" sheep breed located in south-west of France (Rosset *et al.*, 2002) in order to show how various dimensions interact and to what extend it explains the current set up. ). Inspired by sociology of technology theories, we treated, throughout diachronic analysis, the various elements: stakeholders, animals, technical devices, *etc.*

## What can we learn from the "Landais" case?

Three steps are identified: (i) the setting-up of the flock and of a management plan in an eco-museum, (ii) the splitting of this flock into three separate flocks, (iii) the transfer of the management responsibility to a regional association, and an increasing number of breeders. The number of stakeholders has risen from about 3 to 15 and the exchanges of breeding stocks have increased. In the same time, the population size of the breed increased from 150 to about 800 ewes. The distribution of the livestock and the exchanges of rams are unbalanced, with breeders being mostly suppliers or mostly users (Figure 1). This cannot be only associated with the fact that the breeders are distributed in a wide area. Analyzing the breeders' motivations and their perceptions of the population, we can also notice controversies on phenotypical selection criteria as well as the genetic management rules. After having increased during a few years, the collective dimension and the institutional forms of support are now decreasing.

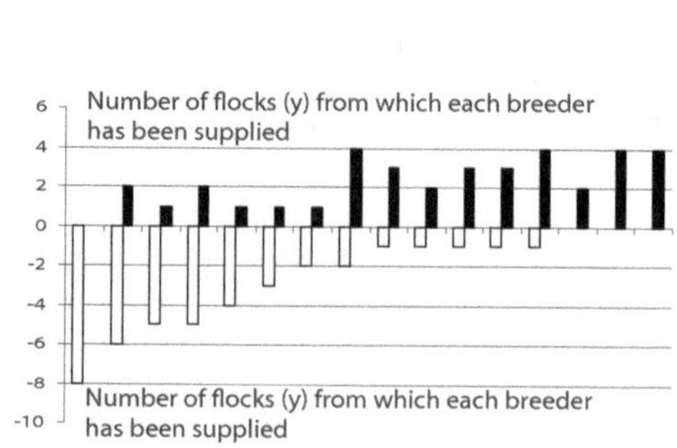


*Figure 1. Rams flows for each breeder (x).*

## Conclusion

All the elements of a rare breed management set up are interrelated and it can be relevant not to give too much emphasis to a particular one. Such an approach should be checked; it could be applied to other breeds, representing a large variety of situations and contexts.

## Acknowledgements

Study realised thanks to the "Institut National de la Recherche Agronomique"(SAD), the "Bureau des Ressources Génétiques", and the french Ministry in charge of agriculture.

## References

Rosset, O., L. Reveleau and C. Guintard, 2002, Population ovine landaise: Historique, situation actuelle et caractérisation morpho-biométrique, Ethnozoot. 69, 75-134.

# Final section – Perspectives for participative systems oriented research

# Introduction to final section – Perspectives for participative systems oriented research

*Jan Verhagen[1], Hans Langeveld[1] and John Dixon[2]*
*[1]Wageningen University and Research Centre, The Netherlands; [2]CIMMYT, Mexico*

Rural areas and more specifically, but not exclusively, agriculture are facing serious challenges. Ecological services are rapidly eroding and economic conditions are becoming increasingly more difficult, in many areas leading to deteriorating living conditions and social environments. Responses to the symptoms of climate change (*e.g.*, increasing temperatures, erratic rainfall patterns, *etc.*), to large-scale outbreaks of animal diseases, and to increasing competition for land (*e.g.*, for industrial purposes, housing or nature development) and water often are inadequate. General support for current land use practices is declining. In many countries, an increasing proportion of the population is living in urban centres and the contradictions between the highly populated and economically better-off urban areas, on the one hand, and depopulating, poor rural areas, on the other, are becoming more evident.

While it is clear that change is needed, it is less clear what the available options are, which opportunities lay ahead, what adaptations must be made, and how research can play a role in realising the potential of rural areas. Participative systems-oriented research, integrating multi-disciplinary knowledge at different scale levels, may be a useful instrument in defining viable solutions. Knowledge alone may not be sufficient to support decision-making and problem-solving in multiple-stakeholder arenas and integrative, participative processes may be required to transfer conflict into concerted action.

To have an impact on the improvement and development of rural areas, research needs to involve different stakeholders, including farmers, civilians and policy makers. They need to participate in problem formulation, to identify effective policies and farming practices, and assist in the adoption of such practices. Although agricultural research traditionally focused largely on intensification and expansion, it is increasingly taking on board rural and environmental issues, often combining input reductions with participative processes and technological innovations. Another re-orientation is the shift from a single component to a systems-oriented approach integrating knowledge of biophysical and socio-economic disciplines. While regional differences exist in foci of agricultural policy and farmers' objectives, in many cases the major emphasis is on the farm household as the level where crucial decision making level in agricultural systems is taking place.

In his presentation delivered at the 17$^{th}$ IFSA symposium in Florida, David Norman (2002) presents a historical overview of the development of a large number of systems-oriented, participative research approaches. But more approaches exist, two of which will be discussed here briefly.

The Nucleus Pilot Farm Research Approach (NUPFRA) has been developed in environmental research projects in The Netherlands. It focuses on ecological and economic performance of farming systems, using ecological (nitrogen surplus, groundwater nitrate concentration and residual mineral soil nitrogen), and economic (farm income, additional costs made for improvement of the system) indicators. NUPFRA combines process research on experimental farms with applied research on commercial pilot farms, using results of an extended monitoring system in the guidance of the pilot farmers (Langeveld *et al.*, 2005).

NUPFRA describes land use and farming activities in detail, defining farm compartments and internal and external material flows, and applying indicators in an integrative, multi-disciplinary assessment of farm performance. It is a data-intensive approach in which land use activities are sequentially evaluated in an intensive, participative process involving farmers, advisors and researchers. An application is discussed in the paper 'Improving nitrogen management on arable farms in a participative project in The Netherlands' by Langeveld *et al.* (section 5).

Farmer Field Schools (FFS), a less formal participatory training approach, is found at the other end of the spectrum. FFS utilises participatory methods to improve analytical skills, critical thinking, and creativity of farmers', and helps in the decisions making process. Trainers are viewed as facilitators rather than instructors, and interactive learning and field-experimentation is used to teach farmers to experiment and improve their problem-solving capacities (Godtland *et al.*, 2003). FFS was developed to introduce knowledge-intensive integrated pest management (IPM) practices for rice, but has since then been adapted to work with other crops and diseases and other problems such as human health, and has spread rapidly across Asia, Africa, and Latin America. See Tripp *et al.* (2005) for a review of its potentials. An application of FFS is discussed in the paper by Vaarst et al (this section).

The approaches discussed here are only two in a range of alternatives. Notwithstanding their differences, both are useful to discuss alternatives in land use practices at the farm and at the regional level. They facilitate analysis of actual and future systems, using a range of disciplines, and relatively simple indicators that appeal to farmers, advisors, researchers and policy makers and thereby allow exchanges of ideas and insights during a process of strategic cooperation.

## References

Godtland, E., E. Sadoulet, A. De Janvry, R. Murgai and O. Ortiz, 2003, The Impact of Farmer-Field-Schools on Knowledge and Productivity: A Study of Potato Farmers in the Peruvian Andes. http://www.farmerfieldschool.net/document_en/Knowledge-Nov03.pdf.

Langeveld, J.W.A., H. Van Keulen, J.J. De Haan, B.M.A. Kroonen-Bakbier and J. Oenema, 2005. The nucleus and pilot farm research approach: examples from The Netherlands. Agricultural Systems 84, pp.227-252.

Norman, D.W., 2002. The Farming Systems approach: A historical Perspective. Presentation at the 17th Symposium of the International Farming Systems Association, Lake Buena Vista, Florida, November 17th-20th, 2002. http://conference.ifas.ufl.edu/ifsa/papers/invite/Norman.doc

Tripp, R., M. Wijeratne and V. Hiroshini Piyadasa, 2005. What Should We Expect from Farmer Field Schools? A Sri Lanka Case Study. World Development Vol. 33, No. 10, pp. 1705–1720.

# A Farmer Field School approach for learning and exchanging knowledge on animal health and production: Results from action research in Kenya, Uganda, Vietnam and Denmark

*M. Vaarst[1], J.C. Riise[2], P. Rasmussen[3] and C.E.S. Larsen[4]*
*[1]Danish Institute of Agricultural Sciences, PO Box 50, DK-8830 Tjele, Denmark; [2]Network for Smallholder Poultry Development, KVL, Dyrlægevej 2, DK-1870 Frederiksberg C, Denmark; [3]The Danish Institute of Agricultural Information and Education, Vesterbrogade 4 A,1, DK - 1620 Copenhagen V, Denmark; [4]Small Livestock Component, ASPS 405, 406 B9, Bo NN&PTNT, So 2 Ngoc Ha, Ba Dinh, Hanoi, Vietnam*

The concept of Farmer Field Schools (FFS) provides a set of methods for common learning, network building and exchange of experiences among farmers, which have the potential to empower the farmers on individual and community level to improve their farms and lives. A FFS group consists of a number of farmers and a facilitator, who can be a public extension worker or from an independent organisation (*e.g.* a farmer organisation, a church or a charity organisation). Although a common framework exists for this FFS concept, the context and conditions for farming in a given community and related to the enterprise, region, market conditions, infrastructure *etc.* necessarily will influence the individual goals of the group of farmers, as well as the way in which a Farmer Field School can operate and work as a part of the farmers' daily life.

The FFS approach is successful in many countries, but also faces difficulties in implementation and sustainability. One of the reasons for this may be the lack of profound and sufficient analysis of the context, in which the FFS is being implemented in each individual case. It is obvious that FFS may not be implemented in the same way in peri-urban Hanoi and rural Uganda. In the presentation, we argue for the necessity of a profound analysis of the existing nature-given, social and cultural context including an identification of the constraints and possible pitfalls.

An important part of this context analysis addresses the farmers' need in terms of which skills and abilities they want to improve (the 'technical' issues) and in which way their personal and community human capacities and environment can be supported in order to take responsibility and control over their own lives as farmers, humans and citizens. This also involves an evaluation of their background and existing knowledge and skills, *e.g.* literacy, educational background, family structure and position in the society. The outcome of an analysis of the context will direct the need for facilitation. In Denmark, where farmers have been through a long education, the farmers are capable of identifying problem areas in their herds, and the common learning process can be mostly that of a mutual advisory session where everybody contributes to identify risk factors and possible solution in a given problem area in a herd owned by one of the group members, where the session takes place. In Uganda, where a farmer group may contain both literate and illiterate farmers, some of whom have neither had any formal school education nor any specific knowledge about livestock animals, the common learning process can be an increased belief and control over the situation at the farm and a move towards a practical experience of gradually trusting ones own abilities and knowledge and get acknowledgement in the surrounding society as a resource person.

## Author index

## Agrosystems Research

### *Science for sustainable agro-ecosystems*

Agrosystems Research is a scientific business unit of Plant Research International, a research institute of Wageningen University and Research Centre, the Netherlands. Total staff of Agrosystems Research is 95 of which 75 are scientists.

Agrosystems has built a reputation on the design, development and testing of systems for sustainable agriculture and land use in The Netherlands, Europe and developing countries. Our scientific research ranges from studies of nutrient and herbicide management strategies, and water and energy use to studies of adaptation to climate change, food security, multifunctional land use and green care.

We study agricultural systems in the arable, horticulture (inclusive of greenhouse production) and dairy farming sectors. The development of innovative mixed cropping systems is also included in our research programme, as well as the development of integrated animal, plant and energy production systems.

Integration of knowledge is our strength. We do this by means of desk and model studies and experimental research. The individual farm is our reference level from which we may go down to the field or process level or up to regional, national or even global level.

Agrosystems Research performs research for national and local governments, the European Commission, industry and non-governmental organisations.

*Dr. Jacques Neeteson*
*Business Unit Manager Agrosystems Research*
*Plant Research International, PO Box 16, 6700 AA Wageningen, The Netherlands*
*E-mail: Jacques.neeteson@wur.nl*

## TransForum

The Netherlands traditionally have a strong agrifood sector. The post-war build-up of knowledge was aimed at the development of a highly productive agricultural sector. After many decades of success this has now led to over-specialisation, an increased environmental burden and encroachment on public spaces. A switch is required to more sustainable production and rural development or, in short, to more sustainable agriculture, based on the three P's: people, planet and prosperity.

TransForum was established in November 2004 in order to promote innovation-oriented knowledge that can contribute towards the realisation of that goal. Such knowledge is largely generated together with businesses in Innovative Practical Projects, thereby ensuring that it provides an answer to the questions thrown up by industry. In addition TransForum develops Scientific Projects in order to look in greater depth at the issues arising out of the practical projects. Finally there are Knowledge Projects in which the knowledge, methods and skills that have been developed are embedded in society, for example by means of education, training, new arrangements or new institutions. Under this approach links are created and new knowledge is developed that can contribute to more sustainable production and use of the landscape.

In order to achieve this TransForum is working on a new network of knowledge institutes, government agencies, industry and community organisations. TransForum is forging new alliances between entrepreneurs, financiers, special interest organisations and knowledge developers. Parties that have not cooperated in this way with one another in the past are brought together in working towards TransForum's objective. Together with TransForum, these partners invest in projects in which innovation-oriented knowledge is built up and shared. In this way a cross-fertilisation is achieved between scientific and practical knowledge. The transmission and recording of new insights is at least as important. This gives rise to a process of continual renewal, leading in turn to the desired turnaround.

TransForum is a temporary organisation. Up to and including 2010 it will be working towards its objective in 50 projects and various conferences, by means of publications and learning moments. After 2010 the approach must become embedded in the way that everyone involved in sustainable agriculture and vital rural areas thinks and acts.

www.ingramcontent.com/pod-product-compliance
Lightning Source LLC
LaVergne TN
LVHW060616300926
849162LV00008B/1633
*9789086860029*